Mechanics of Flexible and Stretchable Electronics

Mechanics of Flexible and Stretchable Electronics

Edited by Yong Zhu and Nanshu Lu

WILEY-VCH

Editors

Prof. Yong Zhu
North Carolina State University
1840 Entrepreneur Dr.
Raleigh
NC 27695-7910
USA

Prof. Nanshu Lu
University of Texas
2617 Wichita St., C0600
Austin
TX 78712-1221
USA

Cover Image: © Eugene Mymrin/Getty Images

Library of Congress Card No.: applied for

British Library Cataloguing-in-Publication Data
A catalogue record for this book is available from the British Library.

Bibliographic information published by the Deutsche Nationalbibliothek The Deutsche Nationalbibliothek lists this publication in the Deutsche Nationalbibliografie; detailed bibliographic data are available on the Internet at <http://dnb.d-nb.de>.

Print ISBN: 978-3-527-35219-7
ePDF ISBN: 978-3-527-84228-5
ePub ISBN: 978-3-527-84229-2
oBook ISBN: 978-3-527-84230-8

Typesetting Straive, Chennai, India

Printed and bound by
CPI Group (UK) Ltd, Croydon, CR0 4YY

C9783527352197_300824

Contents

Preface

Flexible and stretchable electronics represent an emerging technology with the potential to significantly impact a wide array of areas including healthcare, energy, robotics, manufacturing, and defense. Unlike rigid electronics, this technology enables novel applications ranging from wearable or implantable devices to soft robots. Although many interdisciplinary challenges exist in this field, a fundamental challenge is achieving mechanical flexibility, stretchability, or surface conformability while maintaining the functionalities and performance of the electronics.

The mechanics of materials, structures, and interfaces plays a crucial role in addressing this challenge. This book aims to showcase some successful and nascent approaches including soft functional materials (such as composites, nanomaterials, organic semiconductors, hydrogels, liquid metals, and so on), soft structures (like origami/kirigami, serpentines, wrinkles, buckles, architected materials, etc.), and advanced bio-electronics adhesives, among others. The focus throughout is on the mechanistic understandings and/or the mechanics-driven designs of such materials, structures, interfaces, and systems.

Recent books in the field of flexible and stretchable electronics largely focus on materials, manufacturing, electronics, and applications. However, the mechanics underlying these advancements have been instrumental in the rapid progress of this field. We believe that a book summarizing the recent developments in this crucial aspect is both timely and necessary.

Our primary audience includes senior undergraduate students, graduate students, and early-stage researchers who are interested about the mechanics of soft electronics or planning to take it on as their research directions. Additionally, instructors teaching undergraduate and graduate courses on the topic will find valuable resources within these chapters. Anticipated readers may come from a wide range of disciplines including, but not limited to, mechanics, physics, materials science, chemical engineering, electronics, and biomedicine, reflecting the multidisciplinary nature of flexible and stretchable electronics.

Authored by leading experts, this book is divided into three parts, comprising 15 chapters. Part I (Chapters 1–5) addresses the mechanics of a variety of materials commonly used in flexible and stretchable electronics such as hydrogels, 1-D and 2-D nanomaterials, liquid metals, and organic semiconductors. Part II (Chapters 6–10) explores the mechanics involved in the design and manufacturing of these electronics, covering diverse structural designs like mechanics-guided 3D assembly and wrinkling/buckling instabilities, and manufacturing processes including laser-based methods and electrospinning. Part III (Chapters 11–15) delves into mechanics issues relevant to applications, such as the indentation behavior of soft electronics, wet adhesion, and the integration of mechanics in applications like tactile sensors, haptics with curved origami, and transient electronics. These 15 chapters provide a snapshot of the recent advancements in the mechanics of flexible and stretchable electronics.

We extend our sincere gratitude to all contributing authors, whose dedication and hard work made this book possible. We are also indebted to the editorial team at Wiley-VCH for their exceptional technical assistance.

June 2024

Yong Zhu and Nanshu Lu
Raleigh and Austin

Part I

Materials

1

Extreme Mechanics of Hydrogels Toward *In Situ* Hydrogel Bioelectronics

Tsz H. Wong[1], *Xuanhe Zhao*[2,3], *and Shaoting Lin*[4]

[1] Michigan State University, Department of Biomedical Engineering, 775 Woodlot Dr., East Lansing, MI 48824, USA
[2] Massachusetts Institute of Technology, Department of Mechanical Engineering, 77 Massachusetts Avenue, Cambridge, MA 02139, USA
[3] Massachusetts Institute of Technology, Department of Civil and Environmental Engineering, 77 Massachusetts Avenue, Cambridge, MA 02139, USA
[4] Michigan State University, Department of Mechanical Engineering, 474 S Shaw Ln, East Lansing, MI 48824, USA

1.1 Introduction

In-situ bioelectronics, a rapidly evolving field focusing on the development of electronic devices that can operate within the body for on-site sensing, stimulation, and powering, holds great promise for revolutionizing the field of medicine in a variety of ways (Figure 1.1a) [1, 2]. For example, the development of biosensors that can achieve on-site quantification of biomarkers closely related to the development and progression of colorectal cancer (CRC) would enable early CRC detection preventing CRC from progressing, thereby increasing the five-year relative survival rate up to 90% [3]. The development of neural probes that can form intimate interfaces with neurons without provoking a severe foreign body response would enable chronic neuron stimulation and recording, facilitating fundamental understanding of neural activities and offering long-term treatment for neuropsychiatric disorders, traumatic injuries, and inflammatory conditions [4–6]. The advancement of energy harvesting devices that can provide sustainable power supply to cardiac pacemakers could prolong their lifespan and help maintain or restore a normal heart rhythm with electrical impulses [7–9]. However, the key challenge faced by *in-situ* bioelectronics stems from the fundamentally contradictory properties between electronic components and biological systems, which induce foreign body responses due to mechanical, chemical, and biological mismatches. Specifically, electronic components are typically made of metals, silicon, glass, ceramics, and plastics that are hard, dry, and abiotic; in contrast, biological systems are composed of living tissues that are soft, wet, and dynamic.

Hydrogels, as polymer networks infiltrated with water, exhibit intriguing multiphysics phenomena associated with mechanical, electrical, chemical, and thermal

Mechanics of Flexible and Stretchable Electronics, First Edition. Edited by Yong Zhu and Nanshu Lu.

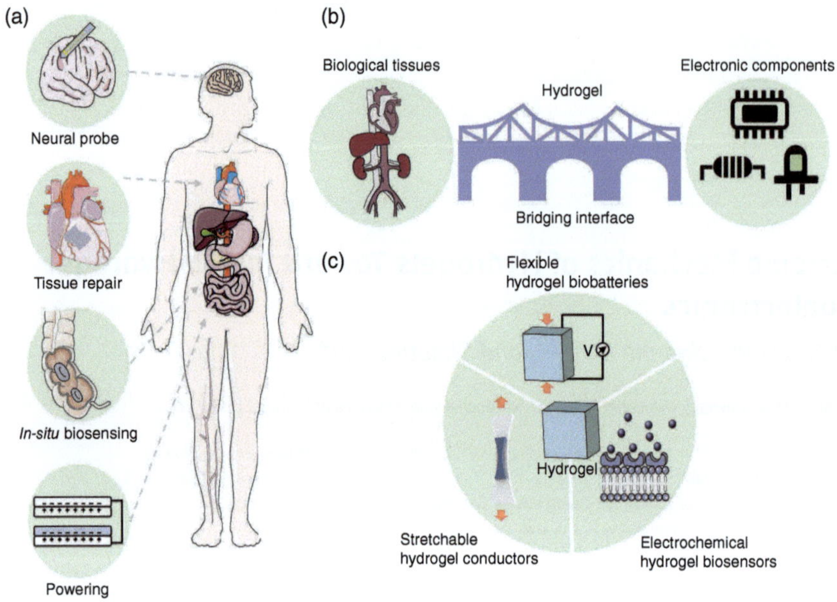

Figure 1.1 Overview of the *in-situ* hydrogel electronics. (a) *In-situ* bioelectronics revolutionized the field of medicine in a variety of ways. (b) Hydrogels form long-term, high-efficacy, multi-modal bridging interfaces between electronic components and biological systems. (c) Three key components of *in-situ* hydrogel bioelectronics include stretchable hydrogel conductors, electrochemical hydrogel biosensors, and flexible hydrogel biobatteries.

couplings [10, 11]. Hydrogels are typically constituted of two phases: one phase of interconnected polymer networks giving the solid-like network elasticity, and the other phase of infiltrated water molecules endowing the fluid-like transport property [12–15]. Due to the unique combination of polymer networks and water molecules, hydrogels show their superior softness, wetness, responsiveness, biocompatibility, and bioactivity, therefore having been regarded as an ideal material candidate to form long-term, high-efficacy, multi-modal bridging interfaces between electronic components and biological systems (Figure 1.1b) [16, 17].

Recently, a nascent field named hydrogel bioelectronics has rapidly evolved, exploiting hydrogels as key components for electronic devices that seamlessly interact with biological systems. The generic idea for the design of hydrogel bioelectronic device is to embed functional electronic components such as conductors, microchips, transducers, resistors, and capacitors inside or attached to the surface of a highly stretchable and tough hydrogel matrix, providing a soft, wet, and biocompatible environment interfacing with biological tissues [18]. As the hydrogel bioelectronic device is stretched, flexible electronic components can deform together with the device while rigid electronic components maintain their undeformed shapes, which involves large deformation of hydrogels around rigid electronic components [19, 20]. Therefore, to maintain reliability and functionality of the device, the hydrogel matrix and the interface between hydrogel and functional components need to be tough and robust. Pioneered by Gong et al., hydrogels with fracture toughness higher than 10^4 J/m^2 have been widely available [21–23]; additionally, initiated in the past 10 years, hydrogel adhesions to diverse substrates can successfully achieve interfacial fracture toughness above 10^3 J/m^2, on the same order as their biological counterparts [24–29]. The recent development of tough hydrogels and tough hydrogel adhesions as well as their synthesis and fabrication techniques have enabled various soft material technologies in the form of diverse hydrogel bioelectronics [18, 30, 31].

More recently, an emerging class of *in-situ* bioelectronics that combines hydrogel technologies with electronic components to create devices that can interact with harsh environments within the body, which we define as *in-situ hydrogel bioelectronics*, have been recently proposed with great promise in potentially addressing the limitations faced by existing *in-situ* bioelectronics [16, 31, 32]. Despite the promise, the complicated chemical, biological, and mechanical factors in the physiological environment pose significant challenges to the reliability and functionality of hydrogel electronic devices when operating within the body. What are the key properties of hydrogels and how to rationally design these properties for developing electronic devices that can operate in the physiological environment for reliable on-site sensing, stimulation, and powering? These are unanswered questions, even considering a growing number of reviews on hydrogel bioelectronics [16, 33–37].

This chapter aims to provide an overview of the design principles, implementation mechanisms, and manufacturing/fabrication techniques, particularly centering on extreme mechanics of hydrogels, for developing three key components of *in-situ* hydrogel bioelectronics (Figure 1.1c): (i) stretchable hydrogel conductors, (ii) electrochemical hydrogel biosensors, and (iii) flexible hydrogel biobatteries. The chapter is organized as follows. Section 1.2 will discuss polymer mechanics for rationally designing extreme mechanical and physical properties of hydrogels crucial for the development of *in-situ* hydrogel bioelectronics. Section 1.3 will discuss the multiscale orthogonal design of stretchable hydrogel conductors and strategies for implementing the orthogonal design. Section 1.4 will discuss the principles for achieving high-specificity and high-sensitivity of electrochemical hydrogel biosensors, specifically focusing on selective transport design in hydrogels and electrochemical sensing performance at hydrogel–electrode interfaces. Section 1.5 will briefly review the recent efforts in developing flexible hydrogel biobatteries by harvesting mechanical energy, chemical energy, and thermal energy. Section 1.6 will conclude the chapter with a set of future opportunities by integrating interdisciplinary efforts in ingestible sensors, neural interfaces, miniature robots, and data analytics.

1.2 Extreme Properties of Hydrogels by Polymer Network Design

Due to the unique combination of solid-like polymer networks and fluid-like water molecules, hydrogels exhibit superior softness, wetness, responsiveness,

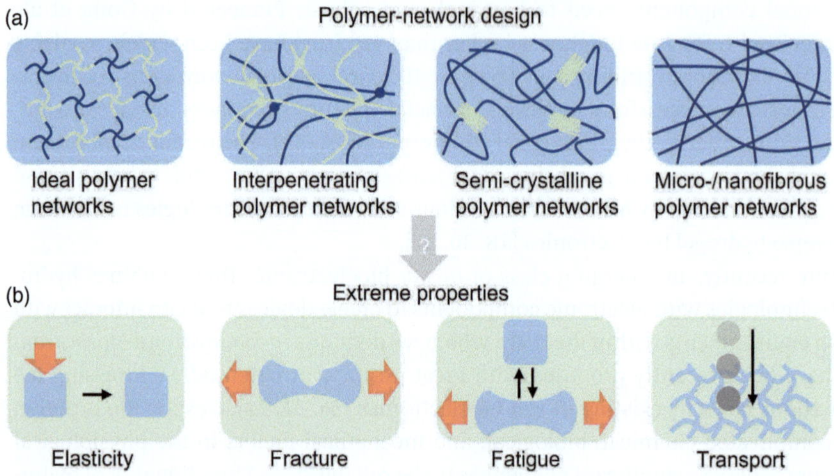

Figure 1.2 Extreme properties of hydrogels by polymer network design. (a) Schematics of unconventional polymer networks such as ideal polymer networks, interpenetrating polymer networks, semi-crystalline polymer networks, micro-/nanofibrous polymer networks. (b) Schematics of mechanical and physical properties critical for the development of *in-situ* hydrogel bioelectronics, including elasticity, fracture, fatigue, and mass transport.

biocompatibility, and bioactivity, therefore being exploited as key material candidates for developing *in-situ* hydrogel bioelectronics. These properties have been intensively studied by understanding the nonlinear elasticity [38], swelling [39–43], poroelasticity [44, 45], viscoelasticity [46–50], fracture [51–53], and fatigue [54–57] of hydrogels, following the pioneering work in the field of polymers and soft materials [58–69]. Despite these unique mechanical and physical properties of common hydrogels, the development of *in-situ* hydrogel bioelectronics also requires hydrogels to possess extreme mechanical and physical properties, such as tunable elastic modulus, extremely high values of fracture toughness and fatigue threshold, and tunable molecular transport. In this section, we will discuss the polymer mechanics to rationally push the limits of mechanical and physical properties of hydrogels including elastic modulus, fracture toughness, fatigue threshold, and mass transport that are crucial for the development of *in-situ* hydrogel bioelectronics (Figure 1.2).

1.2.1 Elastic Modulus

Elastic modulus is one of the most important properties of hydrogels used in biomedical applications, as it governs their ability to withstand deformations while maintaining compliance without damaging the surrounding soft tissues or causing adverse foreign body responses. There are several methods available to measure the elastic modulus of hydrogels, including tensile/compression, rheology, and indentation tests, as illustrated in Figure 1.3a. The shear elastic modulus of a hydrogel G can be measured by identifying the initial slope of the stress–strain curve in tensile or compression tests E, namely $G = \alpha E$, where α is a dimensionless perfector

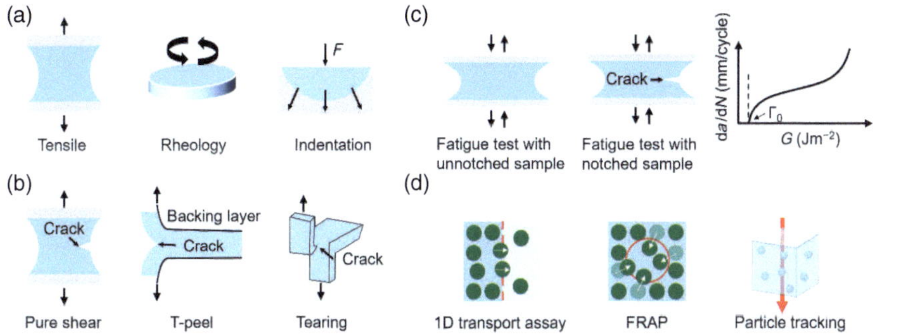

Figure 1.3 Common experimental methods to characterize mechanical and physical properties of hydrogels. (a) Tensile, rheology, and indentation tests to measure the elastic modulus of hydrogels. (b) Pure shear, T-peel, and tearing tests to measure the fracture toughness of hydrogels. (c) Pure shear fatigue tests to measure the fatigue threshold of hydrogels Γ_0 by plotting the crack extension da/dN versus the applied energy release rate G. (d) 1D transport assay, fluorescence recovery after photobleaching (FRAP), and particle tracking methods to measure the diffusivity of particles (e.g. ions, monomers, proteins, viruses) in hydrogels.

depending on samples' dimensions and material incompressibility [70]. Rheology test is the other method to measure the shear elastic modulus of a hydrogel, which is often preferred for soft hydrogels. Rheology tests can provide information on both storage and loss modulus across a range of frequencies, allowing for a quantitative decomposition of elastic and viscous contributions during deformation [71]. Indentation tests can also be used to characterize the shear elastic modulus, but the results may be affected by water migration and stress state. Therefore, a theoretical analysis is necessary to accurately interpret the indentation results and decouple nonlinear elasticity, viscoelasticity, and poroelasticity in hydrogels [72].

Assuming the polymer network of a hydrogel is fully amorphous with negligible molecular entanglements and crystalline domains, the shear elastic modulus of a hydrogel G can be theoretically predicted following the classical affine and phantom network theories of elasticity [12, 66],

$$\frac{G}{kT} = C\phi^{1/3}n \tag{1.1}$$

where ϕ is the volume fraction of dry polymers, n is the number of elastically active polymer chains per unit volume of dry polymers, k is the Boltzmann constant, T is the absolute temperature, and C is a constant that has a value of 1 when the polymer network of the hydrogel follows affine deformation and $1 - \frac{2}{f}$ when the polymer network of the hydrogel follows phantom deformation with f being the functionality of the polymer network (i.e. the number of chains connected to a junction point). By substituting the typical values of $C = 1$, $\phi = 0.1$, $n = 10^{24} - 10^{26}\ \text{m}^{-3}$, and $kT = 4.11 \times 10^{-21}$ J, the shear elastic modulus of a hydrogel is estimated on the order of 1–100 kPa.

Recent experiments have shown that the measured shear modulus of a hydrogel is consistently below the theoretical predictions by Eq. (1.1) due to the presence of molecular defects in a real hydrogel. Given the capability of counting the numbers of various orders of molecular defects in hydrogels, Olsen and Johnson and coworkers [73] developed a real elastic network theory to quantify the impact of molecular defects on shear elastic modulus,

$$\frac{G}{kT} = \frac{f-2}{f}\sum_i \varepsilon_i n_i \tag{1.2}$$

where n_i is the number of the type i polymer chains per unit volume of the hydrogel associated with a specific type of molecular defects, ε_i is the elastic effectiveness of the type i polymer chains. ε_i accounts for the elastic contribution of each polymer chain, having a value of 1 if the polymer chain is an ideal chain with no impact from defects and a value smaller than 1 if the polymer chain is a defective chain affected by surrounding molecular defects. The real elastic network theory suggests the critical role of polymer network architecture on hydrogel's elastic modulus. The design principle for achieving tunable elastic modulus of hydrogels is to engineer the type and density of molecular defects, ubiquitous in synthetic hydrogels and biological tissues [74] (Figure 1.4a).

1.2.2 Fracture Toughness

Fracture toughness, the energy required to propagate a unit area of crack surface under monotonic load, defines the ability of a material to resist crack extension under stress. Fracture toughness of a hydrogel is crucial for maintaining reliability of *in-situ* hydrogel bioelectronics [10, 11]. As illustrated in Figure 1.3b, the fracture toughness of a hydrogel can be measured through pure shear, T-peel, and tearing tests, originally proposed by Rivlin and Thomas for measuring the fracture toughness of rubbers [75]. In a pure shear test, the fracture toughness of a hydrogel Γ is determined by the critical energy release rate applied on a notched sample G_c, $\Gamma = G_c = W(\lambda_c)H$, where W is the strain energy density stored in the unnotched sample with λ_c being the critical stretch for crack extension in the notched sample, and H is the sample height [53]. T-peel tests can also be used to measure the fracture toughness of hydrogels [76], where a hydrogel layer with a precut is sandwiched between two inextensible backing layers. The fracture toughness of the hydrogel is then calculated via $\Gamma = 2F_{SS}/b$, where F_{SS} is the steady-state plateau force, and b is the specimen's thickness. Tearing tests, also known as trouser tests, are another commonly adopted method for measuring hydrogel toughness [77]. Unlike pure shear and T-peel tests with the crack deformed in Mode I, the crack in a tearing test is deformed in Mode III (out-of-plane shear loading). The fracture toughness of a hydrogel in a tearing test is determined by $\Gamma = 2F_{SS}/b$, where F_{SS} is the steady-state plateau force, and b is the specimen's thickness.

Common hydrogels are intrinsically brittle [78]. The intrinsic fracture energy of a hydrogel Γ_0 can be calculated following the classical Lake–Thomas theory [12, 61, 79],

$$\Gamma_0 = \phi^{2/3} \cdot n\sqrt{N}b \cdot NU = \phi^{2/3} nbN^{3/2}U \tag{1.3}$$

where ϕ is the volume fraction of dry polymers, $n\sqrt{N}b$ is the number of elastically active polymer chains per unit area of crack surface, with n being the number of

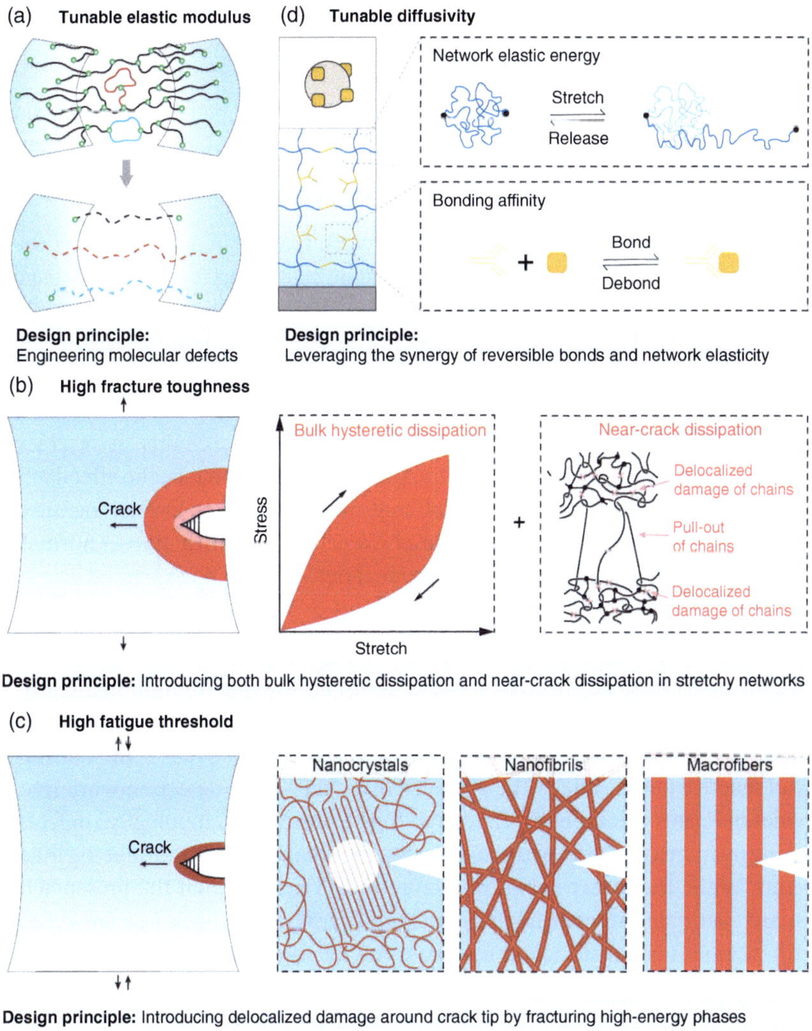

Figure 1.4 Design principles for extreme properties of hydrogels. (a) Achieving tunable elastic modulus by engineering type and density of molecular defects in hydrogels. (b) Achieving high fracture toughness of hydrogels by introducing both bulk hysteretic dissipation and near-crack dissipation in stretchy polymer networks. (c) Achieving high fatigue threshold of hydrogels by introducing high-energy phases to impinge fatigue crack extension. (d) Achieving tunable diffusivity in hydrogels by leveraging the synergy of reversible bonds and network elasticity.

elastically active polymer chains per unit volume of dry polymers, N being the number of Kuhn monomers in each polymer chain, b being the length of each Kuhn monomer, NU is the energy required to fracture a polymer chain, with U being the energy required to fracture a single Kuhn monomer. It is commonly assumed that the dry polymers of a hydrogel satisfy the volume conservation following $Nnb^3 = 1$ with b^3 being the volume of a Kuhn monomer. By imposing the volume conservation, the intrinsic fracture energy of a hydrogel can be calculated via,

$$\Gamma_0 = \phi^{2/3} b^{-2} N^{1/2} U \tag{1.4}$$

By substituting the typical values of $\phi = 0.1$, $b = 10^{-9}$ m, $N = 100-10,000$, and $U = 100\text{kT} = 4.11 \times 10^{-19}$ J, the intrinsic fracture energy of a hydrogel Γ_0 is estimated as low as 1–10 J/m^2, orders of magnitude lower than biological tissues around 1000 J/m^2.

Lin and Zhao recently developed a defect-network fracture model to predict the intrinsic fracture energy of polymer networks containing various types of topological defects including cyclic loops and dangling chains [80]. The defect-network fracture model is inspired by the real elastic network model [73] or, more generally, the phantom network model [66, 81] discussed in Section 1.2.1. Analogous to the previous elastic models, the key idea of this fracture model is to introduce effectively longer fractured chains to account for the energy of the fractured chains on the crack path. However, physically different from the previous elastic models, the effectively longer chains in the fracture model do not change the density of the layer of fractured chains. The intrinsic fracture energy of the polymer network with defects normalized by that of the corresponding defect-free ideal network can be expressed as:

$$\overline{\Gamma} = \sum_X (\gamma_X - 1) n_{\text{X}}^{\text{affected}} - \sum_X n_{\text{X}}^{\text{inactive}} + 1 \tag{1.5}$$

where $\gamma_X \geq 1$ is the fracture effectiveness to account for the contribution to intrinsic fracture toughness by a single polymer chain affected by defect X, $n_{\text{X}}^{\text{affected}}$ is the number density of affected chains due to the presence of defect X, $n_{\text{X}}^{\text{inactive}}$ is the number density of inactive chains due to the presence of defect X. The defect-network fracture model indicates a competing effect due to the presence of topological defects: toughening by increasing effective chain length and weakening by introducing inactive chains. While the defect-network fracture model predicts that the presence of defects can potentially enhance intrinsic fracture energy of hydrogels by a few times, such enhancement is not sufficient to ensure the reliable use of hydrogels in engineering applications.

In the past 20 years, the fracture toughness of hydrogels has been significantly enhanced above 1000 J/m^2, making tough hydrogels key load-bearing components for devices and machines [82, 83]. The generic toughening mechanism of hydrogels is to incorporate two physical processes: one is the scission of a layer of polymer chains on the crack tip, and the other is the bulk hysteretic dissipation around the crack tip such as Mullins effect and viscoelasticity [22, 23, 51, 78, 84–91]. The first process defines the intrinsic fracture energy Γ_0 as discussed in Eqs. (1.3)–(1.5), and the second process defines the bulk hysteretic dissipation's contribution $\Gamma_{\text{D}}^{\text{bulk}}$. Conceptually, the total fracture toughness of a tough hydrogel can be expressed as:

$$\Gamma = \Gamma_0 + \Gamma_{\text{D}}^{\text{bulk}} \tag{1.6}$$

which is often named the bulk dissipation model. The value of $\Gamma_{\text{D}}^{\text{bulk}}$ can be estimated by $\Gamma_{\text{D}}^{\text{bulk}} = U_{\text{D}} L_{\text{D}}$ with U_{D} being the energy for breaking sacrificial bonds per unit volume of the material and L_{D} being the length of the process zone around crack tip where breaking sacrificial bonds occurs. U_{D} can be estimated by the bond energy of one sacrificial bond times the number of sacrificial bonds per unit volume of the material, on the order of 10^6 J/m^3. Since the value of L_D can reach 1 mm [92], the value of Γ_D^{bulk} can reach 1000 J/m^2, with the orders of magnitude larger than Γ_0.

The bulk dissipation model has been widely used to qualitatively explain the toughening mechanisms in diverse soft tough materials [21, 23, 93–95], but recent study shows that the bulk dissipation model significantly underestimates the toughness enhancement of tough hydrogels [93]. The missing term is attributed to a near-crack dissipation that does not rely on bulk hysteresis [96–98]. To account for both bulk hysteretic dissipation and near-crack dissipation in soft tough materials, an extreme toughening model was recently proposed, indicating that the total fracture toughness of a tough hydrogel exhibiting both bulk hysteretic dissipation and near-crack dissipation can be expressed as:

$$\Gamma = \Gamma_0 + \Gamma_D^{bulk} + \Gamma_D^{tip} \tag{1.7}$$

where Γ_0 is the intrinsic fracture energy, Γ_D^{bulk} is the bulk hysteretic dissipation's contribution to fracture toughness, and Γ_D^{tip} is the near-crack dissipation's contribution to fracture toughness. A governing equation for the extreme toughening model can be further derived as:

$$\frac{\Gamma}{\Gamma_0} = \frac{\beta}{1 - \alpha h_m} \tag{1.8}$$

where $\beta = \left(\Gamma_0 + \Gamma_D^{tip}\right)/\Gamma_0 \geq 1$ is a dimensionless number to account for the near-crack dissipation due to molecular entanglements, $0 \leq \alpha \leq 1$ is a dimensionless number depending on the stretch-dependent hysteresis of the bulk materials ($\alpha = 1$ for highly stretchable materials), and $0 \leq h_m < 1$ is the maximum stress-stretch hysteresis of the bulk material. While the cause of the near-crack dissipation is not fully understood, the reported experiments have shown the potential for achieving high values of β up to 10, suggesting the crucial toughening role by the near-crack dissipation. To summarize, the design principle for achieving high fracture toughness in hydrogels is to introduce both bulk hysteretic dissipation (e.g. Mullins effect and viscoelasticity) and near-crack dissipation (e.g. molecular entanglements) in stretchy polymer networks, as depicted in Figure 1.4b.

1.2.3 Fatigue Threshold

Fatigue threshold, the energy required to propagate a unit area of crack surface under cyclic load, defines the ability of a material to resist fatigue crack extension under stress [55, 99, 100]. Fatigue threshold of hydrogels is critical for achieving longevity of *in-situ* hydrogel bioelectronics. As illustrated in Figure 1.3c, the measurement of fatigue threshold of hydrogels requires cyclic loading of unnotched and notched hydrogel specimens. The fatigue tests for hydrogels are typically performed in a chamber with controlled humidity or water bath to ensure the sample reaches an equilibrium state [54, 56]. By monitoring the crack extension Δa at a controlled energy release rate G, one can identify a critical energy release rate at the intersection of abscissa axis as the measured fatigue threshold Γ_0.

While hydrogels have been made tough with high toughness above 1000 J/m^2, as discussed in Section 1.2.2, these tough hydrogels still suffer from fatigue fracture when subjected to prolonged cyclic loading [54, 79, 97, 101]. The experimental findings conclude that the resistance to fatigue crack propagation after prolonged cycles of loads is the energy required to fracture a single layer of polymer chains (i.e. the intrinsic fracture energy of the hydrogel), which is unaffected by the additional dissipation mechanisms introduced in tough hydrogels [56]. To address the challenge of fatigue failures in conventional tough hydrogels, we and others have proposed a general design principle for fatigue-resistant hydrogels (Figure 1.4c) – inducing delocalized damage around the crack tip by fracturing high-energy phases, such as nanocrystals [28, 56, 100, 102–104], micro–/nanofibers [105], and macro-fibers [92, 106] in hydrogels. Additionally, hierarchical molecular structure design such as introducing bi-continuous phase networks can suppress fatigue-induced crack advance [57, 107].

The fatigue threshold of fatigue-resistant hydrogels containing high-energy phases can be qualitatively calculated by modifying the Lake–Thomas theory:

$$\Gamma_0 = \phi^{2/3} n l_d N U \tag{1.9}$$

where ϕ is the volume fraction of dry polymers, n is the number of elastically active polymer chains per unit volume of dry polymers, NU is the energy required to fracture a polymer chain with N being the number of Kuhn monomers in each polymer, U being the energy required to fracture a single Kuhn monomer, and l_d is the length scale of the delocalized damage and understood as the crack processing zone length at the threshold. In conventional hydrogels, l_d is equal to the length of a single layer of polymer chains, i.e. $l_d = \sqrt{N}b$; in fatigue-resistant hydrogels, l_d is larger than the length of a single layer of polymer chains, i.e. $l_d > \sqrt{N}b$. Since the parameters ϕ, N, and U are intrinsic properties of hydrogels with little room to engineer, the strategies to enhance fatigue threshold of hydrogels mainly rely on the mechanisms that can significantly enlarge the value of l_d, which can be achieved by introducing intrinsically high-energy phases above mentioned.

1.2.4 Mass Transport

Diffusion in a hydrogel, movement of particles (e.g. ions, monomers, proteins, and viruses) through the hydrogel, is a ubiquitous phenomenon in nature and a fundamental process that governs the working principles in diverse applications. For example, the distinct diffusion of different biomarkers in hydrogels determines the sensing sensitivity and sensing specificity in electrochemical hydrogel biosensors discussed in Section 1.4; the stress-induced diffusion of ions in nano channels of hydrogels governs the stress-voltage coupling in flexible hydrogel batteries discussed in Section 1.5. The diffusivity of particles in hydrogels can be commonly characterized by 1D transport assay, fluorescence recovery after bleaching (FRAP), and particle tracking methods (Figure 1.3d).

The mode of diffusion in a hydrogel is determined by the mesh size of the hydrogel. When the hydrogel's mesh size is much larger than the size of the substance, the substance moves as Brownian diffusion, the diffusivity of which is governed by the viscosity of solvent (i.e. water) in the hydrogel [108]:

$$D = \frac{kT}{3\pi\eta d} \tag{1.10}$$

where k is the Boltzmann constant, T is the absolute temperature, η is the dynamic viscosity of water, and d is the diameter of the substance. By substituting the typical values of $kT = 4.21 \times 10^{-21}$ J, $\eta = 8.9 \times 10^{-4}$ Pa · s, and $d = 1$ nm, the diffusivity of the substance in a hydrogel with large mesh size is estimated as 10^{-10} m/s^2. When the hydrogel's mesh size is on the same order as the size of the substance, the diffusivity of the substance in a hydrogel is governed by the hydrogel's mesh size ξ [109]:

$$\frac{D}{D_0} \sim \exp\left(-\frac{d}{\xi}\right) \tag{1.11}$$

where D_0 is the diffusivity of the substance in water. For a hydrogel with its mesh size two times the substance diameter (i.e. $d/\xi = 2$), the diffusivity of the substance reduces by 10 times on the order of 10^{-11} m/s^2 .

Existing efforts have been mostly focused on Brownian diffusion of particles in a fluid, the diffusivity of which is governed by the viscosity of fluids. In contrast, when particles diffuse in the polymer networks of a hydrogel, the transport of particles is a discrete process of particles making stochastic hops between neighboring sites, namely hopping diffusion, the diffusivity of which is governed by the energy required to overcome the elasticity of polymer networks [110]. Such network elasticity can be readily tuned by mechanical deformation applied on the hydrogel. This implies the potential of harnessing mechanical deformation as a new design space to program particle transport in hydrogels. Specifically, for a particle with size d confined in a cage formed by polymer chains with mesh size ξ slightly smaller than particle size (i.e. $\xi < d$), the particle can still escape from one cage to the other neighboring cage by overcoming the energy barrier due to network elasticity. The probability for such escape P is determined by the energy barrier E_e via $P = \int_d^\infty \frac{e^{-E_e/kT}}{\xi} dx$ where k is the Boltzmann constant and T is the absolute temperature, resulting in a reduced hopping diffusivity [110],

$$D_{network} = D_0 P \tag{1.12}$$

where D_0 is the particle diffusivity in a hydrogel free of network elasticity. The fundamental understanding of how mechanical deformation modulates particle transport in soft materials remains an unexplored research topic but will potentially lead to multiple previously inaccessible technologies that rely on mechano-transport design in hydrogels, including but not limited to force-sensitive cargo for on-demand drug delivery, and strain-programmable tissue adhesive for prolonged tissue repair.

In addition to network elasticity, hopping diffusion in hydrogels can also be induced by reversible interactions between particles and functional groups in hydrogels [111]. During a hopping event, multipoint particle attachment results in caging. As one or more of the attached points can be released, the particle can escape from one cage to the other neighboring cage, resulting in hopping diffusivity,

$$D_{\text{reversible}} = \sum_{j=2}^{N} k_{\text{off}} P_j D_0. \tag{1.13}$$

where P_j is the probability of a particle maintaining j number of bonding sites, N is the number of bonding sites on the particle, D_0 is the particle diffusivity with one bonding site ($N = 1$), and k_{off} is the dissociate rate of the reversible interaction. Harnessing hopping diffusion by reversible interactions enables selective transport of chemical and biological cargoes based on reversible cargo-barrier interactions [112, 113]. Overall, the principle to tune particle diffusivity in hydrogels is to leverage the synergy of network elasticity and reversible bonds (Figure 1.4d).

1.3 Stretchable Hydrogel Conductors

Hydrogel-based electronic materials that can conduct electricity while being able to stretch and deform without fracture, known as stretchable hydrogel conductors, hold great potential as an alternative to traditional metallic conductors in the development of *in-situ* hydrogel bioelectronics due to their unique combination of tissue-like properties and electrical conductivity [18, 114]. Despite the promising potential, existing stretchable hydrogel conductors face a technical challenge in reconciling superior mechanical toughness and high electrical conductivity, limiting their use in *in-situ* hydrogel bioelectronics that require both mechanical and electrical properties [115–117]. Currently, most stretchable hydrogel conductors consist of a mixture of electrically conductive fillers in stretchable hydrogel matrices, resulting in low electrical conductivity due to low connectivity between electrical phases in the hydrogel matrices. Although increasing the amount of conductive filler can improve the electrical conductivity by forming a percolation network, this approach significantly compromises the hydrogel's mechanical properties by reducing stretchability and fracture toughness.

1.3.1 Multiscale Orthogonal Design

The design principle of stretchable hydrogel conductors is depicted in Figure 1.5, where the integration of electrical and mechanical phases is accomplished by a bi-continuous structure with an orthogonal design of each phase. To achieve high electrical conductivity, various electrical fillers such as ions [118, 119], conductive monomers [120, 121], conductive polymers [122–124], conductive nanoparticles [125, 126], and conductive macro fillers [127, 128] have been used. On the other hand, to achieve superior mechanical elasticity, stretchable hydrogels with interpenetrating polymer networks [21, 23, 129], hybrid crosslinkers [130–132], high-functionality crosslinkers [133–135], monodispersed polymer chains [136, 137], and meso-/macro composites [138–140] have been designed. While high electrical conductivity and superior mechanical elasticity have been

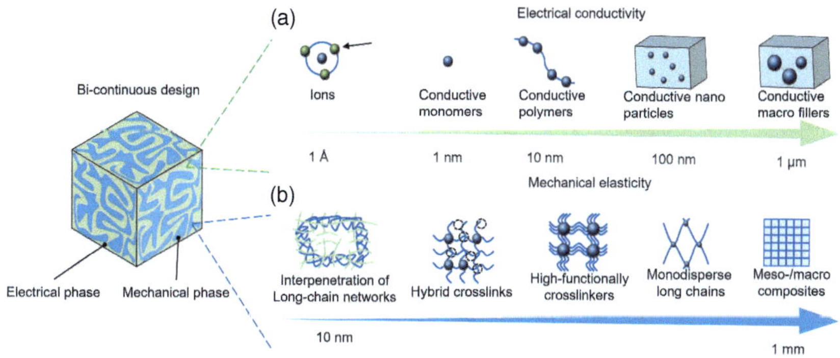

Figure 1.5 Design of stretchable hydrogel conductors relying on a bi-continuous integration of electrical and mechanical phases. (a) The high electrical conductivity in the electrical phase can be achieved by introducing electrical fillers including ions, conductive monomers, conductive polymers, conductive nanoparticles, and conductive macro fillers. (b) The superior mechanical elasticity in the mechanical phase can be achieved by designing stretchable hydrogels with interpenetrating polymer networks, hybrid crosslinkers, high-functionality crosslinkers, monodispersed polymer chains, and meso-/macro composites. Source: Reproduced with permission from Ref. [78]. Copyright 2014 The Royal Society of Chemistry.

demonstrated individually, the integration of these two phases is challenging due to their significant mechanical mismatch. The elastic modulus of the electrical phase (~1 GPa) is typically much larger than that of the mechanical phase (hydrogels, ~1–100 kPa), leading to stress concentrations at their interface and surface delamination when highly deformed. Furthermore, the fracture toughness of the electrical phase (~10 J/m^2) is typically much lower than that of the mechanical phase (tough hydrogels, 1000 J/m^2), resulting in microcrack formation and decreased electrical conductivity. Despite numerous individual efforts to enhance electrical conductivity and mechanical elasticity, implementing the orthogonal design through the bi-continuous integration of the electrical and mechanical phases remains a challenge.

1.3.2 Implementations of the Orthogonal Design

Figure 1.6 illustrates the typical ways in which the orthogonal design of stretchable hydrogel conductors can be implemented. A biocompatible, highly stretchable, and robust hydrogel provides a soft and wet matrix for electronic components, drastically different from dry elastomer/polymer matrices used in conventional stretchable electronics. Electronic components including stretchable, rigid, and brittle conductors are either embedded inside or attached to the surface of the hydrogel. When the stretchable hydrogel conductor is stretched, stretchable electronic components can deform together with the conductor, while rigid electronic components maintain their undeformed shapes, which requires robust interfaces between electronic components and hydrogel

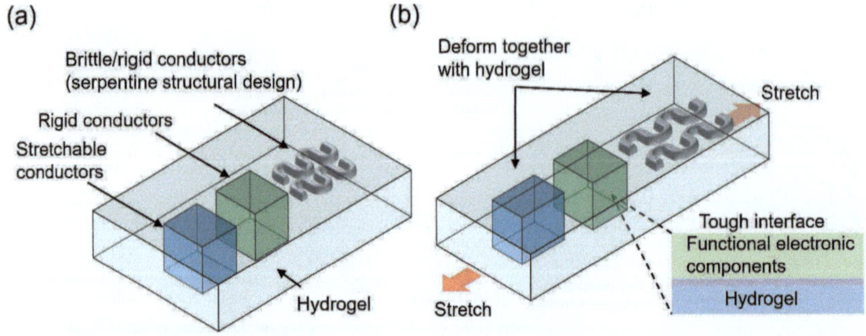

Figure 1.6 General strategies to implement the orthogonal design of stretchable hydrogel conductors. (a) Electronic components including stretchable, rigid, and brittle conductors are embedded inside or attached on the surface of the hydrogel. (b) As the stretchable hydrogel conductor is stretched, stretchable electronic components can deform together with the conductor, but rigid electronic components maintain their undeformed shapes, which requires robust interfaces between electronic components and hydrogel matrix. Additionally, brittle electronic components require serpentine structural designs to reduce the energy for driving the crack formation in electronic components. Source: Reproduced with permission from Ref. [18]. Copyright 2015 John Wiley & Sons, Inc.

matrix. In addition, brittle electronic components require serpentine structural designs to minimize the energy for driving crack formation in the electronic components.

While a stretchable hydrogel conductor with intrinsically stretchable electronic components does not require special interfacial or structural design, it requires material innovation for the design of intrinsically stretchable conductive materials. Conventional conductive materials have difficulty forming long-term stable and conformable interfaces with both biological and synthetic hydrogels due to severe modulus mismatch. For example, carbon-based conductive materials like graphene have moduli of 1 TPa; metallic conductive materials like silicon have moduli of 50 GPa; even conducting polymers such as poly(3,4-ethylenedioxythiophene) polystyrene sulfonate (PEDOT: PSS), polyaniline (PANI), and polypyrrole (PPy), possessing superior flexibility compared to inorganic conductors, typically have moduli on the order of 1 GPa [141]. PEDOT: PSS-based hydrogels [13, 142], owing to their favorable compatibility, processability, and reproducibility, represent one of the most common material candidates for intrinsically stretchable electronic components. Figure 1.7a summarizes the recent efforts in reconciling electrical conductivity and mechanical stretchability in PEDOT: PSS hydrogels through interpenetrating polymer networks [142], dispersing conductive nanofillers [148], and engineering PEDOT: PSS nanofibrils [13, 149]. Liquid metal, a metal alloy including gallium, indium, and tin, is another class of intrinsically stretchable conductive materials, recently embedded in hydrogels [150, 151]. Although liquid metal shows unique properties such as high conductivity, excellent thermal stability, and ease of processing, it is also corrosive, reactive with water or oxygen, and requires safety precautions.

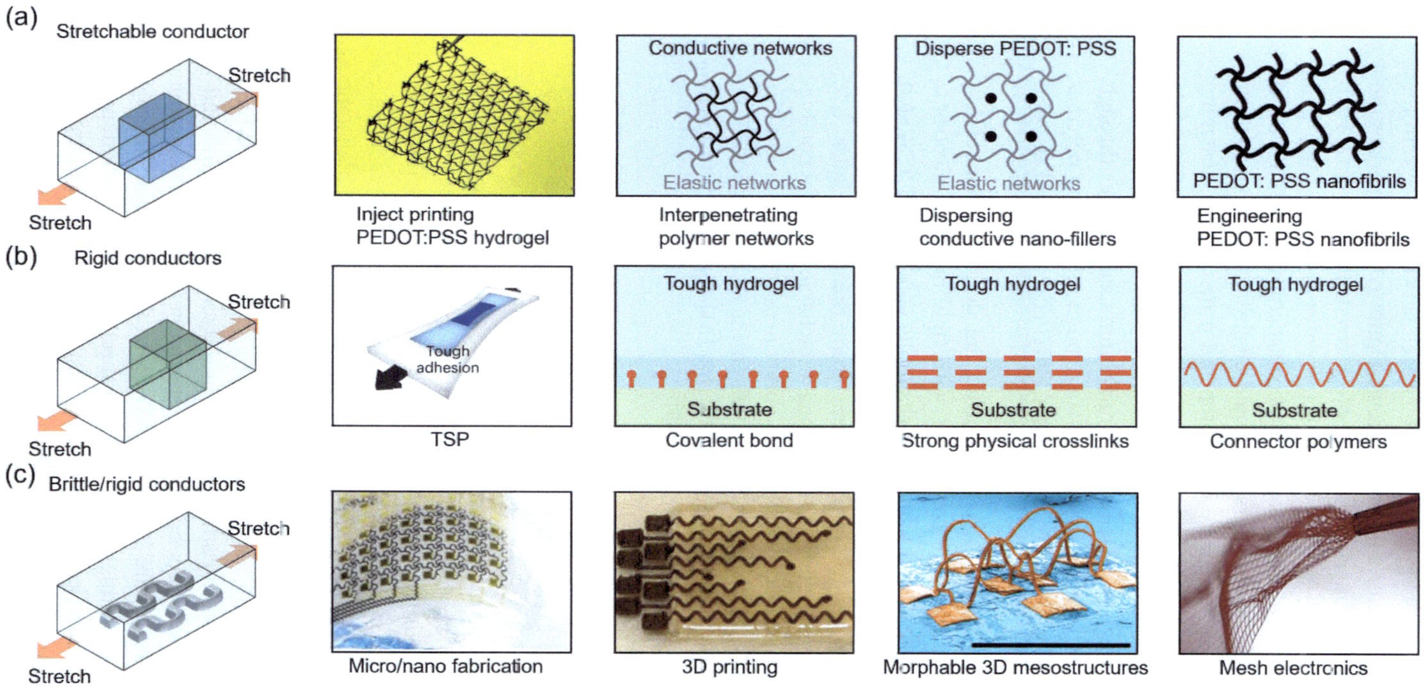

Figure 1.7 Representative examples to implement the orthogonal design of stretchable hydrogel conductors. (a) The implementation of a stretchable hydrogel conductor with stretchable conductors requires material innovations for the design of intrinsically stretchable conductive materials such as PEDOT: PSS-based hydrogels shown in the left image (Source: Ref. [13]/Springer Nature/CC BY 4.0). The design of PEDOT: PSS hydrogels typically relies on interpenetrating polymer networks, dispersing conductive nanofillers, and engineering PEDOT: PSS nanofibrils. (b) The implementation of a stretchable hydrogel conductor with rigid conductors requires strong and tough interfaces between the electronic components and the hydrogel matrix such as TSP recently developed by Jin and Bao and coworkers shown in the left image (Source: Reproduced with permission from Ref. [143]. Copyright 2022 Springer Nature). The design of strong and tough interfaces typically relies on covalent bonds, strong physical crosslinks, and connector polymers. (c) The implementation of a stretchable hydrogel conductor with brittle/rigid conductors also requires serpentine structural design, which typically relies on advanced manufacturing techniques such as micro-/nanofabrication (Source: Ref. [144]), 3D printing (Source: Ref. [145]), morphable 3D meso-structures (Source: Ref. [146]), and syringe-injectable mesh electronics (Source: Ref. [147]).

When incorporating rigid or nonstretchable electronic components into a stretchable hydrogel, the design of strong and tough interfaces between the electronic components and the hydrogel matrix becomes crucial. Figure 1.7b summarizes the recent efforts to develop strong interfacial linkages (e.g. covalent bonds [152, 153], strong physical crosslinks [154–156], and connector polymers [157, 158]) and introduce bulk dissipation mechanisms (such as viscoelasticity and Mullins effect) to create stretchable hydrogel conductors with rigid electronic components. For instance, Jin and Bao and coworkers recently proposed a method for designing a tough interface between a brittle semiconducting thin film and a stretchable substrate that effectively delayed microcrack formation in the film when stretched [143]. They achieved the tough interface using strong interfacial linkages through surface chemistry and dynamic bulk dissipations via a tough self-healing polymer matrix (TSP) that repeatedly dissipates energy via dynamic bond breakage and reformation.

In addition to interfacial design, serpentine structural design is another key feature of many stretchable hydrogel conductors with rigid or brittle electronic components. This undulating design allows the electronic components to stretch and deform together with the hydrogel matrix while minimizing the strain on the electronic components and preventing them from breaking. Figure 1.7c provides an overview of notable efforts in designing stretchable hydrogel conductors with serpentine-structured brittle electronic components. Implementing the serpentine structure design requires advanced manufacturing techniques, including micro-/nanofabrication [144, 159], 3D printing [145, 160, 161], stereographic design [146, 162, 163], and syringe injection [147, 164, 165].

1.4 Electrochemical Hydrogel Biosensors

Electrochemical biosensors have tremendous potential for medical diagnostics as they detect and measure the presence of biological molecules by converting biological events into electrical signals [166, 167]. However, previous efforts have primarily focused on graphene-based sensing materials and metallic electrodes, which have limitations when targeting *in-situ* detection. For instance, the presence of high ionic strength in body fluids produces pronounced screening of electrical fields, significantly reducing the sensitivity of the electrical chemical detection [168]. Additionally, biomarker concentrations in body fluids are typically about 10 ng/ml, which is too low to be detected by most existing electrical biosensors [169]. Recent studies show that introducing hydrogels into existing electrochemical biosensors can provide two unique benefits (Figure 1.8a) [170, 171]. First, hydrogels can act as a protective and selective layer for highly sensitive and specific *in-situ* detection, preventing interferences from biological environments (Figure 1.8b). Second, hydrogels can serve as surface modification to potentially push the limits of electrochemical properties of sensing materials, including Debye length, surface capacitance, and band gap (Figure 1.8c).

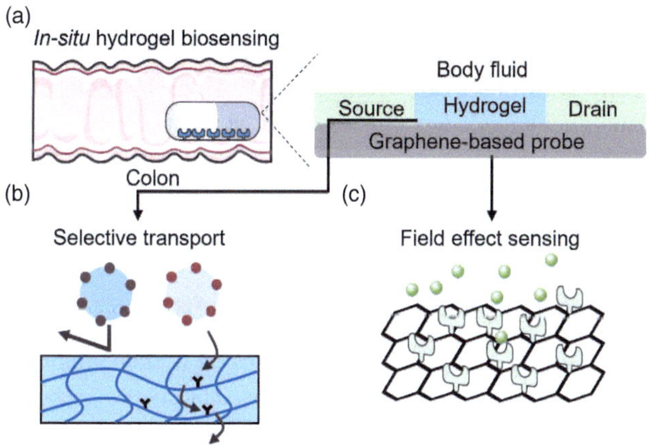

Figure 1.8 Design of electrochemical hydrogel biosensors. (a) Illustration of an electrochemical hydrogel biosensor for *in-situ* detection biomolecules. (b) Hydrogels act as a protective and selective layer for sensing materials. (c) Hydrogels serve as surface modification to potentially push the limits of electrochemical properties of sensing materials.

1.4.1 Selective Transport Design of Hydrogels

Efforts to tailor the transport properties of hydrogels mainly rely on two approaches: filtering by size (Figure 1.9a) and filtering by covalent interaction (Figure 1.9b) [172]. However, neither approach is suitable for selectively controlling the transport of biomolecules in body fluids. Filtering by size is impractical due to the small size difference between most biomolecules in body fluid; while filtering by covalent interaction is possible to distinguish the transport between target and nontarget biomolecules, such covalent interaction is only limited at the outmost surfaces of hydrogels and very challenging to achieve an enhanced transport throughout hydrogels due to the resistance from these covalent interactions. Recent studies have shown that introducing reversible interactions between target biomolecules and functional groups in hydrogels has the potential to achieve selective transport of target biomolecules across hydrogels [111, 112, 173, 174] (Figure 1.9c). The dissociation and reforming rates of the reversible interactions determine the mode and speed of biomolecular transport across hydrogels. Additionally, the network

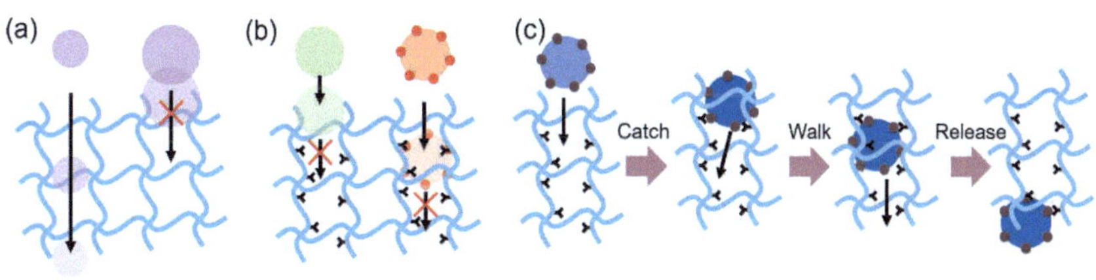

Figure 1.9 Selective transport design in hydrogels. (a) Filtering by size. (b) Filtering by permanent interaction. (c) Filtering by reversible interactions.

elasticity of hydrogels plays an important role in controlling transport. Future opportunities may focus on the synergistic role of network elasticity and reversible interaction in independently controlling the transport of various biomolecules in hydrogels. Such fundamental knowledge may also aid in understanding the selective transport mechanisms employed in biological systems, such as sperm selection by mucus layers.

1.4.2 Electrochemical Design of Hydrogel-2D Material Interfaces

The other challenge faced by *in-situ* electrochemical biosensors is their inability to detect biological events in high ionic strength solutions, ubiquitous in body fluids [166, 175]. The presence of 0.9 wt% mobile ions (Na^+, K^+, Cl^-) in body fluids leads to significant screening of electrical fields, which greatly reduces the sensitivity of electrochemical detection. For instance, the Debye length for 2D sensing materials in body fluids is typically below 1 nm, making it almost impossible to detect proteins (typically around 10 nm) above the Debye length [168]. A potential solution is to engineer the interface between the hydrogel and 2D material to enhance the field-effect sensing performance. Recent studies have shown that surface-functionalized 2D materials can achieve a 10-fold increase in Debye length from 0.82 to 9.6 nm in high ionic strength solutions [170]. One possibility is to graft crosslinked polymer networks of hydrogels onto existing sensing materials and explore the potential of optimizing the network topology and charge density in hydrogels to push the limits of electrochemical properties such as Debye length, surface capacitance, and band gap.

1.5 Flexible Hydrogel Biobattery

A flexible biobattery is a device that converts low-grade energy within the human body into usable energy and is essential for developing self-powered *in-situ* bioelectronic devices [176–179]. Hydrogels are a promising material for use in these devices because they are porous and can modulate ion and electron transport while minimizing potential leakage compared to traditional organic aqueous electrolytes. Furthermore, hydrogels are biocompatible and soft, which reduces damage to surrounding tissues. However, the power outputs of flexible hydrogel biobatteries are still relatively low and do not meet the power requirements of most *in-situ* bioelectronic devices. Figure 1.10a shows the power range and operation time requirements of common biomedical devices [180]. To develop the next generation of self-sustaining *in-situ* bioelectronic devices, high-performing hydrogels must be leveraged to harness various forms of energy within the human body (Figure 1.10b). This potential remains largely unexplored but is highly desirable. This section will discuss recent efforts in exploring the potential of powering *in-situ* bioelectronic devices through mechanical, chemical, and thermal energy harvesters (Figure 1.10c).

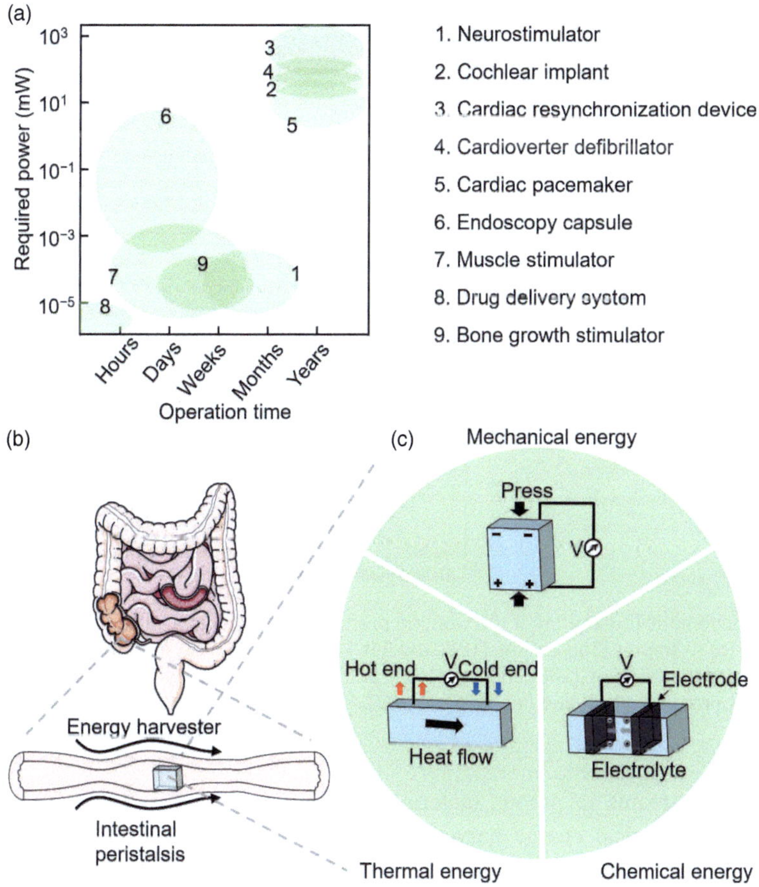

Figure 1.10 Flexible hydrogel batteries to self-power *in-situ* bioelectronic devices. (a) Examples of common implantable medical devices, with their required power supply and operation time (Source: Ref. [180]/Oxford University Press). (b) Schematic illustrations of the energy harvester working inside the intestine and (c) three forms of energy harvesting, including mechanical, chemical, and thermal energy, generated within the human body.

1.5.1 Mechanical Energy Harvester

Mechanical energy harvesters are devices that can convert the mechanical energy of the human body or organs into electrical energy for powering *in-situ* bioelectronic devices. One prime example of a mechanical energy harvester is the triboelectric nanogenerators (TENGs) [181–183], which generates an electrostatic potential difference between two materials of diverse polarities due to the triboelectric effect, causing a transfer of charges and the formation of an electric potential difference between them (Figure 1.11a) [177, 190–192]. Figure 1.11b presents a representative example of using ultrasound to induce vibrations and harness triboelectricity for in-body powering [184, 185]. Another way to harvest mechanical energy in the body is by leveraging the fluid-electro-mechanical coupling

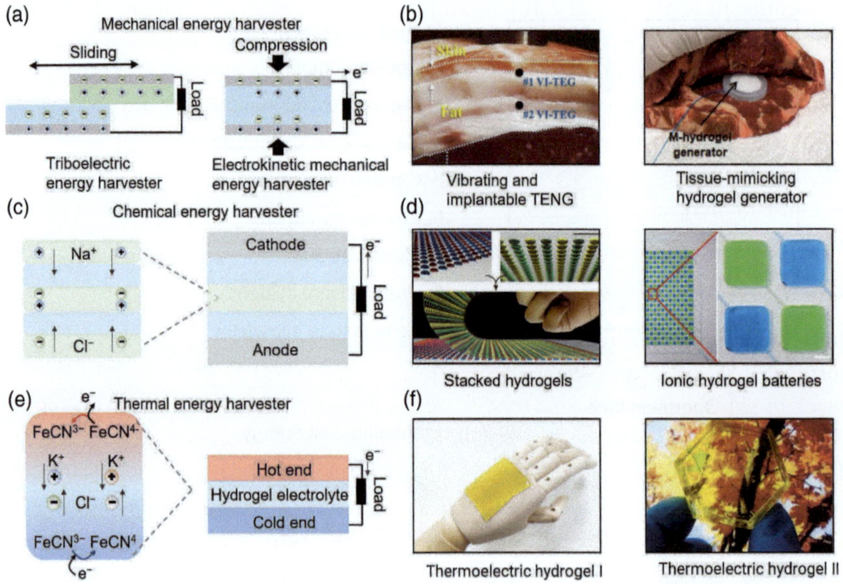

Figure 1.11 Schematic illustration of the working mechanism of (a) mechanical energy harvester with (b) examples (Source: Ref. [184] and Ref. [185]), (c) chemical energy harvester with (d) examples (Source: Ref. [186] and Ref. [187]), and (e) thermal energy harvester with (f) examples (Source: Ref. [188] and Ref. [189]. © 2022/Elsevier).

of electrokinetic streams in porous materials to generate electricity [193, 194]. Electrokinetic mechanical energy harvesters typically involve applying external forces, such as pressure, to drive the movement of micro-/nanofluidic water across a porous membrane, thereby causing the motion of ions to produce electricity. Unlike conventional TENGs, electrokinetic mechanical energy harvester can harvest low-frequency body motions while potentially producing high power output [195].

1.5.2 Chemical Energy Harvesters

Chemical energy harvesters are devices that convert chemical energy into electrical energy through chemical reactions in an electrolyte (Figure 1.11c) [196, 197]. These devices usually have two electrodes and an electrolyte, where the electrolyte acts as a mediator for the chemical reactions. The chemical reactions that occur between the positive and negative electrodes cause the flow of electrons within the electrolyte, thus generating electrical energy that can be used in the circuit. Hydrogel, with its watery nature, is an excellent carrier for chemicals, and acts as the electrolyte [198–200]. Drawing inspiration from the electric eel's power generation mechanism, Yang and Mayer and coworkers harnessed the gradients of ions in hydrogels to develop soft, flexible, transparent, and biocompatible hydrogel biobatteries, generating 110 V at open circuit or 27 mW/m^2 per hydrogel cell, which is a significant achievement in the realm of electrochemical energy harvesting via hydrogels (Figure 1.11d) [186, 187]. Despite the abundant chemical reactions that occur within the human body (such as those that occur during digestion resulting in pH differences), there has been no significant progress in *in-situ* chemical energy harvesting. The main barriers are related to unstable ion concentrations and uncontrolled ion types in body fluids. The stability and size of the device are still the main challenges.

1.5.3 Thermal Energy Harvesters

Thermal energy harvesters using thermoelectric materials (Figure 1.11e), such as thermoelectric hydrogels, offer the potential to harvest low-grade body heat and power *in-situ* bioelectronic devices. These soft and biocompatible thermoelectric hydrogels are regarded as favorable alternatives to conventional thermoelectric materials [201–205]. Recent studies by Chen and Liu and coworkers have demonstrated a giant positive thermopower of 17 mV/K in an ionic thermoelectric hydrogel by harnessing synergistic thermo diffusion and thermo-galvanic effects. The thermos-diffusion effect is dominated by the presence of ions (KCl, NaCl, and KNO_3), while the thermo-galvanic effect is governed by a redox couple ($Fe(CN)_6^{4-}$)/($Fe(CN)_6^{3-}$), also adopted in other thermoelectric hydrogels (Figure 1.11f) [188, 189]. While thermoelectric hydrogels have significant potential for *in-situ* bioelectronics, the low power output and poor mechanical properties of these materials remain key limitations. Overcoming these limitations through further research and development will be crucial to fully exploit the unique advantages of thermoelectric hydrogels in *in-situ* bioelectronics.

1.6 Concluding Remarks

Over the past few years, we have seen many exciting advances and examples in the field of *in-situ* hydrogel bioelectronics that suggest great potential of high-performing hydrogels for many important applications. We will conclude this chapter with a set of opportunities by integrating interdisciplinary efforts in various areas of *in-situ* hydrogel bioelectronics, including ingestible sensors, neural interfaces, miniature robots, and data analytics. Ingestible sensors are one area where hydrogel-based bioelectronics can be leveraged (Figure 1.12a). These sensors can be designed to be swallowed and pass through the gastrointestinal tract, allowing for noninvasive monitoring of various biomarkers in real time. With the integration of hydrogel-based sensors, these devices can provide more accurate and reliable data, as hydrogels can respond to changes in pH, temperature, and other environmental factors [32]. Neural interface technology is another area where hydrogel-based bioelectronics can be applied (Figure 1.12b) [206–209]. By using hydrogels as a platform for neural interfaces, these devices can be made more biocompatible and less invasive, reducing the risk of rejection or other complications. With the integration of hydrogel-based sensors and actuators, these

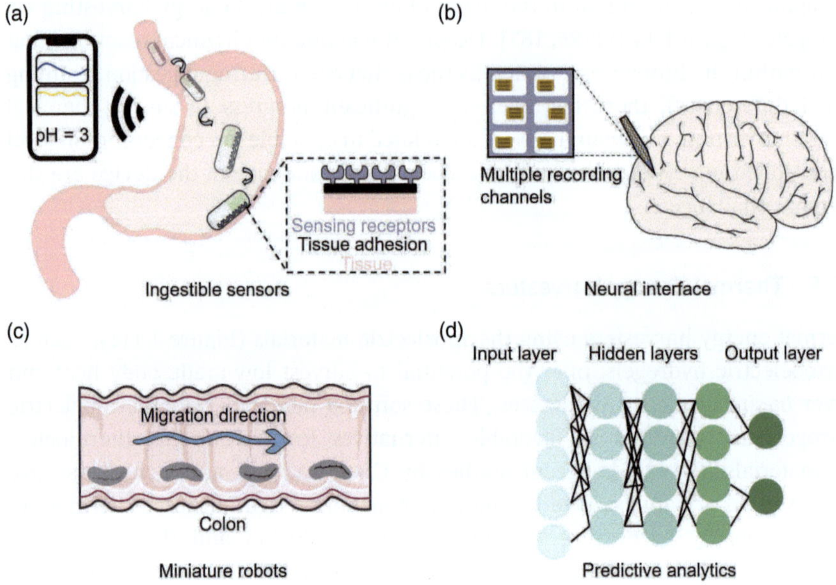

Figure 1.12 Future opportunities of *in-situ* hydrogel bioelectronics. Schematic illustrations of (a) ingestible sensors, (b) neural interfaces, (c) miniature robots, and (d) predictive analytics.

interfaces can provide more accurate and precise control over prosthetic devices or assistive technologies, allowing for more natural movements and interactions with the environment. Miniature robots are also an exciting area of research for *in-situ* hydrogel bioelectronics (Figure 1.12c) [210, 211]. With the use of hydrogels, these robots can be made more flexible and compliant, allowing for safer and more effective integration with the body. By incorporating hydrogel-based sensors and actuators, these robots can be controlled and manipulated to perform targeted drug delivery, tissue engineering, and surgical procedures. Finally, the integration of data analytics and machine learning algorithms is crucial for unlocking the full potential of *in-situ* hydrogel bioelectronics (Figure 1.12d) [212, 213]. With the vast amounts of data generated by these devices, there is a need for advanced analytics tools to help interpret and make sense of the data. By leveraging these tools, we can gain new insights into biological systems and develop more effective treatments and therapies.

Acknowledgments

We acknowledge the startup funds from the College of Engineering at Michigan State University. T. H. W is supported by financial support from the Biomedical Engineering Department at MSU.

References

1 Yu, Y., Nyein, H.Y.Y., Gao, W. et al. (2020). Flexible electrochemical bioelectronics: the rise of in situ bioanalysis. *Advanced Materials* 32 (15): 1902083.

2 Kim, D.W., Song, K.I., Seong, D. et al. (2022). Electrostatic–mechanical synergistic in situ multiscale tissue adhesion for sustainable residue-free bioelectronics interfaces. *Advanced Materials* 34 (5): 2105338.

3 Rawla, P., Sunkara, T., and Barsouk, A. (2019). Epidemiology of colorectal cancer: incidence, mortality, survival, and risk factors. *Gastroenterology Review/Przegląd Gastroenterologiczny* 14 (2): 89–103.

4 Tang, X., Shen, H., Siyuan, Z. et al. (2023). Flexible brain–computer interfaces. *Nature Electronics* 6: 1–10.

5 Zhao, S., Tang, X., Tian, W. et al. (2023). Tracking neural activity from the same cells during the entire adult life of mice. *Nature Neuroscience* 26: 1–15.

6 Steinmetz, N.A., Aydin, C., Lebedeva, A. et al. (2021). Neuropixels 2.0: a miniaturized high-density probe for stable, long-term brain recordings. *Science* 372 (6539): eabf4588.

7 Zheng, Q., Shi, B., Fan, F. et al. (2014). In vivo powering of pacemaker by breathing-driven implanted triboelectric nanogenerator. *Advanced Materials* 26 (33): 5851–5856.

8 Ryu, H., Park, H., Kim, M.K. et al. (2021). Self-rechargeable cardiac pacemaker system with triboelectric nanogenerators. *Nature Communications* 12 (1): 4374.

9 Ouyang, H., Liu, Z., Li, N. et al. (2019). Symbiotic cardiac pacemaker. *Nature Communications* 10 (1): 1–10.

10 Yuk, H., Wu, J., and Zhao, X. (2022). Hydrogel interfaces for merging humans and machines. *Nature Reviews Materials* 7 (12): 935–952.

11 Liu, X., Liu, J., Lin, S. et al. (2020). Hydrogel machines. *Materials Today* 36: 102–124.

12 Zhao, X., Chen, X., Yuk, H. et al. (2021). Soft materials by design: unconventional polymer networks give extreme properties. *Chemical Reviews* 121 (8): 4309–4372.

13 Lu, B., Yuk, H., Lin, S. et al. (2019). Pure pedot: Pss hydrogels. *Nature Communications* 10 (1): 1043.

14 Lin, X., Zhao, X., Xu, C. et al. (2022). Progress in the mechanical enhancement of hydrogels: fabrication strategies and underlying mechanisms. *Journal of Polymer Science* 60 (17): 2525–2542.

15 Kuang, X. et al. (2022). Functional tough hydrogels: design, processing, and biomedical applications. *Accounts of Materials Research* 4: 101–104.

16 Yuk, H., Lu, B., and Zhao, X. (2019). Hydrogel bioelectronics. *Chemical Society Reviews* 48 (6): 1642–1667.

17 Shi, Z., Arıcan, M.O., Zhou, T. et al. (2016). Electroconductive natural polymer-based hydrogels. *Biomaterials* 111: 40–54.

18 Lin, S., Yuk, H., Zhang, T. et al. (2016). Stretchable hydrogel electronics and devices. *Advanced Materials* 28 (22): 4497–4505.

19 Hu, L., Chee, P.L., Sugiarto, S. et al. (2022). Hydrogel-based flexible electronics. *Advanced Materials* 35: 2205326.

20 Guo, X., Li, J., Wang, F. et al. (2022). Application of conductive polymer hydrogels in flexible electronics. *Journal of Polymer Science* 60 (18): 2635–2662.

21 Gong, J.P., Katsuyama, Y., Kurokawa, T. et al. (2003). Double-network hydrogels with extremely high mechanical strength. *Advanced Materials* 15 (14): 1155–1158.

22 Sun, T.L., Kurokawa, T., Kuroda, S. et al. (2013). Physical hydrogels composed of polyampholytes demonstrate high toughness and viscoelasticity. *Nature Materials* 12 (10): 932–937.

23 Sun, J.-Y., Zhao, X., Illeperuma, W.R.K. et al. (2012). Highly stretchable and tough hydrogels. *Nature* 489 (7414): 133–136.

24 Yang, J., Bai, R., Chen, B. et al. (2020). Hydrogel adhesion: a supramolecular synergy of chemistry, topology, and mechanics. *Advanced Functional Materials* 30 (2): 1901693.

25 Yang, H., Li, C., Tang, J. et al. (2019). Strong and degradable adhesion of hydrogels. *ACS Applied Bio Materials* 2 (5): 1781–1786.

26 Steck, J., Yang, J., and Suo, Z. (2019). Covalent topological adhesion. *ACS Macro Letters* 8 (6): 754–758.

27 Yuk, H., Zhang, T., Lin, S. et al. (2016). Tough bonding of hydrogels to diverse non-porous surfaces. *Nature Materials* 15 (2): 190–196.

28 Liu, J., Lin, S., Liu, X. et al. (2020). Fatigue-resistant adhesion of hydrogels. *Nature Communications* 11 (1): 1071.

29 Li, J., Celiz, A.D., Yang, J. et al. (2017). Tough adhesives for diverse wet surfaces. *Science* 357 (6349): 378–381.

30 Wirthl, D., Pichler, R., Drack, M. et al. (2017). Instant tough bonding of hydrogels for soft machines and electronics. *Science Advances* 3 (6): e1700053.

31 Deng, J., Yuk, H., Wu, J. et al. (2021). Electrical bioadhesive interface for bioelectronics. *Nature Materials* 20 (2): 229–236.

32 Liu, X., Steiger, C., Lin, S. et al. (2019). Ingestible hydrogel device. *Nature Communications* 10 (1): 493.

33 Li, S., Cong, Y., and Fu, J. (2021). Tissue adhesive hydrogel bioelectronics. *Journal of Materials Chemistry B* 9 (22): 4423–4443.

34 Xie, C., Wang, X., He, H. et al. (2020). Mussel-inspired hydrogels for self-adhesive bioelectronics. *Advanced Functional Materials* 30 (25): 1909954.

35 Jia, M. and Rolandi, M. (2020). Soft and ion-conducting materials in bioelectronics: from conducting polymers to hydrogels. *Advanced Healthcare Materials* 9 (5): 1901372.

36 Dechiraju, H., Jia, M., Luo, L. et al. (2022). Ion-conducting hydrogels and their applications in bioelectronics. *Advanced Sustainable Systems* 6 (2): 2100173.

37 Yu, Q., Zheng, Z., Dong, X. et al. (2021). Mussel-inspired hydrogels as tough, self-adhesive and conductive bioelectronics: a review. *Soft Matter* 17 (39): 8786–8804.

38 Storm, C., Pastore, J.J., MacKintosh, F.C. et al. (2005). Nonlinear elasticity in biological gels. *Nature* 435 (7039): 191–194.

39 Ganji, F., Vasheghani, F.S., and Vasheghani, F.E. (2010). Theoretical description of hydrogel swelling: a review. *Iranian Polymer Journal* 19 (5): 375–398.

40 Kim, S.W., Bae, Y.H., and Okano, T. (1992). Hydrogels: swelling, drug loading, and release. *Pharmaceutical Research* 9: 283–290.

41 Holback, H., Yeo, Y., and Park, K. (2011). Hydrogel swelling behavior and its biomedical applications. In: *Biomedical Hydrogels* (ed. S. Rimmer), 3–24. Elsevier.

42 Chen, J., Park, H., and Park, K. (1999). Synthesis of superporous hydrogels: hydrogels with fast swelling and superabsorbent properties. *Journal of Biomedical Materials Research: An Official Journal of The Society for Biomaterials, The Japanese Society for Biomaterials, and the Australian Society for Biomaterials* 44 (1): 53–62.

43 Canal, T. and Peppas, N.A. (1989). Correlation between mesh size and equilibrium degree of swelling of polymeric networks. *Journal of Biomedical Materials Research* 23 (10): 1183–1193.

44 Hong, W., Zhao, X., Zhou, J. et al. (2008). A theory of coupled diffusion and large deformation in polymeric gels. *Journal of the Mechanics and Physics of Solids* 56 (5): 1779–1793.

45 Chester, S.A. and Anand, L. (2010). A coupled theory of fluid permeation and large deformations for elastomeric materials. *Journal of the Mechanics and Physics of Solids* 58 (11): 1879–1906.

46 Zhao, X., Huebsch, N., Mooney, D.J. et al. (2010). Stress-relaxation behavior in gels with ionic and covalent crosslinks. *Journal of Applied Physics* 107 (6): 063509.

47 Long, R., Mayumi, K., Creton, C. et al. (2014). Time dependent behavior of a dual cross-link self-healing gel: theory and experiments. *Macromolecules* 47 (20): 7243–7250.

48 Guo, J., Long, R., Mayumi, K. et al. (2016). Mechanics of a dual cross-link gel with dynamic bonds: steady state kinetics and large deformation effects. *Macromolecules* 49 (9): 3497–3507.

49 Mao, Y., Lin, S., Zhao, X. et al. (2017). A large deformation viscoelastic model for double-network hydrogels. *Journal of the Mechanics and Physics of Solids* 100: 103–130.

50 Grindy, S.C., Learsch, R., Mozhdehi, D. et al. (2015). Control of hierarchical polymer mechanics with bioinspired metal-coordination dynamics. *Nature Materials* 14 (12): 1210–1216.

51 Zhang, T., Lin, S., Yuk, H. et al. (2015). Predicting fracture energies and crack-tip fields of soft tough materials. *Extreme Mechanics Letters* 4: 1–8.

52 Qi, Y., Caillard, J., and Long, R. (2018). Fracture toughness of soft materials with rate-independent hysteresis. *Journal of the Mechanics and Physics of Solids* 118: 341–364.

53 Long, R. and Hui, C.-Y. (2016). Fracture toughness of hydrogels: measurement and interpretation. *Soft Matter* 12 (39): 8069–8086.

54 Bai, R., Yang, Q., Tang, J. et al. (2017). Fatigue fracture of tough hydrogels. *Extreme Mechanics Letters* 15: 91–96.

55 Bai, R., Yang, J., and Suo, Z. (2019). Fatigue of hydrogels. *European Journal of Mechanics-A/Solids* 74: 337–370.

56 Lin, S., Liu, X., Liu, J. et al. (2019). Anti-fatigue-fracture hydrogels. *Science Advances* 5 (1): eaau8528.

57 Li, X., Cui, K., Sun, T.L. et al. (2020). Mesoscale bicontinuous networks in self-healing hydrogels delay fatigue fracture. *Proceedings of the National Academy of Sciences* 117 (14): 7606–7612.

58 Lake, G. and Lindley, P. (1965). The mechanical fatigue limit for rubber. *Journal of Applied Polymer Science* 9 (4): 1233–1251.

59 Biot, M.A. (1941). General theory of three-dimensional consolidation. *Journal of Applied Physics* 12 (2): 155–164.

60 Flory, P.J. (1953). *Principles of Polymer Chemistry*. Cornell University Press.

61 Lake, G. and Thomas, A. (1967). The strength of highly elastic materials. *Proceedings of the Royal Society of London. Series A. Mathematical and Physical Sciences* 300 (1460): 108–119.

62 Treloar, L.G. (1975). *The Physics of Rubber Elasticity*. OUP Oxford.

63 De Gennes, P.-G. and Gennes, P.-G. (1979). *Scaling Concepts in Polymer Physics*. Cornell University Press.

64 Gent, A.N. (1996). A new constitutive relation for rubber. *Rubber Chemistry and Technology* 69 (1): 59–61.

65 Boyce, M.C. and Arruda, E.M. (2000). Constitutive models of rubber elasticity: a review. *Rubber Chemistry and Technology* 73 (3): 504–523.

66 Rubinstein, M. and Colby, R.H. (2003). *Polymer Physics*, vol. 23. New York: Oxford University Press.

67 Mark, J.E. and Erman, B. (2007). *Rubberlike Elasticity: A Molecular Primer*. Cambridge University Press.

68 Argon, A.S. (2013). *The Physics of Deformation and Fracture of Polymers*. Cambridge: New York.

69 Creton, C. and Ciccotti, M. (2016). Fracture and adhesion of soft materials: a review. *Reports on Progress in Physics* 79 (4): 046601.

70 Lin, S., Mao, Y., Radovitzky, R. et al. (2017). Instabilities in confined elastic layers under tension: fringe, fingering and cavitation. *Journal of the Mechanics and Physics of Solids* 106: 229–256.

71 Baby, D.K. (2020). Rheology of hydrogels. In: *Rheology of Polymer Blends and Nanocomposites* (ed. S. Thomas, C. Sarathchandran, and N. Chandran), 193–204. Elsevier.

72 Hu, Y., Zhao, X., Vlassak, J.J. et al. (2010). Using indentation to characterize the poroelasticity of gels. *Applied Physics Letters* 96 (12): 121904.

73 Zhong, M., Wang, R., Kawamoto, K. et al. (2016). Quantifying the impact of molecular defects on polymer network elasticity. *Science* 353 (6305): 1264–1268.

74 Gu, Y., Zhao, J., and Johnson, J.A. (2019). A (macro) molecular-level understanding of polymer network topology. *Trends in Chemistry* 1 (3): 318–334.

75 Rivlin, R. and Thomas, A.G. (1953). Rupture of rubber. I. Characteristic energy for tearing. *Journal of Polymer Science* 10 (3): 291–318.

76 Wang, Y., Yin, T., and Suo, Z. (2021). Polyacrylamide hydrogels. III. Lap shear and peel. *Journal of the Mechanics and Physics of Solids* 150: 104348.

77 Bai, R., Chen, B., Yang, J. et al. (2019). Tearing a hydrogel of complex rheology. *Journal of the Mechanics and Physics of Solids* 125: 749–761.

78 Zhao, X. (2014). Multi-scale multi-mechanism design of tough hydrogels: building dissipation into stretchy networks. *Soft Matter* 10 (5): 672–687.

79 Tang, J., Li, J., Vlassak, J.J. et al. (2017). Fatigue fracture of hydrogels. *Extreme Mechanics Letters* 10: 24–31.

80 Lin, S. and Zhao, X. (2020). Fracture of polymer networks with diverse topological defects. *Physical Review E* 102 (5): 052503.

81 James, H.M. and Guth, E. (1943). Theory of the elastic properties of rubber. *The Journal of Chemical Physics* 11 (10): 455–481.

82 Cao, J., Li, J., Chen, Y. et al. (2018). Dual physical crosslinking strategy to construct moldable hydrogels with ultrahigh strength and toughness. *Advanced Functional Materials* 28 (23): 1800739.

83 Hu, Z. and Chen, G. (2014). Novel nanocomposite hydrogels consisting of layered double hydroxide with ultrahigh tensibility and hierarchical porous structure at low inorganic content. *Advanced Materials* 26 (34): 5950–5956.

84 Tsukeshiba, H., Huang, M., Na, Y.H. et al. (2005). Effect of polymer entanglement on the toughening of double network hydrogels. *The Journal of Physical Chemistry B* 109 (34): 16304–16309.

85 Gong, J.P. (2010). Why are double network hydrogels so tough? *Soft Matter* 6 (12): 2583–2590.

86 de Gennes, P.-G. (1996). Soft adhesives. *Langmuir* 12 (19): 4497–4500.

87 Chen, Q. et al. (2016). Engineering of tough double network hydrogels. *Macromolecular Chemistry and Physics* 217 (9): 1022–1036.

88 Li, J., Suo, Z., and Vlassak, J.J. (2014). Stiff, strong, and tough hydrogels with good chemical stability. *Journal of Materials Chemistry B* 2 (39): 6708–6713.

89 Wang, Z., Zheng, X., Ouchi, T. et al. (2021). Toughening hydrogels through force-triggered chemical reactions that lengthen polymer strands. *Science* 374 (6564): 193–196.

90 Luo, F., Sun, T.L., Nakajima, T. et al. (2014). Crack blunting and advancing behaviors of tough and self-healing polyampholyte hydrogel. *Macromolecules* 47 (17): 6037–6046.

91 Haque, M.A., Kurokawa, T., and Gong, J.P. (2012). Super tough double network hydrogels and their application as biomaterials. *Polymer* 53 (9): 1805–1822.

92 Chen, C., Wang, Z., and Suo, Z. (2017). Flaw sensitivity of highly stretchable materials. *Extreme Mechanics Letters* 10: 50–57.

93 Lin, S., Londono, C.D., Zheng, D. et al. (2022). An extreme toughening mechanism for soft materials. *Soft Matter* 18 (31): 5742–5749.

94 Sun, T.L., Londono, C.D., Zheng, D. et al. (2017). Bulk energy dissipation mechanism for the fracture of tough and self-healing hydrogels. *Macromolecules* 50 (7): 2923–2931.

95 Liu, C., Morimoto, N., Jiang, L. et al. (2021). Tough hydrogels with rapid self-reinforcement. *Science* 372 (6546): 1078–1081.

96 Kim, J., Zhang, G., Shi, M. et al. (2021). Fracture, fatigue, and friction of polymers in which entanglements greatly outnumber cross-links. *Science* 374 (6564): 212–216.

97 Zheng, D., Lin, S., Ni, J. et al. (2022). Fracture and fatigue of entangled and unentangled polymer networks. *Extreme Mechanics Letters* 51: 101608.

98 Norioka, C., Inamoto, Y., Hajime, C. et al. (2021). A universal method to easily design tough and stretchable hydrogels. *NPG Asia Materials* 13 (1): 34.

99 Zhang, E., Bai, R., Morelle, X.P. et al. (2018). Fatigue fracture of nearly elastic hydrogels. *Soft Matter* 14 (18): 3563–3571.

100 Lin, S., Liu, J., Liu, X. et al. (2019). Muscle-like fatigue-resistant hydrogels by mechanical training. *Proceedings of the National Academy of Sciences* 116 (21): 10244–10249.

101 Bai, R., Yang, J., Morelle, X.P. et al. (2018). Fatigue fracture of self-recovery hydrogels. *ACS Macro Letters* 7 (3): 312–317.

102 Liang, X., Chen, G., Lin, S. et al. (2022). Bioinspired 2D isotropically fatigue-resistant hydrogels. *Advanced Materials* 34 (8): 2107106.

103 Hua, M., Wu, S., Ma, Y. et al. (2021). Strong tough hydrogels via the synergy of freeze-casting and salting out. *Nature* 590 (7847): 594–599.

104 Guo, X., Dong, X., Zou, G. et al. (2023). Strong and tough fibrous hydrogels reinforced by multiscale hierarchical structures with multimechanisms. *Science Advances* 9 (2): eadf7075.

105 Ni, J., Lin, S., Qin, Z. et al. (2021). Strong fatigue-resistant nanofibrous hydrogels inspired by lobster underbelly. *Matter* 4 (6): 1919–1934.

106 Xiang, C., Wang, Z., Yang, C. et al. (2020). Stretchable and fatigue-resistant materials. *Materials Today* 34: 7–16.

107 Li, X., Cui, K., Kurokawa, T. et al. (2021). Effect of mesoscale phase contrast on fatigue-delaying behavior of self-healing hydrogels. *Science Advances* 7 (16): eabe8210.

108 Edward, J.T. (1970). Molecular volumes and the Stokes-Einstein equation. *Journal of Chemical Education* 47 (4): 261.

109 Moncure, P.J., Simon, Z.C., Millstone, J.E. et al. (2022). Relationship between gel mesh and particle size in determining nanoparticle diffusion in hydrogel nanocomposites. *The Journal of Physical Chemistry B* 126 (22): 4132–4142.

110 Cai, L.-H., Panyukov, S., and Rubinstein, M. (2015). Hopping diffusion of nanoparticles in polymer matrices. *Macromolecules* 48 (3): 847–862.

111 Yang, Y.J., Mai, D.J., Dursch, T.J. et al. (2018). Nucleopore-inspired polymer hydrogels for selective biomolecular transport. *Biomacromolecules* 19 (10): 3905–3916.

112 Gu, Y., Distler, M.E., Cheng, H.F. et al. (2021). A general DNA-gated hydrogel strategy for selective transport of chemical and biological cargos. *Journal of the American Chemical Society* 143 (41): 17200–17208.

113 Maguire, L., Stefferson, M., Betterton, M.D. et al. (2019). Design principles of selective transport through biopolymer barriers. *Physical Review E* 100 (4): 042414.

114 Cheng, T., Zhang, Y.Z., Wang, S. et al. (2021). Conductive hydrogel-based electrodes and electrolytes for stretchable and self-healable supercapacitors. *Advanced Functional Materials* 31 (24): 2101303.

115 Deng, Z., Yu, R., and Guo, B. (2021). Stimuli-responsive conductive hydrogels: design, properties, and applications. *Materials Chemistry Frontiers* 5 (5): 2092–2123.

116 Shen, Z., Zhang, Z., Zhang, N. et al. (2022). High-stretchability, ultralow-hysteresis conducting polymer hydrogel strain sensors for soft machines. *Advanced Materials* 34 (32): 2203650.

117 Fu, F., Wang, J., Zeng, H. et al. (2020). Functional conductive hydrogels for bioelectronics. *ACS Materials Letters* 2 (10): 1287–1301.

118 Huang, H., Han, L., Li, J. et al. (2020). Super-stretchable, elastic and recoverable ionic conductive hydrogel for wireless wearable, stretchable sensor. *Journal of Materials Chemistry A* 8 (20): 10291–10300.

119 Huang, H., Han, L., Fu, X. et al. (2020). Multiple stimuli responsive and identifiable zwitterionic ionic conductive hydrogel for bionic electronic skin. *Advanced Electronic Materials* 6 (7): 2000239.

120 Shi, Y., Ma, C., Peng, L. et al. (2015). Conductive "smart" hybrid hydrogels with PNIPAM and nanostructured conductive polymers. *Advanced Functional Materials* 25 (8): 1219–1225.

121 Mondal, S., Das, S., and Nandi, A.K. (2020). A review on recent advances in polymer and peptide hydrogels. *Soft Matter* 16 (6): 1404–1454.

122 Zhang, C., Wang, M., Jiang, C. et al. (2022). Highly adhesive and self-healing γ-PGA/PEDOT: PSS conductive hydrogels enabled by multiple hydrogen bonding for wearable electronics. *Nano Energy* 95: 106991.

123 Shi, J., Chen, X., Li, G. et al. (2019). A liquid PEDOT: PSS electrode-based stretchable triboelectric nanogenerator for a portable self-charging power source. *Nanoscale* 11 (15): 7513–7519.

124 Guzinski, M., Jarvis, J.M., Perez, F. et al. (2017). PEDOT (PSS) as solid contact for ion-selective electrodes: the influence of the PEDOT (PSS) film thickness on the equilibration times. *Analytical Chemistry* 89 (6): 3508–3516.

125 Sun, X., Yao, F., and Li, J. (2020). Nanocomposite hydrogel-based strain and pressure sensors: a review. *Journal of Materials Chemistry A* 8 (36): 18605–18623.

126 Han, L., Yan, L., Wang, M. et al. (2018). Transparent, adhesive, and conductive hydrogel for soft bioelectronics based on light-transmitting polydopamine-doped polypyrrole nanofibrils. *Chemistry of Materials* 30 (16): 5561–5572.

127 Ohm, Y., Liao, J., Luo, Y. et al. (2022). Reconfigurable electrical networks within a conductive hydrogel composite. *Advanced Materials* 35: 2209408.

128 Hui, Y., Yao, Y., Qian, Q. et al. (2022). Three-dimensional printing of soft hydrogel electronics. *Nature Electronics* 5: 1–11.

129 Yasuda, K., Gong, J.P., Katsuyama, Y. et al. (2005). Biomechanical properties of high-toughness double network hydrogels. *Biomaterials* 26 (21): 4468–4475.

130 Kong, H.J., Wong, E., and Mooney, D.J. (2003). Independent control of rigidity and toughness of polymeric hydrogels. *Macromolecules* 36 (12): 4582–4588.

131 Tuncaboylu, D.C., Sari, M., Oppermann, W. et al. (2011). Tough and self-healing hydrogels formed via hydrophobic interactions. *Macromolecules* 44 (12): 4997–5005.

132 Abdurrahmanoglu, S., Can, V., and Okay, O. (2009). Design of high-toughness polyacrylamide hydrogels by hydrophobic modification. *Polymer* 50 (23): 5449–5455.

133 Seitz, M.E., Martina, D., Baumberger, T. et al. (2009). Fracture and large strain behavior of self-assembled triblock copolymer gels. *Soft Matter* 5 (2): 447–456.

134 Haraguchi, K. and Takehisa, T. (2002). Nanocomposite hydrogels: a unique organic–inorganic network structure with extraordinary mechanical, optical, and swelling/de-swelling properties. *Advanced Materials* 14 (16): 1120–1124.

135 Haraguchi, K. and Li, H.J. (2005). Control of the coil-to-globule transition and ultrahigh mechanical properties of PNIPA in nanocomposite hydrogels. *Angewandte Chemie International Edition* 44 (40): 6500–6504.

136 Sakai, T., Matsunaga, T., Yamamoto, Y. et al. (2008). Design and fabrication of a high-strength hydrogel with ideally homogeneous network structure from tetrahedron-like macromonomers. *Macromolecules* 41 (14): 5379–5384.

137 Sakai, T., Akagi, Y., Matsunaga, T. et al. (2010). Highly elastic and deformable hydrogel formed from tetra-arm polymers. *Macromolecular Rapid Communications* 31 (22): 1954–1959.

138 Liao, I.C., Moutos, F.T., Estes, B.T. et al. (2013). Composite three-dimensional woven scaffolds with interpenetrating network hydrogels to create functional synthetic articular cartilage. *Advanced Functional Materials* 23 (47): 5833–5839.

139 Lin, S., Cao, C., Wang, Q. et al. (2014). Design of stiff, tough and stretchy hydrogel composites via nanoscale hybrid crosslinking and macroscale fiber reinforcement. *Soft Matter* 10 (38): 7519–7527.

140 King, D.R., Sun, T.L., Huang, Y. et al. (2015). Extremely tough composites from fabric reinforced polyampholyte hydrogels. *Materials Horizons* 2 (6): 584–591.

141 Nezakati, T., Seifalian, A., Tan, A. et al. (2018). Conductive polymers: opportunities and challenges in biomedical applications. *Chemical Reviews* 118 (14): 6766–6843.

142 Feig, V.R., Tran, H., Lee, M. et al. (2018). Mechanically tunable conductive interpenetrating network hydrogels that mimic the elastic moduli of biological tissue. *Nature Communications* 9 (1): 2740.

143 Kang, J., Mun, J., Zheng, Y. et al. (2022). Tough-interface-enabled stretchable electronics using non-stretchable polymer semiconductors and conductors. *Nature Nanotechnology* 17: 1–7.

144 Han, M., Chen, L., Aras, K. et al. (2020). Catheter-integrated soft multilayer electronic arrays for multiplexed sensing and actuation during cardiac surgery. *Nature Biomedical Engineering* 4 (10): 997–1009.

145 Asulin, M., Michael, I., Shapira, A. et al. (2021). One-step 3D printing of heart patches with built-in electronics for performance regulation. *Advanced Science* 8 (9): 2004205.

146 Yan, Z., Zhang, F., Liu, F. et al. (2016). Mechanical assembly of complex, 3D mesostructures from releasable multilayers of advanced materials. *Science Advances* 2 (9): e1601014.

147 Hong, G., Yang, X., Zhou, T. et al. (2018). Mesh electronics: a new paradigm for tissue-like brain probes. *Current Opinion in Neurobiology* 50: 33–41.

148 Choi, S., Han, S.I., Kim, D. et al. (2019). High-performance stretchable conductive nanocomposites: materials, processes, and device applications. *Chemical Society Reviews* 48 (6): 1566–1595.

149 Yuk, H., Lu, B., Lin, S. et al. (2020). 3D printing of conducting polymers. *Nature Communications* 11 (1): 1604.

150 Vallem, V., Roosa, E., Ledinh, T. et al. (2021). A soft variable-area electrical-double-layer energy harvester. *Advanced Materials* 33 (43): 2103142.

151 Shay, T., Velev, O.D., and Dickey, M.D. (2018). Soft electrodes combining hydrogel and liquid metal. *Soft Matter* 14 (17): 3296–3303.

152 Mao, S., Zhang, D., Zhang, Y. et al. (2020). A universal coating strategy for controllable functionalized polymer surfaces. *Advanced Functional Materials* 30 (40): 2004633.

153 Wei, K., Chen, X., Zhao, P. et al. (2019). Stretchable and bioadhesive supramolecular hydrogels activated by a one-stone–two-bird postgelation functionalization method. *ACS Applied Materials & Interfaces* 11 (18): 16328–16335.

154 Liu, Q., Nian, G., Yang, C. et al. (2018). Bonding dissimilar polymer networks in various manufacturing processes. *Nature Communications* 9 (1): 846.

155 Chen, H., Liu, Y., Ren, B. et al. (2017). Super bulk and interfacial toughness of physically crosslinked double-network hydrogels. *Advanced Functional Materials* 27 (44): 1703086.

156 Fan, H., Wang, J., Tao, Z. et al. (2019). Adjacent cationic–aromatic sequences yield strong electrostatic adhesion of hydrogels in seawater. *Nature Communications* 10 (1): 5127.

157 Ji, H. and De Gennes, P. (1993). Adhesion via connector molecules: the many-stitch problem. *Macromolecules* 26 (3): 520–525.

158 Raphael, E. and De Gennes, P. (1992). Rubber-rubber adhesion with connector molecules. *The Journal of Physical Chemistry* 96 (10): 4002–4007.

159 Wong, T.H., Liu, Y., Li, J. et al. (2021). Triboelectric nanogenerator tattoos enabled by epidermal electronic technologies. *Advanced Functional Materials* 32: 2111269.

160 Kamyshny, A. and Magdassi, S. (2019). Conductive nanomaterials for 2D and 3D printed flexible electronics. *Chemical Society Reviews* 48 (6): 1712–1740.

161 Shin, S.R., Farzad, M.R., Tamayol, A. et al. (2016). A bioactive carbon nanotube-based ink for printing 2D and 3D flexible electronics. *Advanced Materials* 28 (17): 3280–3289.

162 Cheng, X. and Zhang, Y. (2019). Micro/nanoscale 3D assembly by rolling, folding, curving, and buckling approaches. *Advanced Materials* 31 (36): 1901895.

163 Zhao, H., Kim, Y., Wang, H. et al. (2021). Compliant 3D frameworks instrumented with strain sensors for characterization of millimeter-scale engineered muscle tissues. *Proceedings of the National Academy of Sciences* 118 (19): e2100077118.

164 Liu, J. (2018). Syringe injectable electronics. In: *Biomimetics Through Nanoelectronics* (ed. J. Liu), 65–93. Springer.

165 Dai, X. et al. (2018). Mesh nanoelectronics: seamless integration of electronics with tissues. *Accounts of Chemical Research* 51 (2): 309–318.

166 Ronkainen, N.J., Halsall, H.B., and Heineman, W.R. (2010). Electrochemical biosensors. *Chemical Society Reviews* 39 (5): 1747–1763.

167 Mahshid, S.S., Flynn, S.E., and Mahshid, S. (2021). The potential application of electrochemical biosensors in the COVID-19 pandemic: a perspective on the rapid diagnostics of SARS-CoV-2. *Biosensors and Bioelectronics* 176: 112905.

168 Kesler, V., Murmann, B., and Soh, H.T. (2020). Going beyond the debye length: overcoming charge screening limitations in next-generation bioelectronic sensors. *ACS Nano* 14 (12): 16194–16201.

169 Menon, S., Mathew, M.R., Sam, S. et al. (2020). Recent advances and challenges in electrochemical biosensors for emerging and re-emerging infectious diseases. *Journal of Electroanalytical Chemistry* 878: 114596.

170 Piccinini, E., Alberti, S., Longo, G.S. et al. (2018). Pushing the boundaries of interfacial sensitivity in graphene FET sensors: polyelectrolyte multilayers strongly increase the Debye screening length. *The Journal of Physical Chemistry C* 122 (18): 10181–10188.

171 Gao, N., Gao, T., Yang, X. et al. (2016). Specific detection of biomolecules in physiological solutions using graphene transistor biosensors. *Proceedings of the National Academy of Sciences* 113 (51): 14633–14638.

172 Lieleg, O. and Ribbeck, K. (2011). Biological hydrogels as selective diffusion barriers. *Trends in Cell Biology* 21 (9): 543–551.

173 Kim, M., Chen, W.G., Kang, J.W. et al. (2015). Artificially engineered protein hydrogels adapted from the nucleoporin Nsp1 for selective biomolecular transport. *Advanced Materials* 27 (28): 4207–4212.

174 Yang, Y.J., Mai, D.J., Li, S. et al. (2021). Tuning selective transport of biomolecules through site-mutated nucleoporin-like protein (NLP) hydrogels. *Biomacromolecules* 22 (2): 289–298.

175 Purwidyantri, A., Domingues, T., Borme, J. et al. (2021). Influence of the electrolyte salt concentration on DNA detection with graphene transistors. *Biosensors* 11 (1): 24.

176 Kishore, R.A. and Priya, S. (2018). A review on low-grade thermal energy harvesting: materials, methods and devices. *Materials* 11 (8): 1433.

177 Fan, F.-R., Tian, Z.-Q., and Wang, Z.L. (2012). Flexible triboelectric generator. *Nano Energy* 1 (2): 328–334.

178 Zhu, G., Pan, C., Guo, W. et al. (2012). Triboelectric-generator-driven pulse electrodeposition for micropatterning. *Nano Letters* 12 (9): 4960–4965.

179 Kim, S., Choi, S.J., Zhao, K. et al. (2016). Electrochemically driven mechanical energy harvesting. *Nature Communications* 7 (1): 10146.

180 Stauss, S. and Honma, I. (2018). Biocompatible batteries—materials and chemistry, fabrication, applications, and future prospects. *Bulletin of the Chemical Society of Japan* 91 (3): 492–505.

181 Zheng, Q., Shi, B., Li, Z. et al. (2017). Recent progress on piezoelectric and triboelectric energy harvesters in biomedical systems. *Advanced Science* 4 (7): 1700029.

182 Song, Y., Shi, Z., Hu, G.H. et al. (2021). Recent advances in cellulose-based piezoelectric and triboelectric nanogenerators for energy harvesting: a review. *Journal of Materials Chemistry A* 9 (4): 1910–1937.

183 Luo, J. and Wang, Z.L. (2020). Recent progress of triboelectric nanogenerators: from fundamental theory to practical applications. *EcoMat* 2 (4): e12059.

184 Hinchet, R., Yoon, H.J., Ryu, H. et al. (2019). Transcutaneous ultrasound energy harvesting using capacitive triboelectric technology. *Science* 365 (6452): 491–494.

185 Lee, K.H., Zhang, Y.Z., Jiang, Q. et al. (2020). Ultrasound-driven two-dimensional $Ti_3C_2T_x$ MXene hydrogel generator. *ACS Nano* 14 (3): 3199–3207.

186 He, P., He, J., Huo, Z. et al. (2022). Microfluidics-based fabrication of flexible ionic hydrogel batteries inspired by electric eels. *Energy Storage Materials* 49: 348–359.

187 Schroeder, T.B., Guha, A., Lamoureux, A. et al. (2017). An electric-eel-inspired soft power source from stacked hydrogels. *Nature* 552 (7684): 214–218.

188 Fu, X., Zhuang, Z., Zhao, Y. et al. (2022). Stretchable and self-powered temperature–pressure dual sensing ionic skins based on thermogalvanic hydrogels. *ACS Applied Materials & Interfaces* 14 (39): 44792–44798.

189 Bai, C., Li, X., Cui, X. et al. (2022). Transparent stretchable thermogalvanic PVA/gelation hydrogel electrolyte for harnessing solar energy enabled by a binary solvent strategy. *Nano Energy* 100: 107449.

190 Liu, W., Wang, Z., Wang, G. et al. (2019). Integrated charge excitation triboelectric nanogenerator. *Nature Communications* 10 (1): 1426.

191 Wang, J., Wu, C., Dai, Y. et al. (2017). Achieving ultrahigh triboelectric charge density for efficient energy harvesting. *Nature Communications* 8 (1): 88.

192 Niu, S. and Wang, Z.L. (2015). Theoretical systems of triboelectric nanogenerators. *Nano Energy* 14: 161–192.

193 Wang, Y., Guo, T., Tian, Z. et al. (2022). MXenes for energy harvesting. *Advanced Materials* 34 (21): 2108560.

194 Yang, G., Lei, W., Chen, C. et al. (2020). Ultrathin $Ti_3C_2T_x$ (MXene) membrane for pressure-driven electrokinetic power generation. *Nano Energy* 75: 104954.

195 Cang, Y., Liu, J., Ryu, M. et al. (2022). On the origin of elasticity and heat conduction anisotropy of liquid crystal elastomers at gigahertz frequencies. *Nature Communications* 13 (1): 5248.

196 Wu, H., Yu, G., Pan, L. et al. (2013). Stable Li-ion battery anodes by in-situ polymerization of conducting hydrogel to conformally coat silicon nanoparticles. *Nature Communications* 4 (1): 1943.

197 Kim, T., Song, W., Son, D.Y. et al. (2019). Lithium-ion batteries: outlook on present, future, and hybridized technologies. *Journal of Materials Chemistry A* 7 (7): 2942–2964.

198 Wang, Z., Li, H., Tang, Z. et al. (2018). Hydrogel electrolytes for flexible aqueous energy storage devices. *Advanced Functional Materials* 28 (48): 1804560.

199 Chan, C.Y., Wang, Z., Jia, H. et al. (2021). Recent advances of hydrogel electrolytes in flexible energy storage devices. *Journal of Materials Chemistry A* 9 (4): 2043–2069.

200 Chen, Z., To, J.W.F., Wang, C. et al. (2014). A three-dimensionally interconnected carbon nanotube–conducting polymer hydrogel network for

high-performance flexible battery electrodes. *Advanced Energy Materials* 4 (12): 1400207.

201 Tian, Y. and Zhao, C.-Y. (2013). A review of solar collectors and thermal energy storage in solar thermal applications. *Applied Energy* 104: 538–553.

202 Pu, S., Liao, Y., Chen, K. et al. (2020). Thermogalvanic hydrogel for synchronous evaporative cooling and low-grade heat energy harvesting. *Nano Letters* 20 (5): 3791–3797.

203 Ding, T., Zhou, Y., Wang, X.Q. et al. (2021). All-soft and stretchable thermogalvanic gel fabric for antideformity body heat harvesting wearable. *Advanced Energy Materials* 11 (44): 2102219.

204 Lee, S.W., Yang, Y., Lee, H.W. et al. (2014). An electrochemical system for efficiently harvesting low-grade heat energy. *Nature Communications* 5 (1): 3942.

205 Han, C.-G., Qian, X., Li, Q. et al. (2020). Giant thermopower of ionic gelatin near room temperature. *Science* 368 (6495): 1091–1098.

206 Song, E., Li, J., Won, S.M. et al. (2020). Materials for flexible bioelectronic systems as chronic neural interfaces. *Nature Materials* 19 (6): 590–603.

207 Zhang, M., Tang, Z., Liu, X. et al. (2020). Electronic neural interfaces. *Nature Electronics* 3 (4): 191–200.

208 Shi, J. and Fang, Y. (2019). Flexible and implantable microelectrodes for chronically stable neural interfaces. *Advanced Materials* 31 (45): 1804895.

209 Park, Y., Franz, C.K., Ryu, H. et al. (2021). Three-dimensional, multifunctional neural interfaces for cortical spheroids and engineered assembloids. *Science Advances* 7 (12): eabf9153.

210 Son, D., Gilbert, H., and Sitti, M. (2020). Magnetically actuated soft capsule endoscope for fine-needle biopsy. *Soft Robotics* 7 (1): 10–21.

211 Erin, O., Alici, C., and Sitti, M. (2021). Design, actuation, and control of an MRI-powered untethered robot for wireless capsule endoscopy. *IEEE Robotics and Automation Letters* 6 (3): 6000–6007.

212 Mohan, S., Thirumalai, C., and Srivastava, G. (2019). Effective heart disease prediction using hybrid machine learning techniques. *IEEE Access* 7: 81542–81554.

213 Uddin, S., Khan, A., Hossain, M.E. et al. (2019). Comparing different supervised machine learning algorithms for disease prediction. *BMC Medical Informatics and Decision Making* 19 (1): 1–16.

2

Multiscale Mechanics of Metal Nanowire-Based Stretchable Electronics

Shuang Wu and Yong Zhu

North Carolina State University, Department of Mechanical and Aerospace Engineering, 1840 Entrepreneur Dr, Raleigh, NC 27695, USA

2.1 Introduction

Stretchable electronics is a technology that builds electronic circuits on top of a stretchable substrate or embedded in a stretchable matrix [1–10]. It is an emerging technology that has enabled a wide range of applications such as soft wearable sensors, electronic skins, bioelectronics, and soft robotics [11–16]. A stretchable conductor or an electrode is the most fundamental building block of stretchable electronics [17, 18]. For example, it is a common approach to build stretchable electronics based on stiff active device components and stretchable conductors or interconnects. Different types of materials have been used to fabricate stretchable conductors such as metal thin films, conductive polymers, and a variety of nanomaterials including metal nanoparticles, metal nanowires (NWs), carbon nanotubes, metal flakes, and two-dimensional (2D) materials (e.g. graphene) [17–26]. Among all these materials, metal NWs have shown superior performances in terms of electrical conductivity and mechanical stretchability, owing to the outstanding electrical and mechanical properties of NWs as well as the percolation network structure that the NWs form.

On the other hand, the operation and reliability of the NW-based devices call for a thorough understanding of the mechanical behavior of individual NWs and NW networks that can be quite different from their bulk counterparts. Indeed, mechanics of NWs have seen exciting advances in both experiments and simulations [27–32]. It is known that surface effects become dominant for these small-volume materials as their characteristic dimension approaches sub-100 nm. Size effects on elastic modulus have been reported [31, 33, 34]. Surface dislocation nucleation becomes dominant in the case of metal NWs, in contrast to bulk dislocation dynamics [35–37]. NWs are known to have fewer defects such as dislocations when synthesized, but they often possess well-defined internal (in contrast to surface) defects such as twinning boundaries (TBs) and stacking faults, which makes their mechanical properties

Mechanics of Flexible and Stretchable Electronics, First Edition. Edited by Yong Zhu and Nanshu Lu.

even more interesting. For stretchable electronics, NWs are either placed on top of polymeric substrates or embedded in polymeric matrices. Therefore, in addition to the mechanics of individual NWs, interfacial mechanics between NWs and polymeric substrates or matrices is of critical relevance for the application of stretchable electronics.

This chapter aims to provide an overview of the mechanics issues of metal NWs of relevance to their application in stretchable electronics, including mechanics of individual NWs, interfacial mechanics between NWs and polymer matrix, and mechanics of NW/polymer composite in stretchable structures. It is organized as follows. Section 2.2 provides an overview of the design of NW-based flexible and stretchable electrodes. Following the percolation theory, the electric conductivity of the network electrode depends on the NW length and the network density. Section 2.3 presents mechanics of individual metal NWs of different crystalline structures (single-crystalline, bi-twinned, and penta-twinned). Interesting mechanical behaviors pertinent to stretchable electronics are discussed such as recoverable plasticity, strain hardening, large ductility, and insensitivity to surface flaws imparted by the TBs. Section 2.4 investigates the NW–polymer interfacial mechanics under either van der Waals (vdW) interactions or chemical bonding. The NWs can be positioned on top of a polymeric substrate or embedded in a polymeric matrix. Both *in-situ* experiments and shear-lag modeling will be discussed. Section 2.5 presents the mechanical design of stretchable structures using NW/polymer composites. A variety of stretchable structures exist including wrinkle, serpentine, Kirigami, and buckle-delamination. The fracture behavior of NW/polymer composite is discussed including crack propagation, opening, and blunting. Section 2.6 concludes the chapter with challenges in metal NW-based stretchable conductors from individual NW properties to the performances of stretchable conductors. Of note is that a number of metal NWs such as Ag, Au, and Cu have been used in stretchable conductors. This chapter focuses on Ag NWs while the mechanics understanding can be extended to other metal NWs.

2.2 Metal NW-Based Flexible and Stretchable Electronics

Metal NWs have been widely used in flexible transparent electrodes and stretchable electrodes [38, 39] (Figure 2.1). The resistance of an NW network has the contributions from both the NWs themselves and the junctions between them. For an as-prepared network (without annealing or welding), the junction resistance is typically much higher and is the limiting factor for the conductivity of the network. According to the percolation theory [40, 41], for a random 2D conducting network of filaments, the critical density (N_c) to maintain electron transport in the network, for a given filament (NW in this case) length L, can be determined by:

$$N_c L^2 = 5.71 \tag{2.1a}$$

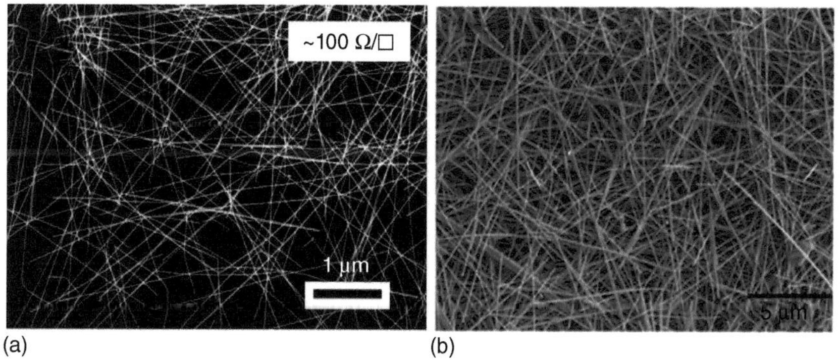

Figure 2.1 Ag NW-based (a) transparent electrode and (b) stretchable electrode. Source: (a) Reproduced with permission of Ref. [38]; Copyright 2010, American Chemical Society. (b) Reproduced with permission of Ref. [39]; Copyright 2012, John Wiley & Sons.

It can be easily seen that the longer the NWs, the lower the critical density (N_c). The conductivity of the network is given by:

$$\sigma \propto (N - N_c)^t \tag{2.2a}$$

where N is the density of the conducting filaments/NWs.

For transparent electrodes, the tradeoff between electric conductivity and optical transmittance is the key. It is desirable to increase electrical conductivity or optical transmittance or both. For a given density (N) of the NWs, the longer the NWs, the higher the conductivity that can be achieved while maintaining the same optical transmittance. For stretchable electrodes, it is important to achieve high electrical conductivity while optical transmittance is often less important. Longer NWs and higher density (N) of the NWs can both lead to higher electric conductivity. Longer NWs, according to Eq. (2.1a), can reduce the critical density and thus, at the same NW density, increase the electrical conductivity. Furthermore, longer NWs can lead to more stable electrical conductivity under deformation [42, 43], which is of important relevance to both flexible transparent electrodes and stretchable electrodes. For these applications, multiscale mechanics – including mechanics of individual NWs, mechanics of NW network, interfacial mechanics between NWs and polymer matrix, and mechanics of NW/polymer composite in stretchable structures – plays a critically important role.

2.3 Mechanics of Individual NWs

Section 2.3 presents mechanics of individual metal NWs of different crystalline structures (single-crystalline, bi-twinned, and penta-twinned). Interesting mechanical behaviors pertinent to stretchable electronics are discussed such as recoverable plasticity, strain hardening, large ductility, and insensitivity to surface flaws. These behaviors are particularly pronounced in penta-twinned NWs as a result of the TB-dislocation interactions.

2.3.1 Overview of Mechanics of Metal NWs

Metal NWs have been synthesized by a number of methods, including the template method [44], hydrothermal method [45], electrochemical deposition [46], chemical vapor deposition [47], physical vapor deposition [48, 49], and polyol method (solution phase) [50–53]. The last two methods are known to produce high-quality NWs with high crystallinity, uniform diameter, smooth surfaces, and well-defined defect structures if any (e.g. TBs).

NWs have been found to exhibit size-dependent elasticity [29, 31, 54–58]. The surface atoms can be elastically stiffer or softer than the bulk atoms [59]. The softening effect is primarily due to the bond loss (i.e. loss of neighboring atoms on the surface), while the stiffening effect is attributed to the electron redistribution (so-called bond saturation) [59, 60]. Another mechanism for size-dependent elasticity, especially metallic nanomaterials, is bulk (or core) nonlinear elasticity [61]. That is, the interior compressive stress caused by the tensile surface stress in metallic nanomaterials is large enough to induce a nonlinear increase in Young's modulus of the bulk atoms. Which mechanism plays a more dominant role depends on a number of factors such as bonding type, NW diameter, axial orientation, and side surface facets [30, 59, 61]. Loading type can also play an important role in the measured size effect in Young's modulus [34, 62]. In general, metal NWs exhibit a stiffening effect in Young's modulus as the NW diameter decreases.

For metal NWs, dislocation nucleation from free surfaces is the dominant mechanism for plasticity [28–30, 35, 63–68]. By contrast, internal dislocation interaction is the dominant mechanism for plasticity of bulk materials. As a result, yield strength is expected to show sensitive temperature and strain-rate dependence and increase modestly with the decreasing NW diameter [37]. For example, an increase in the yield strength from 200 MPa – bulk value of the yield strength – to as high as 8 GPa, with a reduction of diameter from 300 to 40 nm, was reported [69–73].

As a result of size reduction, a transition from ductile to brittle fracture in metallic NWs was found [48, 74]. In one study, Cu NWs were synthesized by physical vapor deposition with the <110> axial direction. When the Cu NW diameter decreased from 300 to 75 nm, the ductile to brittle transition was clearly observed in the stress–strain curve and the fracture surface, while the tensile strength increased from 1 to 7 GPa. Note that the size effect is pronounced when comparing the strength data of the Cu NWs and the Cu micro-whiskers tested by Brenner back in the 1950s [75].

2.3.2 Mechanics of Single-Crystalline Metal NWs

High-quality single-crystalline metal NWs are typically synthesized by physical vapor deposition in ultrahigh vacuum [48, 49, 76]. Single-crystalline Au NWs exhibited large plastic deformation as a result of coherent twin propagation as revealed by *in-situ* scanning electron microscopy (SEM) tensile tests [72, 73, 77]. The flow stress increased with the decreasing NW diameter [78]. For single-crystalline face-centered cubic (FCC) metal NWs, two types of mechanical behaviors have been

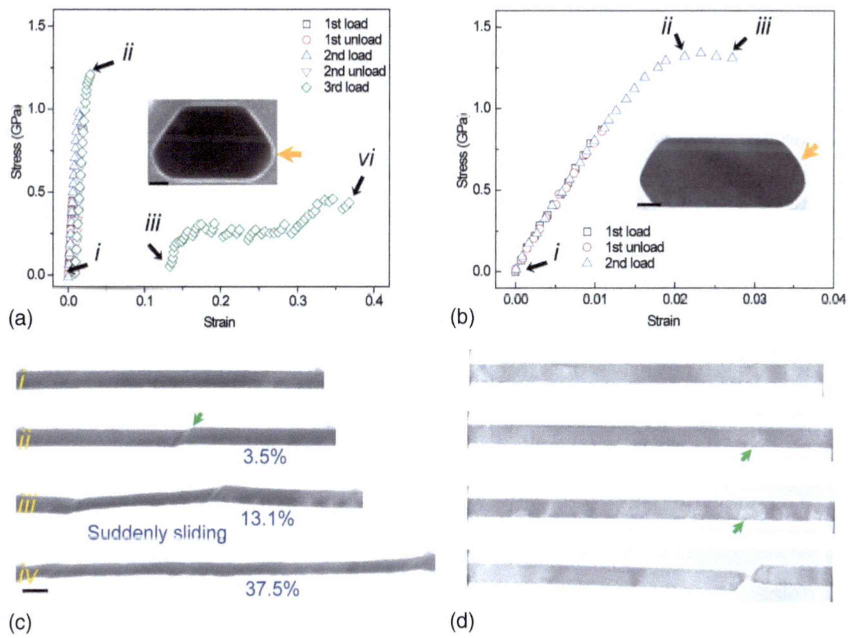

Figure 2.2 Mechanical behaviors of single-crystalline Ag NWs under *in-situ* transmission electron microscopy (TEM) tensile testing. (a,b) Engineering stress–strain curves of two single-crystalline Ag NWs with different aspect ratios. Insets in (a) and (b) are the corresponding cross-sectional images of the tested NWs (sectioned from the undeformed part after the test). Scale bar: 20 nm. (c,d) Snapshots of microstructure evolutions corresponding to (a) and (b). The four snapshots in each case correspond to the stresses and strains marked in (a) and (b). Planar sliding is labeled by green arrows. Scale bars: 100 nm. Source: Reproduced with permission of Ref. [79]; Copyright 2019, American Chemical Society.

observed: high ductility with coherent twin propagation and limited ductility with localized dislocation slip [73]. The competition between these mechanisms can be understood depending on factors like loading direction, stacking fault energies, and Schmid factor [28]. Yin et al. recently reported an additional factor, cross-sectional shape, that can affect the competition (Figure 2.2) [79]. For a truncated rhombic cross section, which is typical for bottom-up synthesized nanocrystals, the extent of truncation determines the competition. A transition from twinning to localized dislocation slip occurs with an increasing extent of truncation; more specifically, a cross section with large aspect ratio leads to slip while that with small aspect ratio does twinning. Their simulations and analyses reveal that the energy barriers for twinning and dislocation slip depend on the cross-sectional shape of the NW. The energy barrier for twinning is proportional to the change of surface energy associated with the twinning, while the energy barrier for dislocation slip is nearly constant. Thus, the transition of the two deformation modes can be attributed to the change of surface energy during twinning as a function of the cross-sectional shape.

2.3.3 Mechanics of Bi-Twinned Metal NWs

During the bottom-up synthesis, it is common to produce NWs with TBs. The interactions between the TBs and the surface-nucleated dislocations lead to interesting mechanical behavior of NWs [80–83]. Cheng et al. recently reported the observation and quantification of dislocation nucleation, interaction with TBs, and retraction in so-called bi-twinned Ag NWs with a single TB along the NW length direction [76]. In the physical vapor deposition process used to synthesize single-crystalline metal NWs [48], such bi-twinned NWs are often synthesized too. Cheng's results have shown that leading partial dislocations nucleated from the free surface can be hindered by the TB, and upon unloading all or part of the leading partials can retract due to the repulsive force from the TB, leading to full or partial plastic strain recovery (Bauschinger effect), respectively (Figure 2.3). The bi-twinned Ag NWs can undergo stress relaxation as a result of dislocation nucleation. This work illustrated that the internal TBs in NWs can interact with surface-nucleated dislocations, leading to time-dependent plastic strain recovery.

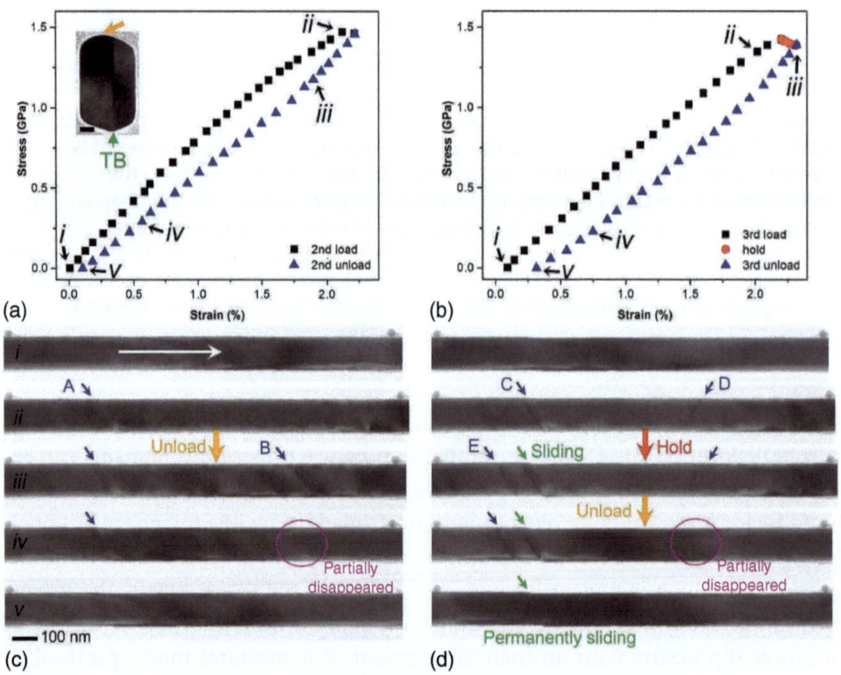

Figure 2.3 Mechanical behaviors of a bi-twinned Ag NW under *in-situ* TEM tensile testing. (a,b) Stress–strain curves for the bi-twinned NW without or with a holding step, respectively. The relaxation step took five minutes. Inset in (a) are the corresponding cross-sectional images of the tested NW. Scale bar, 20 nm. (c,d) Snapshots of microstructure evolution during loading–unloading and loading–holding–unloading processes, respectively. The location of partial dislocations and planar slip in the NW are marked by blue and green arrows, respectively. Scale bar, 100 nm. The viewing direction in (c,d) is from the $<1\bar{1}0>$ zone axis as marked by the orange arrow in the inset in (a).
Source: Reproduced with permission of Ref. [76]; Copyright 2020, Elsevier.

For such bi-twinned NWs, a transition of the deformation mechanism from localized dislocation slip to delocalized plasticity via an anomalous tensile detwinning mechanism was reported [84]. The anomalous detwinning under tension starts with detwinning of a segment of the pre-existing TB, even under nominally zero resolved shear stress. The detwinned segment forms a pair of new TBs and grain boundaries (with the existing twin variants), which propagate along the NW length, leading to a large plastic deformation. This propagation is like the coherent twin propagation observed in the single-crystalline NWs [72, 79]. The observed transition of the deformation mechanism is found to depend on the volume ratio between the two twin variants and the cross-sectional aspect ratio.

2.3.4 Mechanics of Penta-Twinned Metal NWs

The polyol method is based on solution phase synthesis, which can produce NWs in very large quantities [50, 52, 53, 85–87]. Metal NWs that are commonly used in stretchable electronics, e.g. Ag, Au, and Cu, can be synthesized by the polyol method. The synthesized NWs possess the so-called penta-twinned structure with five twin segments joined along a common quintuple line in the axial direction of <110> and five surface facets of {100} planes. Below we will discuss the mechanics of penta-twinned metal NWs in detail.

For penta-twinned Ag NWs, a dislocation-mediated, time-dependent, and fully reversible plastic behavior was reported [88]. Figure 2.4a shows a typical tensile stress–strain response of a penta-twinned Ag NW that consists of four steps: loading, relaxation, unloading, and recovery. During the relaxation step, the stress decreases with time while the strain increases. After unloading, the strain in the NW completely recovers in about the same amount of time as the relaxation. The stress relaxation is attributed to the surface nucleation of (partial) dislocations, while the plastic strain recovery is to the retraction of the dislocations. The TBs and the intrinsic stress field associated with the penta-twinned structure promote retraction of the dislocations. As a result of the same dislocation/TB interaction,

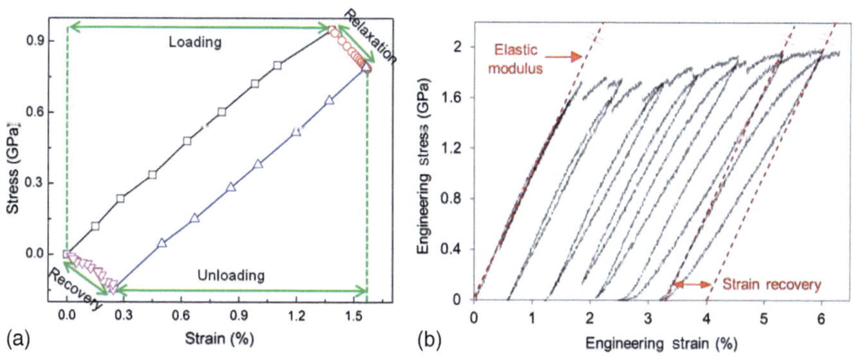

Figure 2.4 (a) Recoverable plasticity and (b) Bauschinger effect for penta-twinned Ag NWs. (a) Source: Reproduced with permission of Ref. [88]/Springer Nature/CC BY 4.0. (b) Reproduced with permission of Ref. [89]; Copyright 2015, American Chemical Society.

Bauschinger effect was observed for the penta-twinned Ag NWs under cyclic loading [89] (Figure 2.4b). Bauschinger effect has been reported in thin films and NWs due to several different mechanisms such as heterogeneous distribution of grain size [90, 91], dislocation–grain boundary interaction [92], surface passivation [93, 94], or twinning–detwinning transition [95]. Note that for the observed recoverable plasticity, the applied strain to the NWs was relatively small (<2%) and hence the dislocations did not transmit across the TBs, which is essential for the recoverable plasticity [88].

If the dislocations transmit across the TBs, which can occur under larger strain and/or sustained loading, they cannot retract and, as a result, the associated plastic strain cannot recover. In this case, interaction between the dislocations and the pre-existing TBs in the penta-twinned NWs can lead to pronounced strain hardening [96, 97] (Figure 2.5). The strain hardening originated from the obstruction of surface-nucleated dislocations by the TBs. More specifically, so-called stacking fault

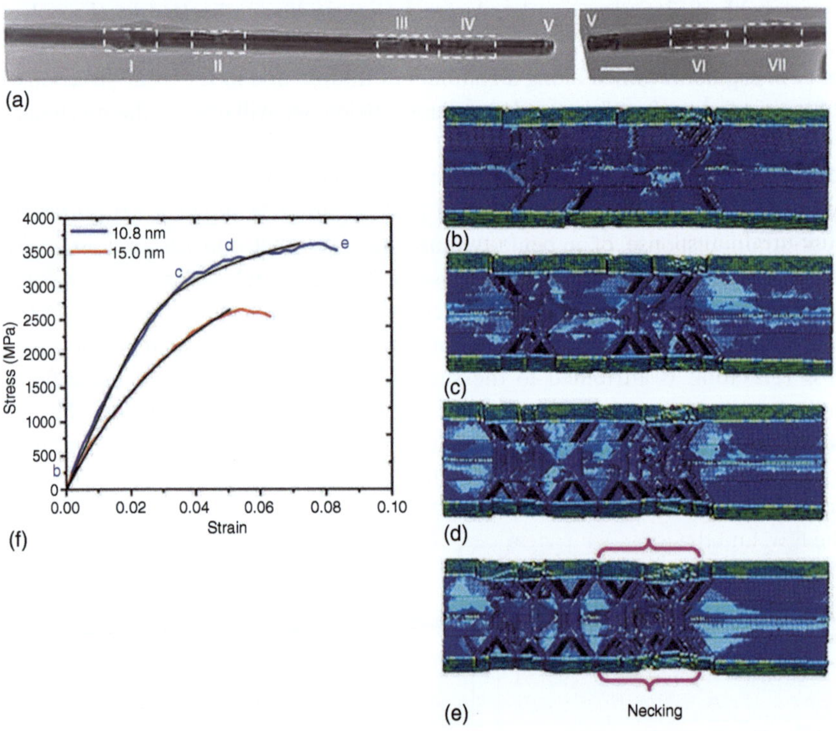

Figure 2.5 (a) TEM image of an Ag NW after tensile testing exhibiting multiple distributed plastic zones along the NW. Scale bar: 100 nm. Source: Reproduced with permission of Ref. [97]; Copyright 2012, John Wiley & Sons. (b–e) Atomic configurations of the 10.8 nm diameter NW at various strain levels as marked in (a). The onset of necking is marked by pink brackets in (e). (f) Stress–strain curves for the tensile loading of defective fivefold twinned Ag NWs of 10.8 and 15 nm in diameter, respectively, showing size-dependent strain hardening. The red and blue curves are the molecular dynamics (MD) data, while the black curves are the Ramberg–Osgood fitting. Source: Reproduced with permission of Ref. [98]; Copyright 2015, American Chemical Society.

decahedrons (SFDs) are formed due to the TB obstruction and the fivefold symmetry. Formation of SFDs locally hardens the NW and promotes subsequent nucleation of SFDs at other locations, which leads to enhanced ductility of the NW as manifested in multiple distributed plastic zones along the NW length (Figure 2.5a). By confining dislocation activity to SFD chain propagation, the internal twin boundaries cause local hardening of thin NWs, resulting in defect-insensitive structures with significantly enhanced flow stress, ductility, and strength. Strain hardening was found to be size-dependent – thinner NWs achieved more hardening and higher ductility. To put in perspective, a penta-twinned NW with a small diameter (~41 nm) exhibited an unprecedented combination of high strength, ductility, and strain hardening with a strain-to-failure of ~14% and strain hardening rate of 11.7 ± 0.4 GPa. This work showed that the statistical variation of source strengths of surface-nucleated dislocations plays an important role in mechanical properties of small-sized structures [98].

2.4 Interfacial Mechanics of the NW–Polymer Interface

Section 2.4 presents interfacial mechanics of individual metal NWs on an elastomeric substrate. For stretchable electronics applications, metal NW network is typically deposited on top of a polymer substrate or embedded in a polymer matrix. Under deformation (tension or bending), the interaction between the NWs and the polymer substrate (or matrix) – in the form of shear stress transfer – is essential for the mechanical and electrical behaviors of the NW network. This section will discuss mechanical analysis of such an interaction. The mechanical analysis is similar for the cases with a substrate and a matrix, except that the contact areas in the two cases are different. Therefore, when we describe the interaction in general, we will call it NW–polymer matrix interaction.

2.4.1 Classic Theory of Shear-Lag Analysis

According to the classic shear-lag analysis, the relative displacement, the axial strain in the NW, and the interfacial shear stress can be obtained as, respectively,

$$\delta(x) = \frac{\varepsilon_{\mathrm{m}}}{\beta} \frac{\sinh(\beta x)}{\cosh(\beta L)} \tag{2.1b}$$

$$\varepsilon(x) = \varepsilon_{\mathrm{m}} \left(1 - \frac{\cosh(\beta x)}{\cosh(\beta L)}\right) \tag{2.1c}$$

$$\tau(x) = \frac{K_0 \varepsilon_{\mathrm{m}}}{\beta} \frac{\sinh(\beta x)}{\cosh(\beta L)} \tag{2.1d}$$

where ε_{m} is the strain applied to the substrate, L is half of the NW length in the x-direction (loading direction) with $x = 0$ at the center of the NW, K_0 is a stiffness constant, and $\beta = \sqrt{\frac{K_0}{E_{\mathrm{NW}} h}}$ is the shear-lag parameter with h as the effective height (defined as the ratio between the cross-sectional area and the side width of the cross section), and E_{NW} as Young's modulus of the NW [99–101]. In the classic shear-lag analysis, τ is assumed to be proportional to δ, i.e. $\tau = K_0\delta$.

2.4.2 Shear-Lag Analysis Considering Bonding Mechanisms

The NW–matrix interaction typically involves different bonding mechanisms, such as vdW interactions and chemical bonding. Nonlinear shear-lag and bilinear cohesive shear-lag models were found to appropriately account for the vdW interactions and chemical bonding, respectively [101–105]. For the nonlinear shear-lag model, the interface could consist of two zones: a bonded zone near the center ($x < L_1$) and a sliding zone emerging from the edge ($L_1 < x < L$), as shown in Figure 2.6a,b. For the bilinear cohesive shear-lag model, the interface could consist of three zones: a bonded zone ($x < L_1$), a damaged zone ($L_1 < x < L_2$), and a debonded zone ($L_2 < x < L$), as shown in Figure 2.6c,d. The key results of both nonlinear and bilinear cohesive shear-lag models are summarized below. For more details, please refer to our previous work [107–109].

For the nonlinear shear-lag model, the traction–separation law includes two independent interface parameters: interfacial stiffness K_0 and shear strength τ_c. The critical matrix strain for onset of sliding is given by:

$$\varepsilon_{c1} = \frac{\tau_c}{\beta E_{NW} h} \coth(\beta L) = \delta_1 \beta \coth(\beta L) \tag{2.2b}$$

where δ_1 is the critical relative displacement between NW and matrix as shown in Figure 2.6b. When $\varepsilon_m < \varepsilon_{c1}$, the entire NW is bonded to the matrix, and the axial strain in the NW is given by Eq. (2.1c).

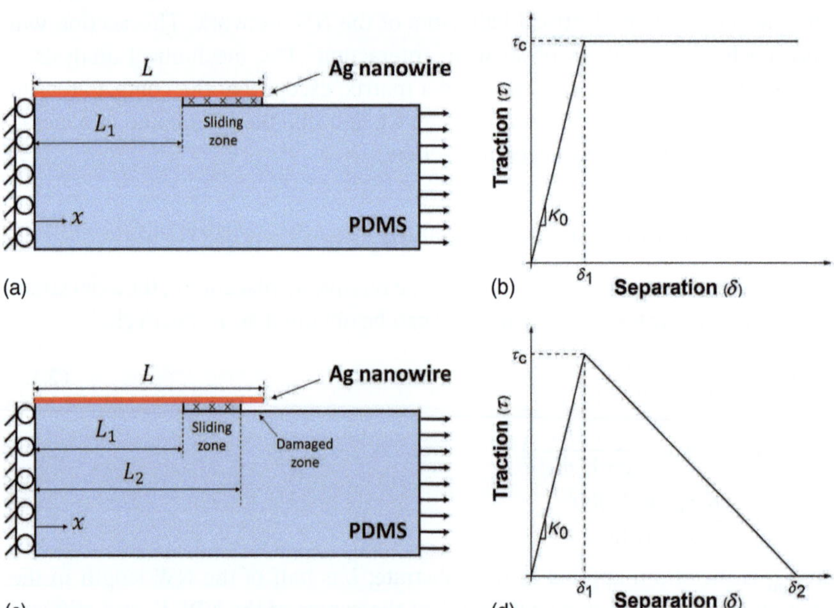

Figure 2.6 (a) Schematic of the nonlinear shear-lag model between an Ag NW and a PDMS substrate and (b) the corresponding traction–separation law; (c) schematic of the cohesive shear-lag model between an Ag NW and a PDMS substrate and (d) the corresponding traction–separation law. Source: Reproduced with permission of Ref. [106]; Copyright 2020, Elsevier.

When $\varepsilon_{\mathrm{m}} > \varepsilon_{\mathrm{c1}}$, the interface consists of two zones: a bonded zone and a sliding zone (Figure 2.6a). The axial strain in the NW, in the bonded zone and sliding zone, are given by, respectively,

$$\varepsilon(x) = \varepsilon_{\mathrm{m}} - \delta_1 \beta \frac{\cosh(\beta x)}{\sinh(\beta L_1)} \tag{2.3a}$$

$$\varepsilon(x) = \frac{\tau_{\mathrm{c}}}{E_{\mathrm{NW}} h}(L - x) = \beta^2 \delta_1 (L - x) \tag{2.3b}$$

L_1 can be obtained by solving the following:

$$\coth(\beta L_1) + \beta(L - L_1) = \frac{\beta E_{\mathrm{NW}} h \varepsilon_{\mathrm{m}}}{\tau_{\mathrm{c}}} \tag{2.4}$$

The average strain in the NW is given by:

$$\bar{\varepsilon} = \frac{\varepsilon_{\mathrm{m}} L_1}{L} + \frac{\left(\frac{\beta^2 (L-L_1)^2}{2} - 1\right)\tau_{\mathrm{c}}}{\beta^2 L E_{\mathrm{NW}} h} \quad \text{when } L_1 > 0 \tag{2.5a}$$

$$= \frac{\tau_{\mathrm{c}}}{2E_{\mathrm{NW}} h} L \quad \text{when } L_1 = 0 \tag{2.5b}$$

For the bilinear cohesive shear-lag model, the traction–separation law is shown in Figure 2.6d, with three independent interface parameters, interfacial stiffness K_0, shear strength τ_{c}, and mode II fracture toughness G. Two critical substrate strains, $\varepsilon_{\mathrm{c1}}$ and $\varepsilon_{\mathrm{c2}}$, correspond to the onset of the damage and the debonding, respectively.

The first critical substrate strain $\varepsilon_{\mathrm{c1}}$ is still given by Eq. (2.2b). When $\varepsilon_{\mathrm{m}} < \varepsilon_{\mathrm{c1}}$, the axial strain in the NW (all bonded to the substrate) is given by Eq. (2.1c). When $\varepsilon_{\mathrm{c1}} < \varepsilon_{\mathrm{m}} < \varepsilon_{\mathrm{c2}}$, the axial strain in the NW in the bonded zone is still given by Eq. (2.3a), while that in the damaged zone is given by:

$$\varepsilon(x) = \varepsilon_{\mathrm{m}} + A\alpha \sin(\alpha(x - L_1)) - B\alpha \cos(\alpha(x - L_1)) \tag{2.6}$$

where $\alpha = \sqrt{\tau_{\mathrm{c}}/(E_{\mathrm{NW}} h(\delta_2 - \delta_1))}$, $A = \delta_2 - \delta_1$, and $B = \delta_1 \beta \coth(\beta L_1)/\alpha$. L_1 can be obtained by solving the following:

$$-\varepsilon_{\mathrm{m}} = A\alpha \sin(\alpha(L - L_1)) - B\alpha \cos(\alpha(L - L_1)) \tag{2.7}$$

At the second critical substrate strain $\varepsilon_{\mathrm{c2}}$, we have:

$$A \cos(\alpha(L - L_1)) + B \sin(\alpha(L - L_1)) = 0 \tag{2.8}$$

This equation gives a critical value of L_1, at which $\varepsilon_{\mathrm{c2}}$ can be determined by Eq. (2.7).

When $\varepsilon_{\mathrm{m}} > \varepsilon_{\mathrm{c2}}$, the axial strain in the NW in the bonded zone and the damaged zone still take the same form of Eqs. (2.3a) and (2.6), respectively. In the debonded part, the axial strain in the NW becomes zero (Figure 2.6c). The length of L_1 and L_2 can be obtained simply by solving Eqs. (2.7) and (2.8) simultaneously by substituting L with L_2.

The average strain in the NW is given by:

$$\overline{\varepsilon} = \frac{\varepsilon_m L_2 - \delta_1}{L} + \frac{A(1-\cos(\alpha(L_2-L_1))) - B\sin(\alpha(L_2-L_1))}{L} \quad \text{when } L_1 > 0 \tag{2.9a}$$

$$= \frac{\varepsilon_m L_2}{L} + \frac{A(1-\cos(\alpha L_2)) - B\sin(\alpha L_2)}{L} \quad \text{when } L_1 = 0 \text{ and } L_2 > 0 \tag{2.9b}$$

$$= 0 \quad \text{when } L_2 = 0 \tag{2.9c}$$

Figure 2.7a,c show the results for the nonlinear shear-lag model, while Figure 2.7b,d show those for the bilinear shear-lag model. Individual NWs can be placed on top of a polymer substrate or embedded in a polymer matrix; the polymer can be either as-prepared or ultraviolet–ozone (or plasma) treated. VdW interactions exist between metal NWs and as-prepared polymer, while chemical bonding does between metal NWs and treated polymer. Accordingly, the nonlinear shear-lag model and bilinear shear-lag model were found to well describe the NW behavior (e.g. average strain) on the as-prepared and ultraviolet–ozone treated

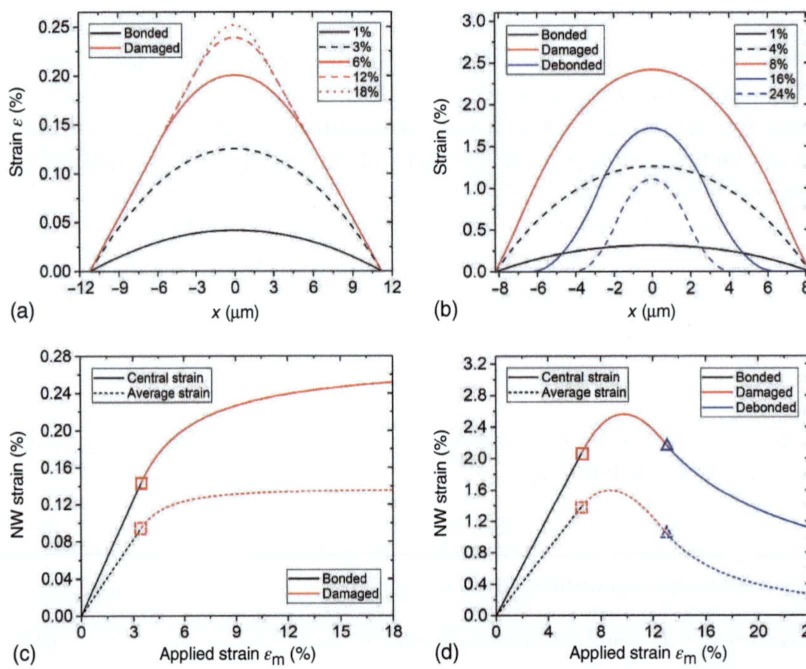

Figure 2.7 Nonlinear shear-lag model results for ε_m varying from 1% to 18% with $L = 11.2$ μm, $a = 10$ nm, and $\beta = 0.026$/m: (a) NW axial strain, (c) NW center strain and average strain versus substrate strain (square dots mark the bonded–damaged transition). Bilinear shear-lag model results for ε_m varying from 1% to 24% when $L = 8.2$ μm, $a = 10$ nm, and $\beta = 0.113$/m: (b) NW axial strain, and (d) NW center strain and average strain versus substrate strain (square dots mark the bonded–damaged transition and triangle dots mark the damaged-debonded transition). Source: Reproduced with permission of Ref. [107]; Copyright 2019, Elsevier.

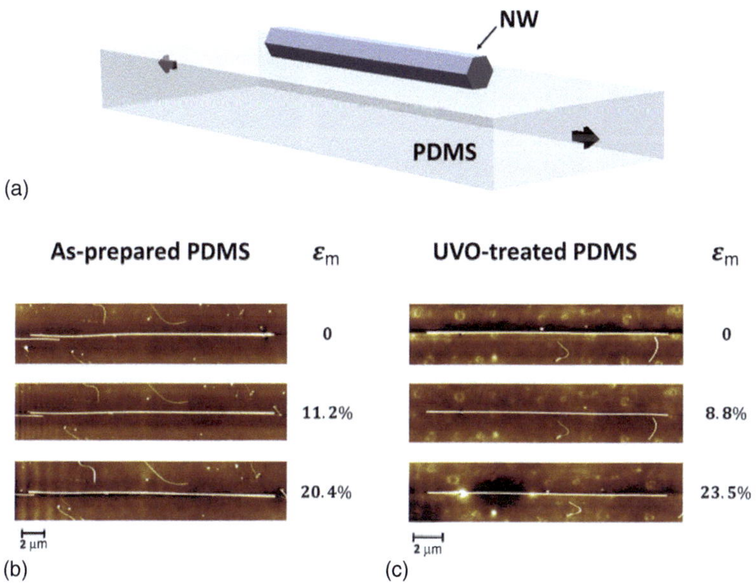

Figure 2.8 (a) Schematic of the experimental setup with a Si NW on top of a PDMS substrate, (b) AFM images of a Si NW on top of as-prepared PDMS, and (c) AFM images of a Si NW on top of UVO-treated PDMS. Source: Reproduced with permission of Ref. [107]; Copyright 2019, Elsevier.

polymer substrates, respectively. Figure 2.8 shows an example of the experimental setup, where an individual Si NW is placed on top of the as-prepared or ultraviolet–ozone treated polydimethylsiloxane (PDMS) substrate [107]. Atomic force microscopy (AFM) imaging is used to obtain the average strain in the NW. Optical imaging can be used to obtain the average strain too with the aid of digital image correlation [106].

2.4.3 Fracture of NWs Due to Shear Stress Transfer

As a result of interfacial shear stress transfer, when stretching the polymer substrate, NWs on top of or embedded in the polymer can be stretched. When placing Ag NWs on top of as-prepared PDMS, the vdW interactions were found to be insufficient to fracture them. When embedded in an as-prepared PDMS matrix, a larger contact area exists between the Ag NWs and PDMS (five side surfaces versus one side surface), which leads to a more effective interfacial stress transfer from PDMS to Ag NWs and could fracture the NWs [106]. As shown in Eq. (2.5b), with the increasing applied strain to the substrate, the average NW strain approaches a plateau value, given by $\frac{\tau_c}{2E_{NW}h}L$. Figure 2.9 shows the stretching until fracture of a single Ag NW embedded in a PDMS matrix. As it can be seen in Figure 2.9b, the average strain in the NW approached a plateau value, which can be used to obtain the interfacial shear strength τ_c. Before sliding occurs, the average strain in the NW, as given in Eq. (2.1c), is proportional to the applied strain with the proportional

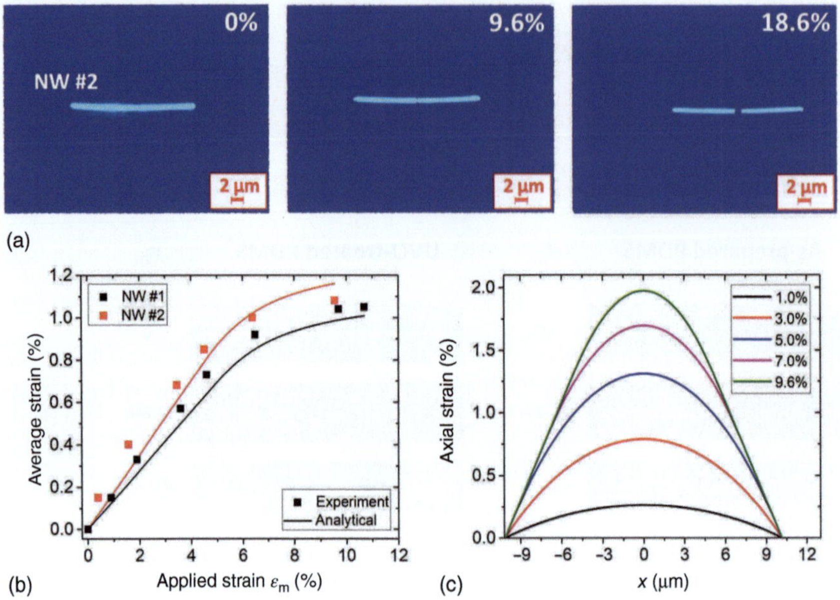

Figure 2.9 Results of Ag NWs embedded in as-prepared PDMS. (a) Optical images of NW #2 (20.3 μm long) during the experiment. (b) Fitting experimental data of average nanowire strain for the two Ag NWs using the nonlinear shear-lag model. (c) Predicted normal strain along the Ag NW length for NW #2 for the applied PDMS strain ranging from 1.0% to 9.6%. Source: Reproduced with permission of Ref. [106]; Copyright 2020, Elsevier.

factor of $1 - \frac{\sin h(\beta L)}{\beta L \cos h(\beta L)}$, which should be able to yield the interfacial stiffness K_0 [107]. With K_0 and τ_c known, the axial strain in the embedded NW can be fully predicted. Figure 2.9c shows the axial strain distribution in the NW. The fracture strain can thus be obtained, i.e. the maximum strain along the NW length (Table 2.1). Therefore, this method offers a simple and scalable method to measure the fracture strain of NWs.

Ultraviolet–ozone treatment of PDMS can also be used to enhance the bonding with Ag NWs as a result of the formation of chemical bonds between PDMS and Ag NWs. In this case, Ag NWs can be fractured when placed on top of treated PDMS substrate. The bilinear shear-lag model should be used to analyze the strain distribution in the NW. Three interfacial parameters – interfacial stiffness K_0, shear strength τ_c, and mode II fracture toughness G – can be identified by fitting the model to the experimental data [107]. After that, the axial strain distribution in the NW can be predicted using the bilinear shear-lag model. Hence the fracture strain can be obtained similar to the above case, NWs embedded in as-prepared PDMS matrix (the fracture strain is still in the middle of the NW as shown in Figure 2.7b).

Note that Ag NWs exhibit size-dependent Young's modulus [62, 96]. The Ag NWs used in the reported study had diameters of ~60 nm, corresponding to Young's modulus of ~90 GPa [106]. Ag NWs were synthesized by the polyol method with a pentagonal cross section. When on top of the substrate, the NW/substrate interface is estimated to be one side surface of the NW cross section, corresponding

Table 2.1 Fracture strains of Ag NWs calculated from the shear-lag analysis.

		NW length (μm)	Fracture strain (%)
Embedded in as-prepared PDMS	NW #1	17.6	1.72
	NW #2	20.3	1.97
	NW #3	9.7	1.84
	NW #4	13.3	1.82
	NW #5	16.5	1.82
On top of UVO-treated PDMS	NW #1	11.6	1.62
	NW #2	14.4	2.07 (1st)
		7.3	1.18 (2nd)
	NW #3	15.1	2.13

Source: Ref. [106]/with permission of Elsevier.

to $h = \frac{1}{4}\sqrt{5(5+2\sqrt{5})}a$ with a as the side width; when embedded in the substrate (matrix), the interface is the entire side surface of the NW cross section, corresponding to $h = \frac{1}{20}\sqrt{5(5+2\sqrt{5})}a$.

2.4.4 Elastoplastic Analysis of Metal NWs

Most composites are for structural applications, where high strength and often brittle fillers such as ceramics are embedded in ductile matrices such as polymer or metal. The fillers are typically considered to be elastic. Indeed, in all the analyses discussed above, the NWs are assumed to be elastic, which is not true in the case of metal NWs. Therefore, it is of relevance to treat metal NWs elastoplastic; Figure 2.10 shows a typical stress–strain curve of penta-twinned Ag NWs. Zhu and coworkers recently presented the analysis for the interfacial shear stress transfer between an elastoplastic fiber (more specifically, NW) and an elastic substrate/matrix [110].

Since the shear-lag model of NW on untreated substrate includes two parts: the bonded zone and the sliding zone, there may be three cases with respect to the plastic zone, l_p, as shown in Figure 2.11. Below we will briefly provide the analytical solution for each case. More details can be found elsewhere [110].

Case 1:

$$\begin{cases} \delta = B_2 \sinh(\beta_2 x), \text{when } 0 < x < l_p \\ \delta = A_1 \cosh(\beta_1 x) + B_1 \sinh(\beta_1 x), \text{when } l_p < x < l \end{cases} \tag{2.10}$$

Constants A_1, B_1, B_2, and l_p cannot be solved analytically, so numerical solution is pursued.

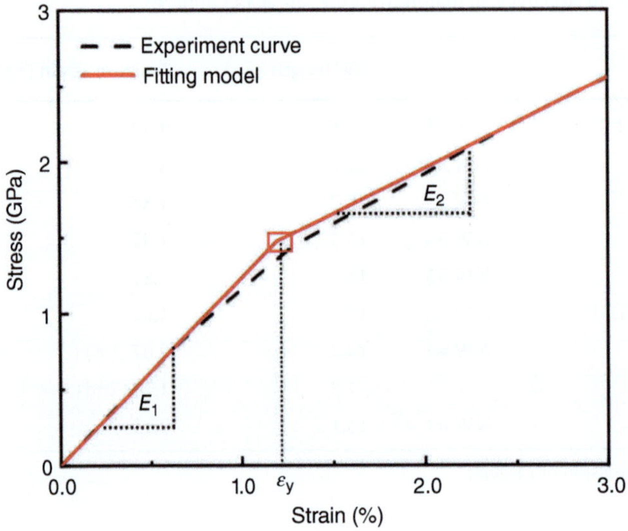

Figure 2.10 Typical experimental stress–strain curve of silver NWs and the bilinear fitting (note that the square dot marks the yield strain, ε_y). Source: Reproduced with permission of Ref. [110]; Copyright 2023, Elsevier.

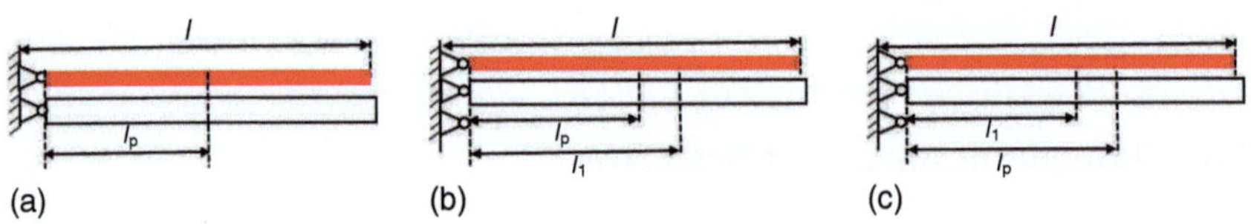

Figure 2.11 Distribution of the plastic zone on the untreated substrate surface: (a) The entire NW is in the bonded zone and plastic zone. (b) The NW contains bonded zone and sliding zone, and plastic zone $l_p < l_1$ and (c) NW contains bonded zone and sliding zone, and plastic zone $l_p > l_1$. Source: Reproduced with permission of Ref. [110]; Copyright 2023, Elsevier.

Case 2:

$$\begin{cases} \delta = B_2 \sinh(\beta_2 x), \text{when } 0 < x < l_p \\ \delta = A_1 \cosh(\beta_1 x) + B_1 \sinh(\beta_1 x), \text{when } l_p < x < l_1 \\ \delta = \frac{\tau_c}{2E_1 h}(x + l_1 - 2l)(x - l_1) + \varepsilon_m (x - l_1) + \delta_1, \text{when } l_1 < x < l \end{cases} \tag{2.11}$$

Constants A_1, B_1, B_2, l_p, and l_1 will be solved numerically.

Case 3:

$$\begin{cases} \delta = B_2 \sinh(\beta_2 x), \text{when } 0 < x < l_1 \\ \delta = \frac{\tau_c}{2E_2 h}x^2 + C_2 x + D_2, \text{when } l_1 < x < l_p \\ \delta = \frac{\tau_c}{2E_1 h}(x + l_1 - 2l)(x - l_1) + \varepsilon_m (x - l_1) + \delta_1, \text{when } l_p < x < l \end{cases} \tag{2.12}$$

Constants B_2, C_2, D_2, l_p, and l_1 will be solved numerically.

Table 2.2 Different shear-lag models and their respective dimensionless parameters. Note that the substrate or matrix is elastic in all cases.

	Elastic fiber	**Elastoplastic fiber**
Classic shear-lag model	c_1	—
Nonlinear shear-lag model (vdW interactions)	c_1, c_2	c_1, c_2, c_3, c_4
Bilinear shear-lag model (Chemical bonding)	c_1, c_2, c_5	c_1, c_2, c_3, c_4, c_5

Source: Ref. [110]/with permission of Elsevier.

The shear-lag analysis on the treated substrate considering the electroplastic fiber is more complicated. The interface between NW- and ultraviolet-ozone (UVO)-treated surfaces consist of three parts: bonded zone, damaged zone, and debonded zone. It can include six cases with respect to the plastic zone, l_p. The analysis will not be presented here. Interested readers are referred to Shao et al.

To generalize the shear-lag analysis, five dimensionless parameters are proposed as follows:

$$c_1 = \beta_1 l, c_2 = \frac{\tau_c l}{E_1 h}, c_3 = \varepsilon_y, c_4 = \frac{E_2}{E_1}, c_5 = \frac{\delta_2}{\delta_1} \tag{2.13}$$

It is interesting to note that in the classic shear-lag model, only one dimensionless parameter c_1 is used. In the shear-lag model that considers an interfacial shear strength (for vdW interactions), two parameters c_1 and c_2 are used. When considering plasticity in the fiber or NW, two additional parameters c_3 and c_4 must be used. In the shear-lag model that considers the damage (for chemical bonding), three parameters c_1, c_2, and c_5 are used. Similarly, when considering plasticity in the fiber or NW, two additional parameters c_3 and c_4 must be used. Table 2.2 summarizes the necessary parameters for the different cases. While the case of classic shear-lag model with elastoplastic fiber is not analyzed in this work, it is reasonable to predict that three parameters, c_1, c_3, c_4, will be needed.

The maximum strain in the NW versus the applied strain of the substrate is shown in Figure 2.12, from which the dependence of interfacial shear stress transfer between them on the parameters c_1–c_4 can be observed. Note that the circle dot marks the bonded-sliding transition. Figure 2.12a shows the significant effect of c_1 on the interfacial stress transfer; the maximum strain in the NW increases with c_1 in both the fully bonded case and the bonded-sliding case. The larger c_1 is, the earlier the bonded-sliding transition occurs. Figure 2.12b shows that while c_2 does not affect the interfacial stress transfer in the fully bonded case, it affects the bonded-sliding case significantly. The higher c_2 is, the slower the bonded-sliding transition occurs. Figure 2.12c shows the yield strain, ε_y, of the NW has only a modest effect on the interfacial stress transfer, and has nearly no influence on the bonded-sliding transition. Figure 2.12d illustrates that the smaller c_4 is, the larger the maximum strain in the NW is without the effect on the bonded-sliding transition. In Figure 2.12, c_3 and c_4 appear to have no influence on the interfacial shear stress transfer for the fully bonded case.

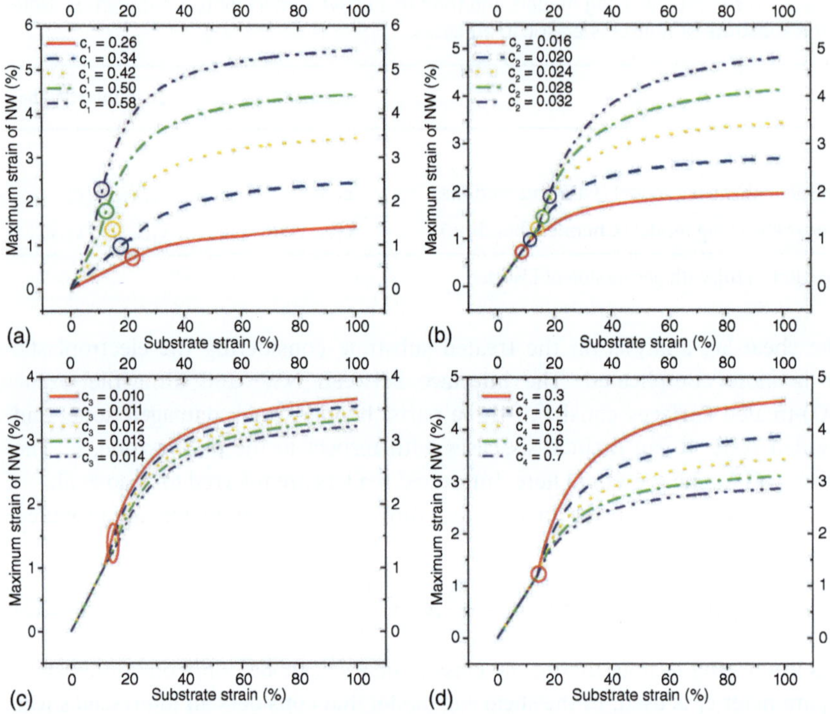

Figure 2.12 The effect of the dimensionless parameters (a) c_1, (b) c_2, (c) c_3, and (d) c_4, on the interfacial shear stress transfer. Note that the default values $c_1 = 0.42$; $c_2 = 0.02$; $c_3 = 0.012$; $c_4 = 0.5$. The circle dot marks the bonded-sliding transition. Source: Reproduced with permission of Ref. [110]; Copyright 2023, Elsevier.

2.5 Mechanical Design of Stretchable Structures

Two main strategies have been exploited to develop stretchable devices – using nanocomposites with synthesized nanomaterials dispersed into a polymer matrix and mechanics-guided structural design to reduce the effective stress on the structure. Both strategies have seen remarkable progress in developing highly conductive and stretchable devices [1, 5, 111]. However, so far, the mechanical design strategy has been mainly applied to thin films deposited on a substrate. It would be interesting to exploit this strategy for nanocomposites. Here we present two examples – delaminated buckle and surface crack – for developing stretchable Ag NW devices.

2.5.1 Buckle-Delamination Enabled Stretchable Silver Nanowire Conductors

To develop stretchable conductors, a common method is to use nanocomposites with synthesized nanomaterials dispersed into a polymer matrix [39, 42, 112–114]. The key to this strategy is to maintain the conductive percolation pathways when the substrate or matrix is stretched. On the other hand, the structural design strategy can significantly reduce the effective strain on the conductors, leading to nearly constant conductivity irrespective of the applied strain. Controlled buckling and delamination of thin films on a compliant substrate has attracted much attention for applications ranging from micro/nanofabrication to flexible and stretchable electronics to bioengineering [115–117]. However, so far this strategy has been mainly applied to the deposited thin films, relying on costly microfabrication processes.

A highly conductive and stretchable conductor was recently reported by attaching a polymer composite film (with a thin layer of Ag NWs embedded below the surface of the polymer matrix) on top of a pre-stretched elastomer substrate followed by releasing the prestrain (Figure 2.13a) [118]. A partially delaminated wavy geometry of the polymer film is created (Figure 2.13b). During the evolution of the buckle-delamination, the blisters pop up randomly but self-adjust into a uniform distribution, which effectively reduces the local strain in the Ag NWs (Figure 2.13c). The resistance change of the conductor is less than 3% with the applied strain up to 100% (Figure 2.13d). An ideal stretchable conductor should maintain constant conductance irrespective of the applied strain, which is however very challenging because of a tradeoff between stretchability and quality factor (Q, the percent strain divided by the percent resistance change). Figure 2.13e shows the quality factor with respect to the stretchability of representative nanomaterial-enabled stretchable conductors. In one demonstration, an integrated Ag NW/elastomer sensing module is designed and fabricated to show the electrical stability of the interconnection between sensor and data acquisition module under large bending strain (Figure 2.13f,g). In another demonstration, a stretchable thermochromic device is developed and tested with multiple stretching and releasing cycles (up to 100% strain) (Figure 2.13h,i). The cycling test does not affect the heating temperature because the buckle-delamination structure offers excellent stretchability and minimal resistance change. This work highlights the important relevance of mechanics-based design in nanomaterial-enabled stretchable devices.

2.5.2 A Highly Sensitive, Stretchable, and Robust Strain Sensor Based on Crack Advancing and Opening

Soft and stretchable strain sensors have been attracting significant attention [8, 119–121]. However, the tradeoff between the sensitivity (gauge factor) and the sensing range has been a major challenge. A soft stretchable strain sensor with an unusual combination of high sensitivity, large sensing range, and high robustness has been recently reported [122]. The sensor is made of metal NW network embedded below the surface of an elastomeric matrix (Figure 2.14a). Periodic mechanical cuts are applied to the top surface of the sensor (Figure 2.14b). Under the applied strain, the resistance increases as the crack propagates but remains constant as the crack reaches the cut length and only opens (the reversible range) (Figure 2.14c,d).

To study the electrical performance of the sensors, three major geometrical parameters – ratio of the cut depth to the sample thickness d_c/t, ratio of the pitch between the cracks to the specimen width p/w, and ratio of the cut length to the

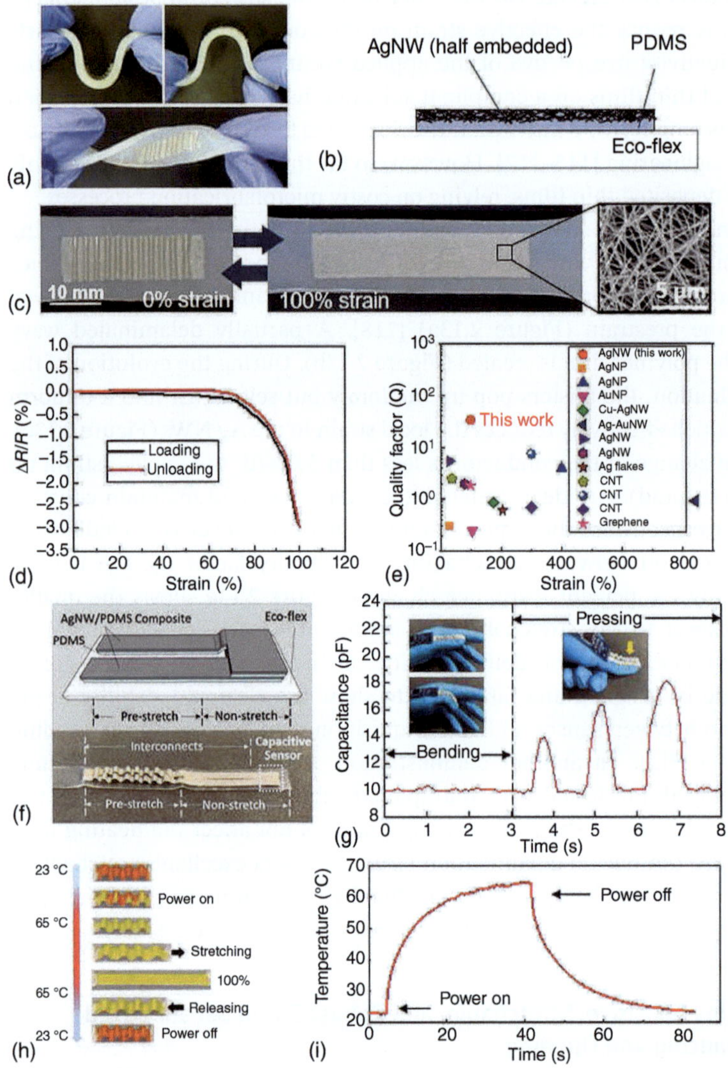

Figure 2.13 (a) Photographs of the stretchable Ag NW conductor bent and twisted. (b) Schematic cross-sectional view of the stretchable conductor showing the Ag NWs just below the PDMS surface (not to scale). (c) Photographs of the stretchable conductor stretched from 0% to 100% in the longitudinal direction and the SEM image showing the top surface of the conductor. (d) Theoretically calculated resistance change compared to experimental measurements of resistance change with respect to applied stretching strain. (e) A comparison of this work to recent work in stretchable conductors. (f) Schematic illustration and photograph of integrated sensor–conductor module. (g) Capacitance changes of the integrated sensor–conductor module attached on a thumb that bends and presses. (h) Photographs of the stretchable thermochromic conductor changing color: First heat up the stretchable conductor with 2 V direct current resulting temperature rise from 23 to 65 °C; then stretch the conductor to 100% strain and release; finally turn off the power and cool down to original room temperature. (i) Temperature change of the stretchable thermochromic conductor when powered with 2 V and then turning off the power. Source: Reproduced with permission of Ref. [118]; Copyright 2020, American Chemical Society.

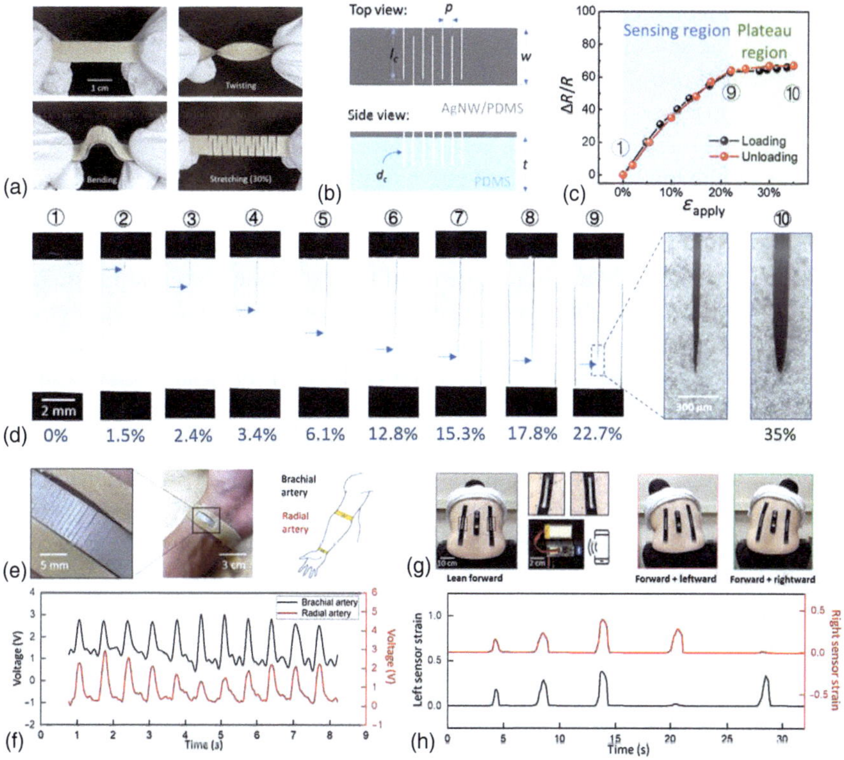

Figure 2.14 (a) The undeformed crack-based strain sensor and the sensor under twisting, bending, and stretching. (b) Schematic of the sensor showing the pitch between cuts p, crack length l_c, specimen width w, cut depth d_c, and specimen thickness t. (c) Resistance change of the sensor corresponding to snapshots in (d). (d) Snapshots of the top surface of the sensor under applied strain with magnified photographs showing the crack tip opening and blunting. (e) The sensor integrated with a wristband for monitoring blood pulse. (f) Sensing results of two blood pulse sensors placed on the radial artery on the wrist and the brachial artery on the arm. (g) Photographs of the sensor integrated with an athletic tape attached side by side along the spine for monitoring lower back strains. Inset shows the two sensors and the Bluetooth evaluation board for data collection and transmission. (h) Sensing results of the two sensor system detecting different bending motions of the lower back. Source: Reproduced with permission of Ref. [122]; Copyright 2022, American Chemical Society.

specimen width l_c/w. 1000 stretching/unloading cycles were applied to show the excellent repeatability of the sensor for long-term use. This sensor overcame the limitation of most existing strain sensors and offered an unprecedented combination of gauge factor (GF), strain sensing range, and robustness. A large GF of 290.1 was achieved with a sensing range over 22%. Finite element analysis was conducted to validate the electrical performance and predict the mechanical damage, agreeing very well with the experimental results [122].

The stretchable strain sensor can be integrated into wearable systems for monitoring physiological functions and body motions associated with different levels of strain, such as blood pressure and lower back health. To demonstrate the versatile applicability, the sensors were first applied on the wrist to detect the pulse wave, which represents one of the most delicate strain signals on human skin. Figure 2.14e shows the setup consisting of a rubber band with a slot filled with the strain sensor. When the blood travels through the vein, the strain sensor is slightly bent, opening the cracks on the top surface. The red curve in Figure 2.14f shows the pulse wave captured from the radial artery on the wrist. Similarly, another strain sensor can be placed on the brachial artery on the arm, recording another pulse wave simultaneously. By measuring the distance between the two pulse areas and taking the average of the time gap between peaks of the two pulse waves, the averaged pulse wave velocity can be measured. The other demonstration is aimed at monitoring the large strains on the lower back, which is a critical signal for metabolic syndrome and spine issues. The sensor was integrated onto an athletic tape by fixing the two ends of the sensor on the tape using a PDMS precursor (Figure 2.14g inset). Two sensor/tape setups were attached on the lower back side by side in parallel with the spine. An Arduino Bluetooth board/tape was attached in the middle area of the lower back to collect and transmit the sensing signals to a smartphone. The results measured by the sensor and by the fiducial markers showed excellent agreement (Figure 2.14h) [122].

2.6 Concluding Remarks

In this chapter, we reviewed the various mechanics issues of metal NWs of relevance to their application in stretchable electronics, including mechanics of individual NWs, interfacial mechanics between NWs and polymer matrix, and mechanics of NW/polymer composite in stretchable structures. The metal NWs are generally in the form of nanocomposites composed of metal NWs in polymer matrices. So, the mechanics issues discussed here are typical for nanocomposites. Metal NWs with internal structures like TBs possess excellent mechanical properties such as recoverable plasticity, strain hardening, large ductility, and insensitivity to surface flaws, owing to the interactions between TBs and surface-nucleated dislocations. These properties are of great value to stretchable electronics. The NW–polymer interfacial interactions typically include vdW interactions and chemical bonding. The interactions determine the interfacial shear stress transfer, which imparts the strain in the polymer matrix to the NWs. Mechanics models accounting for the different interfacial interactions and the plastic deformation of the metal NWs have been developed, in agreement with the experiments. Mechanical design has been applied to the NW/polymer nanocomposites to develop highly stretchable structures. As an example, buckle-delamination was used to develop a highly stretchable conductor with nearly constant resistance up to 100% strain. Fracture behavior of NW/polymer composite is discussed including crack propagation, opening, and blunting, for the development of a highly stretchable, sensitive, and robust strain sensor.

NW-enabled stretchable electronics have seen exciting progress in the last decade. For further development in this promising area, several challenges lie ahead across multiple length scales. At the individual NW level, effect of cyclic loading on the mechanical behavior has been investigated for penta-twinned Ag NWs, exhibiting the pronounced Bauschinger effect [89]. However, the number of loading cycles is limited. It would be interesting to study the high-cycle fatigue behavior. For stretchable electronics, the NWs undergo combined mechanical loading and electrical loading (i.e. current). Current can significantly alter the mechanical behaviors, especially combined with mechanical stress, through effects like electroplasticity [123–125] and electromigration [126–128], which have only been rarely investigated in NWs. Temperature can modify the dislocation nucleation and interactions, which is particularly pronounced at the nanoscale [129]. Electrical current can lead to Joule heating; thus, temperature increases to the NWs. It would be interesting yet challenging to study the combined effect of mechanical stress, electrical current, and temperature, especially under cyclic mechanical stress and electrical current.

Interfacial mechanics between NWs and polymer matrix plays a critical role in the morphology of the NW percolation network under deformation. According to the percolation theory, the longer the NW, the higher the electrical conductivity. Efforts have been pursued to fabricate very long NWs for flexible transparent electronics and stretchable electronics [43]. However, according to the shear-lag theory, an NW can fracture above a certain length; the critical length depends on the interfacial mechanical properties (and thus the bonding types) [106]. It would be interesting to optimize the electrical and mechanical properties of an NW network by tailoring the NW length and/or the bonding strength.

Mechanics of network materials have been well-studied [130]. The network can have different types of interactions between the fibers – crosslinking, topological interactions between tortuous fibers, and surface interactions such as cohesion. Often multiple types of interactions coexist. If the NW network is not welded, the second and third types of interactions could be both present. If welded, then the first type of interaction is present too. For stretchable electronics, the NWs are typically not bonded with each other. For flexible transparent electronics, they are typically welded. And cold welding has been reported for metal NWs [131, 132]. While the mechanics framework of network materials is established, its application to the NW network has not been explored considering the size-dependent NW behavior, NW–NW interactions, and NW–matrix interactions. In addition, it is important to understand the electron transport across the NW junctions for the different types of interactions between the NWs.

Many other nanomaterials including their hybrids have been used in stretchable electronics including carbon nanotubes, metal flakes, and 2D materials [17–23]. The fundamental mechanics learned from the metal NWs for stretchable electronics could be applied to other nanomaterials.

Acknowledgments

We acknowledge the financial support from the National Science Foundation (NSF) under Award Nos. 1929646, 2122841, and 2134664.

References

1 Rogers, J.A., Someya, T., and Huang, Y.G. (2010). Materials and mechanics for stretchable electronics. *Science* 327 (5973): 1603–1607.

2 Someya, T., Bao, Z., and Malliaras, G.G. (2016). The rise of plastic bioelectronics. *Nature* 540 (7633): 379–385.

3 Yao, S., Swetha, P., and Zhu, Y. (2018). Nanomaterial-enabled wearable sensors for healthcare. *Advanced Healthcare Materials* 7 (1): 1700889.

4 Kim, D.-H., Lu, N., Ma, R. et al. (2011). Epidermal electronics. *Science* 333 (6044): 838–843.

5 Sekitani, T., Noguchi, Y., Hata, K. et al. (2008). A rubberlike stretchable active matrix using elastic conductors. *Science* 321 (5895): 1468–1472.

6 Sekitani, T. and Someya, T. (2010). Stretchable, large-area organic electronics. *Advanced Materials* 22 (20): 2228–2246.

7 Lipomi, D.J., Tee, B.C.K., Vosgueritchian, M., and Bao, Z.N. (2011). Stretchable organic solar cells. *Advanced Materials* 23 (15): 1771–1775.

8 Lipomi, D.J., Vosgueritchian, M., Tee, B.C.-K. et al. (2011). Skin-like pressure and strain sensors based on transparent elastic films of carbon nanotubes. *Nature Nanotechnology* 6 (12): 788–792.

9 Daeneke, T., Khoshmanesh, K., Mahmood, N. et al. (2018). Liquid metals: fundamentals and applications in chemistry. *Chemical Society Reviews* 47 (11): 4073–4111.

10 Zhu, S., So, J.-H., Mays, R. et al. (2013). Ultrastretchable fibers with metallic conductivity using a liquid metal alloy core. *Advanced Functional Materials* 23 (18): 2308–2314.

11 Trung, T.Q. and Lee, N.E. (2016). Flexible and stretchable physical sensor integrated platforms for wearable human-activity monitoring and personal healthcare. *Advanced Materials* 28 (22): 4338–4372.

12 Yang, J.C., Mun, J., Kwon, S.Y. et al. (2019). Electronic skin: recent progress and future prospects for skin-attachable devices for health monitoring, robotics, and prosthetics. *Advanced Materials* 31 (48): 1904765.

13 McCoul, D., Hu, W., Gao, M. et al. (2016). Recent advances in stretchable and transparent electronic materials. *Advanced Electronic Materials* 2 (5): 1500407.

14 Bao, Z.N. and Chen, X.D. (2016). Flexible and stretchable devices. *Advanced Materials* 28 (22): 4177–4179.

15 Park, J., You, I., Shin, S., and Jeong, U. (2015). Material approaches to stretchable strain sensors. *ChemPhysChem* 16 (6): 1155–1163.

16 Takei, K., Takahashi, T., Ho, J.C. et al. (2010). Nanowire active-matrix circuitry for low-voltage macroscale artificial skin. *Nature Materials* 9 (10): 821–826.

17 Matsuhisa, N., Kaltenbrunner, M., Yokota, T. et al. (2015). Printable elastic conductors with a high conductivity for electronic textile applications. *Nature Communications* 6 (1): 1–11.

18 Yao, S. and Zhu, Y. (2015). Nanomaterial-enabled stretchable conductors: strategies, materials and devices. *Advanced Materials* 27 (9): 1480–1511.

19 Gong, S. and Cheng, W. (2016). One-dimensional nanomaterials for soft electronics. *Advanced Electronic Materials* 3 (3): 1600314.

20 Xia, Y.N., Yang, P.D., Sun, Y.G. et al. (2003). One-dimensional nanostructures: synthesis, characterization, and applications. *Advanced Materials* 15 (5): 353–389.

21 McAlpine, M.C., Friedman, R.S., Jin, S. et al. (2003). High-performance nanowire electronics and photonics on glass and plastic substrates. *Nano Letters* 3 (11): 1531–1535.

22 De Volder, M.F.L., Tawfick, S.H., Baughman, R.H., and Hart, A.J. (2013). Carbon nanotubes: present and future commercial applications. *Science* 339 (6119): 535–539.

23 Ge, J., Yao, H.B., Wang, X. et al. (2013). Stretchable conductors based on silver nanowires: improved performance through a binary network design. *Angewandte Chemie International Edition* 52 (6): 1654–1659.

24 Jung, D., Lim, C., Shim, H.J. et al. (2021). Highly conductive and elastic nanomembrane for skin electronics. *Science* 373 (6558): 1022–1026.

25 Jung, D., Kim, Y., Lee, H. et al. (2023). Metal-like stretchable nanocomposite using locally-bundled nanowires for skin-mountable devices. *Advanced Materials* 35 (44): 2303458.

26 Park, M., Im, J., Shin, M. et al. (2012). Highly stretchable electric circuits from a composite material of silver nanoparticles and elastomeric fibres. *Nature Nanotechnology* 7 (12): 803–809.

27 Diao, J., Gall, K., and Dunn, M.L. (2003). Surface-stress-induced phase transformation in metal nanowires. *Nature Materials* 2 (10): 656–660.

28 Weinberger, C.R. and Cai, W. (2012). Plasticity of metal nanowires. *Journal of Materials Chemistry* 22 (8): 3277–3292.

29 Park, H.S., Cai, W., Espinosa, H.D., and Huang, H. (2009). Mechanics of crystalline nanowires. *MRS Bulletin* 34 (3): 178–183.

30 Zhu, Y. (2017). Mechanics of crystalline nanowires: an experimental perspective. *Applied Mechanics Reviews* 69 (1): 010802.

31 Agrawal, R., Peng, B., Gdoutos, E.E., and Espinosa, H.D. (2008). Elasticity size effects in ZnO nanowires—a combined experimental-computational approach. *Nano Letters* 8 (11): 3668–3674.

32 Zhu, Y. (2016). In situ nanomechanical testing of crystalline nanowires in electron microscopes. *JOM* 68: 84–93.

33 He, J. and Lilley, C.M. (2008). Surface effect on the elastic behavior of static bending nanowires. *Nano Letters* 8 (7): 1798–1802.

34 Xu, F., Qin, Q.Q., Mishra, A. et al. (2010). Mechanical properties of ZnO nanowires under different loading modes. *Nano Research* 3: 271–280.

35 Park, H.S. and Zimmerman, J.A. (2005). Modeling inelasticity and failure in gold nanowires. *Physical Review B* 72: 54106.

36 Park, H.S., Gall, K., and Zimmerman, J.A. (2005). Shape memory and pseudoelasticity in metal nanowires. *Physical Review Letters* 95: 255504.

37 Zhu, T., Li, J., Samanta, A. et al. (2008). Temperature and strain-rate dependence of surface dislocation nucleation. *Physical Review Letters* 100: 25502.

38 Hu, L.B., Kim, H.S., Lee, J.Y. et al. (2010). Scalable coating and properties of transparent, flexible, silver nanowire electrodes. *ACS Nano* 4 (5): 2955–2963.

39 Xu, F. and Zhu, Y. (2012). Highly conductive and stretchable silver nanowire conductors. *Advanced Materials* 24 (37): 5117–5122.

40 Pike, G.E. and Seager, C.H. (1974). Percolation and conductivity: a computer study. I. *Physical Review B* 10 (4): 1421.

41 Kasry, A., Kuroda, M.A., Martyna, G.J. et al. (2010). Chemical doping of large-area stacked graphene films for use as transparent, conducting electrodes. *ACS Nano* 4 (7): 3839–3844.

42 Lee, P., Lee, J., Lee, H. et al. (2012). Highly stretchable and highly conductive metal electrode by very long metal nanowire percolation network. *Advanced Materials* 24 (25): 3326–3332.

43 Lee, J., Lee, P., Lee, H. et al. (2012). Very long Ag nanowire synthesis and its application in a highly transparent, conductive and flexible metal electrode touch panel. *Nanoscale* 4 (20): 6408.

44 Bera, D., Kuiry, S.C., and Seal, S. (2004). Synthesis of nanostructured materials using template-assisted electrodeposition. *JOM* 56 (1): 49–53.

45 Liu, Z., Yang, Y., Liang, J. et al. (2003). Synthesis of copper nanowires via complex-surfactant-assisted hydrothermal reduction process. *The Journal of Physical Chemistry. B* 107: 12658–12661.

46 Tian, M., Wang, J., Kurtz, J. et al. (2003). Electrochemical growth of single-crystal metal nanowires via a two-dimensional nucleation and growth mechanism. *Nano Letters* 3 (7): 919–923.

47 Kim, C., Gu, W., Briceno, M. et al. (2008). Copper nanowires with a five-twinned structure grown by chemical vapor deposition. *Advanced Materials* 20 (10): 1859–1863.

48 Richter, G., Hillerich, K., Gianola, D.S. et al. (2009). Ultra high strength single crystalline nanowhiskers grown by physical vapor deposition. *Nano Letters* 9: 3048–3052.

49 Yoo, Y., Seo, K., Han, S. et al. (2010). Steering epitaxial alignment of Au, Pd, and AuPd nanowire arrays by atom flux change. *Nano Letters* 10 (2): 432–438.

50 Sun, Y., Gates, B., Mayers, B., and Xia, Y. (2002). Crystalline silver nanowires by soft solution processing. *Nano Letters* 2 (2): 165–168.

51 Wiley, B., Sun, Y., Mayers, B., and Xia, Y. (2005). Shape-controlled synthesis of metal nanostructures: the case of silver. *Chemistry – A European Journal* 11 (2): 454–463.

52 Murphy, C.J. and Jana, N.R. (2002). Controlling the aspect ratio of inorganic nanorods and nanowires. *Acta Materialia* 14 (1): 80–82.

53 Wiley, B., Sun, Y., and Xia, Y. (2007). Synthesis of silver nanostructures with controlled shapes and properties. *Accounts of Chemical Research* 40 (10): 1067–1076.

54 Chen, C.Q., Shi, Y., Zhang, Y.S. et al. (2006). Size dependence of the Young's modulus of ZnO nanowires. *Physical Review Letters* 96: 75505.

55 Duan, H.L., Wang, J., Huang, Z.P., and Karihaloo, B.L. (2005). Size-dependent effective elastic constants of solids containing nano-inhomogeneities with interface stress. *Journal of the Mechanics and Physics of Solids* 53: 1574–1596.

56 Park, H.S. (2008). Surface stress effects on the resonant properties of silicon nanowires. *Journal of Applied Physics* 103: 123504.

57 Sharma, P., Ganti, S., and Bhate, N. (2003). Effect of surfaces on the size-dependent elastic state of nano-inhomogeneities. *Applied Physics Letters* 82 (4): 535–537.

58 Zhu, Y., Xu, F., Qin, Q. et al. (2009). Mechanical properties of vapor-liquid-solid synthesized silicon nanowires. *Nano Letters* 9 (11): 3934–3939.

59 Zhou, L.G. and Huang, H.C. (2004). Are surfaces elastically softer or stiffer? *Applied Physics Letters* 84 (11): 1940–1942.

60 Shim, H.W., Zhou, L.G., Huang, H., and Cale, T.S. (2005). Nanoplate elasticity under surface reconstruction. *Applied Physics Letters* 86: 151912.

61 Liang, H.Y., Upmanyu, M., and Huang, H.C. (2005). Size-dependent elasticity of nanowires: nonlinear effects. *Physical Review B* 71 (24): 241403.

62 Chang, T.-H., Cheng, G., Li, C., and Zhu, Y. (2016). On the size-dependent elasticity of penta-twinned silver nanowires. *Extreme Mechanics Letters* 8: 177–183.

63 Zhu, T. and Li, J. (2010). Ultra-strength materials. *Progress in Materials Science* 55 (7): 710–757.

64 Chen, L.Y., He, M.-R., Shin, J. et al. (2015). Measuring surface dislocation nucleation in defect-scarce nanostructures. *Nature Materials* 14: 707–713.

65 Yue, Y., Liu, P., Zhang, Z. et al. (2011). Approaching the theoretical elastic strain limit in copper nanowires. *Nano Letters* 11 (8): 3151–3155.

66 Zheng, H., Cao, A., Weinberger, C.R. et al. (2010). Discrete plasticity in sub-10-nm-sized gold crystals. *Nature Communications* 1: 144.

67 Sun, J., He, L., Lo, Y.-C. et al. (2014). Liquid-like pseudoelasticity of sub-10-nm crystalline silver particles. *Nature Materials* 13 (11): 1007–1012.

68 Wang, J., Zeng, Z., Weinberger, C.R. et al. (2015). In situ atomic-scale observation of twinning-dominated deformation in nanoscale body-centred cubic tungsten. *Nature Materials* 14 (6): 594–600.

69 Wu, B., Heidelberg, A., and Boland, J.J. (2005). Mechanical properties of ultrahigh-strength gold nanowires. *Nature Materials* 4 (7): 525–529.

70 Wang, J., Sansoz, F., Huang, J. et al. (2013). Near-ideal theoretical strength in gold nanowires containing angstrom scale twins. *Nature Communications* 4: 1742.

71 Lu, Y., Song, J., Huang, J.Y., and Lou, J. (2011). Surface dislocation nucleation mediated deformation and ultrahigh strength in sub-10-nm gold nanowires. *Nano Research* 4 (12): 1261–1267.

72 Seo, J.-H., Yoo, Y., Park, N.-Y. et al. (2011). Superplastic deformation of defect-free Au nanowires via coherent twin propagation. *Nano Letters* 11 (8): 3499–3502.

73 Sedlmayr, A., Bitzek, E., Gianola, D.S. et al. (2012). Existence of two twinning-mediated plastic deformation modes in Au nanowhiskers. *Acta Materialia* 60 (9): 3985–3993.

74 Peng, C., Zhan, Y., and Lou, J. (2012). Size-dependent fracture mode transition in copper nanowires. *Small* 8 (12): 1889–1894.

75 Brenner, S.S. (1956). Tensile strength of whiskers. *Journal of Applied Physics* 27 (12): 1484–1491.

76 Cheng, G., Yin, S., Li, C. et al. (2020). In-situ TEM study of dislocation interaction with twin boundary and retraction in twinned metallic nanowires. *Acta Materialia* 196: 304–312.

77 Hwang, B., Kang, M., Lee, S. et al. (2015). Effect of surface energy on size-dependent deformation twinning of defect-free Au nanowires. *Nanoscale* 7 (38): 15657–15664.

78 Seo, J.H., Park, H.S., Yoo, Y. et al. (2013). Origin of size dependency in coherent-twin-propagation-mediated tensile deformation of noble metal nanowires. *Nano Letters* 13 (11): 5112–5116.

79 Yin, S., Cheng, G., Richter, G. et al. (2019). Transition of deformation mechanisms in single-crystalline metallic nanowires. *ACS Nano* 13 (8): 9082–9090.

80 Wang, J. and Huang, H. (2006). Novel deformation mechanism of twinned nanowires. *Applied Physics Letters* 88: 203112.

81 Deng, C. and Sansoz, F. (2009). Enabling ultrahigh plastic flow and work hardening in twinned gold nanowires. *Nano Letters* 9 (4): 1517–1522.

82 Deng, C. and Sansoz, F. (2010). Repulsive force of twin boundary on curved dislocations and its role on the yielding of twinned nanowires. *Scripta Materialia* 63 (1): 50–53.

83 Sansoz, F., Huang, H., and Warner, D.H. (2008). An atomistic perspective on twinning phenomena in nano-enhanced fcc metals. *JOM* 60 (9): 79–84.

84 Cheng, G., Yin, S., Chang, T.-H. et al. (2017). Anomalous tensile detwinning in twinned nanowires. *Physical Review Letters* 119 (25): 256101.

85 Sun, Y. and Xia, Y. (2002). Large-scale synthesis of uniform silver nanowires through a soft, self-seeding, polyol process. *Advanced Materials* 14 (11): 833.

86 Caswell, K.K., Bender, C.M., and Murphy, C.J. (2003). Seedless, surfactantless wet chemical synthesis of silver nanowires. *Nano Letters* 3 (5): 667–669.

87 Li, X.D., Hao, H.S., Murphy, C.J., and Caswell, K.K. (2003). Nanoindentation of silver nanowires. *Nano Letters* 3 (11): 1495–1498.

88 Qin, Q., Yin, S., Cheng, G. et al. (2015). Recoverable plasticity in penta-twinned metallic nanowires governed by dislocation nucleation and retraction. *Nature Communications* 6: 5983.

89 Bernal, R.A., Aghaei, A., Lee, S. et al. (2015). Intrinsic Bauschinger effect and recoverable plasticity in pentatwinned silver nanowires tested in tension. *Nano Letters* 15 (1): 139–146.

90 Rajagopalan, J., Han, J.H., and Saif, M.T.A. (2007). Plastic deformation recovery in freestanding nanocrystalline aluminum and gold thin films. *Science* 315 (5820): 1831–1834.

91 Rajagopalan, J., Rentenberger, C., Peter Karnthaler, H. et al. (2010). In situ TEM study of microplasticity and Bauschinger effect in nanocrystalline metals. *Acta Materialia* 58 (14): 4772–4782.

92 Mompiou, F., Caillard, D., Legros, M., and Mughrabi, H. (2012). In situ TEM observations of reverse dislocation motion upon unloading in tensile-deformed UFG aluminium. *Acta Materialia* 60 (8): 3402–3414.

93 Xiang, Y. and Vlassak, J.J. (2005). Bauschinger effect in thin metal films. *Scripta Materialia* 53 (2): 177–182.

94 Jennings, A.T., Gross, C., Greer, F. et al. (2012). Higher compressive strengths and the Bauschinger effect in conformally passivated copper nanopillars. *Acta Materialia* 60 (8): 3444–3455.

95 Lee, S., Im, J., Yoo, Y. et al. (2014). Reversible cyclic deformation mechanism of gold nanowires by twinning-detwinning transition evidenced from in situ TEM. *Nature Communications* 5: 3033.

96 Zhu, Y., Qin, Q.Q., Xu, F. et al. (2012). Size effects on elasticity, yielding and fracture of silver nanowires: in situ experiments. *Physical Review B* 85: 45443.

97 Filleter, T., Ryu, S., Kang, K. et al. (2012). Nucleation-controlled distributed plasticity in penta-twinned silver nanowires. *Small* 8 (19): 2986–2993.

98 Narayanan, S., Cheng, G., Zeng, Z. et al. (2015). Strain hardening and size effect in five-fold twinned Ag nanowires. *Nano Letters* 15 (6): 4037–4044.

99 Cox, H.L. (1952). The elasticity and strength of paper and other fibrous materials. *British Journal of Applied Physics* 3 (3): 72–79.

100 Nairn, J.A. (1997). On the use of shear-lag methods for analysis of stress transfer in unidirectional composites. *Mechanics of Materials* 26, 63–80.

101 Chen, B., Wu, P.D., and Gao, H. (2009). A characteristic length for stress transfer in the nanostructure of biological composites. *Composites Science and Technology* 69: 1160–1164.

102 Needleman, A. (1990). An analysis of tensile decohesion along an interface. *Journal of the Mechanics and Physics of Solids* 38 (3): 289–324.

103 Park, K. and Paulino, G.H. (2011). Cohesive zone models: a critical review of traction-separation relationships across fracture surfaces. *Applied Mechanics Reviews* 64 (6).

104 Jiang, L.Y., Huang, Y., Jiang, H. et al. (2006). A cohesive law for carbon nanotube/polymer interfaces based on the van der Waals force. *Journal of the Mechanics and Physics of Solids* 54 (11): 2436–2452.

105 Wei, X., Filleter, T., and Espinosa, H.D. (2015). Statistical shear lag model-unraveling the size effect in hierarchical composites. *Acta Biomaterialia* 18: 206–212.

106 Poblete, F.R., Cui, Z., Liu, Y., and Zhu, Y. (2020). Stretching nanowires on a stretchable substrate: a method towards facile fracture testing and elastic strain engineering. *Extreme Mechanics Letters* 41: 101035.

107 Poblete, F.R. and Zhu, Y. (2019). Interfacial shear stress transfer at nanowire-polymer interfaces with van der Waals interactions and chemical bonding. *Journal of the Mechanics and Physics of Solids* 127: 191–207.

108 Guo, G. and Zhu, Y. (2015). Cohesive-shear-lag modeling of interfacial stress transfer between a monolayer graphene and a polymer substrate. *Journal of Applied Mechanics* 82 (3): 031005.

109 Jiang, T., Huang, R., and Zhu, Y. (2014). Interfacial sliding and buckling of monolayer graphene on a stretchable substrate. *Advanced Functional Materials* 24 (3): 396–402.

110 Shao, L.-H., Qu, X., Wang, T. et al. (2023). Interfacial shear stress transfer between elastoplastic fiber and elastic matrix. *Journal of the Mechanics and Physics of Solids* 173: 105218.

111 Lacour, S.P., Jones, J., Wagner, S. et al. (2005). Stretchable interconnects for elastic electronic surfaces. *Proceedings of the IEEE* 93 (8): 1459–1467.

112 Xu, F., Wang, X., Zhu, Y.T., and Zhu, Y. (2012). Wavy ribbons of carbon nanotubes for stretchable conductors. *Advanced Functional Materials* 22 (6): 1279–1283.

113 Zhu, Y. and Xu, F. (2012). Buckling of aligned carbon nanotubes as stretchable conductors: a new manufacturing strategy. *Advanced Materials* 24 (8): 1073–1077.

114 Hu, W., Niu, X., Li, L. et al. (2012). Intrinsically stretchable transparent electrodes based on silver-nanowire-crosslinked-polyacrylate composites. *Nanotechnology* 23 (34): 344002.

115 Song, J., Jiang, H.Q., Huang, Y., and Rogers, J.A. (2009). Mechanics of stretchable inorganic electronic materials. *Journal of Vacuum Science and Technology A* 27: 1107–1125.

116 Khang, D.Y., Jiang, H.Q., Huang, Y., and Rogers, J.A. (2006). A stretchable form of single-crystal silicon for high-performance electronics on rubber substrates. *Science* 311 (5758): 208–212.

117 Sun, Y.G., Choi, W.M., Jiang, H.Q. et al. (2006). Controlled buckling of semiconductor nanoribbons for stretchable electronics. *Nature Nanotechnology* 1 (3): 201–207.

118 Wu, S., Yao, S., Liu, Y. et al. (2020). Buckle-delamination-enabled stretchable silver nanowire conductors. *ACS Applied Materials & Interfaces* 12 (37): 41696–41703.

119 Yao, S. and Zhu, Y. (2014). Wearable multifunctional sensors using printed stretchable conductors made of silver nanowires. *Nanoscale* 6 (4): 2345–2352.

120 Amjadi, M., Kyung, K.U., Park, I., and Sitti, M. (2016). Stretchable, skin-mountable, and wearable strain sensors and their potential applications: a review. *Advanced Functional Materials* 26 (11): 1678–1698.

121 Amjadi, M., Pichitpajongkit, A., Lee, S. et al. (2014). Highly stretchable and sensitive strain sensor based on silver nanowire–elastomer nanocomposite. *ACS Nano* 8 (5): 5154–5163.

122 Wu, S., Moody, K., Kollipara, A., and Zhu, Y. (2023). Highly sensitive, stretchable, and robust strain sensor based on crack propagation and opening. *ACS Applied Materials & Interfaces* 15 (1): 1798–1807.

123 Zhao, S., Zhang, R., Chong, Y. et al. (2020). Defect reconfiguration in a Ti–Al alloy via electroplasticity. *Nature Materials* 20 (4): 468–472.

124 Dimitrov, N.K., Liu, Y., and Horstemeyer, M.F. (2022). Electroplasticity: a review of mechanisms in electro-mechanical coupling of ductile metals. *Mechanics of Advanced Materials and Structures* 29 (5): 705–716.

125 Conrad, H. (2000). Electroplasticity in metals and ceramics. *Materials Science and Engineering A* 287 (2): 276–287.

126 Ho, P.S. and Kwok, T. (1989). Electromigration in metals. *Reports on Progress in Physics* 52 (3): 301.

127 Tu, K.N. (2003). Recent advances on electromigration in very-large-scale-integration of interconnects. *Journal of Applied Physics* 94: 5451–5473.

128 Zhang, Z., Suo, Z., and He, J. (2005). Saturated voids in interconnect lines due to thermal strains and electromigration. *Journal of Applied Physics* 98 (7): 074501.

129 Cheng, G., Zhang, Y., Chang, T.-H. et al. (2019). In situ nano-thermomechanical experiment reveals brittle to ductile transition in silicon nanowires. *Nano Letters* 19 (8): 5327–5334.

130 Picu, C.R. (2022). *Network Materials: Structure and Properties*. London: Cambridge University Press.

131 Lu, Y., Huang, J.Y., Wang, C. et al. (2010). Cold welding of ultrathin gold nanowires. *Nature Nanotechnology* 5 (3): 218–224.

132 Liu, Y., Zhang, J., Gao, H. et al. (2017). Capillary-force-induced cold welding in silver-nanowire-based flexible transparent electrodes. *Nano Letters* 17 (2): 1090–1096.

3

Liquid Metal-Based Electronics

Carmel Majidi

Carnegie Mellon University, Department of Mechanical Engineering, 5000 Forbes Avenue, Pittsburgh, PA 15213, USA

3.1 Introduction

Liquid metals (LMs) are metals that have a melting point and crystallization temperature that are below room temperature (25 °C). LMs like mercury and gallium-indium-tin (Galinstan) have been used for decades in strain gauges [1], tilt sensors [2], thermometers [3], and other applications that rely on their unique combination of electrical, thermal, and rheological properties [4]. In recent years, mercury has been replaced with gallium-based LM alloys like Galinstan and eutectic gallium-indium (EGaIn) on account of their limited toxicity and negligible vapor pressure [5]. Moreover, there has been increasing interest in applying liquid metals to a wider range of applications – not only for sensing but also in the development of soft and highly elastic and stretchable circuits.

In this chapter, I will review efforts to use gallium-based LM alloys for applications in soft electronics. This will include an overview of material architectures, manufacturing methods, and design principles based on the mechanics of deformation of both the LM inclusions and the surrounding elastomer matrix. Following a brief introduction of Ga-based LMs (Section 3.1.1) and their treatment in the academic literature (Section 3.1.2), I will present an overview of various material architectures that use liquid metal for soft and stretchable circuitry (Section 3.2). This includes a brief review of fabrication methods that have been developed to synthesize these architectures and manufacture them at a large scale. Next, I will discuss some of the relevant mechanics that influence the properties of the material systems and present theoretical models based on these principles that inform device-level design (Section 3.3). Lastly, I will close with a brief discussion of open challenges and future prospects for the field (Section 3.4).

3.1.1 Ga-Based Liquid Metals

Over the past few decades, EGaIn and Galinstan have emerged as popular materials for LM-based sensing and electronics (Figure 3.1a) [5–8]. These Ga-based alloys

Mechanics of Flexible and Stretchable Electronics, First Edition. Edited by Yong Zhu and Nanshu Lu.

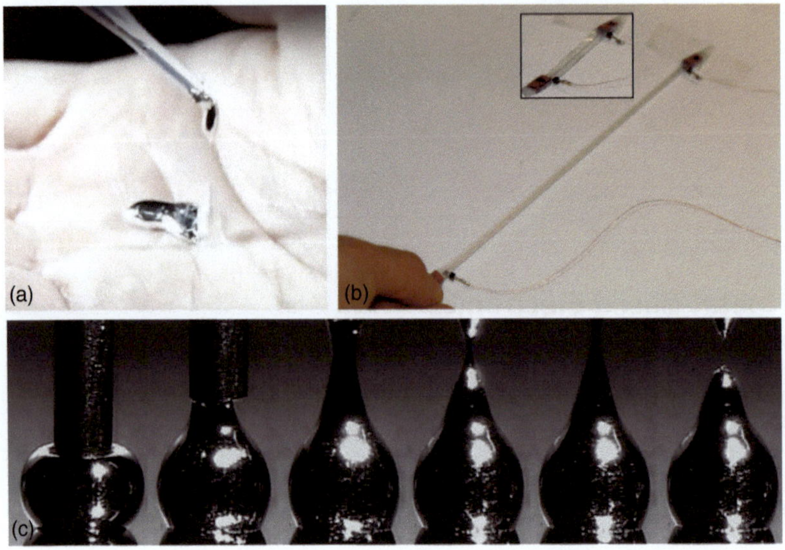

Figure 3.1 (a) Eutectic gallium-indium (EGaIn) liquid metal dispensed from a pipette. Source: From Dr. Michael Ford, Soft Machines Lab, Carnegie Mellon University. (b) Elongation of a strip of silicone elastomer embedded with a microfluidic channel of EGaIn. Source: From Dr. Carmel Majidi, Soft Machines Lab, Carnegie Mellon University (Chapter Author). (c) Despite their low viscosity, EGaIn can hold its shape in air due to the formation of a thin oxide skin. Source: Reproduced with permission of Chiechi et al. [20], © 2008 John Wiley & Sons.

have a high electrical conductivity (3×10^6 S/m) that far exceeds the threshold required to support digital electronic circuitry. Moreover, unlike mercury and other low melting point metals (Francium, Cesium), these alloys have negligible vapor pressure and low toxicity. Lastly, they have a low viscosity (2 mPa/s) – similar to that of water – which allows them to flow with limited mechanical resistance. When encapsulated as a microfluidic channel in a soft elastomer, the metal can flow and elongate with the surrounding elastomer when it is stretched, thereby maintaining electrical connectivity (Figure 3.1b).

Despite being regarded as a nontoxic alternative to mercury, the biocompatibility of Ga-based LM alloys is still an area of open research. Various studies have examined cell cytotoxicity and tissue growth on EGaIn or EGaIn-filled materials under a variety of conditions [9, 10]. While direct exposure to EGaIn can impair cell growth – and even lead to cell death – in many circumstances, current studies suggest that EGaIn may be safe in the presence of biological tissue if it is encapsulated within a polymer matrix [10]. In the case of soft and stretchable circuits, liquid metal will almost always be fully encapsulated within elastomer. Therefore, for the purposes of this chapter, cell cytotoxicity will not be a prime factor in examining the design, applications, and underlying mechanics of LM-based electronics.

An important feature of Ga-based LMs like EGaIn and Galinstan is that they form a nanometer-then oxide layer when exposed to oxygen. This oxide skin functions like a membrane that allows droplets or traces of liquid metal to hold their shape when deposited on a substrate (Figure 3.1c). Such a feature enables patterning of LM on a soft elastomer through direct ink write (DIW) printing [11], 3D printing [12], and microcontact printing [13]. Moreover, the oxide skin functions as a surfactant that allows bulk LM to break up into smaller droplets during shear mixing or ultrasonication within a solvent or polymer liquid. This enables the synthesis of microscale or nanoscale droplets that can be used for printed electronic inks [14, 15] or LM–polymer composites in which a dispersion of LM droplets are suspended within a cured elastomer matrix [16–19]. The role of the oxide skin will be discussed here in the context of LM architectures and fabrication methods.

3.1.2 Relevant Literature

Interest in liquid metal for soft electronics accelerated following seminal papers from the George Whitesides lab at Harvard in 2007 [21] and 2008 [6, 20]. These papers explored the practice of creating stretchable circuits and antennas using microfluidic channels of EGaIn and other low melting point metals embedded in a soft silicone rubber. Since these early papers, there has been a rapid growth in novel material architectures, manufacturing methods, and applications of LM-based soft microfluidics for elastic circuits, sensors, and antennas.

A complete review of all the progress in Ga-based LM electronics is outside the scope of this chapter. Instead, interested readers should refer to past review papers that cover various aspects of the field [4, 5, 7, 8, 22–26]. Together, these review articles represent a comprehensive overview of how liquid metal is used for applications in soft electronics.

3.2 LM Architectures

Early efforts in LM-based electronics focused on microfluidic channels of liquid metal embedded within a soft elastomeric film or substrate. As described below in Section 3.2.1, these early architectures were fabricated using "soft lithography" techniques popularized for microfluidics engineering. Since then, there has been several new manufacturing methods that allow for more scalable manufacturing as well as a wider range of possible circuit geometries. These include thin-film metal traces coated with LM alloy (Section 3.2.2), LM–polymer composites composed of EGaIn droplets suspended in a soft elastomer matrix (Section 3.2.3), and printable conductive inks composed of LM droplets suspended within a solvent or elastomer binder (Section 3.2.4).

3.2.1 Microfluidic LM Channels

Starting with "microsolidics" in 2007 [21], there has been interest in using soft materials to create flexible and stretchable alternatives to conventional printed circuit board (PCB) technologies. Microsolidics involve injection of a liquified metal – typically indium or a low melting point metal solder that is melted – into microchannels embedded within a sheet of polydimethylsiloxane (PDMS) silicone

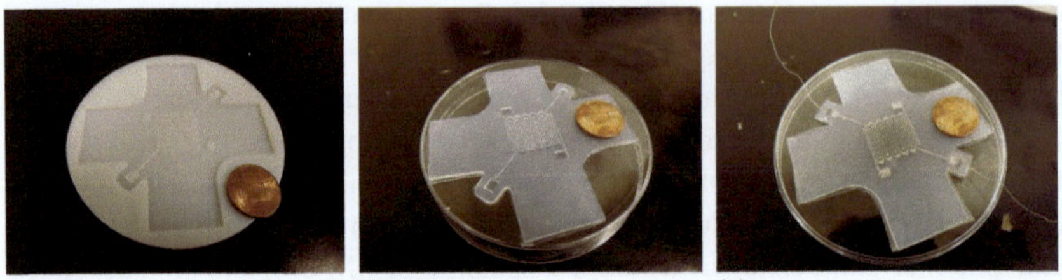

(a) (b) (c)

Figure 3.2 Steps for fabricating a fluidic LM circuit in silicone elastomer. (a) Create a 2.5D mold for casting the elastomer. (b) Peel elastomer from mold and seal the exposed microchannels with a protective layer of additional elastomer. (c) Use a needle and syringe to inject EGaIn into the channels. Source: From Andrew Fassler, Soft Machines Lab, Carnegie Mellon University.

rubber. The microchannels were fabricated by casting the PDMS in a mold that was produced using photolithography and then sealing the molded channels with an additional layer of PDMS. When the metal solidified, the circuit was flexible but not stretchable.

To achieve stretchable functionality, the low melting point metal was subsequently replaced with EGaIn, which flows at room temperature and keeps the fluidic channels intact when the surrounding PDMS is stretched [6]. Early demonstrations of EGaIn-based microfluidic circuits focused on a mechanically-tunable dipole antenna [27], strain gauge [28], patch antenna [29], and pressure sensors [30]. In all cases, the microfluidic architecture is fabricated by casting PDMS into a 2.5D mold that is produced using SU-8 photolithography or 3D printing. Depending on the method of mold fabrication, the cross-section dimensions of the fluidic channels can range from millimeters [31] to 10s of microns [28] and even sub-micron [32]. The complete fabrication process for a fluidic circuit produced using a 3D printed mold is presented in Figure 3.2.

Soft electronic architectures composed of microfluidic channels of EGaIn in PDMS remain a popular approach to creating soft and stretchable circuits and sensors. However, some limitations exist related to their manufacturing. Because fabrication relies on manual elastomer casting and syringe-based EGaIn injection, manufacturing can be laborious and difficult to scale up. Moreover, dependency on liquid metal injection results in a limited range of possible circuit geometries. For example, it is difficult to produce circuits in which the fluidic channel self-intersects. Likewise, it is difficult to create large filled areas, as in the case of a capacitor electrode.

3.2.1.1 Printing-Based Deposition Methods

Efforts to automate manufacturing and broaden the variety of possible fluidic EGaIn geometries have focused on additive manufacturing techniques. This includes DIW printing in which EGaIn is deposited from a nozzle onto a PDMS substrate [11]. The printed EGaIn coating can then be coated with a sealing layer of silicone elastomer. Because of the oxide skin that forms on the surface of the EGaIn, the printed traces can hold their shape when uncured liquid elastomer is deposited on top to form the sealing layer. The oxide skin also allows for droplet-by-droplet 3D printing [12] or microcontact deposition [13, 33] of EGaIn to form patterned geometries that can hold their shape in air or when sealed with a protective elastomer coating.

3.2.1.2 Direct LM Casting

Other approaches to creating microfluidic LM electronics include casting EGaIn in open channels of a molded elastomer. The deposited EGaIn can then be sealed with an additional layer of elastomer, thereby eliminating the need for needle injection [32]. Such an approach exploits the oxidation of the liquid metal, which assists with wetting inside the molded channels. For smaller microscale channels, EGaIn exhibits a preferential affinity for wetting the channels and only residual amounts of gallium-oxide will be observed outside of the channels. This is most pronounced for submicron patterns, where atomic force microscopy images suggest only a thin Ga_2O_3 layer over the exposed portions of the molded PDMS substrate.

Rather than sealing the EGaIn after it has been deposited in the open channels, it can be transferred to another substrate to form free-standing structures. One method for doing this is to freeze the EGaIn and then peel it out of the elastomer mold [34]. This requires cooling the liquid metal to below its crystallization temperature within a humidity controlled chamber. Alternatively, the molded EGaIn can be pressed against another surface and transferred without performing a freezing step [33].

3.2.2 LM-Coated Thin-Film Metal Traces

In addition to microfluidic LM electronics, PDMS and other soft elastomers have also been popular as a substrate for flexible and stretchable circuits composed of ultrathin-films of gold, silver, or copper. Stretchability is typically achieved through wavy or buckled geometries in which the metal thin-film has a naturally sinusoidal shape that flattens out when the underlying elastomer substrate is stretched [35]. In 2016, Stephanie Lacour, an early pioneer of these stretchable thin-film electronics, and her research team presented stretchable "biphasic" architectures in which traces of thin-film gold patterned on an elastic substrate are coated with gallium (Figure 3.3a) [36]. Gallium and gold were observed to form an $AuGa_2$ intermetallic alloy. Because of the low melting point and fluid-like properties of the excess gallium, the resulting traces could remain electrically conductive when the PDMS substrate was stretched.

This concept of biphasic conductors in which a metal thin-film is coated with liquid metal has also been demonstrated on lithographically-patterned Au/Cr bilayers [37], laser patterned Cu/Cr bilayers (Figure 3.3b) [38], and printed silver inks [39]. As demonstrated by Ozutemiz et al. in their 2018 study [40], the EGaIn coating also doubles as a room temperature solder that can be used to mount microelectronic chips and create fully functional digital circuits. This is accomplished by applying a reducing vapor of hydrochloric acid, which removes the oxide skin from liquid metal and allows it to readily wet to the pins of the surface-mounted microelectronic chips.

Recent efforts in biphasic LM circuits have focused on the use of mass scale manufacturing techniques to allow for mass production of soft digital circuits. This

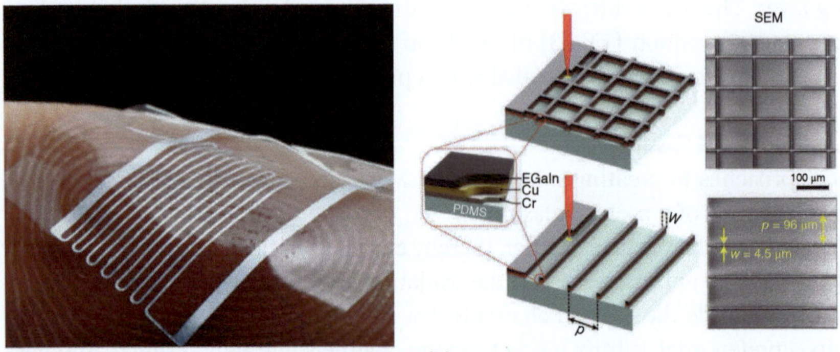

(a) (b)

Figure 3.3 (a) Intrinsically stretchable "biphasic" soft circuit composed of Au traces coated with Ga on a thin PDMS substrate. Source: Reproduced with permission of Hirsch et al. [36], © 2016 John Wiley & Sons. (b) Layup of biphasic EGaIn–Cr–Cu circuit that is patterned using UV laser micromachining. Source: Reproduced with permission of Pan et al. [38], © 2018, John Wiley & Sons.

includes both the use of standard lithographic methods to pattern the thin-film metal template as well as dip-coating techniques for applying the LM coating [41]. Other recent approaches to mass scale production have focused on eliminating the lithographic step and instead producing a patterned circuit template using inkjet printing methods [39, 42, 43].

3.2.3 LM–Polymer Composites

One of the challenges of working with LM microfluidics and biphasic thin-films is the propensity of liquid metal to leak or smear. To overcome this, researchers have explored the development of polymer composites in which a soft elastomer (e.g. PDMS) is embedded with a suspension of microscale droplets of EGaIn (Figure 3.4a) [16–19]. The resulting composite has the elasticity and material integrity of the elastomer matrix while also exhibiting enhanced thermal conductivity [44] or electrical permittivity [45]. For high concentrations of EGaIn filler, the droplets can form a connected, percolating network that results in soft rubbery composites that are electrically conductive (Figure 3.4b) [46–48].

There have been several review papers in recent years that are focused on the synthesis, properties, and applications of these LM–polymer composites [16–19]. In Section 3.3 of this chapter, special attention is focused on the relevant principles of mechanics that are used to predict the thermal, electrical, and mechanical properties of these material systems. Such principles are used to derive theoretical models that describe the influence of LM droplet shape and concentrations on the "effective medium" properties of LM–polymer composites and inform their design.

Once an LM microstructure (i.e. droplet size, shape, density) is selected, composites with tailor-made combinations of elastic, electrical, and thermal properties

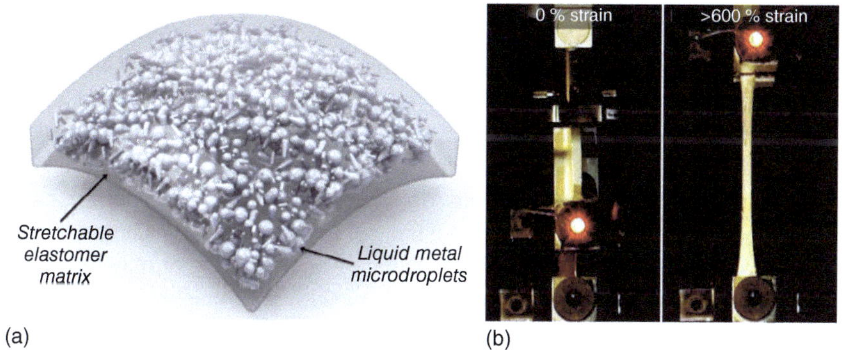

Figure 3.4 (a) LM–polymer composite composed of EGaIn droplets embedded in a soft elastomer matrix. Source: Reprinted with permission from Bartlett et al. [45], Copyright 2016 Wiley-VCH. (b) Electrically conductive EGaIn–silicone composite that maintains low electrical resistance when stretched to >600% strain. Source: Reproduced with permission from Ford et al. [47], © John Wiley & Sons.

can be synthesized using a variety of shear mixing techniques. These methods include planetary centrifugal mixing [49], high speed mixing with an overhead shear mixer [47], and ultrasonication in a liquid dispersion medium using bath or probe sonication [50]. LM droplets can be synthesized by directly shear mixing in a liquid pre-polymer or in a solvent (e.g. ethanol, toluene, hexane), after which the droplets are mixed with the pre-polymer and the solvent is allowed to evaporate. In either case, synthesis results in a liquid emulsion composed of both an EGaIn and pre-polymer phase. As a final step, the emulsion can be molded or deposited as a sheet or thin film and allowed to cure based on the typical curing conditions of the polymer matrix.

As with other LM architectures, the surface oxidation of Ga-based alloys plays a central role in synthesis. In the case of LM–polymer emulsions, the Ga_2O_3 skin formed on the surface of the droplets functions like a surfactant that stabilizes the droplets and prevents them from coalescing back into a bulk phase. Moreover, they improve the binding affinity between the liquid metal and pre-polymer and can act as an initiator for polymerization. Nonetheless, there have been recent efforts to incorporate additional surfactants, monomers, and reactants into the emulsion to provide even tighter control on EGaIn nanodroplet synthesis and elastomer polymerization. A particularly powerful approach to control the synthesis of LM–polymer composites is through the use of atom-transfer radical polymerization (ATRP) [51–53]. Efforts led by the Krzysztof Matyjaszewski lab at Carnegie Mellon University, where the original methods for ATRP synthesis of polymers were pioneered, have resulted in new classes of LM–polymer nanocomposites that are produced using surface initiated ATRP (SI-ATRP) [53]. With these methods, it is possible to create composites with highly monodisperse suspensions of EGaIn droplets that are embedded in a wide range of polymers, including polyacrylates and fluoropolymer elastomers.

3.2.4 Printable LM-Based Inks

Current efforts in LM-based electronics are increasingly focused on EGaIn-filled inks that can be deposited using DIW or inkjet printing. Because of their rapid oxidation in air, it is difficult to directly print Ga-based LM alloys by extrusion through a nozzle. Instead, efforts in printed LM circuits have largely focused on inks composed of EGaIn nanodroplets or blends of LM droplets, liquid-phase polymer, and rigid metal particles (e.g. Ag flakes). Depending on the choice of ink, the printed circuits can be conductive immediately after printing or may require a "sintering" step in which the LM droplets coalesce with their neighbors to form an electrically conductive network.

Printable circuits with LM nanoparticles are typically synthesized using ultrasonication within a solvent (e.g. ethanol) [15]. After depositing on a substrate using DIW printing, high pressure is applied in order to rupture the droplets and form conductive pathways [14]. Printable inks that become conductive through such mechanical sintering has also been demonstrated with EGaIn nanodroplets that are coated in a thin polymer shell [54]. Alternatively, conductivity can be induced by laser ablation with an infrared (10.6 μm wavelength) CO_2 engraver [55] or thermal sintering in a high temperature (900 °C) furnace [56].

Sinter-free conductive inks with EGaIn typically require the addition of rigid metal microparticles, such as Ag flakes. Biphasic combinations of EGaIn and metal powder result in a paste that can be printed through extrusion [16, 57]. Blending EGaIn droplets and Ag flakes with a styrene–isoprene–styrene (SIS) copolymer elastomer results in a conductive ink that becomes a rubber solid when cured after printing [58]. These LM–Ag–SIS inks can be printed on soft thin films and function as electrical interconnects for digital circuits [58, 59] as well as bioelectrodes for wearable health monitoring [60, 61].

3.3 Mechanics and Modeling

Some of the material architectures described in Section 3.2 exhibit electrical, thermal, and/or mechanical properties that can be examined using effective medium theories and principles of elasticity. In this section, I will present theoretical models that have been successfully used to explain empirical observations in the following classes of LM-based materials: (i) microfluidic LM channels for strain and pressure sensing; (ii) electrically insulating LM-polymer composites with enhanced thermal conductivity and electrical permittivity; and (iii) electrically conductive LM-polymer composites that exhibit stable electrical resistance under mechanical strain.

3.3.1 Microfluidic LM Strain Gauge

The earliest examples of LM electronics were highly stretchable strain gauges composed of rubber tubes filled with mercury [1]. In the unstretched state, the strain gauge has an electrical resistance of $R_0 = \rho L_0/A_0$, where ρ is the volumetric resistivity of the liquid metal, L_0 is the natural (unloaded, undeformed) length of the

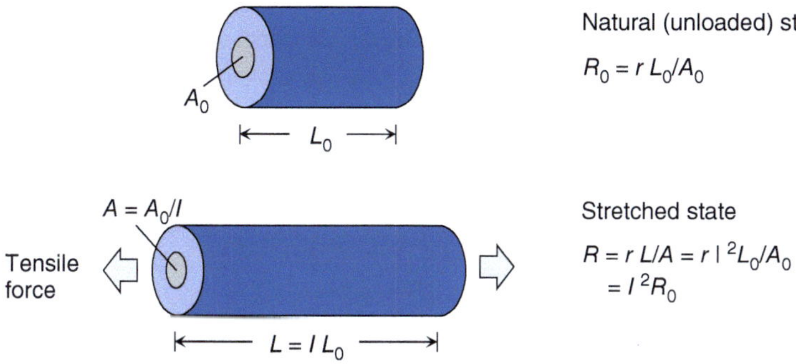

Figure 3.5 Hyperelastic strain gauge composed of an elastomer tube filled with liquid metal. Blue denotes the elastomer and grey corresponds to the embedded LM channel.

rubber tube, and A_0 is the natural cross-sectional area of the LM-filled channel. When strained in tension with a stretch of $\lambda = L/L_0$, where L is the stretched length of the tube, the cross-sectional area reduces to $A = A_0/\lambda$ and the electrical resistance becomes $R = \rho L/A = \lambda^2 R_0$ (Figure 3.5). The λ^2 dependency of the relative electrical resistance (R/R_0) with strain makes the LM-filled tube effective as a "hyperelastic" strain gauge that can measure large deformations. Moreover, the electromechanical dependency is invariant to the channel geometry so long as the channel and tube are prismatic – i.e. the cross section remains uniform along the length.

The above theoretical prediction is based on Pouillet's law and makes the following assumptions. First, it assumes that the liquid metal is incompressible, i.e. its volume remains conserved, and that the tensile strain is uniform along the length of the tube when it is stretched. Moreover, it assumes that the LM channels are straight and prismatic. Studies with zig-zag or diamond-patterned channels show that changing the channel geometry can result in different gauge factors and electromechanical dependencies [28].

3.3.2 Microfluidic LM Pressure Sensor

Another common application of LM microchannels is for pressure sensing [30, 62]. Using principles of elasticity, it is possible to derive simple algebraic predictions for how the electrical resistance of a microfluidic channel of EGaIn increases when pressure is applied to the surrounding elastomer. Several models exist depending on the cross-sectional geometry of the EGaIn microchannel. These can include channels with triangular, rectangular, diamond-shaped, and ellipse-shaped cross sections [63]. Among these, pressure sensors with rectangular cross sections are the most popular since they are the easiest to fabricate. Regardless, in all cases, the change in electrical resistance as a function of pressure can be estimated using plane strain elasticity [63].

Referring to Figure 3.6, applying pressure deforms the elastomer surrounding the channel and reduces the cross-sectional from an original value A_0 to a deformed value A. Normalizing with the original electrical resistance R_0 of the deformed portion of the fluidic channel, the relative resistance is $R/R_0 = A_0/A$. Predictions for A

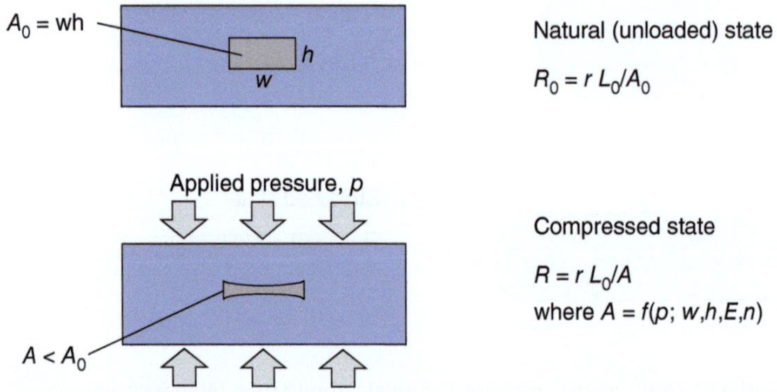

Figure 3.6 Hyperelastic pressure sensor composed of a liquid metal microchannel embedded within an elastomer substrate. Blue denotes the elastomer and gray denotes the embedded LM channel.

as a function of the applied pressure p exerted over a prescribed area of the elastomer can be obtained by solving the field equations for a linear elastic solid under plane strain loading. This can be accomplished numerically using a partial differential equation (PDE) solver or with a finite element method (FEM). Alternatively, for certain cross-sectional geometries, analytic solutions to the governing PDEs can be obtained using complex analysis and the Kolosov–Muskhelishvili formulation [63].

In order to obtain a simplified algebraic expression for R/R_0 as a function of p, it is necessary to approximate the deformation of the channel using the principle of linear superposition. In the case of a channel with a rectangular cross section, this is achieved by superposing uniform deformation of a homogenous Hookean solid with the opening of a slit under far-field stress. For an elastomer with a Poisson's ratio v and Young's modulus E embedded with a microchannel of width w and height h, the cross-sectional area is approximated as

$$A = wh\left\{1 - (1 - v^2)\frac{p}{E}\right\}\left\{1 + v(1 + v^2)\frac{p}{E}\right\} - \frac{4(1 - v^2)\sqrt{2}pw^2}{3E}. \tag{3.1}$$

Dividing by $A_0 = wh$ yields the relative change in resistance R/R_0 as a function of pressure.

Of particular interest is the gauge factor, $G = 1 - A/A_0$, which is related to the slope of the relative change in resistance $\Delta R/R_0 = R/R_0 - 1$ plotted against a normalized pressure $\hat{p} = p/E$. The parameter G is an important metric for electromechanical coupling that relates the "sensitivity" of the resistive strain sensor to applied pressure. In the limit as $p \to 0$, the gauge factor converges to an approximate value of

$$G = 1 - v - 2v^2 + \frac{4(1 - v^2)\sqrt{2}}{3\alpha}, \tag{3.2}$$

where $\alpha = h/w$ is the aspect ratio of the channel cross section. Plots of G versus α obtained from the numerical, analytic, and approximate solutions all suggest that sensors with a smaller height-to-width cross section exhibit the greatest sensitive to pressure [63].

3.3.3 LM–Polymer Composites

Suspensions of LM droplets in a soft elastomer matrix have been modeled using a variety of techniques, including FEM, statistical mechanics, and effective medium theory (EMT). EMT has been especially useful in predicting the influence of LM droplets on the electrical permittivity, thermal conductivity, and effective Young's modulus of composites that are electrically insulating. For the case of electrically conductive LM–polymer composites, FEM and statistical mechanics techniques have been used to estimate how electrical resistance changes as a function of tensile strain.

3.3.3.1 Electrical Permittivity and Thermal Conductivity

Experimental measurements of the thermal conductivity (k) and electrical permittivity (ϵ) of EGaIn–PDMS composites as a function of LM volume fraction (φ) have been shown to be in good agreement with predictions from the Maxwell–Garnett (M–G) EMT formulation [44, 45]. According to M–G,

$$\frac{\epsilon}{\epsilon_m} \approx \frac{k}{k_m} \approx \frac{1+2\varphi}{1-\varphi}. \tag{3.3}$$

Here, the subscript "m" denotes the material properties of the polymer matrix and it is assumed that the metal has relatively infinite permittivity and thermal conductivity [17]. Other assumptions inherent in the M–G formulation are that the LM inclusions are spherical, monodisperse (i.e. have a uniform diameter), and are sufficiently sparse that the heat flux or electrical displacement of inclusions do not influence those of their neighbors. More accurate predictions based on dense suspensions (i.e. high concentrations) and nonspherical shapes have been derived using a Bruggeman EMT formulation [45, 50]. Accounting for nonspherical droplet shapes is of particular importance when the composite is stretched and the LM inclusions elongate along the direction of strain. In the case of predictions for k/k_m, these more accurate models are also modified to account for the finite thermal conductivity of EGaIn [44]. Nonetheless, the simplified M–G prediction in Eq. (3.3) does a reasonable job of capturing the influence of LM volume fraction on increasing the electrical permittivity and thermal conductivity of the composite.

3.3.3.2 Electromechanical Coupling

In the case of electrically conductive LM–polymer composites, modeling has primarily been used to establish the electromechanical coupling between electrical resistance R/R_0 and stretch λ. This is typically accomplished using a combination of finite element analysis to establish the deformation of the connected network of LM inclusions under strain and Pouillet's law for estimating the change in electrical resistance based on changes in the LM microstructure geometry. Detailed FEM simulations of single droplets in contact with their neighbors [64] are in reasonable agreement with experimental measurements by Fassler and Majidi [46], both of which show little to no increase in electrical resistance when the composite is stretched up to 50% strain. Subsequent FEM simulations of larger networks of connected LM droplets show that various parameters of the microstructure can be adjusted to achieve agreement with electromechanical measurements from additional experimental studies [65]. Lastly, Zhao et al. [66] combined FEM with statistical techniques that account for the stochasticity of connectivity between adjacent LM droplets. As with the other studies, this analysis also predicted trends in electromechanical coupling that span the range of measurements – from virtually no coupling to slightly positive or negative coupling between electrical resistance and strain – that have been experimentally observed.

3.3.3.3 Effective Young's Modulus

Because the EGaIn droplets within an LM–polymer composite are liquid-phase, it is intuitive that a larger LM volume fraction (φ) would result in a lower effective Young's modulus E_{eff} of the composite. Such a prediction would be consistent with Eshelby's Theory of Inclusions [67], which – assuming that the LM inclusions have zero elastic modulus – would predict an effective modulus of

$$E_{\text{eff}} = \frac{E_m}{1 + 5\varphi/3}, \quad \text{(Eshelby)} \tag{3.4}$$

where E_m is the Young's modulus of the polymer matrix.

Predictions for E_{eff} based on Eshelby's Theory of Inclusions are consistent with experimental measurements for EGaIn–PDMS composites in which the EGaIn droplet sizes are on the order of ~10–100 μm diameter [50]. However, as shown in Figure 3.7, composites with micron (~1 μm) or sub-micron (~0.1 μm) diameter droplets exhibit *increased* effective modulus with increasing volume fraction φ. Such a phenomenon is not limited to liquid metal inclusions and has also been empirically observed with elastomers embedded with droplets of ionic liquid [68]. For the latter study, the authors introduced a modified Eshelby's Theory of Inclusions that predicts this stiffening effect by accounting for the surface tension γ of the fluidic inclusions [68]. For a nominal droplet radius R and again letting E_m correspond to the modulus of the elastomer matrix, they predict that the

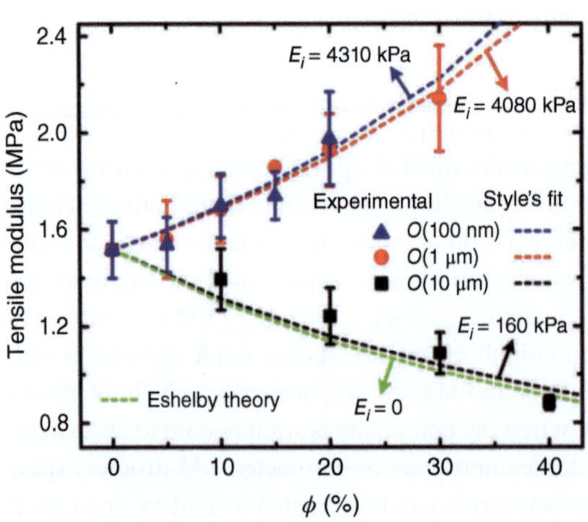

Figure 3.7 Experimental measurements and theoretical fits for the effective Young's modulus of an EGaIn–silicone composite with varying LM volume fraction (φ) and droplet size. Source: Reproduced with permission from Pan et al. [50], Copyright 2019 John Wiley & Sons.

contribution of surface tension results in droplets that are equivalent to elastic inclusions with a stiffness:

$$E_i = \frac{24\gamma/E_m R}{10 + 9\gamma/E_m R} E_m. \tag{3.5}$$

Substituting this into the original Eshelby formulation results in the following prediction for the effective modulus of the composite:

$$E_{\text{eff}} = \frac{1 + 2E_i/3E_m}{1 + 5\varphi/3 + (2/3 - 5\varphi/3)(E_i/E_m)} E_m. \quad \text{(Style)} \tag{3.6}$$

In the case of EGaIn, the Ga_2O_3 oxide skin influences the mechanical resistance to deformation and so Eq. (3.5) is no longer valid. Instead, E_i is treated as a fitting parameter for establishing an empirical relationship between E_{eff} and φ. Again referring to Figure 3.7, the Style et al. model can be used to capture the experimentally observed trend for the case of micron and sub-micron scale droplets while Eshelby's Theory of Inclusions accurately predicts the trend for LM–polymer composites with larger droplet diameters [50].

3.4 Open Challenges and Future Directions

Despite over a decade of progress in LM-based material architectures for soft and stretchable electronics, the field is still in its early stages. Compared to more established methods of flexible and stretchable circuitry, there is still much to learn about the underlying mechanics of LM-based electronics and how they can be utilized in real-world applications for computing, robotics, and healthcare.

One open challenge in the modeling of LM microfluidics and printable conductive inks relates to the electrical failure of LM interconnects under high current [69, 70]. Possible contributions include expansion of air bubbles trapped within the conductive trace, thermally-controlled oxidation, and electromigration. While such failures are not expected to arise in digital circuit applications, where voltages and currents are in the ~1–10 V and ~0.1–1 A range, they can potentially occur for power electronics applications that involve delivering electrical current to a motor or heating element [69]. Electromigration is the leading candidate for explaining why electrical failure occurs in liquid metal at elevated currents [70], although further experimental and theoretical investigation is required.

Theoretical modeling is also important for addressing critical failure modes related to delamination at the interface of the elastomer, LM trace, and embedded microelectronic chip when a circuit is under mechanical load. Such failure is likely caused by the large stiffness mismatch between the rigid circuit component and the surrounding soft material. Under tensile or shear loading, these stiffness mismatches result in large stress concentrations and the propagation of tears or interfacial delamination. Gradients in stiffness within the elastomer could potentially enable greater stiffness matching and reduce these stress concentrations [71]. However, additional studies of the underlying mechanics using FEM simulations and principles in fracture mechanics are required in order to establish predictive models for materials selection and design.

Rupture of microfluidic channels or microscale droplets of LM under extreme external pressure represents another challenge area where mechanics-based modeling techniques can provide valuable insight. In the case of LM–polymer composites, extreme pressure can have a useful role in "mechanical sintering" of embedded droplets to form conductive pathways. However, once a circuit is fabricated, the liquid metal should remain embedded in the elastomer and not leak out if high pressure is applied. There has only been limited empirical study of the role of applied pressure on the rupture of LM inclusions and further experimental and theoretical examination is required to understand how to make embedded LM inclusions resistant to mechanical failure and leakage.

Lastly, more needs to be understood about the electromechanical coupling of LM interconnects within a soft and stretchable circuit. Both experimental measurements and theoretical models show evidence of limited coupling between electrical resistance and strain for circuits composed of LM–polymer composites. However, circuits composed of microfluidic LM channels can exhibit a significant degree of coupling depending on the microchannel geometry. Models that combine elasticity with Pouillet's law for predicting how the electrical resistance of stretchable circuit interconnects changes under mechanical loading could be useful in the development of computational tools for LM circuit design.

In closing, there has been tremendous progress over the past decade in the development of LM-based material architectures for stretchable electronics. This has included not only novel fabrication methods and device-level designs but also theoretical studies that use principles of mechanics to model the interplay of LM microstructure, elasticity, and circuit properties. While several open challenges remain in the modeling of LM-based soft electronics, these challenges can potentially be addressed through further theoretical investigation and represent possible directions for future research.

References

1 Whitney, R.J. (1949). The measurement of changes in human limb-volume by means of a mercury-inrubber strain gauge. *The Journal of Physiology* 109 (1–2): Proc–5.

2 Łuczak, S. and Ekwińska, M. (2021). Electric-contact tilt sensors: a review. *Sensors* 21 (4): 1097.

3 Liu, T., Sen, P., and Kim, C.-J. (2011). Characterization of nontoxic liquid-metal alloy galinstan for applications in microdevices. *Journal of Microelectromechanical Systems* 21 (2): 443–450.

4 Dickey, M.D. (2014). Emerging applications of liquid metals featuring surface oxides. *ACS Applied Materials & Interfaces* 6 (21): 18369–18379.

5 Lin, Y., Genzer, J., and Dickey, M.D. (2020). Attributes, fabrication, and applications of gallium-based liquid metal particles. *Advanced Science* 7 (12): 2000192

6 Dickey, M.D., Chiechi, R.C., Larsen, R.J. et al. (2008). Eutectic gallium–indium (EGaIn): a liquid metal alloy for the formation of stable structures in microchannels at room temperature. *Advanced Functional Materials* 18 (7): 1097–1104.

7 Joshipura, I.D., Ayers, H.R., Majidi, C., and Dickey, M.D. (2015). Methods to pattern liquid metals. *Journal of Materials Chemistry C* 3 (16): 3834–3841.

8 Tang, S.-Y., Tabor, C., Kalantar-Zadeh, K., and Dickey, M.D. (2021). Gallium liquid metal: the Devil's Elixir. *Annual Review of Materials Research* 51: 381–408.

9 Kim, J.-H., Kim, S., So, J.-H. et al. (2018). Cytotoxicity of gallium–indium liquid metal in an aqueous environment. *ACS Applied Materials & Interfaces* 10 (20): 17448–17454.

10 Won, P., Coyle, S., Ko, S.H. et al. (2023). Controlling C2C12 cytotoxicity on liquid metal embedded elastomer (LMEE). *Advanced Healthcare Materials* 12 (17): 2202430

11 Boley, J.W., White, E.L., Chiu, G.T.-C., and Kramer, R.K. (2014). Direct writing of gallium–indium alloy for stretchable electronics. *Advanced Functional Materials* 24 (23): 3501–3507.

12 Ladd, C., So, J.-H., Muth, J., and Dickey, M.D. (2013). 3d printing of free standing liquid metal microstructures. *Advanced Materials* 25 (36): 5081–5085.

13 Tabatabai, A., Fassler, A., Usiak, C., and Majidi, C. (2013). Liquid-phase gallium–indium alloy electronics with microcontact printing. *Langmuir* 29 (20): 6194–6200.

14 Boley, J.W., White, E.L., and Kramer, R.K. (2015). Mechanically sintered gallium–indium nanoparticles. *Advanced Materials* 27 (14): 2355–2360.

15 Hohman, J.N., Kim, M., Wadsworth, G.A. et al. (2011). Directing substrate morphology via self-assembly: ligand-mediated scission of gallium–indium microspheres to the nanoscale. *Nano Letters* 11 (12): 5104–5110.

16 Chen, S., Wang, H.-Z., Zhao, R.-Q. et al. (2020). Liquid metal composites. *Matter* 2 (6): 1446–1480.

17 Kazem, N., Hellebrekers, T., and Majidi, C. (2017). Soft multifunctional composites and emulsions with liquid metals. *Advanced Materials* 29 (27): 1605985

18 Majidi, C., Alizadeh, K., Ohm, Y. et al. (2022). Liquid metal polymer composites: from printed stretchable circuits to soft actuators. *Flexible and Printed Electronics* 7 (1): 013002

19 Malakooti, M.H., Bockstaller, M.R., Matyjaszewski, K., and Majidi, C. (2020). Liquid metal nanocomposites. *Nanoscale Advances* 2 (7): 2668–2677.

20 Chiechi, R.C., Weiss, E.A., Dickey, M.D., and Whitesides, G.M. (2008). Eutectic gallium–indium (EGaIn): a moldable liquid metal for electrical characterization of self-assembled monolayers. *Angewandte Chemie International Edition* 47 (1): 142–144.

21 Siegel, A.C., Bruzewicz, D.A., Weibel, D.B., and Whitesides, G.M. (2007). Microsolidics: fabrication of three-dimensional metallic microstructures in poly (dimethylsiloxane). *Advanced Materials* 19 (5): 727–733.

22 Chen, S., Cui, Z., Wang, H. et al. (2023). Liquid metal flexible electronics: past, present, and future. *Applied Physics Reviews* 10 (2): 021308

23 Dickey, M.D. (2017). Stretchable and soft electronics using liquid metals. *Advanced Materials* 29 (27): 1606425

24 Wang, X. and Liu, J. (2016). Recent advancements in liquid metal flexible printed electronics: properties, technologies, and applications. *Micromachines* 7 (12): 206.

25 Won, P., Jeong, S., Majidi, C., and Ko, S.H. (2021). Recent advances in liquid-metal-based wearable electronics and materials. *Iscience* 24 (7): 102698

26 Zhu, L., Wang, B., Handschuh-Wang, S., and Zhou, X. (2020). Liquid metal–based soft microfluidics. *Small* 16 (9): 1903841

27 So, J.-H., Thelen, J., Qusba, A. et al. (2009). Reversibly deformable and mechanically tunable fluidic antennas. *Advanced Functional Materials* 19 (22): 3632–3637.

28 Kim, H.-J., Son, C., and Ziaie, B. (2008). A multiaxial stretchable interconnect using liquid-alloy-filled elastomeric microchannels. *Applied Physics Letters* 92 (1): 011904

29 Cheng, S. and Wu, Z. (2011). A microfluidic, reversibly stretchable, large-area wireless strain sensor. *Advanced Functional Materials* 21 (12): 2282–2290.

30 Park, Y.-L., Majidi, C., Kramer, R. et al. (2010). Hyperelastic pressure sensing with a liquid-embedded elastomer. *Journal of Micromechanics and Microengineering* 20 (12): 125029

31 Fassler, A. and Majidi, C. (2013). Soft-matter capacitors and inductors for hyperelastic strain sensing and stretchable electronics. *Smart Materials and Structures* 22 (5): 055023

32 Gozen, B.A., Tabatabai, A., Ozdoganlar, O.B., and Majidi, C. (2014). High-density soft-matter electronics with micron-scale line width. *Advanced Materials* 26 (30): 5211–5216.

33 Yalcintas, E.P., Ozutemiz, K.B., Cetinkaya, T. et al. (2019). Soft electronics manufacturing using microcontact printing. *Advanced Functional Materials* 29 (51): 1906551

34 Fassler, A. and Majidi, C. (2013). 3D structures of liquid-phase GaIn alloy embedded in PDMS with freeze casting. *Lab on a Chip* 13 (22): 4442–4450.

35 Rogers, J.A., Someya, T., and Huang, Y. (2010). Materials and mechanics for stretchable electronics. *Science* 327 (5973): 1603–1607.

36 Hirsch, A., Michaud, H.O., Gerratt, A.P. et al. (2016). Intrinsically stretchable biphasic (solid–liquid) thin metal films. *Advanced Materials* 28 (22): 4507–4512.

37 Li, G., Wu, X., and Lee, D.-W. (2015). Selectively plated stretchable liquid metal wires for transparent electronics. *Sensors and Actuators B: Chemical* 221: 1114–1119.

38 Pan, C., Kumar, K., Li, J. et al. (2018). Visually imperceptible liquid-metal circuits for transparent, stretchable electronics with direct laser writing. *Advanced Materials* 30 (12): 1706937

39 Tavakoli, M., Malakooti, M.H., Paisana, H. et al. (2018). EGaIn-assisted room-temperature sintering of silver nanoparticles for stretchable, inkjet-printed, thin-film electronics. *Advanced Materials* 30 (29): 1801852

40 Ozutemiz, K.B., Wissman, J., Ozdoganlar, O.B., and Majidi, C. (2018). EGaIn–metal interfacing for liquid metal circuitry and microelectronics integration. *Advanced Materials Interfaces* 5 (10): 1701596

41 Ozutemiz, K.B., Majidi, C., and Ozdoganlar, O.B. (2022). Scalable manufacturing of liquid metal circuits. *Advanced Materials Technologies* 7 (11): 2200295

42 Lopes, P.A., Gomes, D.V., Marques, D.G. et al. (2019). Soft bioelectronic stickers: selection and evaluation of skin-interfacing electrodes. *Advanced Healthcare Materials* 8 (15): 1900234

43 Silva, A.F., Paisana, H., Fernandes, T. et al. (2020). High resolution soft and stretchable circuits with PVA/liquid-metal mediated printing. *Advanced Materials Technologies* 5 (9): 2000343

44 Bartlett, M.D., Kazem, N., Powell-Palm, M.J. et al. (2017). High thermal conductivity in soft elastomers with elongated liquid metal inclusions. *Proceedings of the National Academy of Sciences of the United States of America* 114 (9): 2143–2148.

45 Bartlett, M.D., Fassler, A., Kazem, N. et al. (2016). Stretchable, high-*k* dielectric elastomers through liquid-metal inclusions. *Advanced Materials* 28 (19): 3726–3731.

46 Fassler, A. and Majidi, C. (2015). Liquid-phase metal inclusions for a conductive polymer composite. *Advanced Materials* 27 (11): 1928–1932.

47 Ford, M.J., Patel, D.K., Pan, C. et al. (2020). Controlled assembly of liquid metal inclusions as a general approach for multifunctional composites. *Advanced Materials* 32 (46): 2002929

48 Markvicka, E.J., Bartlett, M.D., Huang, X., and Majidi, C. (2018). An autonomously electrically self-healing liquid metal–elastomer composite for robust soft-matter robotics and electronics. *Nature Materials* 17 (7): 618–624.

49 Kazem, N., Bartlett, M.D., and Majidi, C. (2018). Extreme toughening of soft materials with liquid metal. *Advanced Materials* 30 (22): 1706594

50 Pan, C., Markvicka, E.J., Malakooti, M.H. et al. (2019). A liquid-metal–elastomer nanocomposite for stretchable dielectric materials. *Advanced Materials* 31 (23): 1900663

51 Malakooti, M.H., Kazem, N., Yan, J. et al. (2019). Liquid metal supercooling for low-temperature thermoelectric wearables. *Advanced Functional Materials* 29 (45): 1906098

52 Wei, Q., Sun, M., Wang, Z. et al. (2020). Surface engineering of liquid metal nanodroplets by attachable diblock copolymers. *ACS Nano* 14 (8): 9884–9893.

53 Yan, J., Malakooti, M.H., Lu, Z. et al. (2019). Solution processable liquid metal nanodroplets by surface-initiated atom transfer radical polymerization. *Nature Nanotechnology* 14 (7): 684–690.

54 Thrasher, C.J., Farrell, Z.J., Morris, N.J. et al. (2019). Mechanoresponsive polymerized liquid metal networks. *Advanced Materials* 31 (40): 1903864

55 Lin, Y., Cooper, C., Wang, M. et al. (2015). Handwritten, soft circuit boards and antennas using liquid metal nanoparticles. *Small* 11 (48): 6397–6403.

56 Liu, S., Shah, D.S., and Kramer-Bottiglio, R. (2021). Highly stretchable multilayer electronic circuits using biphasic gallium–indium. *Nature Materials* 20 (6): 851–858.

57 Daalkhaijav, U., Yirmibesoglu, O.D., Walker, S., and Mengüç, Y. (2018). Rheological modification of liquid metal for additive manufacturing of stretchable electronics. *Advanced Materials Technologies* 3 (4): 1700351

58 Lopes, P.A., Fernandes, D.F., Silva, A.F. et al. (2021). Bi-phasic Ag–in–Ga-embedded elastomer inks for digitally printed, ultra-stretchable, multi-layer electronics. *ACS Applied Materials & Interfaces* 13 (12): 14552–14561.

59 Hajalilou, A., Silva, A.F., Lopes, P.A. et al. (2022). Biphasic liquid metal composites for sinter-free printed stretchable electronics. *Advanced Materials Interfaces* 9 (5): 2101913

60 Zu, W., Ohm, Y., Carneiro, M.R. et al. (2022). A comparative study of silver microflakes in digitally printable liquid metal embedded elastomer inks for stretchable electronics. *Advanced Materials Technologies* 7 (12): 2200534

61 Reis Carneiro, M., Majidi, C., and Tavakoli, M. (2022). Multi-electrode printed bioelectronic patches for long-term electrophysiological monitoring. *Advanced Functional Materials* 32 (43): 2205956

62 Park, Y.-L., Tepayotl-Ramirez, D., Wood, R.J., and Majidi, C. (2012). Influence of cross-sectional geometry on the sensitivity and hysteresis of liquid-phase electronic pressure sensors. *Applied Physics Letters* 101 (19): 191904

63 Ramachandran, V. and Majidi, C. (2018). Deformation of microchannels embedded in an elastic medium. *Journal of Applied Mechanics* 85 (10): 101004

64 Cohen, N. and Bhattacharya, K. (2019). A numerical study of the electromechanical response of liquid metal embedded elastomers. *International Journal of Non-Linear Mechanics* 108: 81–86.

65 Zolfaghari, N., Khandagale, P., Ford, M.J. et al. (2020). Network topologies dictate electromechanical coupling in liquid metal–elastomer composites. *Soft Matter* 16 (38): 8818–8825.

66 Zhao, Y., Khandagale, P., and Majidi, C. (2021). Modeling electromechanical coupling of liquid metal embedded elastomers while accounting stochasticity in 3D percolation. *Extreme Mechanics Letters* 48: 101443

67 Eshelby, J.D. (1957). The determination of the elastic field of an ellipsoidal inclusion, and related problems. *Proceedings of the Royal Society of London. Series A: Mathematical and Physical Sciences* 241 (1226): 376–396.

68 Style, R.W., Boltyanskiy, R., Allen, B. et al. (2015). Stiffening solids with liquid inclusions. *Nature Physics* 11 (1): 82–87.

69 Kent, T.A., Ford, M.J., Markvicka, E.J., and Majidi, C. (2020). Soft actuators using liquid crystal elastomers with encapsulated liquid metal joule heaters. *Multifunctional Materials* 3 (2): 025003

70 Michaud, H.O. and Lacour, S.P. (2019). Liquid electromigration in gallium-based biphasic thin films. *APL Materials* 7 (3): 031504

71 Naserifar, N., LeDuc, P.R., and Fedder, G.K. (2016). Material gradients in stretchable substrates toward integrated electronic functionality. *Advanced Materials* 28 (18): 3584–3591.

4

Mechanics of Two-Dimensional Materials

Olugbenga Ogunbiyi and Yingchao Yang

University of Missouri, Department of Mechanical and Aerospace Engineering, Columbia, MO, USA

4.1 Introduction

Since the groundbreaking discovery of C60 by Richard Errett Smalley in 1985, often hailed as the "Forebear of Nanotechnology" [1], research on carbon-related low-dimensional materials has been irresistibly driving innovation in the development of new smart materials. Six years later, Sumio Iijima pioneered the creation of carbon nanotubes (CNTs) [2], a one-dimensional (1D) material with diverse applications in materials fabrication, ranging from reinforcement in bulk structural materials (nanocomposites) to flexible smart electronic materials. The extensive study of CNTs has been fueled by their broad spectrum of physical and chemical properties [3–9]. Against this backdrop, a breakthrough occurred in 2004 with the discovery of single-layer graphene by Konstantin Novoselov et al. [10], sparking tremendous interest in 2D materials. These materials exist in the form of a single-atom-thick or polyhedral-thick layer, with bonding occurring through covalent or ionic interactions within the plane and van der Waals (vdW) bonding out of the plane [11–16]. In addition to graphene, semiconducting transition-metal dichalcogenides (TMDs) have gained considerable attention from researchers [17–19] due to their exceptional electrical, thermal, optical, mechanical, and magnetic properties, stemming from the unique symmetry of their honeycomb structure. Figure 4.1 provides a schematic illustrating the properties of 2D crystals and their associated applications.

Graphene emerges as a preeminent 2D material, distinguished by its densely packed arrangement of carbon atoms forming a regular hexagonal pattern through sp^2-bonding [21]. Often described as an atomic-scale, one-atom-thick layer of graphite [21], it serves as the foundational structural unit for various carbon-based allotropes, including graphite, fullerenes, charcoal, and CNTs. Typical graphene displays versatility, capable of being manipulated into 0D fullerenes, rolled up to create 1D nanotubes, or stacked to generate 3D graphite structures. Possessing exceptional strength, graphene stands as one of the toughest materials on record, boasting a reported Young's modulus of 1 TPa and intrinsic strength of 130 GPa [22].

Mechanics of Flexible and Stretchable Electronics, First Edition. Edited by Yong Zhu and Nanshu Lu.

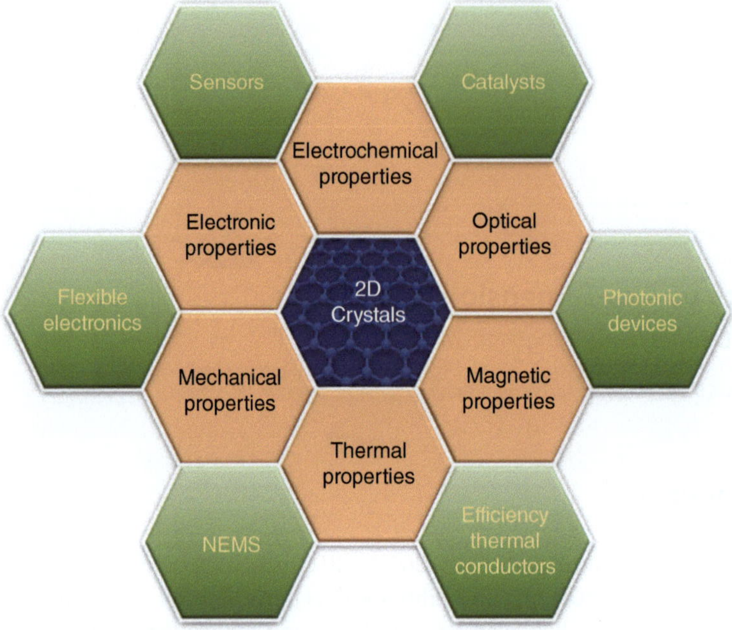

Figure 4.1 Schematic showing the properties and potential uses of 2D crystals. Source: Ref. [20]/Taylor & Francis/CC BY 4.0.

These values surpass those of steel by factors of 5 and 20, respectively, while maintaining a significantly lower weight– only one-third as much [23, 24]. Consequently, graphene sheets hold the potential to reinforce lightweight composites, serving as formidable mechanical reinforcements [25–27].

As a member of the family of layered vdW solids, TMDs, which have been extensively explored, share similarities with other 2D materials [28–31]. The crystalline structure of 2D TMDs follows a development pattern comparable to MX_2 stoichiometry, where M represents a transition element from group IV, V, or VI, and X represents chalcogen classes such as S, Se, or Te [32]. It is noteworthy that layered TMDs are composed of stacked hexagonal lattices, and the interaction between the transition metal and chalcogen atoms is contingent on their structural arrangement. Among the TMDs, molybdenum disulfide (MoS_2) has undergone a thorough examination [33–36]. In MoS_2, a hexagonal layer of molybdenum (Mo) atoms is positioned between two layers of hexagonal sulfur (S) atoms. The MoS_2 in monolayer form has a direct bandgap of 1.9 eV, distinguishing it from graphene [37]. This property is pivotal for semiconductor applications. MoS_2 demonstrates potential in the fabrication of flexible electronics owing to its cost-effectiveness and high performance, rendering it a promising material in this domain [38–40].

Conversely, hexagonal boron nitride (*h*-BN) has a historical application as a lubricant and has recently garnered significant attention due to its appealing properties, including chemical stability, dielectric behavior, anticorrosion capabilities, and intrinsic insulation [41, 42]. The *h*-BN sheet comprises alternating boron and nitrogen atoms, connected by covalent bonds, with layers held together through weak vdW interactions. This gentle interlayer bonding facilitates the exfoliation and synthesis of single- or few-layer *h*-BN crystals. With a wide bandgap of 5.56 eV, *h*-BN exhibits substantial transparency across a diverse range of UV–visible wavelengths [43]. The availability of large-area *h*-BN films has opened avenues for the production of 2D graphene-based electronics [44–47]. These films can serve as supporting substrates or dielectric layers in graphene-based devices, contributing to the advancement of next-generation electronic technologies.

Moreover, the mechanical stability of individual components in 2D materials plays a pivotal role in ensuring the reliability of manufactured devices, such as flexible electronics and energy storage systems. Damage to any component unit during mechanical deformation can result in the malfunction of the entire device. Therefore, studying the mechanical properties of 2D materials is essential, not only for predicting their lifespan in various applications but also for comprehending their unique mechanical behaviors. Extensive efforts, encompassing both theoretical and experimental mechanics, have been dedicated to investigating and quantifying the mechanical properties of 2D materials, including fracture toughness, elastic modulus, and fracture strength [48–52]. Moreover, additional mechanical characteristics, such as work of fracture, sliding friction, and interface adhesion, have been determined using diverse strategies [53–62]. These studies contribute to a comprehensive understanding of the mechanical behavior and performance of 2D materials.

In addition to original research, several insightful review articles have been published, specifically focusing on the mechanics of 2D materials. Zhang et al. [63] conducted a review highlighting the progress made in experimental and theoretical studies concerning the fracture behavior of graphene, covering aspects such as theoretical strength, impact fracture, mode I fracture toughness, mixed mode fracture, and sonication fracture. Similarly, Megra and Suk [64] conducted a review specifically addressing the adhesion properties of 2D materials, concentrating on measurement techniques and the characteristics of adhesion behaviors, with emphasis on graphene and MoS_2 materials. Wei and coworkers [65] provided a review of *in situ* characterization techniques utilizing scanning electron microscopy (SEM), atomic force microscopy (AFM), and transmission electron microscopy (TEM) for investigating the mechanical properties of 2D materials, aiming to present an overview of the application of these techniques in analyzing the mechanical behavior of 2D materials. Additionally, Akinwande et al. [48] conducted a comprehensive review covering both experimental and theoretical studies associated with the mechanics and mechanical properties of 2D materials, discussing various mechanical properties and addressing the coupling between mechanical properties and other physical characteristics of 2D materials.

In this chapter, selected experimental techniques developed for visualizing mechanical behaviors and evaluating the mechanical properties of 2D materials will be presented and discussed. The techniques include nanoindentation, AFM-enabled indentation, micro-/nano-mechanical devices, and pressurized bulge testing. Their working mechanisms, as well as the advantages and disadvantages associated with each technique, will be elaborated using monolayer and multilayer graphene, MoS_2, h-BN, and other emerging 2D structures as case studies. The mechanical properties of interest, such as elastic modulus and fracture strength/toughness, will be summarized in general. This book chapter aims to enhance the understanding of the mechanical behavior and properties of these materials.

4.2 Nanoindentation Method

The advent of ultrathin nanomaterials, encompassing 1D, 2D, and freestanding film materials, has necessitated the development of testing systems for the assessment of their mechanical properties. This imperative arises from the notable differences in structural formation and physical characteristics between these materials and their bulk 3D counterparts, demanding specialized techniques for accurate mechanical evaluation. Nanoindentation analysis, particularly through longitudinal loading, provides a valuable avenue for comprehending the mechanical response of freestanding films to external forces. Standard properties such as hardness and modulus of elasticity can be discerned through direct measurement of load–displacement behavior using nanoindentation [66]. Moreover, theoretical models exist that elucidate the process of deriving properties such as fracture toughness, yield strength, work-hardening exponent, residual stress, and rate-dependent properties from raw data obtained through indentation.

Two prevalent nanoindentation scenarios, illustrated in Figure 4.2a,b, are employed for measuring the elastic properties of freestanding 2D materials: linear strip and clamped membrane geometries. A representative example is depicted in Figure 4.2c, featuring a SEM image of freestanding fitted-end graphene strips [67]. In the case of a doubly clamped linear thin strip with initial dimensions of length (l), width (w), and thickness (t), as presented in Figure 4.2a, a normal load (P) can be applied using a wedge indenter tip. The deflection distance of the strip is denoted by (h), assuming effective clamping on both ends and negligible bending stiffness for the ultra-thin film. Thus, the elastic deflection of the thin strip can be understood as the balance between the work done by the indenter (Ph) and the stretching energy in the strip $\left(EAL\left(\frac{h}{l}\right)^4\right)$, where $A = wt$. This balance arises from the longitudinal external load being countered by the local tension within the strip through deflection [66]. Moreover, the strip typically experiences uniform residual stress σ_r, which is often the biaxial tensile stress across the film in a clamped position. Consequently, the energy balance can be expressed as follows:

$$Ph \sim EAl\left(\frac{h}{l}\right)^4 + \sigma_r Al\left(\frac{h}{l}\right)^2 \tag{4.1}$$

The preparation of linear strips is a more intricate process compared to circular membranes as it necessitates the patterning of graphene. Nevertheless, the outcomes are more straightforward to interpret due to the presence of a linear stress field. In the

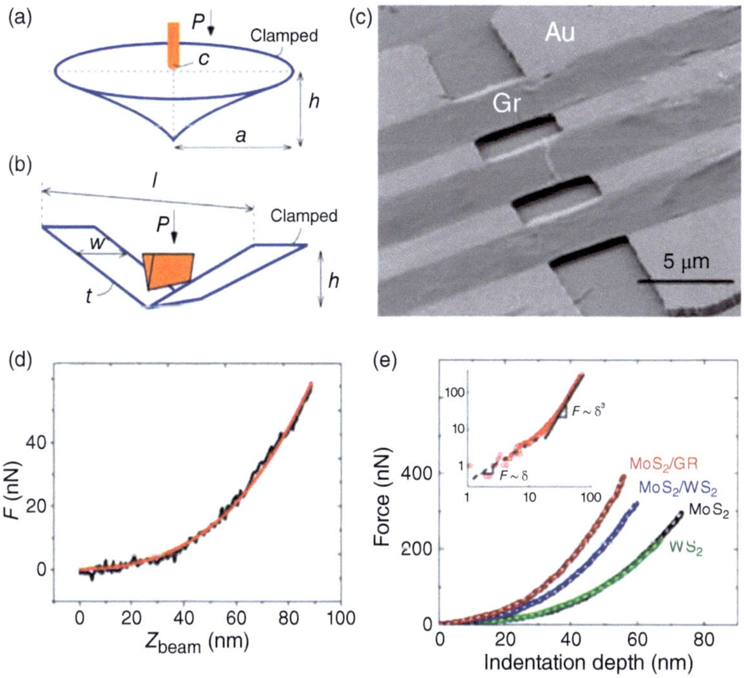

Figure 4.2 Nanoindentation experiments on 2D membranes or ribbons. (a) Schematic of a thin ribbon that is double-clamped and has a wedge tip. (b) Schematic of a clamped circular membrane with a spherical point indenting it. (c) Crossing gold electrodes with freestanding monolayer graphene strips. Source: Ref. [67]/with permission of AIP Publishing LLC. (d) A plot of force against deflection curve. The red line signifies that force is cubic to deflection [67]. (e) Typical force–deflection curves for 2D materials from circular membrane indentation. The linear-cubic membrane behavior is shown in the inset. Source: Ref. [68]/ American Chemical Society.

case of linear graphene strips, an AFM cantilever applies force at the center, inducing out-of-plane deflection. Figure 4.2d,e presents a representative force–deflection curve for a doubly clamped thin strip and typical force–deflection curves for 2D materials from circular membrane indentation. The linear-cubic membrane behavior is illustrated in the inset. The specific relationship between the applied force and the deflection of the thin strip can be determined by referring to Eq. (4.1) or employing force balance principles within the elastic range [66]. Additionally, commonly used equations for determining the elastic modulus and residual stress of 2D materials are compiled in Table 4.1. It is crucial to note that these equations are primarily applicable to long, thin, and narrow strips that primarily undergo stretching for normal deflection rather than significant bending [66].

However, as the number of layers in 2D materials increases, the role of bending becomes more noticeable in strip deformation. Senturia [74] incorporated the bending impact into the force–deflection relationship by assuming a cosine-shaped strip deformation. A commonly employed method for determining the elastic properties of thin film materials involves utilizing a circular drum-like structure, as depicted in

Table 4.1 Load–displacement relationships influencing the elastic response of freestanding thin films by nanoindentation.

Model	Calculation	Materials	Indenter	*E*	σ_r	σ_{max}	Reference
Circular fastened membrane	$P = E^{2D}q^3a\left(\frac{h}{a}\right)^3 + \pi a\sigma_r^{2D}\left(\frac{h}{a}\right)$	Monolayer Gr	AFM tip ($R = 16.5$, 27.5 nm)	342 ±30 N/m	0.07–0.74 N/m	42 ±4 N/m	[22]
		~2-layer *h*-BN	AFM tip ($R = 50$ nm)	223±16 N/m	—	8.8 ± 1.2 N/m	[43]
		Monolayer MoS_2	AFM tip ($R = 12$ nm)	180 ± 60 N/m	0.02–0.1 N/m	15 ± 3 N/m	[69]
		Double- layer MoS_2/Gr	AFM tip	467 ± 48 N/m	0.35 ± 0.05 N/m	—	[68]
		Double- MoS_2		300 ± 13 N/m	—	—	
		Monolayer WS_2		177 ±12 N/m	0.15 ± 0.03 N/m	—	
		Double-MoS_2/WS_2		314 ± 31 N/m	0.25 ± 0.05 N/m	—	
	$P = Eq^3at\left(\frac{h}{a}\right)^3 + \pi at\sigma_r\left(\frac{h}{a}\right)$	Monolayer GO ($t = 0.75$ nm)	AFM tip ($R = 25$ nm)	256.4±28.2 GPa	650±300 MPa	5.3±1.2 GPa	[70]
	$P = E^{2D}q^3a\left(\frac{h}{a}\right)^3 + \frac{2\pi\sigma_r^{2D}}{ln\frac{a}{c}}h$	Monolayer Gr	AFM tip	55 N/m	0.085 N/m	~35 GPa	[71]
	$P = \left[\frac{4\pi E}{3(1-v^2)}\left(\frac{t^3}{a^2}\right)\right]h + (\pi\sigma_r^{2D})\,h + \left(\frac{q^3Et}{a^2}\right)h^3$	~5 layer MoS_2	AFM tip	330 ± 70 GPa	0.13 ± 0.10 N/m	—	[72]
Double -clamped beam/strip	$P = \frac{8wtEh^3}{3l^3} + \frac{4wt\sigma_r h}{l}$	Monolayer Gr	AFM tip	800 GPa	—	—	[67]
	$P = \frac{\pi^4AEh^3}{8l^3} + \frac{\pi^4AEt^2h}{6l^3} + \frac{\pi^2A\sigma_r h}{2l}$	Al/0.5 wt. % Cu thin film ($t = 547$ nm)	Wedge tip	75.3 ± 1.0 GPa	28.7 ± 0.6 MPa	—	[66]
	$P = \frac{8wE^{2D}h^3}{l^3} + \frac{8w\sigma_r^{2D}h}{l}$	Monolayer Gr	Wedge tip	336 ± 20 N/m	0.08–2.4 N/m	—	[73]

P represents the load, *h* is the deflection distance and represents the cross-sectional area. The *l* represents the original length of the strip, σ_r is the residual stress, *w* denotes the width of the strip, *t* represents the film thickness, *a* is the original radius of the membrane, *v* is the Poisson's ratio, and *c* represents the radius of the indenter. *q* is a dimensionless parameter, which is equal to $1/1.05 - 0.15v - 0.16v^2$.

Figure 4.2a. This geometrical configuration is relatively easier to create and test, as a rigid substrate firmly fastens a circular membrane with a radius of a by clamping it along its edge. A sharp indenter tip/AFM cantilever with a curvature radius of c applies a central normal load P while the central deflection h is measured during the reload.

Consequently, to fabricate the freestanding clamped circular membrane structures in experiments, thin film materials are directly exfoliated onto a pre-patterned substrate featuring circular holes [22, 72, 75, 76]. In comparison to thin strips, the influence of bending stiffness can be disregarded for ultra-thin membranes. In such cases, the point-load assumption is employed when $c/a \ll 1$. These assumptions significantly streamline the analysis of the collected data. The deformation of circular drum-like membranes is then determined by balancing the work performed by the indenter, the stretching energy, and the strain energy produced by the clamping mechanism [77].

$$Ph \sim Et\frac{h^4}{a^2} + \sigma_r th^2 \tag{4.2}$$

Table 4.1 enumerates the regular elastic force–deformation relationships for the circular clamped membrane structure. The relations demonstrate the combined linear and cubic responses of the central deflection h to the applied force. For small indentation depths, $\tilde{h} = \frac{h}{a}\sqrt{\frac{E}{\sigma_r}} \ll 1$, where $\tilde{h}$ represents the normalized deflection, the stress in the membrane closely aligns with the preexisting tension from the initial state before indentation, thereby leading to the linear force law [78]. On the other hand, for large indentation depths $\tilde{h} = \frac{h}{a}\sqrt{\frac{E}{\sigma_r}} \gg 1$, it has been observed that the applied force (p) is proportional to h^3 ($P \sim h^3$) [53, 55–57]. This scenario is confirmed from the force–deflection curves depicted in Figure 4.2e, showing the transition from linear to cubic behavior during indentation of 2D materials [68]. The stretching performance predominantly influences the membrane deformation in dealing with monolayer graphene, MoS_2, WS_2, or double-layer heterostructure membranes, including MoS_2/Gr and MoS_2/WS_2. By utilizing Eq. (4.2) and the relations provided in Table 4.1, the elastic modulus and residual stress can be obtained through careful curve fitting of the cubic and linear relations separately. For wider membranes with 5–25 layers (MoS_2), the consequence of bending stiffness becomes significant, introducing an additional linear term $\frac{4\pi E}{3(1-\nu^2)}\left(\frac{t^3}{a^2}\right)h$ to the force–deformation relationship [72, 79].

4.3 AFM-Enabled Nanoindentation

4.3.1 Setup of AFM-Enabled Nanoindentation

AFM belongs to the family of scanning probe microscopy (SPM), offering high-resolution imaging capabilities at the nanoscale level. The resolution of AFM is typically on the order of a fraction of a nanometer, rendering it an invaluable tool for studying nanoscale features and structures. It operates by collecting surface

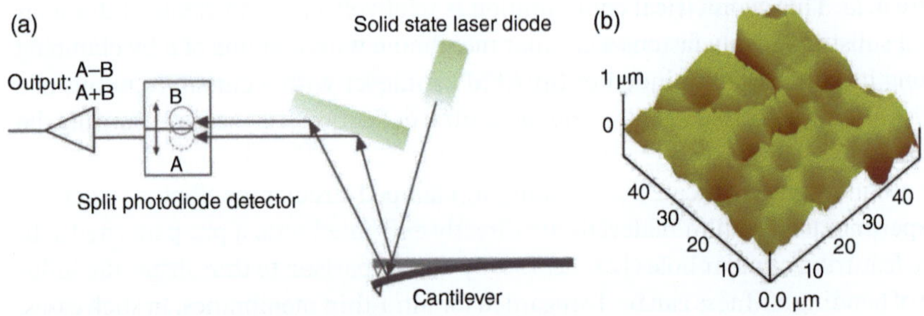

Figure 4.3 (a) Schematic showing the laser's alignment with respect to the probe cantilever and the split photodiode detector and (b) AFM image of the surface of a Faraday plate. Source: Ref. [20]/with permission of Taylor & Francis/CC BY 4.0.

information from a sample using a mechanical probe that interacts with the sample's surface through "feeling" or "touching" it. The probe's movement is controlled by piezoelectric elements, allowing precise and accurate positioning. AFM provides detailed topographic information and can also measure other properties such as mechanical, electrical, and magnetic characteristics. With the capability to manipulate single atoms at ambient conditions, AFM can be utilized for positioning and assembling atoms, offering possibilities for constructing nanoscale devices with precision.

In Figure 4.3a, a sketch illustrates the laser alignment process from a laser diode, the probe cantilever, and the split photodiode detector. The laser diode serves as the light source, directed toward the probe cantilever. The interaction between the probe and the sample causes a deflection in the cantilever, which is detected by the split photodiode. This deflection data is then used to generate the topographic image of the sample.

Generally, AFM proves to be a powerful technique for studying the mechanical properties and imaging the surfaces of 2D materials and other nanoscale structures, with its capabilities extending beyond imaging to potential applications in nanoscale device fabrication [80–82]. Specifically, the imaging capability of AFM relies on recording the probe's height in relation to constant probe–sample interaction while raster scanning the sample surface. The height information is then employed to generate high-resolution 2D and 3D images, which can be displayed using a pseudo-color scheme. Figure 4.3b exemplifies an AFM image of a Faraday plate obtained using a Veeco Dimension 3100 AFM. The image visually represents the sample's surface topography, enabling the observation of fine details and features at the nanoscale level.

Using AFM for imaging offers several advantages. First, AFM allows for the export of both 2D and 3D images, providing additional depth information and the ability to visualize the sample from different observation angles. This flexibility contributes to a more comprehensive understanding of the sample's topography. Second, AFM can image both conducting and insulating materials, making it a versatile tool for a wide range of samples. Unlike SEM techniques, which necessitate sample conductivity, AFM achieves true atomic resolution on conducting and nonconducting surfaces.

In force measurement, AFM quantifies the force between the AFM probe and the sample as a function of their separation distance. When a load, denoted as P, is applied to the AFM probe, it results in a deflection, represented as δ, provided the deflection is relatively small compared to the suspended length of the AFM probe cantilever (e.g. less than 1/6 of the suspended length). In that case, the relationship between the force and deflection can be simplified using Hooke's law:

$$P = k \cdot \delta \tag{4.3}$$

where k represents the spring constant of the AFM probe. The spring constant of commercial AFM probes typically falls within the range of 0.01–450 N/m. This constant describes the stiffness of the probe and determines the relationship between the applied force and resulting deflection. AFM-enabled nanoindentation has proven to be a helpful technique for characterizing the mechanical properties of 1D nanostructures [82–85].

4.3.2 Mechanical Testing of 2D Materials

The mechanical performance of 2D materials can be effectively elucidated at a macroscopic level using continuum elasticity theory. Expanding upon this understanding, the nanoindentation technique enabled by AFM proves highly suitable for measuring the macroscopic mechanical properties of such materials [79, 86–88]. This method has gained recognition as the simplest means of characterizing the mechanical properties of 2D materials, including graphene, MoS_2, h-BN, and others. Notably, the elastic modulus can be determined by analyzing the load–displacement data obtained through AFM nanoindentation. Additionally, AFM imaging allows for the accurate identification of the thickness of a 2D material sheet. However, several challenges are associated with characterizing 2D monolayer materials using AFM. These challenges include ensuring well-defined sample geometries, addressing stress concentration at fixing points, achieving high loading force resolution, accounting for defects, and developing mechanical models to determine stress levels accurately.

4.3.2.1 Mechanical Testing of Graphene

Lee et al. [22] were pioneers in utilizing AFM-enabled indentation on mechanically exfoliated monolayer graphene suspended on a Si wafer containing circular holes (with diameters of 1 and 1.5 μm), as illustrated in Figure 4.4. This configuration offers specific boundary conditions and is unconnected to the substrate, making indentation an ideal method for evaluating graphene's elastic modulus and strength. Moreover, no noticeable slippage occurs during indentation or scanning, ostensibly attributed to the vdW interactions between the graphene and substrate. In the case of monolayer graphene, a 2D material, the strain energy density can be normalized by the area of the graphene sheet. The mechanical performance under tensile loading can be described using 2D stress and elastic constants. The nonlinear elastic response follows a third-order stress–strain relationship:

$$\sigma = E\varepsilon + D\varepsilon^2 \tag{4.4}$$

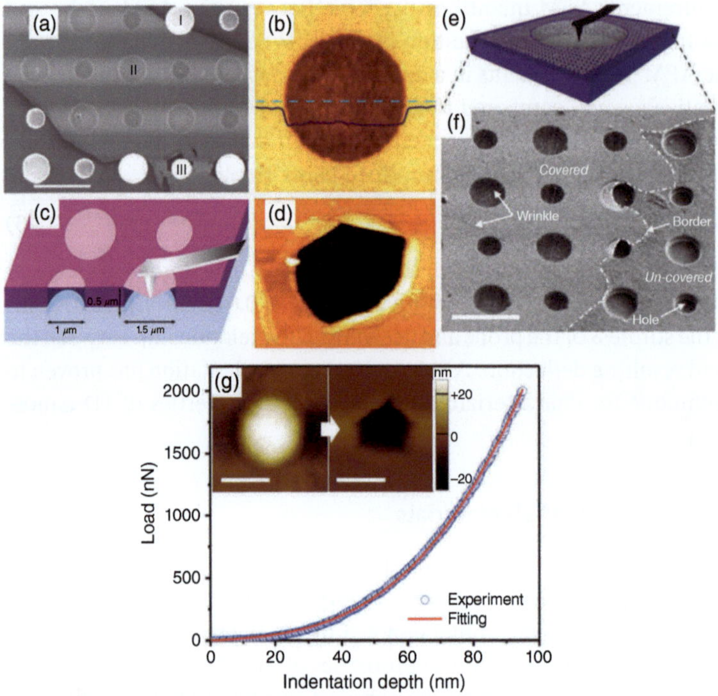

Figure 4.4 Single-crystalline and polycrystalline graphene indentations. (a) SEM image of an exfoliated graphene flake with a series of 1.5- and 1-m-diameter circular holes. (b) Contact-free mode one membrane in an AFM image. (c) Schematic of nanoindentation on suspended graphene membrane. (d) Image from an AFM of a ruptured membrane. Source: Ref. [22]/with permission of American Association for the Advancement of Science - AAAS. (e) Graphene sheet over a hole in a CVD-grown schematic for AFM nanoindentation measurements. (f) SEM image of polycrystalline graphene layer suspended over holes. (g) The polycrystalline graphene film's force–displacement curves during AFM nanoindentation. Insets are (g) the suspended polycrystalline graphene film's AFM topology pictures before and after a fracture. Source: Ref. [89]/with permission of permission of AIP Publishing LLC.

where σ represents the symmetric second Piola–Kirchhoff stress, ε denotes the uniaxial Lagrangian strain, E is Young's modulus, and D represents the third-order elastic modulus. Using a statistical analysis approach, Young's modulus of exfoliated graphene was evaluated to be 1.0 ± 0.1 TPa (corresponding to a 2D elastic constant $E^{2D} = 342$ N/m), assuming that the adequate thickness is 0.335 nm. Lee et al. [22] made a groundbreaking contribution by conducting a comprehensive measurement of the elastic properties of monolayer graphene, marking a significant milestone in the mechanical characterization of 2D crystals. Sequentially, Lee et al. [90] applied AFM-enabled nanoindentation to investigate the mechanical properties of bilayer and trilayer graphene sheets. Notably, the load–displacement behavior during testing remained consistent between loading and unloading, indicating the absence of slippage or plastic deformation. The calculated Young's moduli for bilayer and trilayer graphene sheets were found to be 1.04 and 0.98 TPa, respectively, closely matching the value for monolayer graphene. Palaniapan and coworkers [91] also employed a similar process to determine Young's moduli of graphene sheets. They reported Young's modulus of approximately 1.12 TPa for monolayer graphene, consistent with the findings from Lee et al. [90]. Remarkably, Young's moduli of graphene sheets consisting of two, three, and five layers were observed to increase to 3.2 TPa. This phenomenon was attributed to increased pretension resulting from different adhesion forces between graphene and the substrate, as well as stacking faults in multilayer graphene [92].

Additionally, differences exist in the adhesion force in multilayer graphene sheets compared to monolayer graphene and the substrate [93, 94]. Graphene prepared through mechanical exfoliation exhibits a single crystallinity structure and low defect density. By utilizing AFM-enabled indentation, it is possible to derive the maximum stress as a function of applied load for a clamped, linear elastic, circular membrane subjected to a spherical indenter. Assuming a spherical indenter shape and neglecting nonlinear elasticity, the maximum stress considering this configuration can be developed as follows:

$$\sigma_{\mathrm{m}}^{2\mathrm{D}} = \left(\frac{FE^{2\mathrm{D}}}{4\pi R}\right)^{1/2} \tag{4.5}$$

where $\sigma_{\mathrm{m}}^{2\mathrm{D}}$ represents the maximum stress at the center of the graphene film, F is the maximum load, and R stands for the radius of the indenter. The average breaking strength is determined to be 55 N/m; however, it is important to note that this value might be miscalculated due to the uncertainty associated with nonlinear elasticity. Lee et al. [22] also conducted numerical simulations focusing on the theoretical intrinsic strength by combining principles of nonlinear elasticity. The converted intrinsic strength was determined to be 130 ± 10 GPa using a strain of 25%, which closely aligns with the value obtained using Griffith theory.

Gómez-Navarro and coworkers [95] exploited a comparable technique for measuring the elastic properties of monolayer and bilayer graphene sheets. In contrast to suspending the graphene over a circular hole, they prepared the samples into rectangular shapes, as illustrated in Figure 4.5. The mechanically exfoliated monolayer and bilayer graphene sheets were initially sandwiched between polymethyl-methacrylate (PMMA) layers. The rectangular area was then precisely milled using e-beam exposure, a process that provided complete control over the size and position of the suspended region. Notably, the absence of critical point drying will prevent the collapse of the suspended sheet due to surface tension. Nanoindentation, in combination with an AFM, was utilized for the mechanical characterization of the suspended graphene sheets. The elastic constant (k) (suspended graphene sheet) in the linear region could be employed to estimate the elastic modulus. The valid calculation for a double-clamped beam with the application of a point load is expressed as follows:

$$k = 32Ew\left(\frac{t}{l}\right)^3 + \frac{17T}{l} \tag{4.6}$$

where t represents the thickness of the graphene sheet, w denotes the width, and l corresponds to the suspended length of the graphene. The thickness of monolayer

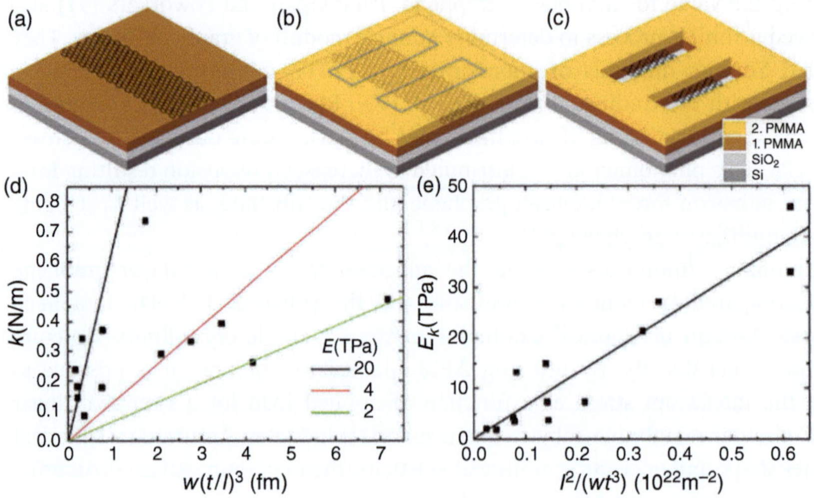

Figure 4.5 Fabrication of suspended graphene and indentation test. (a–c) Schematic of manufactured suspended graphene arranged on a polymer substrate. (d) Elastic constant k in relation to sheet dimensions. (e) Scaled elastic modulus in relation to sheet dimensions. Source: Ref. [95]/IOP Publishing Ltd.

graphene is presumed to be 0.34 nm. T denotes the built-in tension of the sheet. The estimated elastic modulus is determined to be 0.43 TPa, a value less than half the elastic modulus measured by Lee et al. The tested graphene exhibits significantly lower built-in tension compared to exfoliated graphene on a SiO_2 substrate. This lower tension helps to have access to the intrinsic properties of suspended graphene and enables the prediction and reproducibility of resonant frequencies. Several factors may contribute to the variation in elastic modulus, including (i) the crystallographic orientation of graphene in relation to the PMMA windows; (ii) the clamping conditions and point load configuration; (iii) the selection of graphene thickness; (iv) the substrate effect caused by the soft PMMA; and (v) the presence of PMMA residue on the sheets, inducing changes in the local thickness.

While chemical vapor deposition (CVD) allows for the growth of graphene sheets with sizes ranging up to square meters, the method often results in polycrystalline structures [96], characterized by individual dislocations as observed by Hashimoto et al. using TEM [97]. Since the atomic arrangement of graphene grain boundaries is theoretically predicted to give rise to distinctive electronic, magnetic, chemical, and mechanical properties, understanding the mechanical behavior of polycrystalline graphene compared to single graphene is of interest. This interest arises due to the significant size difference between grains and grain boundaries. Huang et al. [98] conducted AFM-enabled indentation tests on CVD-grown polycrystalline graphene membranes (Figure 4.6a). The membranes exhibited failure along the graphene grain boundaries, showing crack propagation (Figure 4.6b). This indicates that grain boundaries mainly establish the mechanical strength of polycrystalline graphene. The failure occurred at loads of ~100 nN, a magnitude lower compared to the usual fracture loads of 1.7 μN described for single-crystal exfoliated graphene. Thus, the

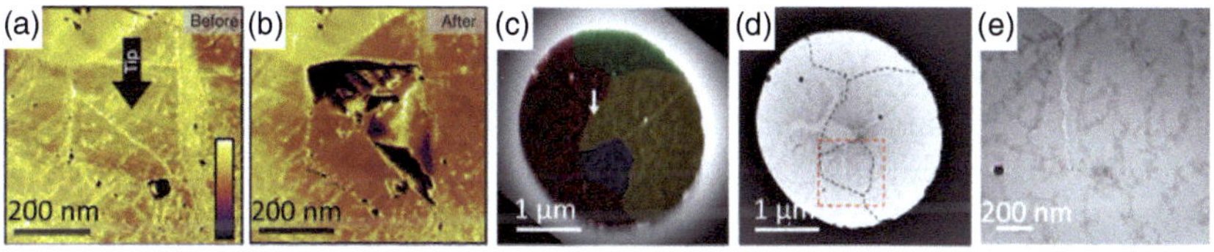

Figure 4.6 Crack path following nanoindentation. AFM phase image representation of a graphene grain prior to (a) and after (b) an indentation measurement. (a) Indentation occurs at a grain's center, as indicated by the arrow. (b) The region indicates crack propagation along grain boundaries following indentation. Source: Ref. [98]/with permission of Springer Nature. (c) False-color dark-field transmission electron microscope (DF-TEM) represents the minute-grain graphene layer placed over a hole before indentation. The white arrow reveals the indentation location. (d) Bright-filed TEM (BF-TEM) appearance after indentation. The black dotted lines reveal grain boundaries. (e) Enlarged BF-TEM image of the red-dashed area of (d). Source: Ref. [75]/with permission of American Association for the Advancement of Science - AAAS.

presence of grain boundaries significantly weakens the mechanical strength of graphene membranes, as also confirmed by the findings of Ruiz-Vargas et al. [71], who performed AFM indentation tests on CVD-grown graphene membranes. They obtained similar results, highlighting the detrimental effect of grain boundaries on the breaking strength of graphene. The upper limit for the in-plane breaking stress of polycrystalline graphene was found to be around 35 GPa, only one-fourth of that observed in single-crystal graphene. Additionally, the presence of out-of-plane ripples in graphene can effectively reduce its in-plane stiffness.

The presence of ripples in graphene, whether free-standing or supported, can lead to an inhomogeneous strain distribution, ultimately degrading the elastic stiffness of graphene. Lee et al. proposed a mechanism for the formation of ripples and developed an improved fabrication technique to minimize their occurrence utilizing CVD. Typically, CVD-grown graphene is synthesized on metal substrates such as copper. In the transfer process, a PMMA support layer is applied on top of the graphene, and the copper substrate is etched out using ferric chloride ($FeCl_3$) solution. Subsequently, the graphene is stamped onto the target substrate, and the PMMA layer is removed. The etching process and PMMA removal steps can introduce ripples in the graphene. However, it is demonstrated that by replacing the copper etchant with ammonium persulfate ($(NH_4)_2S_2O_8$), the air-baking step during PMMA removal can be eradicated, resulting in a significant reduction of ripple formation [99]. Lee et al. [75] also conducted experiments to measure the breaking strength of graphene with large grains, graphene with small grains, and mechanically exfoliated graphene. The results showed that graphene with large grains and exfoliated graphene exhibited an equivalent strength of 118 GPa, while graphene with small grains had a marginally minimal strength of 98.5 GPa. The difference in strength can be ascribed to the presence of defects and grain boundaries. Crack propagation through the grain rather than along the grain boundaries was observed, as shown in Figure 4.6c–e. Atomistic calculations performed by Ruoff and coworkers [100] revealed that graphene sheets with high-density defects in large-angle tilt boundaries exhibited comparable strength to exfoliated graphene and, surprisingly, were even stronger than sheets with low-angle boundaries having few defects. These findings challenge the common intuition that increased defects would minimize material strength. As a result, the discussion regarding the effect of grain boundaries on the strength of graphene continues among researchers [51, 101].

Furthermore, the existence of lattice defects in graphene sheets synthesized through scalable methods can significantly impact their mechanical properties. López-Polín et al. [87] conducted a study where they found that the in-plane Young's modulus initially increases with increasing defect density, reaching almost twice the value of the initial modulus for a vacancy content of ~0.2%. However, once the missing atoms exceed ~0.4%, the elastic modulus decreases significantly due to defect inclusions. Figure 4.7 provides a schematic representation of the indentation process and the measured mechanical behavior of defected graphene. The variation in Young's modulus suggests that the softening effect caused by vacancies becomes less prominent as the vacancy concentration approaches the percolation threshold. At this threshold, the elastic coefficients are expected to reduce linearly with the number of vacancies. Consequently, there is a direct relationship between Young's

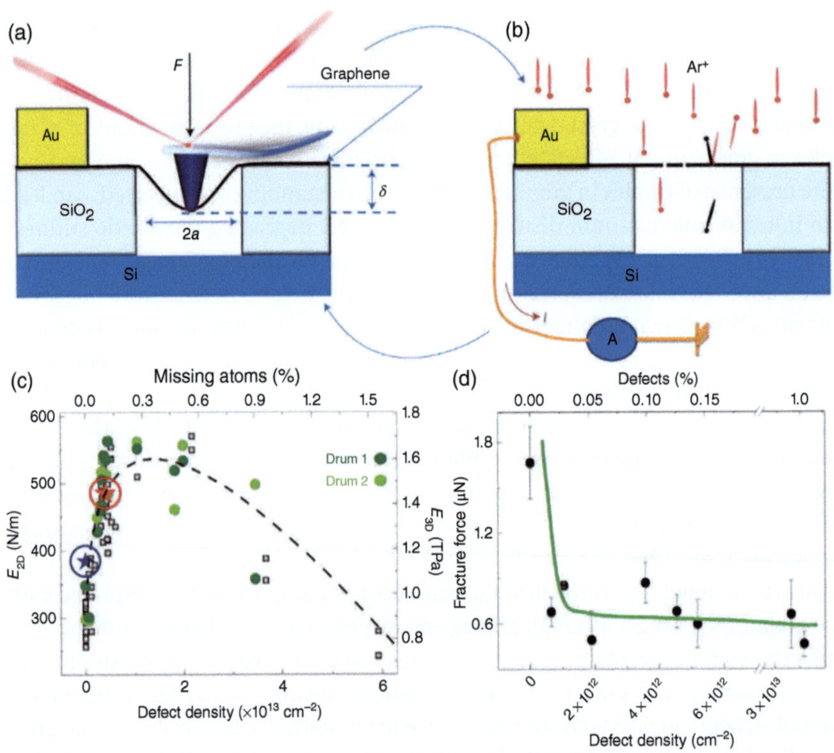

Figure 4.7 Indentation and mechanical behavior of graphene containing defects. (a) Device geometry and an image illustrating nanoindentation framework. (b) Irradiation framework, where the incoming argon ions produce the defects. (c) E_{2D} in response to defect concentration. (d) Breaking force for multiple drumheads in line with defect density generated. Source: Ref. [87]/Springer Nature.

modulus and the density of defects in graphene expressed as follows:

$$E = k\left(b + \frac{1}{l_0^2} + n_i\right)^{\eta/2}\left(1 - c\left(\frac{1}{l_0^2} + n_i\right)\right) \tag{4.7}$$

The equation that comprises the constants k and c, the geometrical factor b, the localization length l_0 for flexural phonons in pristine graphene, and the density of defects induced by irradiation represented by n_i describes the relationship between Young's modulus and defect density [87]. In the relationship between the fracture strength and defect density, the trend is different compared to Young's modulus. The fracture strength continues to decrease as the defect density increases, following the behavior predicted by standard fracture continuum models. Initially, as the defect density increases, the fracture strength drops to ~30% of the original strength. Afterward, it tends to reach a saturation point, where further increases in defect density have a lesser impact on the fracture strength.

4.3.2.2 Mechanical Testing of Graphene Oxide (GO)

The combination of hydrogen and oxygen to carbon atoms in graphene in different proportions and ratios results in the formation of GO compound and is mostly obtained through the treatment of graphite using strong oxidizers constituent [102–104]. Generally, GO processed via this method usually has a layer thickness of 1.1 ± 0.2 nm with oxygen atoms arranged in a rectangular pattern possessing a lattice constant of approximately 0.27 nm × 0.41 nm [105, 106]. At the end of each layer (edges) in the GO bulk materials, the carboxyl and carbonyl groups present enable GO dispersion in the solution. This unique characteristic draws researchers' attention and thus propels the investigation of the mechanical properties of both reduced graphene oxide (rGO) and GO via experimental and theoretical methods.

The multi/monolayer GO membranes prepared by Suk et al. [107] were characterized using AFM in contact mode to assess the Young's modulus and prestress properties, as illustrated in Figure 4.8. The figure shows the transferred membrane coupled with the histogram of the Young's modulus/SAED configurations of GO membranes. The prestress in GO membranes is obtained to be between 39.7 and 76.8 MPa. However, the monolayer GO displays a lower Young's modulus of 207.6 ± 23.4 GPa compared to pristine graphene. In the process of oxidizing graphene to obtain GO, approximately 40% of the original carbon atoms initially bonded in a sp^2 configuration undergo reconstruction to transform into sp^3-bonded atoms through their interaction with oxygen. Due to the large "defected" structure, the GO membrane displays reduced mechanical performance. In the case of the monolayer membrane, the Young's moduli of the bi/tri-layer GO membranes were measured to be 444.8 ± 25.3 and 666.5 ± 34.6 GPa. However, when considering the actual thickness of the membranes, the recalculated Young's moduli for the bi/tri-layer configurations were determined to be 223.9 ± 17.7 and 229.5 ± 27.0 GPa. The similarity observed between the monolayer, bilayer, and trilayer GO membranes suggests that the bonding between the layers is sufficient to prevent interlayer sliding [107]. Therefore, to improve the structural performance of GO, a chemical

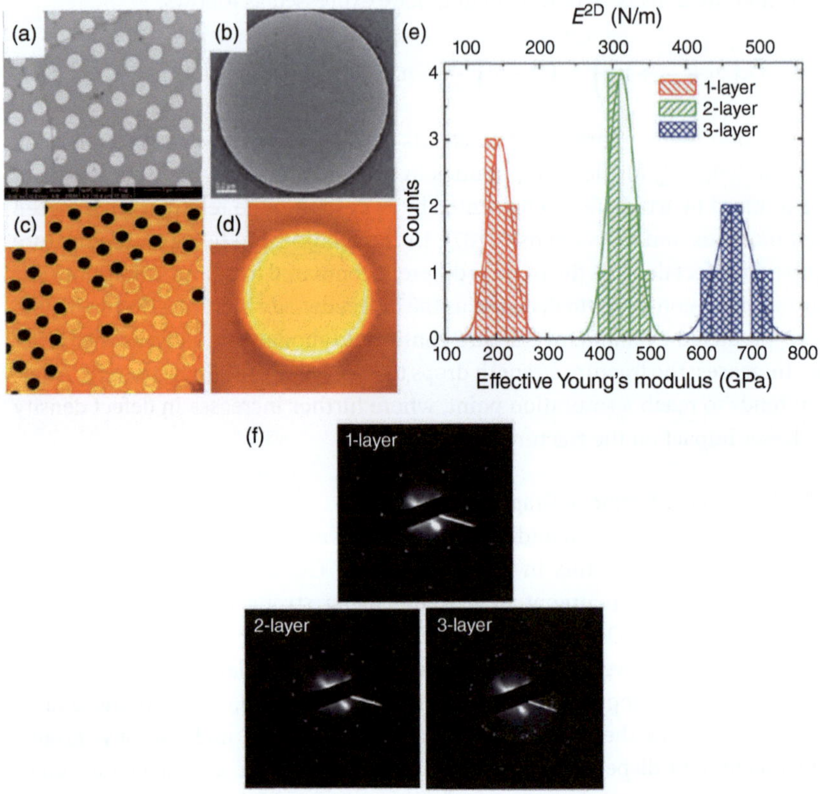

Figure 4.8 Indentation on GO suspended membrane. (a) STEM micrograph and (b) TEM micrograph. (c) AFM topology micrograph in $20 \times 20\,\mu m^2$ scanned location. (d) AFM topology micrograph in $2.2 \times 2.2\,\mu m^2$ (e) Histogram depicting the effective Young's modulus of GO membranes with an assumption that the thickness is one layer for the entirely evaluated membranes. The 2D elastic constant E^{2D} is similarly estimated using the same assumption (monolayer thickness). The solid lines correspond to Gaussian fits to the data. (f) Selective area electron diffraction (SAED) patterns from GO membranes composed of one, two, and three layers. Source: Ref. [107]/with permission of American Chemical Society.

reduction process can be utilized. It involves decreasing the participation of oxygen in the functional group that has oxygen to obtain rGO with a sheet-like structure. For instance, the GO fabricated by Hummers and Offeman [108] was transferred to a Si/SiO_2 substrate with the assistance of hydrogen plasma treatment.

Additionally, a solution-based approach can be employed for GO reduction, offering advantages such as cost-effectiveness and scalability [109, 110]. This method mostly produces sheets with residual oxygenated functional groups because the reduction process is ineffective. To address this challenge, Gómez-Navarro et al. [111] conducted indentation testing to examine the mechanical performance of rGO sheets. In their study, preselected layers of rGO were precisely sectioned using standard e-beam lithography, incorporating Ti/Au electrodes. To achieve the free suspension of the rGO samples, they underwent etching with hydrofluoric acid and were then taken through critical point drying to prevent the sheets from adhering to the substrate. The mechanical performance of the fabricated monolayers for the free-standing rGO was conducted via the indentation method by applying an AFM tip to the center of the suspended region. The measured elastic modulus of rGO monolayers is 0.25 ± 0.15 TPa, which is much less than that of monolayer graphene but comparable to the elastic modulus of GO membranes.

4.3.2.3 Mechanical Testing of MoS_2

MoS_2, a hexagonally structured material composing alternating Mo and S atoms at its corners, is one of the highly investigated TMDs [72, 112–114]. Its incorporation into microelectromechanical systems/nanoelectromechanical systems (MEMS)/NEMS is driven by the fact that compared to graphene, MoS_2 possesses an indirect bandgap of 1.29 eV [37]. Notably, the increasing thickness to a monolayer size enables transitioning from an indirect to a direct bandgap [115]. Thus, MoS_2 exhibits great potential for flexible electronics applications due to its high performance and low cost. To ensure the strong stability and dependability of MoS_2, it is crucial to conduct mechanical properties investigation studies.

According to theoretical calculation, a defect-free material typically displays a fracture strength that approaches an upper boundary of ~10% of its Young's modulus [116]. The in-plane stiffness of a defect-free monolayer MoS_2 is proportional to the effective spring constant of the bonds existing between Mo and S. In the study by Bertolazzi et al. [69], following the existing method (AFM-enabled indentation), the mechanical properties of MoS_2 were evaluated. Figure 4.9 shows the optical and AFM descriptions coupled with the associated indentation testing of the MoS_2 monolayer flake. Prior to analysis, the monolayer MoS_2 was transferred onto a prepared SiO_2 patterned substrate containing an array of spherical holes of 550 nm diameter each. At a thickness of 0.65 nm, Young's modulus obtained was 270 ± 100 GPa, although this value was reduced to 200 ± 60 GPa in bilayer MoS_2. However, this value is still comparable with that of bulk MoS_2 and steel Young's modulus, corresponding to 238 and 210 GPa, respectively. The inconsistency observed in the MoS_2 Young's modulus, with respect to the varied number of layers, is ascribed to the potential effect of interlayer sliding or the possible presence of defects [117]. This observation was supported by the finding of Castellanos-Gomez et al. [72, 118]. The film obtained for the MoS_2 has layers that range from 5 to 25. The measured Young's modulus of 330 ± 70 GPa is close to that of bulk MoS_2. Therefore, there is no size dependence in Young's modulus of MoS_2.

4.3.2.4 Mechanical Testing of WSe_2

Zhang et al. [119] focused on the elastic properties of mechanically exfoliated multilayer suspended WSe_2. The multilayer flakes were transferred to pre-patterned SiO_2/Si substrates designed with hole arrays. The experimental processes are displayed in Figure 4.10, showing the method by which WSe_2 was transferred, including the AFM-enabled indentation test implemented on membranes with different layers. The elastic moduli obtained correspond to 170.3 ± 6.7, 166.3 ± 6.1, 167.9 ± 7.2, and 164.8 ± 5.7 GPa for samples that possess 5, 6, 12, and 14 layers,

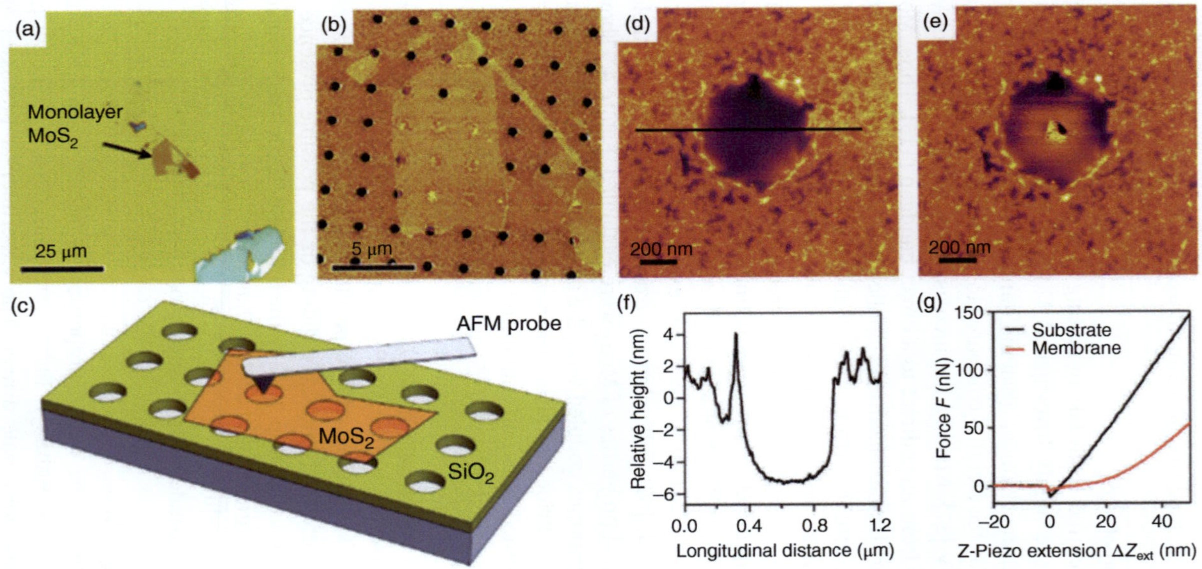

Figure 4.9 Indentation was performed on mechanically exfoliated MoS_2. (a) Optical micrograph of a monolayer MoS_2 flake conveyed onto the pre-patterned SiO_2 substrate having a grouping of spherical holes of 550 nm in diameter. (b) AFM micrograph for the same material (monolayer MoS_2). (c) Illustration showing the indentation experimentation. (d) AFM micrograph of a monolayer MoS_2 flake placed on a hole cavity prior to indentation analysis. (e) A hole cavity could be observed looking at the center of the membrane where the AFM tip punctured. (f) The height sketch of the part highlighted in (d) reveals that the membrane stays to the sidewalls over a vertical distance on the order of 5 nm, causing a pretension between 0.02 and 0.1 N/m. (g) Acquired force against z-piezo extension curves for the hanging membrane and the substrate.Source: Ref. [69]/with permission of American Chemical Society.

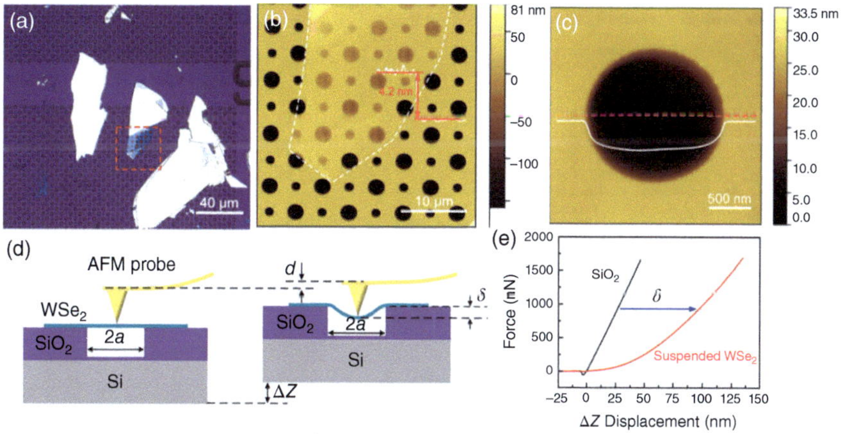

Figure 4.10 Transferred WSe_2 and indentation test. (a) Optical micrograph of WSe_2 flakes conveyed onto pre-patterned SiO_2/Si substrate. (b) AFM micrograph of the corresponding WSe_2 flake in the square region of (a). (c) AFM micrograph of a WSe_2 membrane placed on a hole and an overlaid height profile (along the dotted line) displays a step height of ~30 nm. (d) Schematic of the indentation of experimental analysis on a suspended WSe_2 membrane. (e) Force–displacement curves acquired on a suspended WSe_2 membrane and SiO_2/Si substrate. Source: Ref. [119]/AIP Publishing LLC/CC BY 3.0.

respectively. This observation supports the claim that Young's modulus is not dependent on the thickness of WSe_2. The interlayer interaction in the WSe_2 significantly impeded sliding that could occur between interlayers during testing. Eventually, WSe_2 displays Young's modulus, which is one-sixth of graphene and about two-thirds of those similar 2D semiconducting TMD materials such as WS_2 and MoS_2 [68, 69].

4.3.2.5 Mechanical Testing of *h*-BN

The boron/nitrogen atoms in *h*-BN are bonded through covalent bonds in each layer, while weak vdW interactions hold the layers together. Consequently, weak interlayer bonding allows for easy exfoliation of monolayer and few-layer *h*-BN crystals. Theoretical predictions suggest that *h*-BN exhibits exceptional mechanical properties, similar to graphene [89, 120]. For instance, density functional theory (DFT) simulations have shown that the in-plane elastic constant (E^{2D}) could reach values as high as 275.9 N/m, and the Poisson's ratio is in the range of 0.22 [121, 122]. Song et al. [43] utilized the AFM-enabled nanoindentation technique to evaluate the elastic constants of multilayer *h*-BN. The force–displacement curves are presented in Figure 4.11a–c. The E^{2D} and σ^{2D} for the bilayer *h*-BN analyzed revealed 223 ± 16 N/m and 8.8 ± 1.2 N/m, respectively [43].

The maximum stress taken by monolayer graphene (55 N/m) is significantly higher compared to that in *h*-BN film. Noteworthy is that the 55 N/m maximum stress obtained in monolayer graphene is peculiar to those prepared by mechanical exfoliation. The lower maximum stress is primarily due to the presence of a sizable number of vacancy defects arising during the fabrication of *h*-BN by the CVD

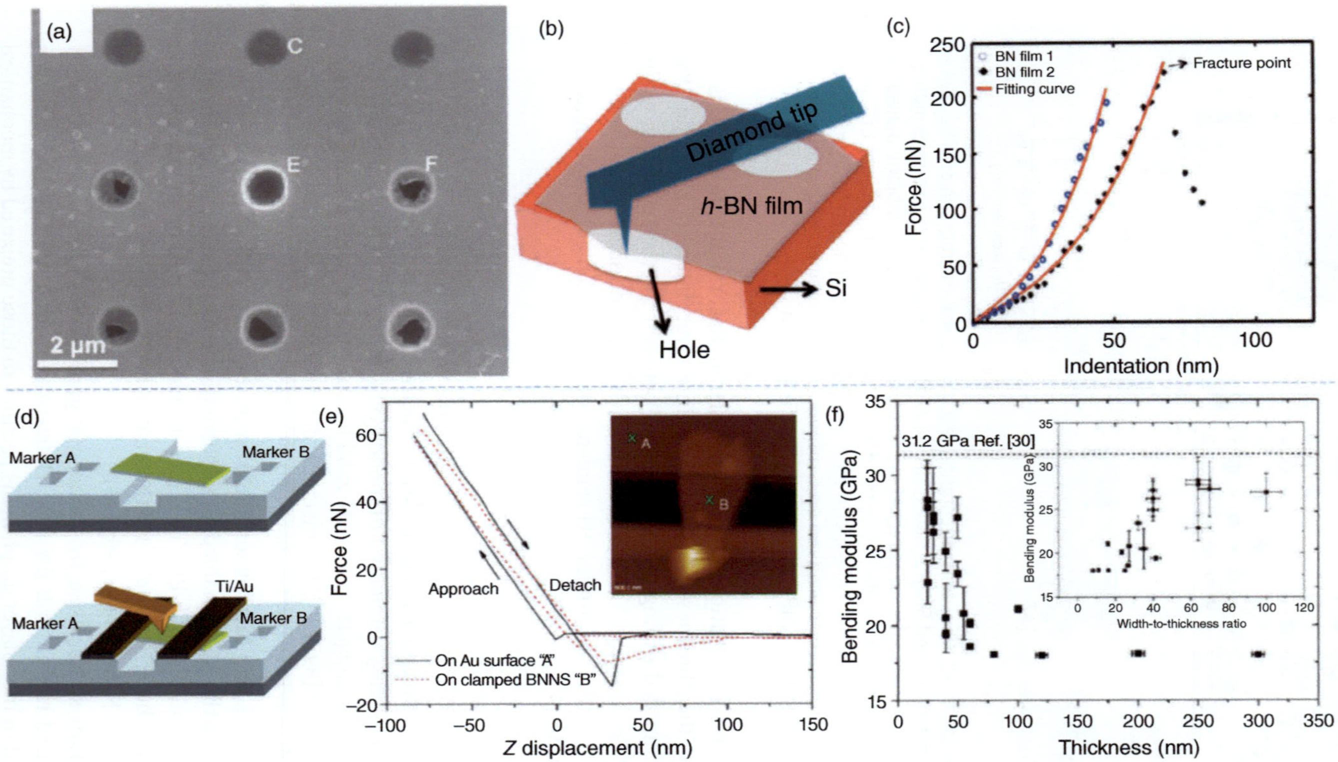

Figure 4.11 Elastic properties and intrinsic strength of *h*-BN film. (a) SEM micrograph of a big *h*-BN film across an array of spherical holes. (b) Graphical image of nanoindentation on draped *h*-BN film. (c) Measured force against displacement curves. (d) Graphical image showing the clamped *h*-BN nanosheet in an AFM bending experiment. Source: Ref. [43]/with permission of American Chemical Society. (e) Comparative experimental advancing and detaching load–displacement curves acquired on the midpoint of a nanosheet and atop the Au substrate. (f) Bending modulus against nanosheet thickness. Source: Ref. [123]/with permission of IOP Publishing. *h*-BN reveals resemblances with MoS_2 regarding its bonding structure.

method. An interesting observation indicates that both E^{2D} and σ^{2D} substantially increase with the increasing thickness of *h*-BN. For example, when the *h*-BN film contains five layers, the measured values of E^{2D} and σ^{2D} are found to be 505 ± 30 and 15.7 ± 1.5 N/m, respectively [43]. This result (E^{2D}) is close to two times higher compared to the theoretically calculated value of 292.1 N/m, and it is significantly higher than the elastic constant of graphene of 342 N/m. The related findings are possibly too sketchy to be fully clarified by the dependence of the in-plane stiffness on inherent defects in the film. Commonly, the materials might have increased defects with an increase in size, indicating that the response of maximum stress in the report could differ. Other factors contributing to the discrepancies in calculating elastic constants and maximum stress could include the dispersion of layers of stacking faults in the films synthesized via CVD, errors due to the estimation of the diameter of the holes present, and the location of the membrane contact with the substrate; however, these are not enough.

To gain further insights, Li et al. [123] focused on three-point bending tests using AFM on boron nitride nanosheets with thicknesses ranging from 25 to 300 nm to elucidate more, and Figure 4.11d–f presents the schematic illustration of the procedure. These are the clamped *h*-BN nanosheets prepared for bending tests. The bending moduli obtained increase as the thickness of the sheet is reduced, corresponding to 31.2 GPa. The significant dependence of the bending modulus on the sheet thickness is associated with the dispersion of the layers of stacking faults present in the nanosheets [124].

4.3.2.6 Mechanical Testing of Black Phosphor (BP)

In the study conducted by Wang et al. [125], nanoindentation analysis was employed to evaluate the mechanical properties of transferred BP nanosheets. Figure 4.12a–c illustrates the suspension of BP nanosheets on spherical holes before and after testing. A continuum mechanics model was introduced to calculate the elastic modulus and pretension of BP nanosheets with thicknesses ranging from 14.3 to 34 nm. The elastic modulus of BP nanosheets was observed to decrease as the thickness decreased, with a maximum elastic modulus of 276.6 ± 32.4 GPa. When the thickness is larger than 30 nm, the modulus remains at 89.7 ± 26.4 GPa, which is equivalent to that of bulk BP.

Moreover, the effective strain of BP varies between 8% and 17%, possessing a breaking strength of 25 GPa. Considering that BP is typically unstable under room conditions, Moreno-Moreno et al. [73] conducted a study to reveal the mechanical properties of nanosheet BP with few layers in a high-vacuum atmosphere. The exposure duration to the room temperature environment was carefully controlled. Figure 4.12d–f represents the mechanical characteristics and landscape of the BP drumheads at room temperature. Mechanical properties were assessed using AFM-enabled nanoindentation on samples positioned on spherical holes (BP flakes) with a thickness between 4 and 30 nm.

In comparison with the study by Wang et al. [125], the elastic modulus and breaking strength for the high-vacuum-tested samples were substantially reduced, yielding values of 46 ± 10 and 2.4 ± 10 GPa, respectively. In contrast to Wang et al.'s

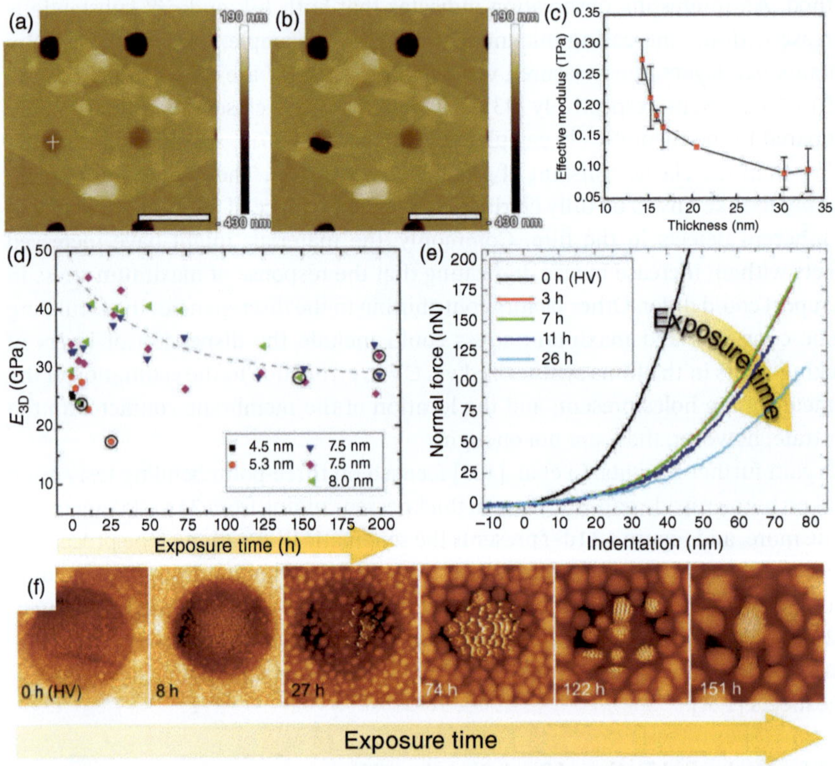

Figure 4.12 AFM micrograph of BP nanosheet prior to (a) and after (b) indentation. (c) Thickness-dependent effectual modulus of BP nanosheets. Source: Ref. [125]/AIP Publishing. (d–f) Development of the mechanical properties and structure of the BP drumheads in ambient conditions. Source: Ref. [50]/IOP Publishing Ltd/CC BY 3.0.

conclusions, both magnitudes were found to be independent of flake thickness. For thick flakes, the elastic modulus exhibited minimal changes with the atmosphere, while for flakes thinner than 6 nm, exposure to air significantly affected the mechanical performance.

4.4 *In Situ* Indentation in SEM

An AFM probe, independent of a photodiode and laser but featuring a cantilever controlled through a manipulator in SEM, can be utilized for force measurement [126]. For instance, a mechanical testing setup, comprising three disconnected linear stages, three picomotors, and an AFM probe, was employed to evaluate the mechanical strength at the CNTs/graphene contact [126]. The tester provides three degrees of freedom (x, y, and z) for aligning, setting, and moving the AFM probe. The spring constant of the attached AFM probe, ranging from 0.01 to 450 N/m, determines the

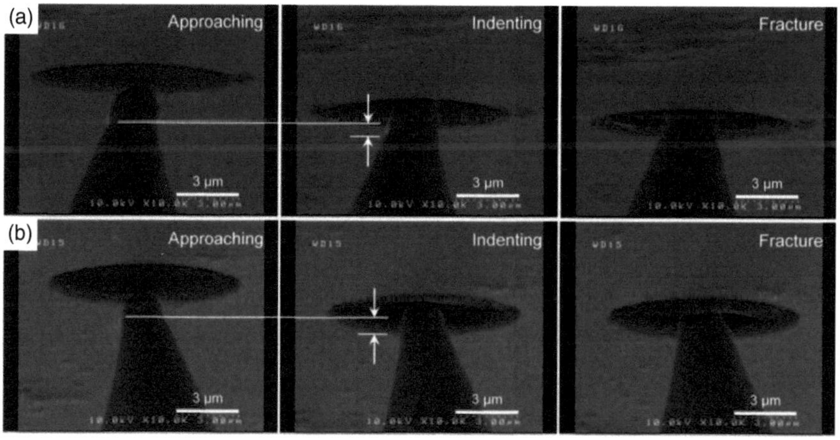

Figure 4.13 Nanoindentation assessments conducted in SEM on graphene membranes (a) with graphene independent of grain boundary and (b) containing small grains. The gap shown by the markers indicates the deflection of the AFM cantilever. Source: Ref. [127]/with permission of John Wiley & Sons.

force resolution. By establishing the relationship between the spring constant and the resulting deflection, the maximum force leading to mechanical failure can be estimated. Consequently, such a method proves valuable for assessing the mechanical behavior of 2D materials. This *in situ* indentation method using SEM facilitates the precise determination of the fracture strength of materials during experimentation by controlling the tip position when placed on the membrane.

Suk et al. [127] conducted indentation tests to scrutinize the fracture behavior of polycrystalline graphene membranes in their study using SEM, as depicted in Figure 4.13. CVD was employed for the fabrication of polycrystalline graphene, which was subsequently transferred onto silicon substrates featuring 9 μm-diameter pores. A nanomanipulator was utilized to apply midpoint loads to the circular graphene membranes, and an AFM cantilever served as the force detector. Raman spectroscopy was employed to observe the grain boundaries of polycrystalline graphene. It is noteworthy that the suspended graphene membrane consisting of a single grain exhibited an average failure strength of approximately 45.4 ± 10.4 GPa, while the average failure strength of suspended graphene membranes comprising various grains was considerably lower at 16.4 ± 5.1 GPa. It was established that the peak fracture strength of polycrystalline graphene membranes did not surpass the lowest failure strength of single-crystal membranes. This indicates that grain boundaries may significantly reduce the fracture strength of graphene, as the failure strength decreases with an increase in the density of grain boundaries. These findings align with previous results from AFM-enabled nanoindentation studies, indicating that grain boundaries negatively impact the breaking strength of graphene [71]. Moreover, the obtained failure strength did not exceed the upper limit of 35 GPa for the in-plane breaking stress of graphene [71].

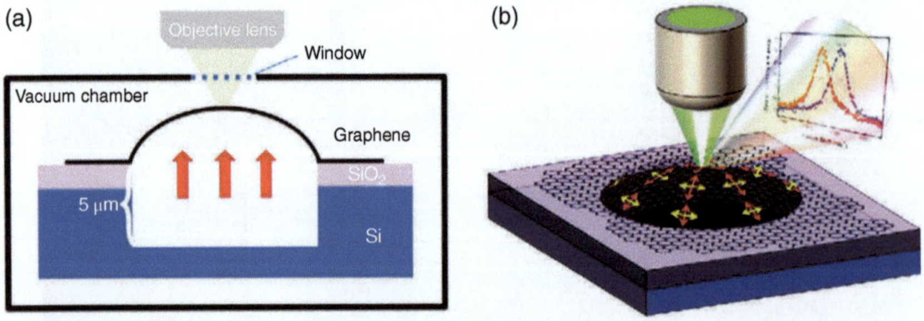

Figure 4.14 (a) Graphics representation of strain-induced pressurized graphene membrane analyzed using Raman spectroscopy. (b) A pressure gradient through the graphene membrane was maneuvered by adjusting the pressure of the vacuum chamber, resulting in an upward bulge. Source: Ref. [128]/IOP Publishing Ltd/CC BY 3.0.

4.4.1 Raman Spectroscopy

Raman spectroscopy stands as a nondestructive experimental technique that scrutinizes the collective phonon modes associated with mechanical and electrical effects [128]. For instance, the elastic modulus of graphene was evaluated utilizing Raman spectroscopy, as illustrated in Figure 4.14a,b [129]. In the context of 2D materials, most Raman investigations have concentrated on high-frequency modes corresponding to intralayer chemical bond vibrations. These high-frequency modes are readily discernible when isolated from Rayleigh background scattering but lack sensitivity to coupling between adjacent layers. Additionally, low-frequency Raman modes (<100/cm), correlating to the interlayer vdW restoring force, consist of two modes: out-of-plane shear and in-plane types [130]. Analyzing these interlayer phonon modes allows for the determination of both the elastic constant in the axial direction (breathing mode) and the elastic constant for the shear mode from the low-frequency Raman spectrum. Assuming that interactions between the nearest neighboring layers are significantly prevalent, the frequency of vibration related to the interlayer breathing mode or shear mode can be determined using a simple linear-chain model [131].

$$\omega_i^2 = \left(\frac{k}{2\mu\pi^2c^2}\right)\left(1 - \cos\left(\frac{i-1}{N}\pi\right)\right) \tag{4.8}$$

where $i = 2, 3, \ldots$ (i=1corresponds to the zero-frequency acoustic mode at Γ point in the Brillouin zone), c is the speed of light in vacuum, K is the in-plane (out-of-plane) force constant, and $\mu = 2.63352 \times 10^{-26}$ kg · Å^{-2} is the mass per unit area of monolayer of 2D materials. The N represents the number of layers present in the 2D material. The in-plane ($K = K_S$) and out-of-plane ($K = K_B$) force constants per unit area can then be obtained by fitting the experimentally obtained peak frequencies of the shear and breathing modes, respectively. This model has been used to effectively envisage the low-frequency phonon modes of different 2D materials, especially graphene [132], MoS_2 [130], WSe_2 [130], and h-BN [133].

4.5 Micro-/Nano-mechanical Devices

Extracting crucial mechanical properties of 2D crystals from local indentation tests using an AFM tip is undeniably challenging due to the generation of nonuniform stress and strain fields by the tip, making it difficult to obtain accurate and representative measurements. Additionally, nanoindentation tests lack the capability to image dynamic fracture processes. To address these challenges, various micro-/nano-mechanical testing tools have been successfully developed, overcoming these limitations and enabling uniform in-plane loading on self-supporting membranes of 2D crystals [134–136]. These devices provide accurate and reliable quantification of mechanical properties once the selected crystals are positioned onto the testing region. *In situ* tensile tests, involving the application of tension to 2D crystals, can be conducted using both SEM and TEM. These tests offer the advantage of real-time visualization of the fracture process, providing a comprehensive understanding of the fracture mechanism of 2D crystals.

4.5.1 Category of Micromechanical Devices

4.5.1.1 Thermal Actuated Micromechanical Devices

Zhu et al. [137] pioneered the development of a micromechanical tool that has gained widespread adoption in the field. This tool features two distinct actuators: a thermal actuator capable of applying controlled thermal stress and an electrostatic (comb drive) actuator that utilizes electrostatic forces for displacement generation. The micromechanical components of this MEMS platform, including an on-chip actuator, an electronic load sensor, and a gap for nanostructure insertion, are illustrated in Figure 4.15a,b [134, 137]. A significant advancement in this device is

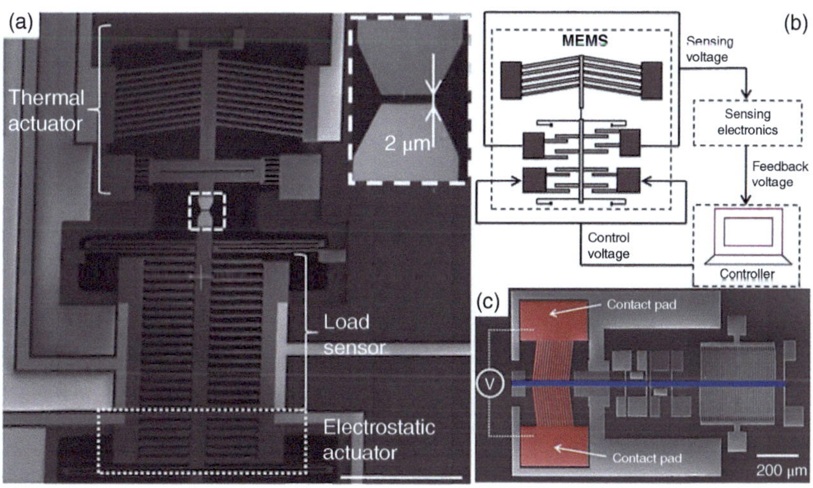

Figure 4.15 (a) Thermal actuated load sensor instability micromechanical devices manufactured by Zhu et al. [137]/with permission of Springer Nature and Pantano et al. [138]. (b) Optimization of MEMS-based tensile testing devices. Source: Adapted from [138] and (c) MEMS devices manufactured by Staufer and coworkers [135]/IOP Publishing Ltd/CC BY 3.0.

the integration of a capacitance load detector, allowing displacement measurements through differential capacitive sensing rather than relying on microscope imaging. This electronic load sensor offers a more precise and efficient method for measuring displacement during testing. The MEMS platforms developed by Espinosa and others are particularly suitable for *in situ* tensile tests conducted in SEM and TEM setups. The thermally actuated micromechanical device is well-suited for testing stiffer materials such as films and nanowires with larger dimensions, while the comb drive micromechanical device is more appropriate for compliant materials, including CNTs and nanowires with reduced dimensions. To characterize the mechanical characteristics of diverse nanostructures, several MEMS platforms based on this architecture have been developed, demonstrating its adaptability and wide range of applications. Figure 4.15c depicts one such MEMS platform [139–142].

4.5.1.2 Micromechanical Devices with Push–Pull Mechanism

Lou and coworkers [142, 143] have recently introduced a micro-fabricated device designed for conducting tensile testing and pullout experiments, as illustrated in Figure 4.16a,b. This innovative device incorporates a "push-pull" mechanism utilizing three movable shuttles interconnected by inclined freestanding beams. The design aims to enable precise and controlled loading as well as displacement

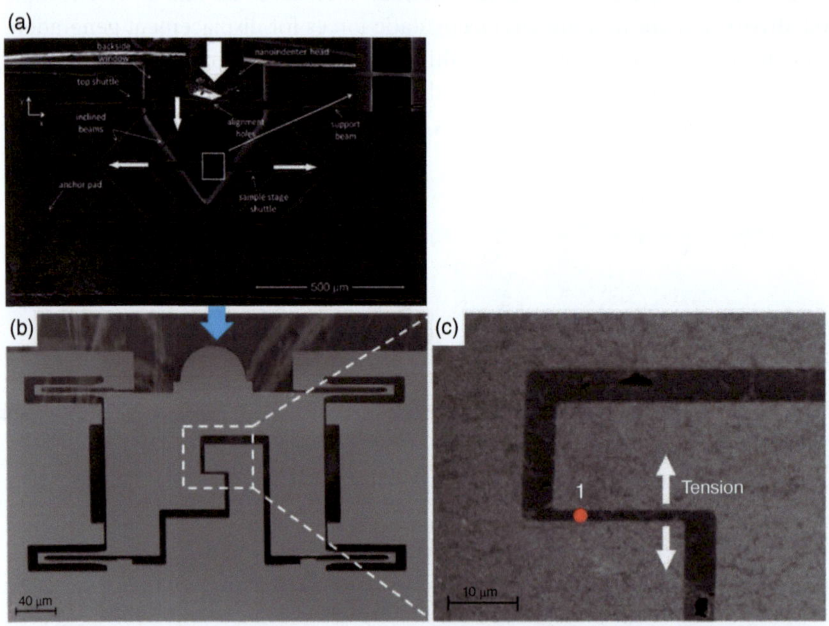

Figure 4.16 Microdevices operating on push–pull mechanisms utilized for *in situ* mechanical testing in SEM and TEM. (a) The micromechanical device invented and manufactured by Lou et al. [143, 144]/with permission of IEEE. (b) The micromechanical device implemented by Pérez-Garza et al. [135]/IOP Publishing Ltd/CC BY 3.0. (c) Shows transferred graphene on the micromechanical device in (a).

measurements during testing. To accurately measure the load and displacement, a quantitative Agilent in SEM nanoindenter is employed, providing independent measurements of load and displacement to minimize potential errors and enhance experimental data reliability.

Actuation is facilitated by the axially applied load (y-direction) on the shuttles top during nanoindentation operation. The sample stage shuttles are translated in 2D across four slanted symmetrical beams, subjecting the clamped sample to uniaxial tension for controlled tensile testing and pullout studies. During these studies, the micro-mechanical device facilitates precise load and displacement measurements. The applied load on the test sample and its elongation can be determined from load and displacement information obtained from the nanoindenter using conversion factors found in finite element simulations or the force reduction method.

With load resolution typically in the range of a few tens of nanonewtons and displacement resolution in the range of a few nanometers, the device provides high-resolution measurements. The stress–strain curve generated by the micro-mechanical device allows for fracture analysis and accurate determination of the elastic modulus alongside fracture strength. This tool is capable of evaluating various 1D nanostructures, including the mechanical testing of transferred graphene, as demonstrated by Liao et al. [145], as illustrated in Figure 4.16c,d.

4.5.2 Development of the "Dry-Transfer" Technique

Transferring thin films at the atomic scale onto a sample stage is crucial for the mechanical assessment of 2D materials. Zhang et al. [146] developed a unique dry transfer method for graphene to overcome the drawbacks of using liquid for the suspended Si functional layer of the stage. The procedure, illustrated in Figure 4.17a–f, involves bonding the as-grown graphene to PMMA which is then lifted by a polydimethylsiloxane (PDMS). The micromechanical device is covered by the suspended graphene/PMMA during transfer because the PDMS block has an open window which is slightly larger than the sample stage. To ensure proper adhesion and avoid wrinkle formation or pre-strain, heat treatment is applied to the PMMA/graphene film, allowing it to cling tightly to the hanging Si layer of the device. The transparent graphene/PMMA film-covered sample stage is then carefully picked up, and the graphene/PMMA film is painstakingly sliced along the PDMS block's open window shape using tweezers. The PMMA is then broken down through calcination in air, leaving a suspended graphene layer covering the entire device.

In contrast to the dry transfer technique developed by Zhang et al. [146], Yang et al. [52] introduced another effective method for transferring individual $MoSe_2$ crystals, demonstrated schematically in Figure 4.17g–j. The 2D crystals and their support were spun and coated with a layer of PMMA (thickness between 200 and 300 nm), followed by heating on a hot plate at 180 °C to promote adhesion. $FeCl_3$ was then utilized to etch away the copper substrate, and NaOH solution was used to remove the Si/SiO_2 substrate. This method was especially effective for moving discontinuous crystals like $MoSe_2$. To retrieve the transferred 2D crystals, a copper TEM grid was used to withdraw the floating film from the etchant solution.

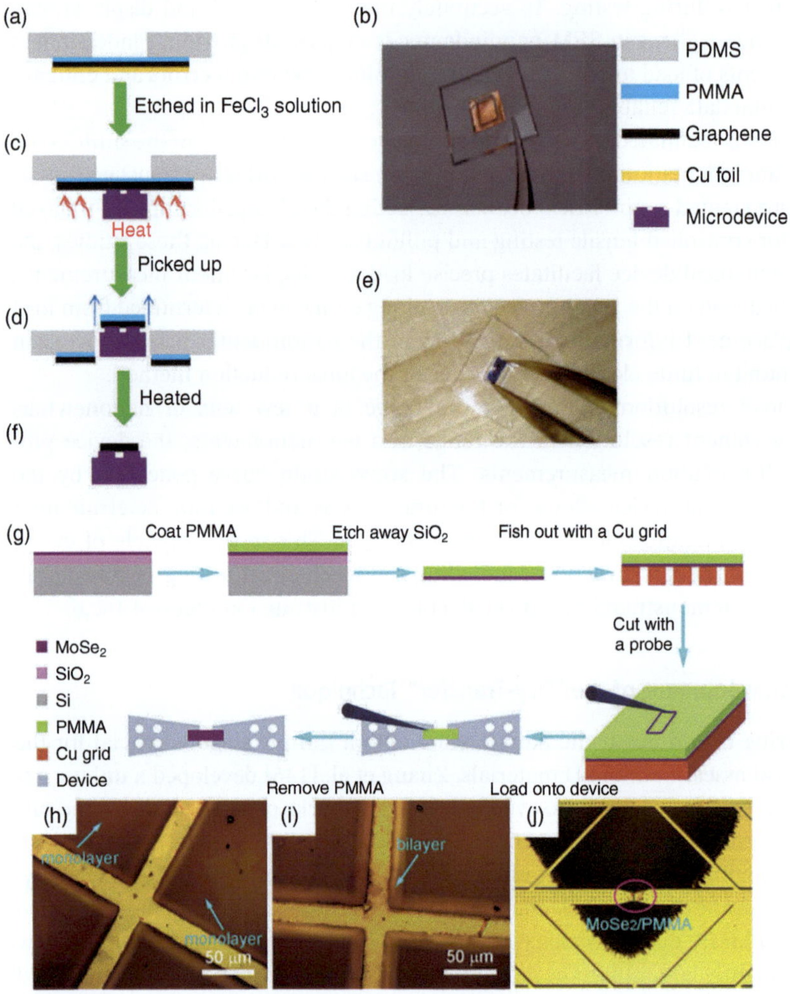

Figure 4.17 Dry transfer of 2D crystals. (a–f) Chronological stages of dry transfer. The two blue arrows in (d) indicate where the device was grasped simultaneously with graphene/PMMA using tweezers. (b,e) indicate the schematic (a,d), respectively. Source: Ref. [146]/with permission of Springer Nature. (g) Schematic sketch of $MoSe_2$ relocation. (h) Monolayer $MoSe_2$/PMMA resting on a copper grid. (i) Bilayer $MoSe_2$/PMMA placed on a copper grid, and (j) $MoSe_2$/PMMA moved to the nanomechanical device. Source: Ref. [52]/with permission of John Wiley & Sons.

The preferred area of PMMA-coated 2D crystals was then precisely sliced using a tungsten micromanipulator probe. The cut 2D material was cautiously put into the test part of the micromechanical device under an optical microscope. Afterward, PMMA was annealed using a gas combination of 90% nitrogen and 10% hydrogen in a CVD tube furnace. In about 40 minutes, the temperature was steadily raised to 400 °C, and it was then kept there for 2 hours with constant gas flow under atmospheric pressure. The mechanical behaviors of 2D crystals with distinct monolayer and bilayer structures, as shown in Figure 4.17h,i, can be compared. As seen in

Figure 4.17j, the efficient "dry-transfer" process makes acquiring suspended monolayer 2D crystals straightforward. This method also facilitates the stacking of different crystals with precise orientations. For instance, prior research has shown that a P–N junction can be created on a piece of *h*-BN by vertically stacking a WSe_2 and a MoS_2 crystal [147].

4.5.3 Mechanical Testing of 2D Materials

4.5.3.1 Mechanical Testing of Graphene

Utilizing the micromechanical apparatus illustrated in Figure 4.18c, Cao et al. [148] conducted *in situ* quantitative tensile testing on freestanding monolayer graphene produced under an SEM. To prepare the graphene samples for testing, robust techniques for sample transfer, shaping, and clamping were employed. The measured Young's modulus of the monolayer graphene was approximately 1 TPa, in good agreement with predictions made by theory and the findings of nanoindentation research [149–152]. Multiple tests revealed that the fracture strength of the graphene samples was ~60 GPa. Although this value is substantially less than the optimal strength of monolayer graphene, estimated to be between 100 and 130 GPa, it still falls within the same order of magnitude. The introduction of edge flaws during the sample-cutting procedure utilizing focused ion beam (FIB) techniques could produce variation in the recorded fracture strengths of graphene samples. These flaws could influence the mechanical characteristics and lower the fracture strength. In contrast to the maximum theoretical strain of graphene, theoretically calculated to be about 35% [153, 154], only about 6% strain was obtained in the experiments. The findings show that freestanding graphene without cracks exhibits

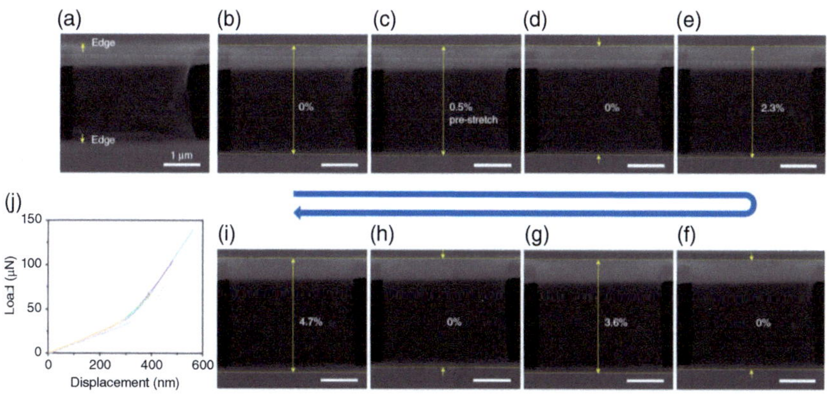

Figure 4.18 Complete restorable elastic straining of suspended monolayer graphene. (a–i) The progression SEM images of the documented video throughout the tensile testing process with the highest indentation displacement of 300 nm (a–c), 400 nm (d,e), 500 nm (f,g), 600 nm (h,i). Initial cycle (b,c) was to pre-stretch the slackened free-standing single-crystalline graphene specimen for observing the initial gauge length. The scale bar within the entire micrograph is 1 µm. (j) The subsequent indenter load–displacement curves for cyclic elastic straining of the freestanding graphene (with unloading segment distinguished with dash line). Source: Ref. [148]/Springer Nature/CC BY 4.0.

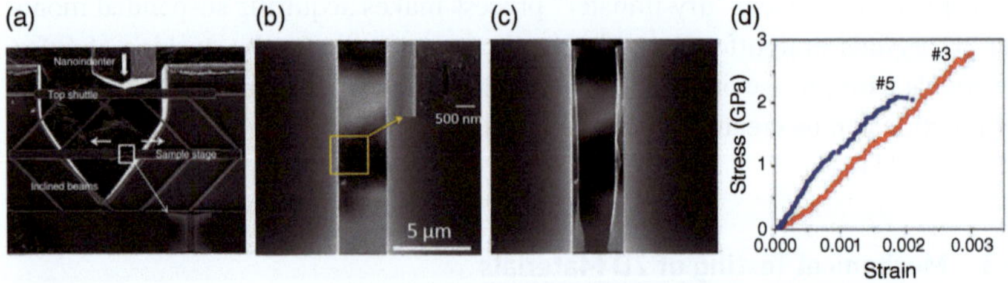

Figure 4.19 Fracture of graphene showcasing a pre-crack. (a) The SEM micrograph demonstrates the *in situ* tensile testing through a microdevice. (b,c) SEM micrograph exhibiting graphene placed on the sample platform before and after tensile analysis. FIB cut the pre-crack (boxed in (b)). (d) Sample of engineering stress–strain curves representing cracked graphene specimen. Source: Ref. [146]/with permission of Springer Nature.

superior elastic stretchability and tensile strength in contrast to earlier work on tensile testing of pre-cracked graphene [146].

Owing to the nanoscale nature of the required testing and the specialized design specifications of the experimental equipment, determining the fracture toughness of monolayer or multilayer graphene poses significant challenges. In the *in situ* tensile testing of suspended graphene conducted by Suk et al. [127], a micromechanical device was integrated into an SEM to investigate its fracture behavior. In its pristine state, perfect graphene exhibits high intrinsic strength as atomic connections are uniformly broken. However, graphene with defects such as single vacancies (SV), stone wales (SW), and slits may exhibit reduced elasticity and weaker fractures. The fracture toughness, which assesses a material's resistance to crack propagation and overall toughness under stress, is a crucial factor in determining the strength of large-area graphene with engineering applications.

Figure 4.19 illustrates nanomechanical testing of graphene fracture with an existing pre-crack. The pre-cracked graphene sample undergoes brittle fracture, displaying sharp edges under tensile loading, occurring at a breaking stress much lower than graphene's intrinsic strength. This brittle fracture characteristic aligns with findings from previous studies [155]. Through a combination of experimental observations and modeling, it has been verified that the classic Griffith theory of brittle fracture is applicable to graphene. The fracture toughness of graphene is quantified as the critical stress intensity, measured to be $4.0 \pm 0.6\,\mathrm{MPa\,m^{1/2}}$.

The obtained fracture toughness of graphene through the tensile testing of notched graphene using the micromechanical device holds significant implications for the strength characterization of large-area graphene. It suggests that, when assessing the mechanical properties of graphene on a larger scale, fracture toughness, rather than intrinsic strength, should be considered. This finding underscores the importance of fracture toughness in determining the overall strength of graphene, encourages further investigations into the fracture toughness of other 2D materials, and addresses the limitation of nanoindentation in measuring the fracture toughness of 2D materials, providing a new avenue for studying the mechanics of these materials.

4.5.3.2 Mechanical Testing of Rebar Graphene

Graphene, when devoid of defects, is recognized as one of the most rigid materials; however, its susceptibility to rupture is less remarkable [146]. The commonly perceived brittle nature of graphene's fracture process, attributable to its 2D structure, is believed to be associated with moderate fracture toughness [52]. Evaluating the mechanical properties of 2D materials is particularly challenging due to their inherent brittleness. The introduction of CNTs into graphene has led to the creation of rebar graphene, which is expected to exhibit enhanced toughness and strength compared to pristine graphene [156]. Hacopian et al. [157] successfully generated freely suspended rebar graphene samples on micromechanical devices, enabling *in situ* tensile testing within an SEM. As illustrated in Figure 4.20, the results unveiled a distinctive zigzag crack propagation pattern in the fracture surface of rebar graphene. This crack behavior is attributed to the presence of CNTs, which redirect crack motion and guide its propagation in a zigzag fashion (Figure 4.20c). This innovative integration of CNTs into graphene offers a potential strategy to augment the fracture toughness and mechanical properties of 2D materials.

In contrast to graphene's direct and brittle fracture behavior, rebar graphene demonstrates a longer fracture path, indicating higher energy absorption during mechanical failure. Experimental results on un-notched rebar graphene samples revealed an average elastic modulus of 466.8 ± 29.95 GPa and a fracture strength of 6.02 ± 1.53 GPa. The stress intensity factor of rebar graphene, measured at 10.50 ± 5.73 MPa/m^2, surpassed that of graphene [146]. The embedded CNTs in rebar graphene guide and redirect the fracture surface, resulting in a distinct zigzag pattern. These observations, supported by theoretical predictions, underscore the improved fracture resistance of rebar graphene compared to pristine graphene.

4.5.3.3 Mechanical Testing of $MoSe_2$

Yang et al. [52] employed a micromechanical device to transfer a monolayer of $MoSe_2$ and conducted real-time observations of its mechanical properties using an SEM. The SEM snapshots depicted in Figure 4.21 capture the progressive tensile deformation of the $MoSe_2$ sample until it undergoes catastrophic failure. The corresponding stress–strain curve, illustrated in Figure 4.21f, offers insights into

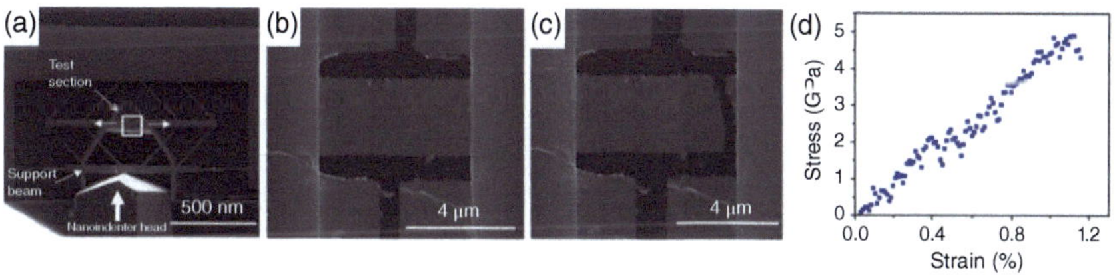

Figure 4.20 *In situ* tensile testing of a rebar graphene specimen. (a) SEM micrograph of experimenting device geometry. (b) SEM micrograph of film in the test region of the device before fracture. (c) SEM micrograph of film in test zone after fracture to reveal zigzag fracture propagation surface (d) Stress–strain curve of specimen described in two micrographs earlier mentioned. Source: Ref. [157]/with permission of American Chemical Society.

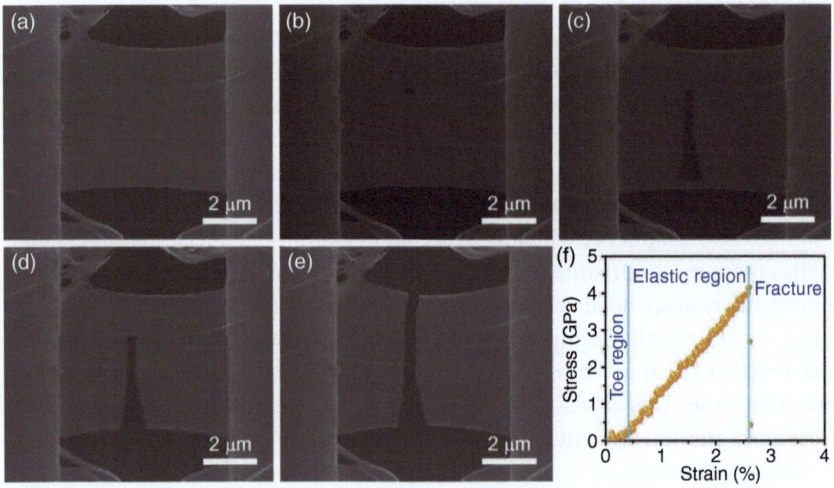

Figure 4.21 *In situ* tensile experimentation of a bilayer $MoSe_2$ sample. (a) The SEM micrograph of $MoSe_2$ before tensile examination. (b–e) SEM micrographs displaying the fracture procedure. (f) Determined stress–strain curve. Source: Ref. [52]/with permission of John Wiley & Sons.

the material's behavior. While the measured average elastic modulus and fracture strength of $MoSe_2$ are 177.2 ± 9.3 and 4.8 ± 2.9 GPa [52], it is noteworthy that the experimental elastic modulus aligns well with the theoretically calculated value of 162.1 GPa. This theoretical value was obtained from DFT calculations [158].

In contrast, the measured fracture strengths of $MoSe_2$ exhibit significant variability, ranging from 2.2 to 9.9 GPa, with a notable standard deviation of 2.9 GPa. This scattering of fracture strengths was attributed to variations in the size of preexisting flaws or cracks within the material. To gain insight into the fracture behavior of $MoSe_2$, the Griffith theory of brittle fracture, which combines *in situ* fracture testing and atomistic modeling, was employed [52]. Accordingly, the application of the Griffith theory enables the estimation of the size of a dominant fracture-producing crack or flaw in 2D $MoSe_2$ [159]. Specifically, for a central crack of length $2a_0$, the Griffith theory of brittle fracture can be utilized to analyze the material's fracture characteristics as follows,

$$a_0 = \frac{2\gamma E}{\pi \sigma_c^2} \tag{4.9}$$

where E stands for elastic modulus and is the surface energy, which may be calculated by dividing the edge energy of a 2D crystal by its thickness, the fracture-producing crack's length was then calculated for each sample examined. With an average value of 33.03 nm, the crack lengths range from 3.6 to 77.5 nm [52]. The abundance of unseen flaws causes the variation in fracture strength.

4.5.3.4 Mechanical Testing of *h*-BN

In a separate mechanical assessment of 2D material, *h*-BN underwent investigation by Yang et al. [160], focusing on intrinsic toughening and the steady propagation

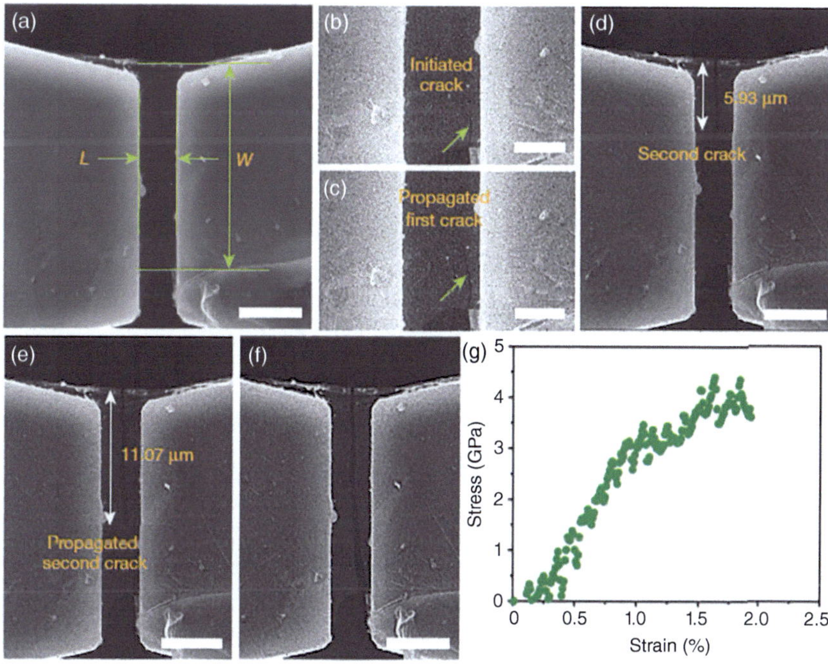

Figure 4.22 *In situ* tensile testing of a monolayer *h*-BN sample independent of pre-crack. (a) Suspended *h*-BN. (b) Crack opening indicated with a pointing arrow in *h*-BN following loading. (c) The first crack developed, indicated by the pointing arrow in *h*-BN. (d), Second crack formation with a 5.93 μm length in *h*-BN. (e) The second crack following propagation to a length of 11.07 μm in *h*-BN. (f) Final failure through fracture. (g) Stress–strain curve of *h*-BN. Source: Yang et al. [160]/with permisison of Springer Nature.

behavior of cracks utilizing a micromechanical "push-pull" device in an SEM. The SEM image depicts the tensile test of a single-layer *h*-BN sample, independent of pre-crack, as shown in Figure 4.22a–f. This illustration delineates the dimensions of *h*-BN, the formation of cracks under stress, crack growth, fracture occurrences, and the corresponding stress–strain plot in Figure 4.22g. In contrast to the catastrophic fracture behavior observed in graphene, this method facilitates a comprehensive understanding of the fracture behavior of CVD-grown single-layer 2D *h*-BN, which exhibits stable crack propagation in contrast to the unstable crack propagation observed in other methodologies.

The initial crack formed at a length of 1.43 μm, subsequently increasing to 2.86 μm with the escalating load. Furthermore, the *h*-BN morphology, as displayed in Figure 4.22a, recorded a high elastic modulus of 413.3 GPa and a fracture toughness of 4.4 GPa, albeit lower than the theoretically predicted values in the literature [43]. Nevertheless, the enhanced mechanical performance, when compared to single-crystal multilayer *h*-BN, is associated with unique toughening characteristics exhibited by the material. The asymmetric edge elastic characteristic displayed at the crack head, along with edge substitution during crack growth, contributes to the intrinsic toughening performance, leading to the formation of steady crack growth in monolayer *h*-BN. Consequently, this intrinsic toughening mechanism results in crack resistance through repeated crack branching and deflection.

4.6 Piezoelectric Tube-Driven Testing in TEM

It is crucial to control the deflection within a small range by applying a higher spring constant AFM cantilever to improve accuracy in AFM-enabled indentation testing. If the deflection is too large, the load–deflection relationship may not be linear, introducing errors in the fracture strength and elastic modulus calculations. Alternatively, a piezoelectric tube-driven manipulator can precisely monitor the load. When a voltage is supplied, the piezoelectric crystal will either expand or contract.

In situ tensile testing was conducted by Wei et al., employing a high-resolution transmission electron microscope (HR-TEM) with a side-entry AFM–TEM holder. This system integrates a force-measuring device and an exceptionally precise piezoelectric tube-driven nanomanipulation system. The experimentation focused on boronitrene and graphene nanosheets, and the setup of the experimental apparatus, as well as TEM images depicting perfect, notched, and fractured graphene nanosheets, are detailed in Figure 4.23.

Fracture toughness values of 12.03 and 5.50 $MPa/m^{1/2}$ were determined for boronitrenes with blunt notches and multilayer graphene, respectively. TEM has been extensively employed for examining the fracture surfaces of 2D materials and assessing their fracture toughness. For example, a study by Wang et al. [34] illustrated that sparse vacancy defects in 2D materials induce crack deflections, while an escalation in defect density leads to a transition from brittle to ductile

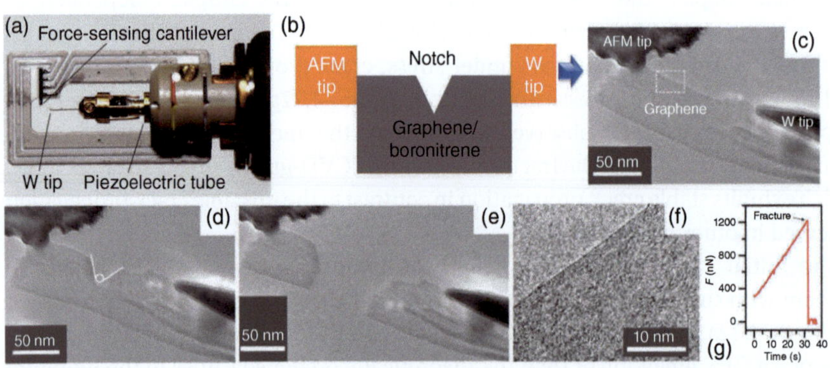

Figure 4.23 Fracture of graphene nanosheets via piezoelectric tube-driven manipulator. (a) Picture of the front view of the AFM–TEM frame [161]. (b) Schematic sketching of fracture toughness experimentation of multilayer graphene and boronitrene nanosheets. (c–f) TEM micrographs of perfect, notched, and fractured graphene nanosheets. (g) Force curve noted in the nanosheet fracture experimentation. Source: Ref. [162]/with permission of American Chemical Society.

fracture behavior through vacancy migration. Similarly, Ly et al. [163] investigated the emission of high-frequency dislocations in MoS_2 during fracture, challenging conventional notions of brittle fracture in this material. Strain analysis further elucidated the close association between dislocation emission and the propagation path of cracks at the nanoscale.

4.7 Bulge Testing

The bulge test, introduced by Beams in 1959 [164], presents several advantages over alternative techniques such as nanoindentation or tensile tests. It effectively eliminates substrate influence and circumvents clamping issues commonly encountered in tensile testing [165]. The mechanical properties of freestanding thin films composed of diverse materials, including metals, semiconductors, and dielectric materials, have been extensively explored through bulge testing [166, 167]. The study of the relationship between applied uniform pressure and the resulting out-of-plane deflection of the bulged film allows for the extraction of valuable mechanical insights.

Various techniques, including optical microscopy with calibrated vertical displacement, laser interferometry, and others, can be employed to measure the out-of-plane deflection of the bulged membrane [168–170]. This deflection, synonymous with the measured bulged membrane's curvature, serves as a means to investigate stress. Consequently, the relationship between pressure, deflection, Young's modulus, and equi-biaxial stress for a pressurized circular membrane can be expressed as follows [165]:

$$P = \frac{2Eh}{(1-\nu)\alpha^4} w_0^3 + \frac{4\sigma_0 h}{\alpha\alpha^2} w_0 \tag{4.10}$$

$$\sigma = \frac{Ew_0^2}{2(1-\nu)a\alpha^2} + \sigma_0 \tag{4.11}$$

where E and ν represent the membrane's Young's modulus and Poisson's ratio, respectively. P stands for pressure and h for sample thickness. The radius of the sample is represented by α, and the maximum deflection is w_0 ($w_0 << \alpha$). Equi-biaxial and residual stress before pressurization are denoted by σ and σ_0, respectively.

4.7.1 Depressurize Inside and Form a Concave Deflection in Film

Recently, Hwangbo et al. [171] utilized the pressured bulging testing apparatus depicted in Figure 4.24 to scrutinize and probe the fracture properties of monolayer CVD-grown graphene. Under room temperature conditions, the monolayer CVD-graphene exhibited environmentally assisted subcritical crack propagation. Once the smallest observable crack appeared, it underwent discontinuous and convoluted expansion, as illustrated in Figure 4.24b. Despite occurring in less than 2 ms, a high-speed camera documented the catastrophic collapse of the CVD-grown

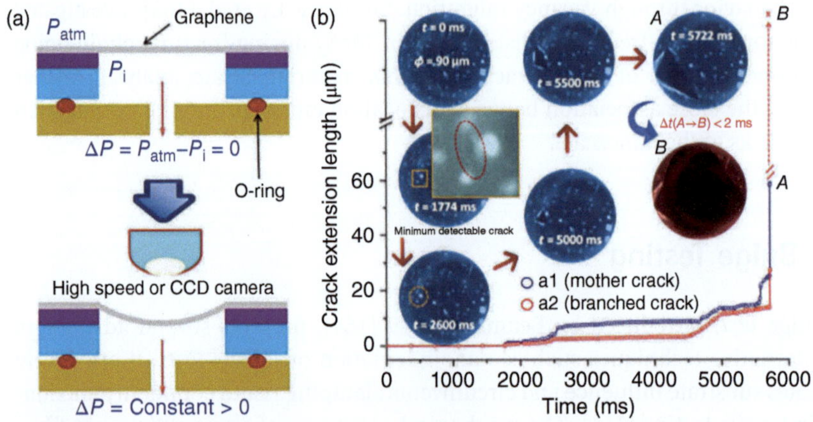

Figure 4.24 Bulge testing using graphene as a case study by depressurizing the inside of the hole. (a) Graphical representation of the bulge test using a freely suspended CVD-grown graphene adhered to SiO_2 substrate as a test sample placed on a cylindrical hole. (b) Crack propagation descriptions of the mother/branched crack with representative images illustrating fracture evolution. Source: Ref. [171]/with permission of Springer Nature.

graphene. Intriguingly, the growth behavior displayed a crack arrest and reinitiation pattern akin to that often observed in 3D bulk materials.

The CVD-grown graphene demonstrated an exceptionally high fracture toughness of $10.7 \pm 3.3\,\text{MPa}\,\text{m}^{1/2}$, significantly surpassing the critical stress intensity of $4.0 \pm 0.6\,\text{MPa}\,\text{m}^{1/2}$ determined through a tensile test utilizing a micromechanical device [146]. Given that the tensile testing was conducted in a high vacuum environment while the pressurized bulge testing took place under ambient conditions, the disparity in fracture toughness may arise from variations in testing conditions.

Indeed, tensile testing confers several advantages for characterizing the fracture toughness of graphene: (i) it provides a comprehensive understanding of the material's mechanical behavior, encompassing deformation, failure, and fracture characteristics; (ii) *in situ* tensile testing conducted in an SEM enables real-time observation of the fracture process, facilitating detailed analysis and characterization. This minimizes experimental and calculation errors, offering valuable insights into crack propagation and failure mechanisms.

Conversely, pressurized bulge testing possesses unique advantages, allowing for the visualization of crack propagation through high-speed cameras. This capability enables the direct observation of crack growth behavior, providing valuable information on crack initiation, propagation, and interaction.

4.7.2 Depressurize Outside and Form a Convex Deflection in Film

Raman spectroscopy has most recently been used to indirectly reveal the mechanical behaviors of 2D crystals. In order to study the intrinsic features of graphene, one can use Raman spectroscopy, which is often used to observe vibrational, rotational, and

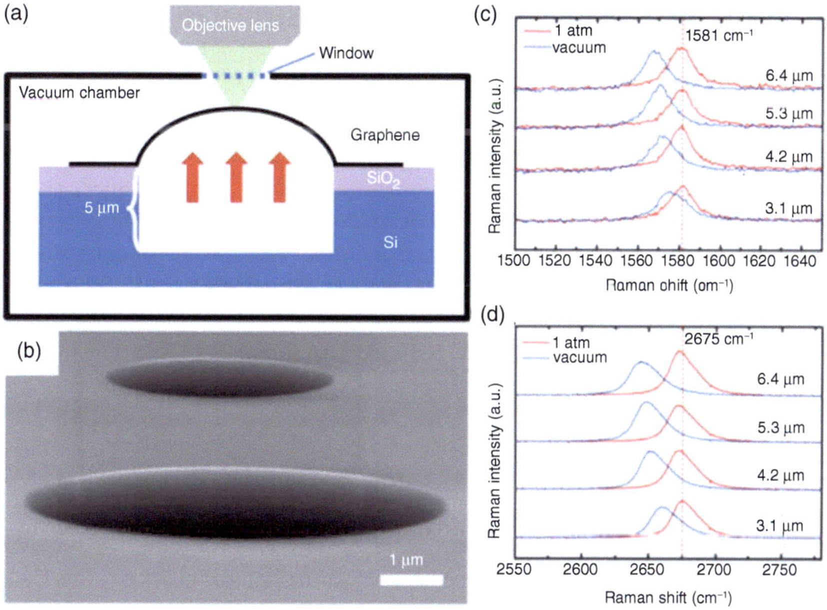

Figure 4.25 Bulge testing of graphene via depressurizing outside of the hole. (a) Schematic illustration of the experimental arrangement. (b) SEM micrograph of a graphene membrane placed in a vacuum, captured at an oblique angle. Two bubbles with varied diameters (3.6 and 6.6 µm), produced via pressure change between the inside and outside of the hole, are illustrated. Relationship of the Raman (c) G band and (d) 2D band spectra, evaluated from a single-layer sample in a vacuum and at room pressure at the midpoint of the holes with 3.1, 4.2, 5.3, and 6.4 µm diameters. Source: Ref. [129]/with permission of American Chemical Society.

other low-frequency modes in a system. This is illustrated in Figure 4.25. These spectral changes provide information about the mechanical deformations experienced by graphene. Experimental evidence of red shift and splitting of the Raman G and 2D bands under uniaxial strain in graphene is available in the literature [172–174]. By utilizing Raman spectroscopy to measure the strain caused by pressurized graphene balloons and comparing the strain with a numerical calculation based on the finite element method, it is possible to determine Young's modulus of graphene layers. The experimental setup for determining Young's modulus of graphene using Raman spectroscopy is schematically depicted in Figure 4.25. The computed Young's modulus of single-layer graphene is 2.40 TPa [129], which is significantly higher than the values that have been published [175, 176]. According to Cheong et al.'s analysis [129], Young's modulus may not be constant across various strains. The calculated Young's modulus would be lower than predicted in the small strain range if graphene exhibits considerable softening at larger strain ranges. The significant uncertainty brought on by trivial Raman peak changes prevents proper quantitative analysis; however, the estimated Young's modulus value is consistently higher than those in the earlier works [22, 75].

4.8 Electrostatic Force Triggered Drum Structure

Frank et al. [174] studied a graphene drum, which was fabricated and electrostatically deflected. An initial scan of the unbiased device was performed to ascertain the drum's initial state, and a second scan of the same region was performed. At the same time, a voltage (V_s) was applied to the back gate to calculate the induced static deflection. The electrostatic force-induced deflection can be calculated by performing profile subtraction. The elastic modulus of deflected graphene can be determined by fitting it to the plate theory. The parallel plate equation was used to determine the electrostatic force F, which was then applied to the model's circular surface as an evenly distributed load. This is expressed as follows:

$$F = \frac{\varepsilon_0 \varepsilon_r A V_S^2}{2(g - \varepsilon d_0)^2} \tag{4.12}$$

where A describes the area of the graphene drum, V_s represents the applied voltage, g describes the initial gap, d_0 represents the peak deflection, $\varepsilon_0 = 8.854 \times 10^{-12}$ F/m represents the permittivity of free space, and ε_r describes the effective relative permittivity. The 2D stress and elastic constant can be determined from the load

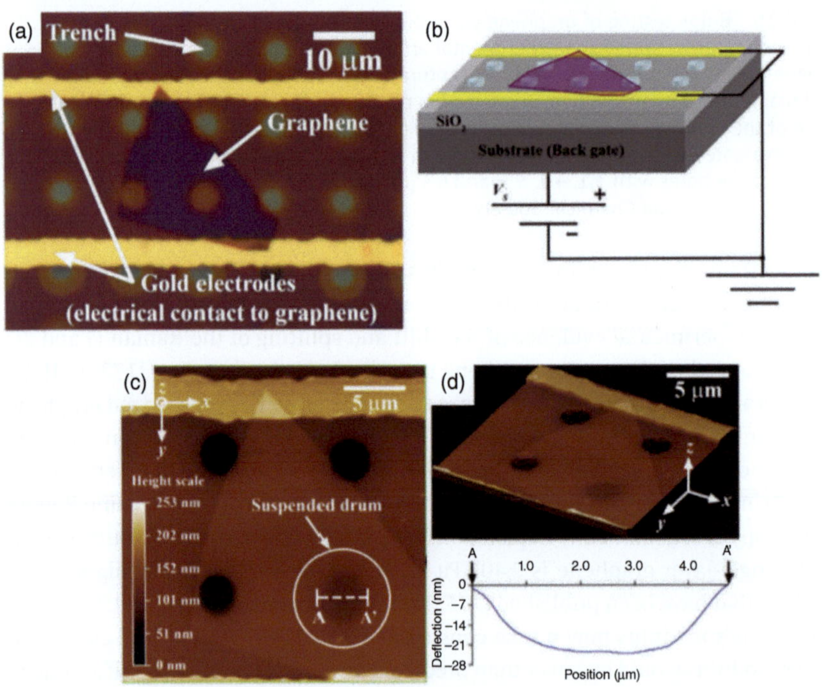

Figure 4.26 Electrostatic force produced drum structure. (a) Optical image of a suspended graphene drum device. (b) Writing of the graphene drum structure for static deflection measurement. A voltage was utilized through the back gate and the graphene. (c) AFM X-ray of the device. (d) A 3D depiction of the scan in (c). (e) A plot of the differences in height at the diameter AA' of the device indicates a preliminary sag of 24 nm. Source: Ref. [175]/IOP Publishing Ltd/CC BY 3.0.

calculation. The optical micrograph of a drum-mounted graphene device is shown in Figure 4.26, and up to 40 layers of graphene are present. Approximately 1 TPa is the calculated Young's modulus. Using finite element modeling, the mechanical behavior of the graphene drum structure was also modeled. It was observed that the computed mechanical characteristics closely matched the findings of the experiment. Likewise, experimental force–deflection data can be used to determine other mechanical characteristics of the drums, such as the linear and nonlinear spring constants. The drum could be actuated to around 18–34% of its thickness before exhibiting nonlinear deflection, and its linear spring constant was discovered to range from 3.24 to 37.4 N/m. Interestingly, it was found that thicker or stiffer devices may exhibit greater resilience to nonlinear behavior. These findings highlight the potential applications of graphene drums as resonators and mass sensors.

4.9 Phonon Dispersion Measurement

Phonon dispersion measurement represents a valuable technique for assessing the mechanical performance of graphene through the application of high-resolution electron energy loss spectroscopy (HREELS). HREELS proves to be an effective tool for investigating surface phonon dispersion, surface structure, and epitaxial growth in materials. Operating in the energy range of 10^{-3} to 1 eV, HREELS is specifically adept at handling minor energy losses. Its key advantages lie in its exceptional surface sensitivity, outstanding resolution in both the energy and momentum domains, and its ability to encompass broad energy and momentum windows.

Phonons in graphene exhibit two main categories: those associated with lattice vibrations within the graphene plane, encompassing transverse and longitudinal acoustic and optical branches, and those linked to out-of-plane lattice vibrations, termed flexural phonons [177]. The acoustic phonons provide insights into the elastic properties of graphene. Furthermore, the determination of the in-plane stiffness and shear modulus of 2D materials can be achieved by analyzing the sound velocities of the transverse and longitudinal acoustic branches.

Politano et al. [177] utilized phonon dispersion measurement to assess Young's modulus and Poisson's ratio of quasi-freestanding graphene samples produced on metal substrates. Based on the scattering angle, HREEL spectra representing monolayer graphene/Pt(111) were revealed in Figure 4.27. The relationship between the graphene's in-plane/branches stiffness (κ) and shear modulus (μ) and the speed of the transverse and longitudinal acoustic branches (v_T and v_L) can be written as:

$$v_L = \left(\frac{\kappa + \mu}{\rho_{2D}}\right)^{1/2} \tag{4.13}$$

$$v_T = \left(\frac{\mu}{\rho_{2D}}\right)^{1/2} \tag{4.14}$$

$$v_L = \frac{\frac{\kappa}{\mu} - 1}{\frac{\kappa}{\mu} + 1} \tag{4.15}$$

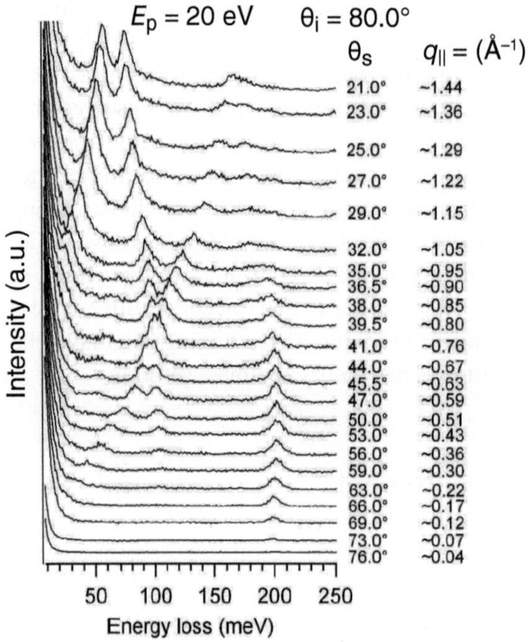

Figure 4.27 HREEL spectra for monolayer graphene/Pt(111) following the scattering angle. The incidence angle is maintained at 80.0° and the impinging energy at 20 eV. Source: Ref. [177]/with permission of Elsevier.

where the 2D mass density ρ_{2D}. Graphene has an in-plane stiffness of 342 N/m and a Poisson's ratio of 0.19. These values match those discovered by Lee et al. using indentation facilitated by an AFM [75, 178]. Thus, phonon dispersion measurement provides a rare chance to conduct analysis for quasi-freestanding and ripple-free graphene sheets.

4.10 Summary

4.10.1 Mechanical Testing Techniques

The study of mechanical characteristics has been addressed through various methods, including AFM-enabled nanoindentation, micro-/nano-mechanical devices, and bulge testing coupled with Raman spectroscopy. AFM-enabled nanoindentation is a widely employed and convenient method for the examination of 2D materials. It entails loading mechanically exfoliated or CVD-grown samples onto holes with diameters up to 9 μm using a wet transfer method. The configuration necessitates no specific substrate and ensures precise boundary conditions. Despite its advantages, it is crucial to acknowledge that local indentation by an AFM tip may result in nonuniform stress and strain distributions within the material, posing challenges in extracting intrinsic mechanical properties. Furthermore, dynamic fracture processes cannot be observed. To address these limitations, *in situ*, indentation in an SEM can complement AFM-enabled nanoindentation experiments by providing precise control of the tip position on the investigated membrane and facilitating real-time observation of fracture processes.

In situ tensile tests conducted using micromechanical devices offer distinct advantages over nanoindentation for evaluating the mechanical properties of 2D materials. Micromechanical devices enable consistent in-plane loading on a freestanding membrane, providing valuable insights into the mechanical properties and fracture mechanisms of 2D materials. Identifying defect categories, distributions, and densities during tensile testing can reveal the impact of defects on mechanical performance. Conducting tensile tests in both SEM and TEM allows for the real-time recording of crack propagation, enhancing our understanding of fracture behavior.

The fracture toughness of suspended 2D membranes can be determined by introducing a notch. Tensile testing of both pristine and notched 2D materials is essential for a comprehensive understanding of their mechanics. While wet transfer techniques are commonly used for nanoindentation tests, dry transfer techniques are necessary for loading 2D materials onto the testing area of micromechanical devices, particularly when liquids are unsuitable for the suspended Si working layer. However, a significant concern lies in the complete removal of the coated PMMA used in the transfer process, as even minor residues can lead to substantial inaccuracies in measured mechanical parameters due to the monolayer or limited number of layers in the tested membrane.

In contrast to nanoindentation and *in situ* tensile tests, bulge testing, electrostatic force-induced drum structures, and phonon dispersion measurement focus on monitoring pressure, voltage, and lattice vibration rather than directly collecting load and displacement data. The mechanical characteristics of tested 2D materials are derived from these parameters. High-speed cameras can be employed to observe crack growth, although quantitative analysis introduces uncertainty, potentially exaggerating Young's modulus.

4.10.2 Mechanical Properties of 2D Materials

Graphene, renowned for its exceptional strength, has garnered considerable attention in the mechanics of materials field. Extensive research has been conducted on the mechanical characteristics of emerging TMDs such as MoS_2 and WSe_2, graphene derivatives including GO and rGO, *h*-BN, and black phosphorus. These materials are obtainable through techniques like mechanical exfoliation and CVD, facilitating the production of samples with variable defect densities, ranging from low to high. Consequently, the impact of defects, such as Stone–Wales defects, vacancies, slits, SV, and grain boundaries, on the mechanical properties of 2D materials has been a focal point of investigation.

Currently, 2D crystals with diverse thicknesses, including monolayer, bilayer, trilayer, and a few nanometers thick, can be obtained through methods such as mechanical exfoliation and liquid dispersion. This allows researchers to explore the relationship between mechanical performance and factors like thickness and interfacial sliding. Thus, elastic modulus, fracture toughness, fracture strength, defect influence on fracture strength, and elasticity have all been extensively discussed in the context of mechanical properties. While the mechanical characteristics and behaviors of graphene and other 2D crystals have been thoroughly examined, the field of 2D material mechanics is still in its nascent stages. To precisely characterize the mechanical properties of 2D materials and gain a deeper understanding of their mechanical behaviors, additional research efforts are warranted.

Ongoing discussions within the mechanics community revolve around mechanical behaviors and qualities, exemplified by the elastic modulus of graphene, experimentally and theoretically ranging from 0.25 to 5.5 TPa. The fracture strength of graphene is reported to range from 55 to 210 GPa. This variability can be attributed to several factors, including differences in experimental techniques, methodologies, sample preparation methods, testing environments, and theoretical approaches used for calculations. Consequently, observed mechanical behaviors may sometimes conflict with mechanical measurements, as evidenced by the significant reduction in the strength of nanocrystalline graphene due to grain boundaries and initial cracks.

The intricate influence of grain boundaries and initial cracks on the mechanical properties of nanocrystalline graphene can be complex and occasionally counterintuitive. Grain boundaries introduce structural discontinuities and defects that can weaken the material, serving as preferential sites for crack initiation and propagation. Additionally, it has been noted that the first crack and grain boundary, regardless of grain size, may be independent of fracture strength. Some fractures follow grain boundaries and traverse through grains. Challenges also arise from the lack of appropriate experimental instruments and sample manipulation options, making it difficult to align experimental findings with theoretical predictions. In contrast to graphene, limited research has been conducted on the mechanical characteristics of MoS_2, *h*-BN, and other 2D crystals, necessitating the development of stronger micro-/nanodevices for comprehensive mechanical characterization. Despite the inherent difficulty in overcoming these challenges, they present opportunities to significantly advance our understanding of 2D materials.

Disclosure Statement

No potential conflict of interest was reported by the authors.

References

1 Kroto, H.W., Allaf, A., and Balm, S. (1991). C60: Buckminsterfullerene. *Chemical Reviews* 91 (6): 1213–1235.

2 Iijima, S. (1991). Helical microtubules of graphitic carbon. *Nature* 354 (6348): 56–58.

3 Li, B., Hahm, M.G., Kim, Y.L. et al. (2011). Highly organized two-and three-dimensional single-walled carbon nanotube–polymer hybrid architectures. *ACS Nano* 5 (6): 4826–4834.

4 Choi, W., Hong, S., Abrahamson, J.T. et al. (2010). Chemically driven carbon-nanotube-guided thermopower waves. *Nature Materials* 9 (5): 423–429.

5 Smith, B.R., Ghosn, E.E.B., Rallapalli, H. et al. (2014). Selective uptake of single-walled carbon nanotubes by circulating monocytes for enhanced tumour delivery. *Nature Nanotechnology* 9 (6): 481–487.

6 Yang, Y., Wang, G., and Li, X. (2011). Water molecule-induced stiffening in ZnO nanobelts. *Nano Letters* 11 (7): 2845–2848.

7 Hao, F., Dong, P., Luo, Q. et al. (2013). Recent advances in alternative cathode materials for iodine-free dye-sensitized solar cells. *Energy & Environmental Science* 6 (7): 2003–2019.

8 Kinloch, I.A., Suhr, J., Lou, J. et al. (2018). Composites with carbon nanotubes and graphene: an outlook. *Science* 362 (6414): 547–553.

9 Wang, Y., Wang, Z., Chen, Y. et al. (2018). Hyperporous sponge interconnected by hierarchical carbon nanotubes as a high-performance potassium-ion battery anode. *Advanced Materials* 30 (32): 1802074.

10 Novoselov, K.S., Geim, A.K., Morozov, S.V. et al. (2004). Electric field effect in atomically thin carbon films. *Science* 306 (5696): 666–669.

11 Deng, D., Novoselov, K.S., Fu, Q. et al. (2016). Catalysis with two-dimensional materials and their heterostructures. *Nature Nanotechnology* 11 (3): 218–230.

12 Low, T., Chaves, A., Caldwell, J.D. et al. (2017). Polaritons in layered two-dimensional materials. *Nature Materials* 16 (2): 182–194.

13 Ahn, G.H., Amani, M., Rasool, H. et al. (2017). Strain-engineered growth of two-dimensional materials. *Nature Communications* 8 (1): 608.

14 Cao, Y., Fatemi, V., Fang, S. et al. (2018). Unconventional superconductivity in magic-angle graphene superlattices. *Nature* 556 (7699): 43–50.

15 Yu, X., Cheng, H., Zhang, M. et al. (2017). Graphene-based smart materials. *Nature Reviews Materials* 2 (9): 1–13.

16 Fujihara, M., Inoue, R., Kurita, R. et al. (2015). Selective formation of zigzag edges in graphene cracks. *ACS Nano* 9 (9): 9027–9033.

17 Zhou, J., Lin, J., Huang, X. et al. (2018). A library of atomically thin metal chalcogenides. *Nature* 556 (7701): 355–359.

18 Browne, M.P., Sofer, Z., and Pumera, M. (2019). Layered and two dimensional metal oxides for electrochemical energy conversion. *Energy & Environmental Science* 12 (1): 41–58.

19 Yang, J., Zeng, Z., Kang, J. et al. (2019). Formation of two-dimensional transition metal oxide nanosheets with nanoparticles as intermediates. *Nature Materials* 18 (9): 970–976.

20 Al-Quraishi, K.K., Qing, H., Wesley, K. et al. (2020). Mechanical testing of two-dimensional materials: a brief review. *International Journal of Smart and Nano Materials* 11 (3): 207–246.

21 Van den Brink, J. (2007). From strength to strength. *Nature Nanotechnology* 2 (4): 199–201.

22 Lee, C., Wei, X., Kysar, J.W. et al. (2008). Measurement of the elastic properties and intrinsic strength of monolayer graphene. *Science* 321 (5887): 385–388.

23 Stankovich, S., Dikin, D.A., Dommett, G.H.B. et al. (2006). Graphene-based composite materials. *Nature* 442 (7100): 282–286.

24 Chawla, N. and Deng, X. (2005). Microstructure and mechanical behavior of porous sintered steels. *Materials Science and Engineering: A* 390 (1–2, 112): 98.

25 Yang, Y., Rigdon, W., Huang, X. et al. (2013). Enhancing graphene reinforcing potential in composites by hydrogen passivation induced dispersion. *Scientific Reports* 3 (1): 2086.

26 Sui, C., Pan, Z., Headrick, R.J. et al. (2018). Aligned-SWCNT film laminated nanocomposites: role of the film on mechanical and electrical properties. *Carbon* 139: 680–687.

27 Huang, Y., Bazarnik, P., Wan, D. et al. (2019). The fabrication of graphene-reinforced Al-based nanocomposites using high-pressure torsion. *Acta Materialia* 164: 499–511.

28 Tanabe, T., Zhao, S., Sato, Y. et al. (2017). Effect of adding Te to layered GaSe crystals to increase the van der Waals bonding force. *Journal of Applied Physics* 122 (16).

29 Li, M., Wan, Y., Tu, L. et al. (2016). The effect of V_{MoS3} point defect on the elastic properties of monolayer MoS_2 with rebo potentials. *Nanoscale Research Letters* 11: 1–7.

30 Wu, J., Hu, Z., Jin, Z. et al. (2016). Spiral growth of $SnSe_2$ crystals by chemical vapor deposition. *Advanced Materials Interfaces* 3 (16): 1600383.

31 Liu, Y., Wu, J., Hackenberg, K.P. et al. (2017). Self-optimizing, highly surface-active layered metal dichalcogenide catalysts for hydrogen evolution. *Nature Energy* 2 (9): 1–7.

32 Yoffe, A. (1990). Electronic properties of low dimensional solids: the physics and chemistry of layer type transition metal dichalcogenides and their intercalate complexes. *Solid State Ionics* 39 (1-2): 1–7.

33 Xu, H., Yi, J., She, X. et al. (2018). 2D Heterostructure comprised of metallic 1T-MoS_2/Monolayer Og-C_3N_4 towards efficient photocatalytic hydrogen evolution. *Applied Catalysis B: Environmental* 220: 379–385.

34 Wang, S., Qin, Z., Jung, G.S. et al. (2016). Atomically sharp crack tips in monolayer MoS_2 and their enhanced toughness by vacancy defects. *ACS Nano* 10 (11): 9831–9839.

35 Zheng, W., Xu, Y., Zheng, L. et al. (2020). MoS_2 Van der Waals p–n junctions enabling highly selective room-temperature NO_2 sensor. *Advanced Functional Materials* 30 (19): 2000435.

36 Wang, X., Tabarraei, A., and Spearot, D.E. (2015). Fracture mechanics of monolayer molybdenum disulfide. *Nanotechnology* 26 (17): 175703.

37 Dolui, K., Pemmaraju, C.D., and Sanvito, S. (2012). Electric field effects on armchair MoS_2 nanoribbons. *ACS Nano* 6 (6): 4823–4834.

38 Lim, Y.R., Song, W., Han, J.K. et al. (2016). Wafer-scale, homogeneous MoS_2 layers on plastic substrates for flexible visible-light photodetectors. *Advanced Materials* 28 (25): 5025–5030.

39 Singh, E., Singh, P., Kim, K.S. et al. (2019). Flexible molybdenum disulfide (MoS_2) atomic layers for wearable electronics and optoelectronics. *ACS Applied Materials & Interfaces* 11 (12): 11061–11105.

40 Kung, Y.C., Hosseini, N., Dumcenco, D. et al. (2019). Air and water-stable n-type doping and encapsulation of flexible MoS_2 devices with SU8. *Advanced Electronic Materials* 5 (1): 1800492.

41 Su, J., Xiao, Y., and Ren, M. (2013). Enhanced thermal conductivity in epoxy nanocomposites with hybrid boron nitride nanotubes and nanosheets. *Physica Status Solidi (a)* 210 (12): 2699–2705.

42 Yi, M., Shen, Z., Zhao, X. et al. (2014). Boron nitride nanosheets as oxygen-atom corrosion protective coatings. *Applied Physics Letters* 104 (14): 143101.

43 Song, L., Ci, L., Lu, H. et al. (2010). Large scale growth and characterization of atomic hexagonal boron nitride layers. *Nano Letters* 10 (8): 3209–3215.

44 Jo, S.H., Kang, D.H., Shim, J. et al. (2016). A high-performance WSe_2/h-BN photodetector using a triphenylphosphine (PPh_3)-based n-doping technique. *Advanced Materials* 28 (24): 4824–4831.

45 Jeong, H., Bang, S., Oh, H.M. et al. (2015). Semiconductor–insulator–semiconductor diode consisting of monolayer MoS_2, h-BN, and GaN heterostructure. *ACS Nano* 9 (10): 10032–10038.

46 Victor, T.W., Easthon, L.M., Ge, M. et al. (2018). X-ray fluorescence nanotomography of single bacteria with a sub-15 nm beam. *Scientific Reports* 8 (1): 13415.

47 Neumaier, D., Pindl, S., and Lemme, M.C. (2019). Integrating graphene into semiconductor fabrication lines. *Nature Materials* 18 (6): 525–529.

48 Akinwande, D., Brennan, C.J., Bunch, J.S. et al. (2017). A review on mechanics and mechanical properties of 2D materials—graphene and beyond. *Extreme Mechanics Letters* 13: 42–77.

49 Zhang, Z., Yang, Y., Penev, E.S. et al. (2017). Elasticity, flexibility, and ideal strength of borophenes. *Advanced Functional Materials* 27 (9): 1605059.

50 Moreno-Moreno, M., Lopez-Polin, G., Castellanos-Gomez, A. et al. (2016). Environmental effects in mechanical properties of few-layer black phosphorus. *2D Materials* 3 (3): 031007.

51 Gludovatz, B., Hohenwarter, A., Thurston, K.V.S. et al. (2016). Exceptional damage-tolerance of a medium-entropy alloy CrCoNi at cryogenic temperatures. *Nature Communications* 7 (1): 10602.

52 Yang, Y., Li, X., Wen, M. et al. (2017). Brittle fracture of 2D $MoSe_2$. *Advanced Materials* 29 (2): 1604201.

53 Huang, L., Li, Y., Wei, Z. et al. (2015). Strain induced piezoelectric effect in black phosphorus and MoS_2 van der Waals heterostructure. *Scientific Reports* 5 (1): 16448.

54 Los, J., Fasolino, A., and Katsnelson, M. (2016). Scaling behavior and strain dependence of in-plane elastic properties of graphene. *Physical Review Letters* 116 (1): 015901.

55 Torres, J., Zhu, Y., Liu, P. et al. (2018). Adhesion energies of 2D graphene and MoS_2 to silicon and metal substrates. *Physica Status Solidi (a)* 215 (1): 1700512.

56 Lin, M.L., Wu, J.B., Liu, X.L. et al. (2018). Probing the shear and layer breathing modes in multilayer graphene by Raman spectroscopy. *Journal of Raman Spectroscopy* 49 (1): 19–30.

57 Bousige, C., Balima, F., Machon, D. et al. (2017). Biaxial strain transfer in supported graphene. *Nano Letters* 17 (1): 21–27.

58 Ren, Y. and Cao, G. (2019). Adhesive boundary effect on free-standing indentation characterization of chemical vapor deposition graphene. *Carbon* 153: 438–446.

59 Fang, L., Liu, D.M., Guo, Y. et al. (2017). Thickness dependent friction on few-layer MoS_2, WS_2, and WSe_2. *Nanotechnology* 28 (24): 245703.

60 Wang, G., Dai, Z., Wang, Y. et al. (2017). Measuring interlayer shear stress in bilayer graphene. *Physical Review Letters* 119 (3): 036101.

61 van Engers, C.D., Cousens, N.E.A., Babenko, V. et al. (2017). Direct measurement of the surface energy of graphene. *Nano Letters* 17 (6): 3815–3821.

62 Lloyd, D., Liu, X., Boddeti, N. et al. (2017). Adhesion, stiffness, and instability in atomically thin MoS_2 bubbles. *Nano Letters* 17 (9): 5329–5334.

63 Zhang, T., Li, X., and Gao, H. (2015). Fracture of graphene: a review. *International Journal of Fracture* 196: 1–31.

64 Megra, Y.T. and Suk, J.W. (2019). Adhesion properties of 2D materials. *Journal of Physics D: Applied Physics* 52 (36): 364002.

65 Li, X., Sun, M., Shan, C. et al. (2018). Mechanical properties of 2D materials studied by in situ microscopy techniques. *Advanced Materials Interfaces* 5 (5): 1701246.

66 Herbert, E.G., Oliver, W.C., De Boer, M.P. et al. (2009). Measuring the elastic modulus and residual stress of freestanding thin films using nanoindentation techniques. *Journal of Materials Research* 24 (9): 2974–2985.

67 Li, P., You, Z., Haugstad, G. et al. (2011). Graphene fixed-end beam arrays based on mechanical exfoliation. *Applied Physics Letters* 98 (25): 253105.

68 Liu, K., Yan, Q., Chen, M. et al. (2014). Elastic properties of chemical-vapor-deposited monolayer MoS_2, WS_2, and their bilayer heterostructures. *Nano Letters* 14 (9): 5097–5103.

69 Bertolazzi, S., Brivio, J., and Kis, A. (2011). Stretching and breaking of ultrathin MoS_2. *ACS Nano* 5 (12): 9703–9709.

70 Wei, X., Mao, L., Soler-Crespo, R.A. et al. (2015). Plasticity and ductility in graphene oxide through a mechanochemically induced damage tolerance mechanism. *Nature Communications* 6 (1): 8029.

71 Ruiz-Vargas, C.S., Zhuang, H.L., Huang, P.Y. et al. (2011). Softened elastic response and unzipping in chemical vapor deposition graphene membranes. *Nano Letters* 11 (6): 2259–2263.

72 Castellanos-Gomez, A., Poot, M., Steele, G.A. et al. (2012). Elastic properties of freely suspended MoS_2 nanosheets. *Advanced Materials* 24 (6): 772–775.

73 Huang, M., Pascal, T.A., Kim, H. et al. (2011). Electronic–mechanical coupling in graphene from in situ nanoindentation experiments and multiscale atomistic simulations. *Nano Letters* 11 (3): 1241–1246.

74 Design Microsystem (2001). *SD Senturia*. Boston: Kluwer Academic Publishers.

75 Lee, G.-H., Cooper, R.C., An, S.J. et al. (2013). High-strength chemical-vapor–deposited graphene and grain boundaries. *Science* 340 (6136): 1073–1076.

76 Lee, J., Wang, Z., He, K. et al. (2013). High frequency MoS_2 nanomechanical resonators. *ACS Nano* 7 (7): 6086–6091.

77 Vella, D. and Davidovitch, B. (2017). Indentation metrology of clamped, ultra-thin elastic sheets. *Soft Matter* 13 (11): 2264–2278.

78 Jennings, R.M., Taylor, J.F., and Farris, R.J. (1995). Determination of residual stress in coatings by a membrane deflection technique. *The Journal of Adhesion* 49 (1–2): 57–74.

79 Castellanos-Gomez, A., Singh, V., van der Zant, H.S.J. et al. (2015). Mechanics of freely suspended ultrathin layered materials. *Annalen der Physik* 527 (1–2): 27–44.

80 Custance, O., Perez, R., and Morita, S. (2009). Atomic force microscopy as a tool for atom manipulation. *Nature Nanotechnology* 4 (12): 803–810.

81 Xu, Z.-H., Yang, Y., Huang, Z. et al. (2011). Elastic modulus of biopolymer matrix in nacre measured using coupled atomic force microscopy bending and inverse finite element techniques. *Materials Science and Engineering: C* 31 (8): 1852–1856.

82 Yang, Y., Xu, Z.H., Pan, Z. et al. (2012). Hydrogen passivation induced dispersion of multi-walled carbon nanotubes. *Advanced Materials* 24 (7): 881–885.

83 Li, X., Gao, H., Murphy, C.J. et al. (2004). Nanoindentation of Cu_2O nanocubes. *Nano Letters* 4 (10): 1903–1907.

84 Gu, Z., Yang, Y., Li, K. et al. (2011). Aligned carbon nanotube-reinforced silicon carbide composites produced by chemical vapor infiltration. *Carbon* 49 (7): 2475–2482.

85 Li, K., Yang, Y., Gu, Z. et al. (2014). Approaching carbon nanotube reinforcing limit in B_4 C matrix composites produced by chemical vapor infiltration. *Advanced Engineering Materials* 16 (2): 161–166.

86 López-Polín, G., Jaafar, M., Guinea, F. et al. (2017). The influence of strain on the elastic constants of graphene. *Carbon* 124: 42–48.

87 López-Polín, G., Gómez-Navarro, C., Parente, V. et al. (2015). Increasing the elastic modulus of graphene by controlled defect creation. *Nature Physics* 11 (1): 26–31.

88 Suk, J.W., Kim, T., and Piner, R.D. (2015). Van der Waals interactions of graphene membranes with a sharp silicon tip. *Journal of the Korean Physical Society* 67: 2003–2006.

89 Tian, Y., Li, Z., Gao, W. et al. (2014). Mechanical properties investigation of monolayer h-BN sheet under in-plane shear displacement using molecular dynamics simulations. *Journal of Applied Physics* 115 (1): 014308.

90 Lee, C., Wei, X., Li, Q. et al. (2009). Elastic and frictional properties of graphene. *Physica Status Solidi (b)* 246 (11–12): 2562–2567.

91 Annamalai, M., Annamalai, M., Mathew, S. et al. (2012). Elastic and nonlinear response of nanomechanical graphene devices. *Journal of Micromechanics and Microengineering* 22 (10): 105024.

92 Huang, R. (2011). Show of adhesive strength. *Nature Nanotechnology* 6 (9): 537–538.

93 Koenig, S.P., Boddeti, N.G., Dunn, M.L. et al. (2011). Ultrastrong adhesion of graphene membranes. *Nature Nanotechnology* 6 (9): 543–546.

94 Gupta, S. and Batra, R. (2010). Elastic properties and frequencies of free vibrations of single-layer graphene sheets. *Journal of Computational and Theoretical Nanoscience* 7 (10): 2151–2164.

95 Traversi, F., Gúzman-Vázquez, F.J., Rizzi, L.G. et al. (2010). Elastic properties of graphene suspended on a polymer substrate by e-beam exposure. *New Journal of Physics* 12 (2): 023034.

96 Bae, S., Kim, H., Lee, Y. et al. (2010). Roll-to-roll production of 30-inch graphene films for transparent electrodes. *Nature Nanotechnology* 5 (8): 574–578.

97 Hashimoto, A., Suenaga, K., Gloter, A. et al. (2004). Direct evidence for atomic defects in graphene layers. *Nature* 430 (7002): 870–873.

98 Huang, P.Y., Ruiz-Vargas, C.S., Van Der Zande, A.M. et al. (2011). Grains and grain boundaries in single-layer graphene atomic patchwork quilts. *Nature* 469 (7330): 389–392.

99 Cao, C., Sun, Y., and Filleter, T. (2014). Characterizing mechanical behavior of atomically thin films: a review. *Journal of Materials Research* 29 (3): 338–347.

100 Grantab, R., Shenoy, V.B., and Ruoff, R.S. (2010). Anomalous strength characteristics of tilt grain boundaries in graphene. *Science* 330 (6006): 946–948.

101 Zhang, T., Li, X., Kadkhodaei, S. et al. (2012). Flaw insensitive fracture in nanocrystalline graphene. *Nano Letters* 12 (9): 4605–4610.

102 Stankovich, S., Dikin, D.A., Piner, R.D. et al. (2007). Synthesis of graphene-based nanosheets via chemical reduction of exfoliated graphite oxide. *Carbon* 45 (7): 1558–1565.

103 Eda, G., Fanchini, G., and Chhowalla, M. (2008). Large-area ultrathin films of reduced graphene oxide as a transparent and flexible electronic material. *Nature Nanotechnology* 3 (5): 270–274.

104 Schniepp, H.C., Li, J.L., McAllister, M.J. et al. (2006). Functionalized single graphene sheets derived from splitting graphite oxide. *The Journal of Physical Chemistry B* 110 (17): 8535–8539.

105 Pandey, D., Reifenberger, R., and Piner, R. (2008). Scanning probe microscopy study of exfoliated oxidized graphene sheets. *Surface Science* 602 (9): 1607–1613.

106 Mkhoyan, K.A., Contryman, A.W., Silcox, J. et al. (2009). Atomic and electronic structure of graphene-oxide. *Nano Letters* 9 (3): 1058–1063.

107 Suk, J.W., Piner, R.D., An, J. et al. (2010). Mechanical properties of monolayer graphene oxide. *ACS Nano* 4 (11): 6557–6564.

108 Hummers, W.S. Jr., and Offeman, R.E. (1958). Preparation of graphitic oxide. *Journal of the American Chemical Society* 80 (6): 1339–1339.

109 Stankovich, S., Piner, R.D., Chen, X. et al. (2006). Stable aqueous dispersions of graphitic nanoplatelets via the reduction of exfoliated graphite oxide in the presence of poly (sodium 4-styrenesulfonate). *Journal of Materials Chemistry* 16 (2): 155–158.

110 Gilje, S., Han, S., Wang, M. et al. (2007). A chemical route to graphene for device applications. *Nano Letters* 7 (11): 3394–3398.

111 Gómez-Navarro, C., Burghard, M., and Kern, K. (2008). Elastic properties of chemically derived single graphene sheets. *Nano Letters* 8 (7): 2045–2049.

112 Chen, J.-R., Odenthal, P.M., Swartz, A.G. et al. (2013). Control of Schottky barriers in single layer MoS_2 transistors with ferromagnetic contacts. *Nano Letters* 13 (7): 3106–3110.

113 Lopez-Sanchez, O., Lembke, D., Kayci, M. et al. (2013). Ultrasensitive photodetectors based on monolayer MoS_2. *Nature Nanotechnology* 8 (7): 497–501.

114 Desai, S.B., Madhvapathy, S.R., Sachid, A.B. et al. (2016). MoS_2 transistors with 1-nanometer gate lengths. *Science* 354 (6308): 99–102.

115 Mak, K.F., Lee, C., Hone, J. et al. (2010). Atomically thin MoS_2: a new direct-gap semiconductor. *Physical Review Letters* 105 (13): 136805.

116 Griffith, A.A. and VI. (1921). The phenomena of rupture and flow in solids. *Philosophical Transactions of the Royal Society of London. Series A, Containing Papers of a Mathematical or Physical Character* 221 (582–593): 163–198.

117 Van Der Zande, A.M., Huang, P.Y., Chenet, D.A. et al. (2013). Grains and grain boundaries in highly crystalline monolayer molybdenum disulphide. *Nature Materials* 12 (6): 554–561.

118 Castellanos-Gomez, A., Poot, M., Steele, G.A. et al. (2012). Mechanical properties of freely suspended semiconducting graphene-like layers based on MoS_2. *Nanoscale Research Letters* 7: 1–4.

119 Zhang, R., Koutsos, V., and Cheung, R. (2016). Elastic properties of suspended multilayer WSe_2. *Applied Physics Letters* 108 (4): 042104.

120 Hernandez, E., Goze, C., Bernier, P. et al. (1998). Elastic properties of C and BxCyNz composite nanotubes. *Physical Review Letters* 80 (20): 4502.

121 Çakır, D., Peeters, F.M., and Sevik, C. (2014). Mechanical and thermal properties of h-MX_2 (M = Cr, Mo, W; X = O, S, Se, Te) monolayers: a comparative study. *Applied Physics Letters* 104 (20): 203110.

122 Duerloo, K.-A.N., Ong, M.T., and Reed, E.J. (2012). Intrinsic piezoelectricity in two-dimensional materials. *The Journal of Physical Chemistry Letters* 3 (19): 2871–2876.

123 Li, C., Bando, Y., Zhi, C. et al. (2009). Thickness-dependent bending modulus of hexagonal boron nitride nanosheets. *Nanotechnology* 20 (38): 385707.

124 Tan, E. and Lim, C. (2004). Physical properties of a single polymeric nanofiber. *Applied Physics Letters* 84 (9): 1603–1605.

125 Wang, J.-Y., Li, Y., Zhan, Z.Y. et al. (2016). Elastic properties of suspended black phosphorus nanosheets. *Applied Physics Letters* 108 (1): 013104.

126 Yang, Y., Kim, N.D., Varshney, V. et al. (2017). In situ mechanical investigation of carbon nanotube–graphene junction in three-dimensional carbon nanostructures. *Nanoscale* 9 (8): 2916–2924.

127 Suk, J.W., Mancevski, V., Hao, Y. et al. (2015). Fracture of polycrystalline graphene membranes by in situ nanoindentation in a scanning electron microscope. *Physica Status Solidi (RRL)–Rapid Research Letters* 9 (10): 564–569.

128 Kim, J.H., Jeong, J.H., Kim, N. et al. (2018). Mechanical properties of two-dimensional materials and their applications. *Journal of Physics D: Applied Physics* 52 (8): 083001.

129 Lee, J.-U., Yoon, D., and Cheong, H. (2012). Estimation of Young's modulus of graphene by Raman spectroscopy. *Nano Letters* 12 (9): 4444–4448.

130 Zhao, Y., Luo, X., Li, H. et al. (2013). Interlayer breathing and shear modes in few-trilayer MoS_2 and WSe_2. *Nano Letters* 13 (3): 1007–1015.

131 Sriv, T., Kim, K. and Cheong, H. Low-frequency Raman spectroscopy of few-layer 2H-SnS_2. *Scientific Reports* 2018; 8: 10194. https://doi.org/10.1038/s41598-018-28569-6.

132 Greiner, M.T., Helander, M.G., Tang, W.M. et al. (2012). Universal energy-level alignment of molecules on metal oxides. *Nature Materials* 11 (1): 76–81.

133 Stenger, I., Schué, L., Boukhicha, M. et al. (2017). Low frequency Raman spectroscopy of few-atomic-layer thick hBN crystals. *2D Materials* 4 (3): 031003.

134 Zhu, Y. and Chang, T.-H. (2015). A review of microelectromechanical systems for nanoscale mechanical characterization. *Journal of Micromechanics and Microengineering* 25 (9): 093001.

135 Pérez-Garza, H.H., Kievit, E.W., Schneider, G.F. et al. (2014). Highly strained graphene samples of varying thickness and comparison of their behaviour. *Nanotechnology* 25 (46): 465708.

136 Li, P., Jiang, C., Xu, S. et al. (2017). In situ nanomechanical characterization of multi-layer MoS_2 membranes: from intraplanar to interplanar fracture. *Nanoscale* 9 (26): 9119–9128.

137 Zhu, Y., Moldovan, N., and Espinosa, H.D. (2005). A microelectromechanical load sensor for in situ electron and x-ray microscopy tensile testing of nanostructures. *Applied Physics Letters* 86 (1): 013506.

138 Pantano, M.F., Bernal, R.A., Pagnotta, L. et al. (2015). Multiphysics design and implementation of a microsystem for displacement-controlled tensile testing of nanomaterials. *Meccanica* 50: 549–560.

139 Yashinski, M. and Muhlstein, C. (2013). The role of deposited layers in the non-linear constitutive behavior of Si nanowires. *Journal of Applied Physics* 114 (19): 193507.

140 Guo, H., Chen, K., Oh, Y. et al. (2011). Mechanics and dynamics of the strain-induced M1–M2 structural phase transition in individual VO_2 nanowires. *Nano Letters* 11 (8): 3207–3213.

141 Cheng, G., Chang, T.H., Qin, Q. et al. (2014). Mechanical properties of silicon carbide nanowires: effect of size-dependent defect density. *Nano Letters* 14 (2): 754–758.

142 Qin, Q., Yin, S., Cheng, G. et al. (2015). Recoverable plasticity in penta-twinned metallic nanowires governed by dislocation nucleation and retraction. *Nature Communications* 6 (1): 5983.

143 Ganesan, Y., Lu, Y., Peng, C. et al. (2010). Development and application of a novel microfabricated device for the in situ tensile testing of 1-D nanomaterials. *Journal of Microelectromechanical Systems* 19 (3): 675–682.

144 Ganesan, Y. and Lou, J. (2009). The mechanical characterization of carbon-nanotube-reinforced polymer-matrix nanocomposites: an unfolding story of interface. *JOM* 61: 32–37.

145 Liao, Z., Sandonas, L.M., Zhang, T. et al. (2017). In-situ stretching patterned graphene nanoribbons in the transmission electron microscope. *Scientific Reports* 7 (1): 211.

146 Zhang, P., Ma, L., Fan, F. et al. (2014). Fracture toughness of graphene. *Nature Communications* 5 (1): 3782.

147 Lu, G., Wu, T., Yang, P. et al. (2017). Synthesis of high-quality graphene and hexagonal boron nitride monolayer In-plane heterostructure on Cu–Ni alloy. *Advanced Science* 4 (9): 1700076.

148 Cao, K., Feng, S., Han, Y. et al. (2020). Elastic straining of free-standing monolayer graphene. *Nature Communications* 11 (1): 284.

149 Kalosakas, G., Lathiotakis, N.N., Galiotis, C. et al. (2013). In-plane force fields and elastic properties of graphene. *Journal of Applied Physics* 113 (13): 134307.

150 Zhang, Y.Y. and Gu, Y. (2013). Mechanical properties of graphene: effects of layer number, temperature and isotope. *Computational Materials Science* 71: 197–200.

151 Wagner, P., Ewels, C.P., Ivanovskaya, V.V. et al. (2011). Ripple edge engineering of graphene nanoribbons. *Physical Review B* 84 (13): 134110.

152 Tsai, J.-L. and Tu, J.-F. (2010). Characterizing mechanical properties of graphite using molecular dynamics simulation. *Materials & Design* 31 (1): 194–199.

153 Reddy, C.D., Rajendran, S., and Liew, K. (2006). Equilibrium configuration and continuum elastic properties of finite sized graphene. *Nanotechnology* 17 (3): 864.

154 Bera, S., Arnold, A., Evers, F. et al. (2010). Elastic properties of graphene flakes: boundary effects and lattice vibrations. *Physical Review B* 82 (19): 195445.

155 Xu, L., Wei, N., and Zheng, Y. (2013). Mechanical properties of highly defective graphene: from brittle rupture to ductile fracture. *Nanotechnology* 24 (50): 505703.

156 Yan, Z., Peng, Z., Casillas, G. et al. (2014). Rebar graphene. *ACS Nano* 8 (5): 5061–5068.

157 Hacopian, E.F., Yang, Y., Ni, B. et al. (2018). Toughening graphene by integrating carbon nanotubes. *ACS Nano* 12 (8): 7901–7910.

158 Mirhosseini, H., Roma, G., Kiss, J. et al. (2014). First-principles investigation of the bulk and low-index surfaces of Mo Se_2. *Physical Review B* 89 (20): 205301.

159 Zhu, T. and Li, J. (2010). Ultra-strength materials. *Progress in Materials Science* 55 (7): 710–757.

160 Yang, Y., Song, Z., Lu, G. et al. (2021). Intrinsic toughening and stable crack propagation in hexagonal boron nitride. *Nature* 594 (7861): 57–61.

161 Lu, D., Chen, H., Wu, J. et al. (2011). Direct measurements of the Young's modulus of a single halloysite nanotube using a transmission electron microscope with a bending stage. *Journal of Nanoscience and Nanotechnology* 11 (9): 7789–7793.

162 Wei, X., Xiao, S., Li, F. et al. (2015). Comparative fracture toughness of multilayer graphenes and boronitrenes. *Nano Letters* 15 (1): 689–694.

163 Ly, T.H., Zhao, J., Cichocka, M.O. et al. (2017). Dynamical observations on the crack tip zone and stress corrosion of two-dimensional MoS_2. *Nature Communications* 8 (1): 14116.

164 Beams, J. (1959). Mechanical properties of thin film of gold and silver. In: *The Structure and Properties of Thin Film.*

165 Xu, D. and Liechti, K. (2010). Bulge testing transparent thin films with moiré deflectometry. *Experimental Mechanics* 50: 217–225.

166 Yu, Z., Xu, H., Chen, H. et al. (2016). Characterization method of thick films using the bulge test technique. *Experimental Mechanics* 56: 881–889.

167 Gruber, P.A., Böhm, J., Onuseit, F. et al. (2008). Size effects on yield strength and strain hardening for ultra-thin Cu films with and without passivation: a study by synchrotron and bulge test techniques. *Acta Materialia* 56 (10): 2318–2335.

168 Tinoco, H.A., Holzer, J., Pikálek, T. et al. (2019). Determination of elastic parameters of Si_3N_4 thin films by means of a numerical approach and bulge tests. *Thin Solid Films* 672: 66–74.

169 Javed, H., Merle, B., Preiß, E. et al. (2016). Mechanical characterization of metallic thin films by bulge and scratch testing. *Surface and Coatings Technology* 289: 69–74.

170 Preiß, E.I., Merle, B., and Göken, M. (2017). Understanding the extremely low fracture toughness of freestanding gold thin films by in-situ bulge testing in an AFM. *Materials Science and Engineering: A* 691: 218–225.

171 Hwangbo, Y., Lee, C.K., Kim, S.M. et al. (2014). Fracture characteristics of monolayer CVD-graphene. *Scientific Reports* 4 (1): 4439.

172 Mohiuddin, T., Lombardo, A., Nair, R.R. et al. (2009). Uniaxial strain in graphene by Raman spectroscopy: G peak splitting, Grüneisen parameters, and sample orientation. *Physical Review B* 79 (20): 205433.

173 Yoon, D., Son, Y.-W., and Cheong, H. (2011). Strain-dependent splitting of the double-resonance Raman scattering band in graphene. *Physical Review Letters* 106 (15): 155502.

174 Frank, O., Mohr, M., Maultzsch, J. et al. (2011). Raman 2D-band splitting in graphene: theory and experiment. *ACS Nano* 5 (3): 2231–2239.

175 Wong, C., Annamalai, M., Wang, Z.Q. et al. (2010). Characterization of nanomechanical graphene drum structures. *Journal of Micromechanics and Microengineering* 20 (11): 115029.

176 Bosak, A., Krisch, M., Mohr, M. et al. (2007). Elasticity of single-crystalline graphite: inelastic x-ray scattering study. *Physical Review B* 75 (15): 153408.

177 Politano, A., Marino, A.R., Campi, D. et al. (2012). Elastic properties of a macroscopic graphene sample from phonon dispersion measurements. *Carbon* 50 (13): 4903–4910.

178 Lee, C.-H., Lee, G.H., Van Der Zande, A.M. et al. (2014). Atomically thin p–n junctions with van der Waals heterointerfaces. *Nature Nanotechnology* 9 (9): 676–681.

5

Mechanics of Flexible and Stretchable Organic Electronics

Abdullah Al Shafe[1] and Brendan T. O'Connor[2]

[1]*North Carolina State University, Department of Materials Science and Engineering and Organic and Carbon Electronic Laboratories (ORaCEL), Engineering Bldg I, 911 Partners Way Room 3002, Raleigh, NC 27606, USA*
[2]*North Carolina State University, Department of Mechanical and Aerospace Engineering and Organic and Carbon Electronic Laboratories (ORaCEL), Engineering Building III, 1840 Entrepreneur Dr, Raleigh, NC 27606, USA*

5.1 Introduction

Since the inception of organic electronic materials, there has been interest in exploiting the potential mechanical advantages of these materials; first in flexible device applications and more recently in stretchable applications [1–5]. Common devices include organic photovoltaic (OPV) cells, organic photodetectors (OPDs), organic light-emitting diodes (OLEDs), organic field effect transistors (OFETs), and organic electrochemical transistors (OECTs). These devices have been targeted for application in foldable displays, deployable power sources, soft robotics, wearable electronics and sensors, electron skin, and implantable electronics among others [6–10]. For successful implementation, the devices are required to maintain functionality under various strains and other mechanical insults. For example, electronics adhered to the skin can require stretchability of 30% or more depending on their placement on the human body [11]. Therefore, a comprehensive understanding of the mechanics of organic electronic devices is needed to create reliable products that can adapt to these diverse applications and environments.

Electrically conducting organic materials rely on conjugated π-electron systems composed of alternating carbon single and double bonds. The pi-orbitals result in electron delocalization and charge transport. Organic conductors can range from metallic to semiconducting and can also be efficient ion conductors resulting in mixed ionic electronic conduction [12]. These conjugated molecules can take the form of small molecules, oligomers, or polymers. To make the molecules soluble in common solvents for solution processing, they often have solubilizing side chains appended to the electronically functional moiety. The design flexibility of organic semiconductors has resulted in a huge library of demonstrated materials with a wide variety of optoelectronic properties. Examples of common small molecules and polymers are provided in Figure 5.1. Historically, molecular design

Mechanics of Flexible and Stretchable Electronics, First Edition. Edited by Yong Zhu and Nanshu Lu.

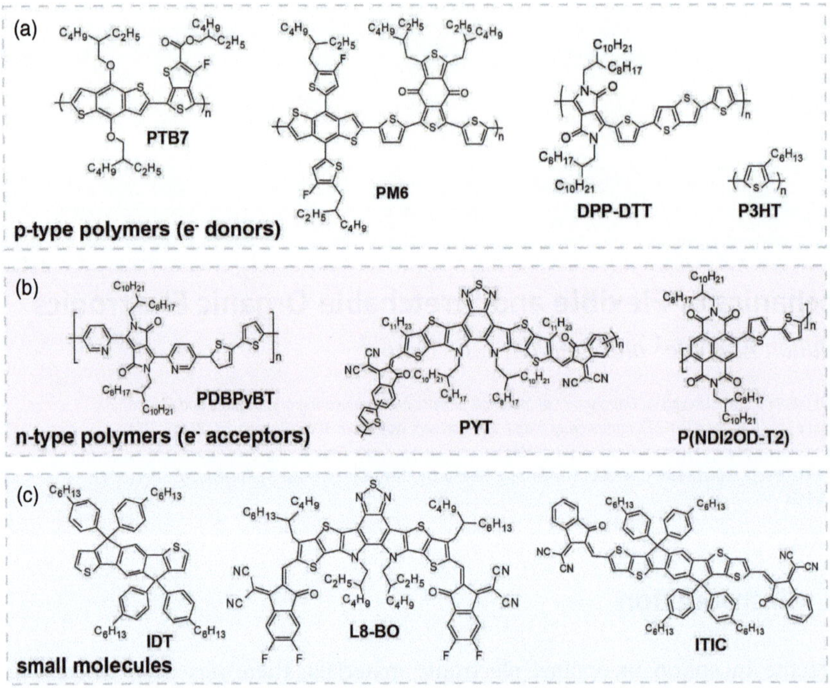

Figure 5.1 Example organic semiconductors used in photovoltaic and transistor applications including (a) p-type polymers, (b) n-type polymers, and (c) small molecules.

considerations have focused on electronic properties, with secondary considerations of processability and molecular packing. However, designing molecules that also target specific mechanical characteristics has been of growing interest [13]. This is a design challenge as optimizing the molecular structure to achieve target optoelectronic and mechanical properties is often in opposition. Beyond the intrinsic properties of the organic electronic materials, the device design of the various layers that compose a complete device must all be optimized to achieve the target mechanical functionality. In this chapter, we consider both the fundamental mechanical properties of organic conductors and device designs. We first outline mechanical properties and related test methods that are considered when probing the mechanical reliability of organic electronics. We then review key molecular features that drive mechanical properties. This is followed by a discussion on device architecture and the role that the various device layers have on mechanical performance. Lastly, applications where flexible and stretchable organic electronic devices have been demonstrated are highlighted.

5.2 Mechanical Characterization Methods

To ensure robust, flexible, and stretchable device operation, there is a need for mechanical characterization that informs operational stability. To gain insight into failure pathways and guide materials development, fundamental mechanical characterization is essential. In this section, we review the mechanical test methods that are commonly applied in organic electronics to assess mechanical properties.

5.2.1 Tensile Tests

From the stress-strain curve of a tensile test, numerous fundamental mechanical properties can be extracted including Young's modulus, modulus of resilience, modulus of toughness, yield stress, tensile strength, and elongation at break, as illustrated in Figure 5.2a. However, organic semiconductor films are typically less than 200 nm thick, making it difficult to handle and test free-standing films. Thicker films can be made but are often cost-prohibitive and may not reflect the mechanical behavior of the thin films due to changes in film morphology [14]. Several test methods have been developed to overcome this limitation. Perhaps the simplest approach has been film on elastomer (FOE) testing. Here, the organic thin film is placed on an elastomer substrate, usually polydimethylsiloxane (PDMS), and the composite is compressed or stretched as illustrated in Figure 5.2b. A number of mechanical properties can be extracted from FOE tests, particularly for brittle films, as reviewed by Stafford et al. [15]. Under compression, the elastic mismatch between the film and substrate results in film wrinkling at a characteristic wavelength that can be used to determine the Young's modulus of the film [16]. The composite can also be strained in tension until film cracking is observed; this is termed crack onset strain (COS). Other properties, including yield strain [17], fracture strength [14], and stress relaxation time constants [18], can also be found using FOE approaches. Lastly, films are often strained on an elastomer and then characterized morphologically to determine microstructural changes that occur with strain [19]. Importantly, the modulus mismatch and interactions between the film and substrate are important considerations in FOE tests. For example, the stiffness mismatch between the film and substrate directly influences the COS [19].

To obtain a more complete view of the stress-strain response of the thin films, a film-on-water (FOW) tensile test technique was introduced by Kim et al. [14]. In this approach, the film of interest is floated off an initial substrate onto water. The film is then gently adhered to the grips of a tensile tester, with a test setup illustrated in Figure 5.2c. Water is chosen to support the film due to its high surface tension and lack of solubility with the organic materials. The FOW is also called pseudo-free-standing tensile test technique because it provides a nearly frictionless environment to the film under test [14]. Recently, the approach has been extended to ionic liquids to allow for temperature-dependent measurements [20]. The desire for a purely free-standing film tensile test remains since water or other liquids could interact with the film in the form of capillary forces, or act as a plasticizer [21]. This has led to the development of tensile tests of free-standing films, or film on air (FOA) tensile tests [21–24]. While free-standing films minimize the influence of neighboring layers, handling the ultrathin films is a challenge and FOW tests continue to be a popular method.

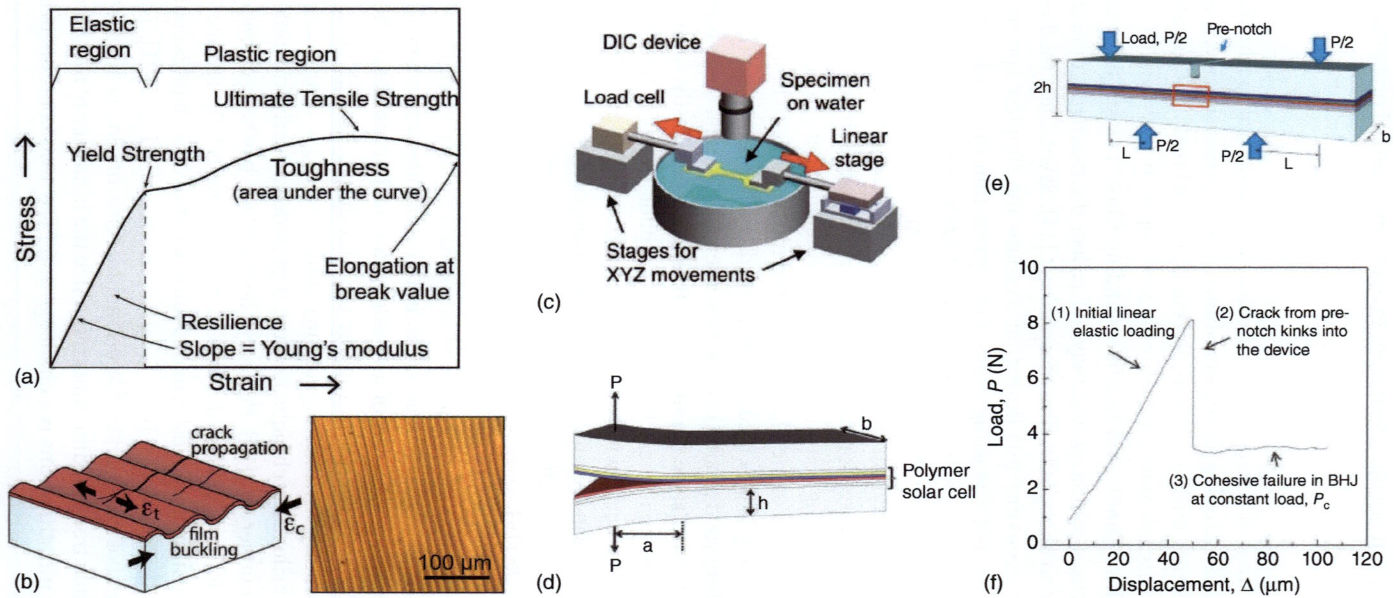

Figure 5.2 (a) Typical stress-strain curve. Source: Adapted from Ref. [25]. (b) Schematic representation of the film on elastomer (PDMS) substrate illustrating the wrinkling instability. On the right is an optical microscope image of wrinkling in P3HT films on PDMS under compression. Source: Ref. [25]/with permission of AIP Publishing. (c) Typical stress-strain measurement setup via film-on-water (FOW) tensile test method. Source: Ref. [14]/Springer Nature. (d) Double cantilever beam (DCB) test specimen. Source: Ref. [26]/with permission of Elsevier. (e) A typical setup of notched four-point bending specimen and (f) associated load displacement curve for cohesive debonding in the P3HT:PCBM bulk heterojunction film. Source: Ref. [27]/with permission of Elsevier.

5.2.2 Fracture Toughness

Primary mechanical failure modes of flexible and stretchable devices are through cohesive fracture of a given layer in the stack or through adhesive failure between functional layers. Capturing the adhesive or cohesive fracture toughness provides a quantitative view of the propensity of the device to fail under certain loading conditions. This includes subcritical fatigue loading that leads to crack propagation [28]. The most common techniques to measure the fracture energy are the double cantilever beam (DCB) test and four-point bend (FPB) test, as illustrated in Figure 5.2d,e [26, 29]. The DCB test relies on a load displacement curve with a known crack length. Measurement of crack length can be difficult, leading to alternative methods being sought. Dauskardt et al. pioneered the FPB technique where the film or interface of interest is sandwiched between two solid elastic substrates and placed between a set of four pins [28]. From the load displacement (Figure 5.2f), the macroscopic fracture work is measured in terms of the release rate of the critical strain energy. In FPB, tests have a couple of advantages over the similar three-point bend test including a higher volume under load and a load that is more evenly distributed [30]. Peel tests are an alternative approach to assess failure with a focus on interfacial toughness. In a peel test, a layer of material is peeled away from the substrate at a controlled angle and speed, while the force required to separate the layers is recorded [32]. A peel test can quantitatively assess the adhesion strength between layers within a device stack. This test helps to identify potential delamination and separation failures, pinpointing the weakest interface among the layers. Finally, facture toughness can be probed through scratch tests [31]. Scratch resistance is important for stable operation, however, these tests are often qualitative in nature.

5.2.3 Thermomechanical Behavior

Thermal phase behavior provides insights into the origin of the mechanical response of the materials. For example, the stiffness and ductility of a polymer depend on the ability of the molecular conformation to reorient under an applied load. The thermal transitions (e.g. glass transition) and related polymer dynamics dictate the ability of the polymer to undergo conformational changes under an applied load. There are a number of tools to measure thermal transitions including differential scanning calorimetry (DSC), dielectric relaxation spectroscopy (DRS), spectroscopic ellipsometry, and dynamic mechanical analysis (DMA) [148, 149]. DMA provides one of the most sensitive tools to measure thermal transitions, and it directly measures the storage and loss modulus of the material. The process involves applying a periodic stress (strain) to a sample while simultaneously measuring its strain (stress) response, as illustrated in Figure 5.3a. DMA has several testing arrangements that typically require relatively large sample specimens, which can be difficult with organic electronic materials as discussed above. To overcome this issue, Andersson and coworkers introduced the approach of

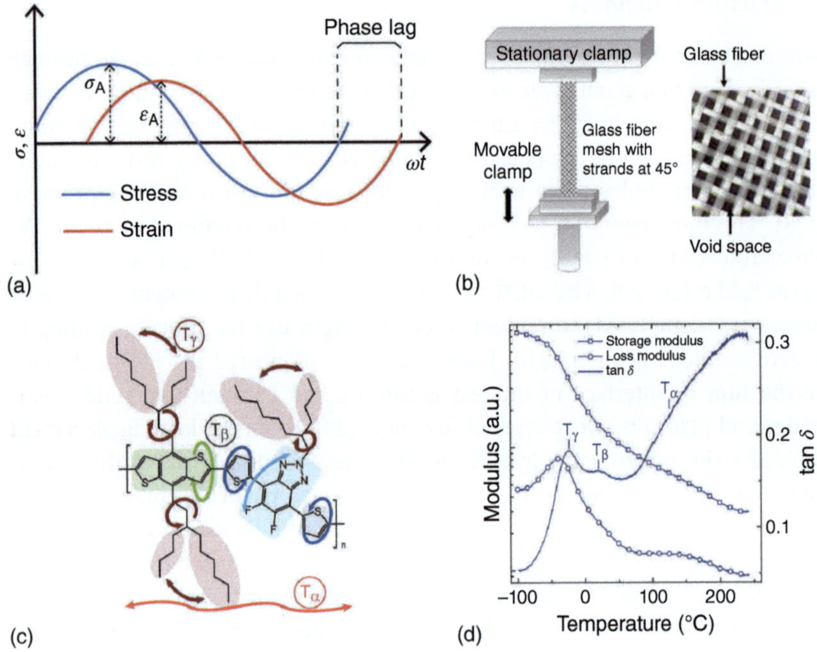

Figure 5.3 (a) Stress-strain application and response in dynamic mechanical analysis. (b) Schematic showing the mounted organic material sample cast on glass mesh, along with an image of the glass mesh support. Source: Ref. [33]/with permission of American Chemical Society. (c) Possible thermal relaxation events of the polymer PBnDT-FTAZ, along with (d) DMA thermogram of PBnDT-FTAZ with thermal transitions labeled. Source: Ref. [35]/John Wiley & Sons.

impregnating a glass mesh with the material of interest [33]. The glass fiber mesh is oriented such that the DMA load is efficiently transferred to the material of interest, as pictured in Figure 5.3b. Alternatively, the material can be placed on a thin substrate support such as polyimide (PI) [34]. However, the stiff PI will diminish the observable behavior of the polymer of interest. To reduce the impact of the substrate, Siddika et al. introduced the strategy of transferring the film of interest onto a cut-patterned PI substrate [44]. The cuts reduce the stiffness of the PI and effectively transfer the load to the polymer of interest, similar to the glass mesh approach. The unique advantage of this approach is that it is capable of measuring the behavior of films processed the same way as those used in devices.

DMA of conjugated polymers often shows several relaxation events, with an example given in Figure 5.3c,d. In many donor–acceptor copolymers, the most prominent relaxation observed has been associated with a side-chain relaxation [35, 36]. There are also additional, less prominent relaxation features, and identifying the nature of these transitions is dependent on the specific polymer [33, 37]. One practical outcome of capturing the thermal relaxation character is that it has been shown that in conjugated polymers, the COS of the film decreases significantly when the temperature of the sample is brought below the side-chain relaxation temperature [35, 37]. Furthermore, small molecule semiconductors added to the polymer, commonly employed in bulk heterojunction solar cells, can suppress the sub-T_g feature of the polymer, and the degree of suppression correlates with the film's elongation at break and modulus of toughness [38]. These observations are consistent with other reports that have highlighted the role of side chains on the ductility of semiconducting polymers [39].

5.3 Material Design

There have been continual advancements in materials tuned specifically for flexible and stretchable applications. The number of demonstrations is growing, and we do not provide a comprehensive summary of all reports. Rather, we summarize a number of common approaches researchers have introduced to address both electronic and mechanical properties, as summarized in Figure 5.4.

5.3.1 Molecular Weight

Small molecule films tend to be brittle and mechanically weak [39]. By moving to polymers, a number of mechanical properties change, including increased ductility and toughness. The large change in mechanical behavior can be brought on solely by the long molecular structure that increases intermolecular interactions and chain entanglements [40]. Thus, in flexible and stretchable organic electronics, significant attention has been given to molecular weight (MW) [40, 41]. It has been shown that when the polymer semiconductors surpass their entanglement MW, there is a large increase in ductility and fracture toughness of the film [16, 40]. According to Bruner and Dauskardt, even a small increase in molecular weight can significantly enhance the cohesion of the active layer film [29]. Figure 5.5a shows the increase in fracture strain and cohesive fracture energy of P(NDI2OD-T2) thin films with increasing MW. Figure 5.5b illustrates the increase in plastic zone at the crack tip with MW and shows the increase of P(NDI2OD-T2) fracture energy and photovoltaic efficiency

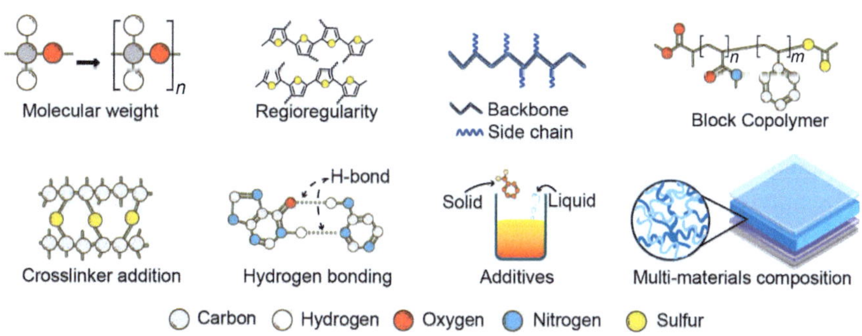

Figure 5.4 Common material design approaches to manage the electrical and mechanical properties of organic semiconductors.

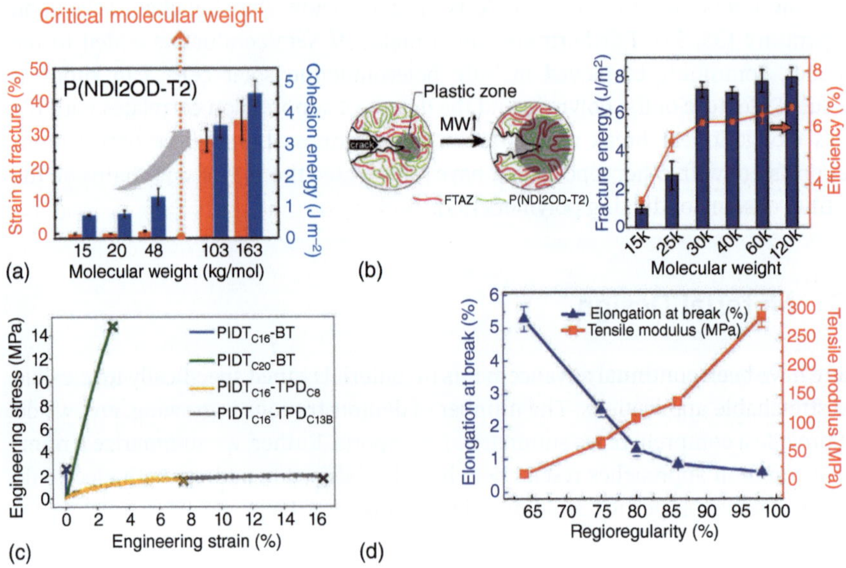

Figure 5.5 (a) Increase in mechanical properties of P(NDI2OD-T2) films at different molecular weight. Source: Ref. [43]/with permission of American Chemical Society. (b) Schematic showing the increase in plastic zone with increasing molecular weight and its impact on the fracture energy and photovoltaic efficiency. Source: Ref. [40]/American Chemical Society.(c) Stress-strain curves of different indacenodithiophene-based copolymers with modified side chains. Source: Ref. [45]/American Chemical Society. (d) Tensile modulus and elongation at break value of P3HT films at different regioregularity. Source: Ref. [46]/American Chemical Society.

with increasing MW [40]. In another work, Zhao et al. demonstrated that when increasing the molecular weight of the polymer IDT-BT from 25 to 1050 kg/mol, the COS increased from less than 10% to over 100% [42]. OFETs based on these ultra-high molecular weight polymers also yielded very high charge mobility. The authors attributed this enhancement to the torsion-free nature of the rigid chains leading to long-range intrachain charge transport of the near amorphous polymers.

5.3.2 Backbone and Side-Chain Design

Both the backbone and side-chain structure are integral to mechanical and electrical properties. First considering side-chain design, research has shown that the stiffness and ductility of the polymer films can be dependent on the type, length, symmetry, and volume of the side chains [47]. The alkyl side-chain length on polythiophene stretchability was considered by Lipomi and coworkers [48]. Their findings revealed that increasing the side-chain length lowered the T_g of the polymer and increased ductility. However, this came at the cost of reduced OPV performance. They found that P3AT with heptyl side chains gave the optimal combination of OPV performance and ductility. Lin et al. investigated the impact of three different asymmetric side chains on isoindigo-bithiophene (PII2T) and isoindigo-difluorobithiophene (PII2TF) polymers on microstructure and stretchability [150]. The three side chains were carbosilane (Si-C8), decyltetradecane (SiO-C8), and decyltetradecane (DT). They found optimum stretchability and mobility in OFETS with asymmetric (Si-C8)/(DT) side chains over the SiO-C8/DT side chains of the backbone fluorinated polymers. The optimal material retained 90% of its field effect charge mobility after 400 stretching cycles to 60% strain, compared to retaining 36% of its nominal charge mobility in the nonfluorinated counterpart. They have ascribed the difference in mechanical response to the mismatched carbon spacer in the side chains that introduces greater chain torsion and disordered chain packing. Recently, Qiu group used diketopyrrolopyrrole-thieno[3,2-*b*]thiophene (DPP-TT) polymer and changed the alkyl length of the side chains with siloxane end groups to modulate stretchability [47]. Here, the side-chain featuring seven carbon spacer groups resulted in an 84% retention of charge mobility under 100% strain. Employing branched side chains also impact mechanical behavior and can increase ductility. However, excess branched side chains can reduce the intermolecular bond strength, resulting in weaker films [49, 50].

The complexity of the polymer backbone structure has made a quantitative framework correlating structure to mechanical behavior difficult to realize; however, there are several design features that have clear impact on mechanical properties [51]. First, conjugation of the backbone typically increases the polymer's persistence length and stiffness. Adding ring structures and reducing single carbon bonds along the backbone further increases stiffness [39, 51]. Disrupting the persistence length is one strategy to reduce stiffness and increase ductility. For example, the Luscombe group has studied the impact of backbone engineering on an indacenodithiophene (IDT) based copolymer by incorporating benzothiadiazole and benzopyrollodione (BPD) for backbone twisting, and substitution of hexadecane (C_{16}) side chains with isocane (C_{20}) side chains [52]. Figure 5.5c shows the stress–strain curves of their IDT copolymers with variable side chains. They have observed a tradeoff between ductility (COS is <5% for 0% BPD to COS >75% for 95% BPD) and hole mobility (3×10^{-2} cm^2/Vs for 0% BPD and 2.5×10^{-4} cm^2/Vs for 95% BPD). BPD increases the average twist angle between monomers, varying the circular distribution of IDT hexadecyl side chains extending away from the backbone, resulting in a drop in intramolecular charge transport and hole mobility.

5.3.3 Regioregularity and Crystallinity

Regioregularity (RR) plays an important role in modulating the ductility and stiffness of the film. One of the most studied materials to understand the effect of RR is the poly(3-alkylthiophenes) (P3ATs). The spatial arrangements of alkyl side chains with the thiophene ring (e.g. head-to-tail) define the RR of these materials. Kim et al. have systematically studied a series of poly(3-hexylthiophene)s (P3HTs) having a range of RR and found that increasing RR decreased the ductility of the films [46, 53]. They attributed the decrease in ductility to an increase in the crystalline domains. It was argued that the amorphous fraction has greater molecular mobility as compared to the crystalline domains. However, decreasing the packing order in the film can come at the cost of charge mobility. To overcome this tradeoff, the same group synthesized regioregular-block-regiorandom P3HT copolymers to exploit the full potential of both blocks to optimize electrical and mechanical performance for OFETs [54]. Figure 5.5d shows the effect of RR on the elongation at break and tensile modulus values of P3HT films. Kim et al. have recently provided a comprehensive review of the impact of RR in conjugated polymers, correlating structure, mechanical properties, and photovoltaic performance [55].

The degree of crystallinity of the polymer will have a large influence on mechanical properties, particularly for materials near or above their T_g [56]. Wang et al. has provided a review of near amorphous materials that have recently emerged as a promising candidate for flexible and stretchable organic electronics [57]. They have attributed the disorder of the polymer chains as an enabling feature for intrinsic stretchability, while charge mobility is maintained owing to efficient intrachain charge transport in the amorphous domains along with occasional intermolecular hopping.

5.3.4 Block Copolymers with Flexible and Stretchable Linkers

Block copolymers have been another approach employed to balance mechanical and electrical properties. Block copolymers tend to show distinct properties of each material allowing it to be both semiconducting and mechanically robust. The flexible block can be insulating, referred to as a conjugated break spacer or linker [58], or a conducting polymer [59]. The insertion of nonconjugated linkers in a DPP-based semiconductor is shown in Figure 5.6a while Figure 5.6b shows the impact of the linker addition on the stretchability of the film. The linkers lower the crystallinity and glass transition temperature of the film, while the pi-conjugated polymer continues to efficiently pack to support charge mobility [60]. In addition, Kim and coworkers synthesized a PDMS-block-PM6 polymer to realize mechanically robust organic solar cells with small molecule acceptor (SMA) L8-BO [61]. The resulting bulk heterojunction thin film structure provided nearly 20% COS while maintaining 18% power conversion efficiency (PCE). They have attributed the high stretchability to the PDMS while the covalent links of the block copolymer limit large-scale phase segregation, thus maintaining device performance. Alternatively, blocks can be made of conjugated polymers. For example, Lipomi and coworkers used blocks of conjugated polymers with each block having a distinct electron acceptor unit, as illustrated in Figure 5.6c [62]. This approach reduced packing order and increased COS relative to the reference polymer.

5.3.5 Crosslinking and Hydrogen Bonding

The addition of a crosslinker has also been applied to improve toughness and ductility of a number of conjugated polymers. Crosslinkers form a three-dimensional covalently bonded network that supports toughness and greater elasticity. At the same time, the resultant polymers can become more ordered, boosting charge mobility [63]. Crosslinking can also aid morphological stability and the lifespan of

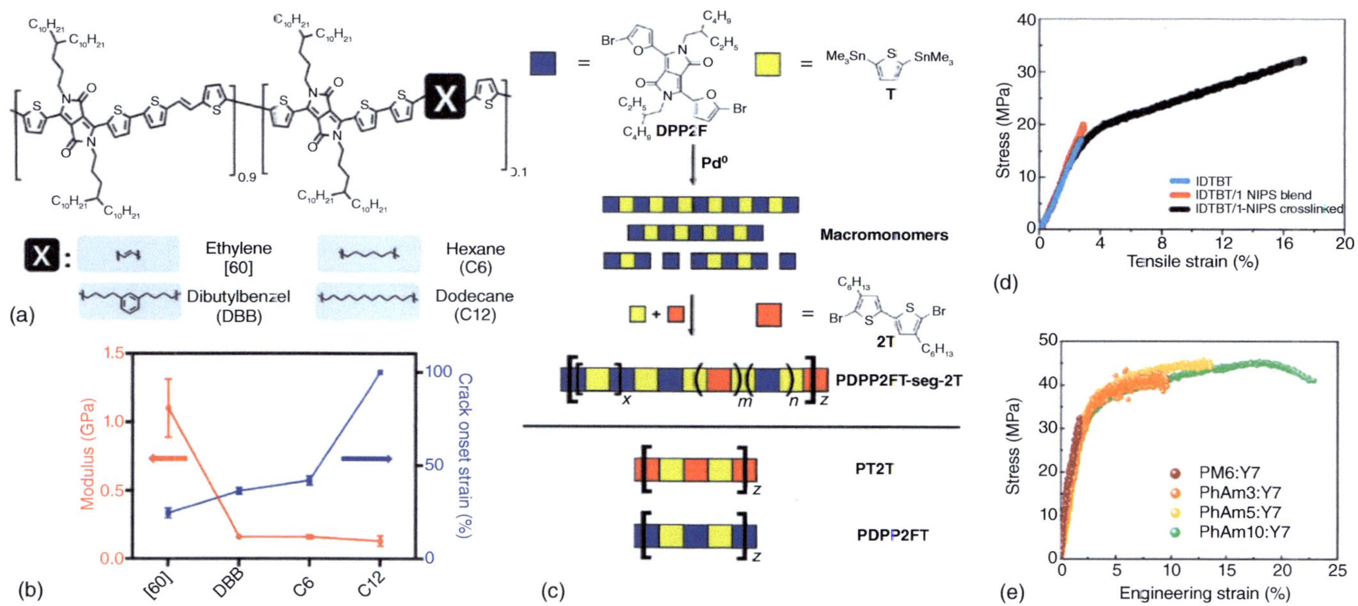

Figure 5.6 (a) Chemical structures of DPP-based semiconducting polymers containing conjugation breakers [60]. (b) Impact of the breakers on the elastic modulus and crack onset strain of the DPP-based organic semiconductors measured by film on elastomer method [60]. (c) Synthesis strategy through segmentation of a DPP-based polymer with different rigid blocks [62] (Royal Society of Chemistry/CC BY 3.0). (d) Impact of nitrene-induced photocrosslinkers on the stress-strain properties of the IDTBT films [65]. (e) Hydrogen-bonded polymers donors (PhAm – 3, 5, 10) yielding higher ductility compared to neat Pm6:Y7 blends [74].

the device [64]. However, certain crosslinkable moieties attached to the polymer backbone can disrupt π–π stacking, resulting in lower charge carrier mobility [65]. Bao and coworkers have demonstrated through several publications that the addition of crosslinkers in conjugated polymers and tuning crosslinker properties (e.g. crystallinity) can enhance the electrical and elastic properties simultaneously [66–69]. Cho's research group recently synthesized a series of nitrene-induced ($-N_3$) photocrosslinkers (NIPSs) with different numbers of ethylene glycol repeating units, to use a crosslinker for indacenodithiophene-benzothiadiazole (IDTBT) polymers, to enhance stretchability without impacting the device's carrier mobility [65]. Figure 5.6d shows the improvement of ductility in IDTBT films crosslinked with NIPS, compared to neat IDTBT and IDTBT–NIPS blend. The IDTBT–NIPS crosslinked OFETs retained 96% mobility at 80% strain, compared to 2.5% of the neat IDTBT OFETs.

Several groups have also developed durable and ductile organic semiconductors by exploiting hydrogen bonding between polymers. Hydrogen bonds exhibit stronger interaction than other noncovalent bonds and can spontaneously and reversibly form between adjacent polymer chains. The hydrogen-bonded chains assist in increasing yield strength and also dissipate stress by breaking the bonds that are then reformed to negate fracture. The bond dynamics improve ductility without degrading the morphology and device performance [70, 71]. For example, Ocheje et al. has synthesized a DPP-based organic semiconductor that incorporated amide-containing alkyl side chains leading to hydrogen bonding that improved COS from 50% for the control film to 75% when 10% hydrogen bonding moieties were added [72]. Bao and coworkers fabricated stretchable transistors using 2,6-pyridine dicarboxamide as a hydrogen bonding unit with a DPP-based polymer backbone [73]. The transistors showed a reduced drop in charge mobility when strained to 100% compared to the reference material without H-bonding and no decrease in OFET performance after 500 cycles when strained to 25%. Recently, Kim and coworkers introduced flexible spacers that involve hydrogen bonding in OPV cells and found over 80% retention of PCE at 32% strain [74]. Figure 5.6e shows the stress-strain curves of active layer films when employing increasing hydrogen bonding active site density.

5.3.6 Additives and Blends

Additives in conjugated polymer films have also been explored to manage mechanical behavior. Particular attention has been given to PEDOT:PSS (poly(3,4-ethylenedioxythiophene)): poly(styrene sulfonate)), a common organic conductor used as electrodes and as an ionic electronic conductor in OECTs. Additives have been used in PEDOT:PSS solutions to improve surface wetting, adhesion, and increase ductility. The Lipomi group added 5% DMSO and 10% Zonyl fluorosurfactant (Zonyl) to PEDOT:PSS to improve strain-sensing properties [75]. The Zonyl acted as a plasticizing agent resulting in an increase in COS from 2% to 28%. Zonyl was also added to PEDOT:PSS by Bao and coworkers, followed by exposing the solution to ultraviolet/ozone (UV/O_3) treatment, which improved the stretchability and boosted the adhesion between PDMS and the PEDOT:PSS [76]. Use of solvent additives has also been considered to manage the film morphology and resulting mechanical behavior. For example, Kim et al. found that the addition of 1,8-diiodoctane (DIO) to chlorobenzene solvent had a significant impact on the mechanical properties of PTB7:PC_{71}BM bulk heterojunction films [77]. The DIO decreased the roughness of the films and prevented material aggregation resulting in a finer distribution of PC_{71}BM. However, with PC_{71}BM loading greater than 1 : 1 ratio by weight, the effect of the DIO was limited.

Blending an organic semiconductor film with a polymer that improves mechanical behavior has been a simple and effective approach to improving mechanical resilience. The blends typically include all organic semiconductors or an organic semiconductor blended with a mechanically robust insulating polymer. Stingelin and coworkers first considered blending P3HT with PE, showing a large increase in ductility of the composite from 13% for neat P3HT to over 660% for 35 : 65 of P3HT:PE blend [78]. Blending two conducting polymers, with one focused on improving ductility and the other on supporting electronic properties has also been considered [79, 80]. For example, the O'Connor group blended a rigid high-performance polymer, PCDTPT, with a more ductile polymer semiconductor, P3HT, and found that the COS of the blend film increased from 4% to over 75% while maintaining the charge mobility of the high-performance polymer semiconductor [79]. Blending elastomers with organic semiconductors has been a particularly effective approach to improve ductility and toughness of the films. The addition of elastomers has been considered across broad loading fractions. At low loading (under 20 wt%), the addition of poly(ethylene-*co*-butylene)-block-polystyrene (SEBS) or silicone elastomers has been shown to increase the mechanical robustness in OPV cells [81–83]. According to these studies, an optimum amount of elastomer can also manage local morphology to not only improve mechanical resilience but also be used as a processing aid to support high-performance devices. For instance, with the addition of 10 wt% SEBS in a polymer, small molecule blend has been shown to have fracture toughness and elongation at break that increased by 200% and 120% compared to the control film [84]. The fracture toughness of the SEBS additive films was found to be comparable to all-polymer (AP) solar cell active layers. Blend films with a large fraction of elastomer have been attractive for stretchable organic electronic applications [85, 86]. One particularly successful demonstration was reported by Bao and coworkers where blending SEBS with a DPP-based polymer resulted in segregated nano-scale fibrils of DPP [87]. They referred to the film as a conjugated polymer/elastomer phase separation–induced elasticity (CONPHINE) and resulted in a film that had a stretchability of 100% without impacting the mobility of the thin film. The mobility was maintained due to the interconnected semiconductor fibril network. The large-scale phase separation of the semiconductor and elastomer was exploited to achieve high-performance transistors. However, in other applications such as OPV cells, large-scale segregation can be detrimental to performance. Yet, by carefully tuning the morphology of SEBS added to an AP active layer, Li et al. showed that SEBS can be added at 40 wt% and still maintain >10% OPV cell efficiency while also increasing the elongation at break to >50% [88].

5.3.7 Organic Photovoltaic Considerations

Significant research attention has been placed on managing the mechanical properties of OPV active layers due to the opportune lightweight, thin, flexible, and stretchable power sources. Thus, we consider some of the unique material considerations associated with OPVs. To start, the photoactive layer is typically composed of a bulk heterojunction of an electron donor and electron acceptor semiconductor. The electron donor has primarily been a polymer while the electron acceptor has been a small molecule semiconductor. The SMA can embrittle the active layer, making mechanically robust devices a challenge [38]. There have been several strategies to manage this, including molecular design of the donor polymer and the use of additives as outlined above. In addition, replacing the SMA with an electron conductive polymer to form an AP active layer has received significant attention [89–93]. For example, Balar et al. have shown an enhancement in fracture energy when replacing the SMA with the polymer acceptor P(NDI2OD-T2) [40]. Later, Lee et al. synthesized P(NDI2OD-T2) of three different molecular weights and found that the polymer acceptor with the highest molecular weight maintains a high crack onset strain (30%) and toughness ($8\,MJ/m^3$), regardless of the donor-acceptor ratio [95]. The polymer improves intermolecular coupling through entanglements and increases the plastic zone near the crack tip leading to tougher films. Polymerization of high-efficiency SMAs has recently gained popularity to improve operational stability and mechanical properties. Yang and Zhan have provided a comprehensive review of the development of the polymerized SMAs for high-performance OPV cells [94]. Lee et al. synthesized a polymerized small molecule by incorporating siloxane flexible spacer into its backbone and fabricated 13.5% efficient organic solar cell that can withstand 15.2% strain [95]. They claimed that the resultant polymerized acceptors provided pathways for better mechanical stress. However, the degree of polymerization can be limited due to solubility limitations of the polymerized SMAs. A higher degree of polymerization should continue to improve mechanical performance of these material systems [96]. While AP OPV cells have increased in popularity, SMAs may still be an attractive system for synthetic reproducibility and device performance. Taking into account the SMA–polymer interactions provide insight into the observable mechanical behavior [38]. The SMA will act as an anti-plasticizing agent in the blend, and reducing the heterointeractions between the SMA and polymer was found to reduce the negative impact of the SMA addition on film toughness [38]. Thus, designing SMAs and polymers with minimally attractive heterointeractions should support mechanical durable polymer: SMA blends. Lastly, OPV active layers now commonly employ ternary and even quaternary semiconductor blends that improve performance and stability [97, 98]. The use of multicomponent systems that incorporate multiple polymer semiconductors will reduce SMA loading and improve mechanical stability.

5.4 Device Design

The device design will dictate the stresses encountered under various mechanical loading conditions. In this section, we review device designs that support mechanically resilient operation.

5.4.1 Neutral Axis and Ultrathin Devices

The typical flexible organic electronic devices comprise an organic semiconductor layer with multiple interface layers, electrodes, substrate, superstrate, and encapsulation. To improve mechanical stability, the weaker layers can be placed near the neutral axis of the device. The neutral axis is the plane at which the stress and strain go to zero when the film is placed in flexure, as illustrated in Figure 5.7a. The neutral axis will be dictated by the stiffness and thickness of each layer in the stack [99]. By placing the weak layers near the neutral axis, the stress and strain on these layers are minimized. By using ultrathin substrate and superstrate, the strain in the entire device can be small even under extremely small radius of curvature. For example, Sekitani et al. has shown an ultrathin transistor with a parylene C top encapsulation layer and a PI substrate with a total thickness of 27 μm [100]. The device was maintained functionally after 60,000 bending cycles to a bending radius of 2 mm. There are numerous other examples of using ultrathin substrates that demonstrate that the use of these substrates minimizes the strain in the active layers, allowing for very small bending radii [101–103]. Figure 5.7b shows one such example of an ultraflexible organic solar cell [101].

5.4.2 Film Thickness

Film thickness can impact a number of mechanical properties including ductility and fracture toughness. It has been shown that increasing film thicknesses can increase fracture toughness, although the dependence is often material-specific [19, 29]. The fracture energy of various films as a function of thickness is given in Figure 5.7c, showing the range in observed behavior. It was observed that the more brittle polymers have a lower thickness dependence. In addition, Figure 5.7d shows that the COS linearly correlates with $(G_c/t)^{0.5}$, where t is the film thickness, and G_c is the cohesive fracture energy of the semiconductor films [19]. Thus, films with greater ductility (i.e. higher COS) are shown to be tougher for a given film thickness. This is partially attributed to a larger strain energy distribution around the crack tip with increased film thickness [104, 105].

Reducing the film thickness to approximately under 50 nm can also change mechanical behavior relative to their bulk counterpart due to nanoconfinement effects [107]. This phenomenon becomes evident when the thickness of the film

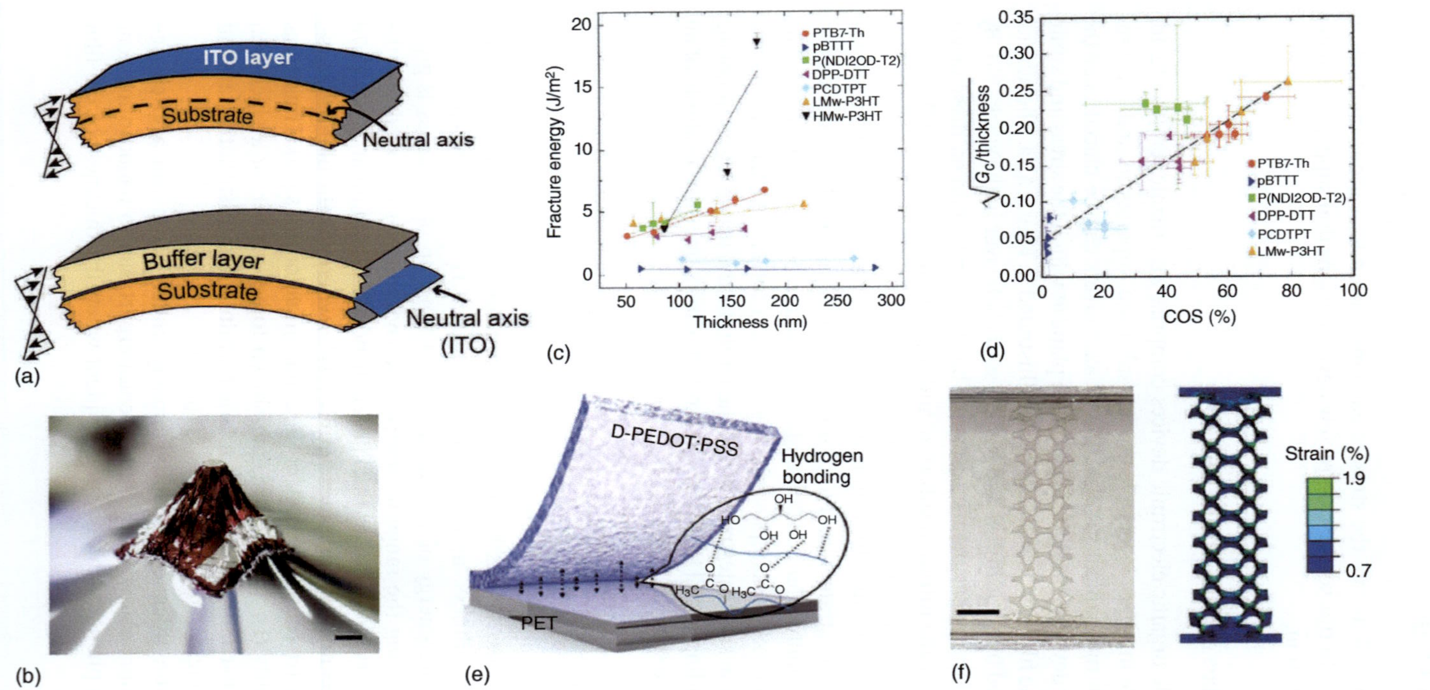

Figure 5.7 (a) Schematic illustration of neutral axis engineering to distribute the stress away from the brittle layer. (b) Structure of ultraflexible organic solar cell on ultraflexible PET substrate. Source: Ref. [101]/with permission of Springer Nature. (c) Thickness dependent fracture energy of various organic semiconductors. Source: Ref. [19]/with permission of American Chemical Society. (d) Relation between crack onset strain (COS) and (Fracture energy, G_c/thickness, t)$^{0.5}$ for different organic semiconductor. Source: Ref. [19]/with permission of American Chemical Society. (e) Interfacial bonding of the PET substrate with the D-PEDOT:PSS electrode layer. Source: Ref. [105]/with permission of Royal Society of Chemistry. (f) Kirigami cut patterns of a PI film at 50% strain. The finite element analysis data shows the local strain below 2%. The scale bar is 5 mm. Source: Ref. [106]/with permission of John Wiley & Sons.

approaches the polymer's radius of gyration, which may be thicknesses found in some organic electronic devices [108]. In such situations, the increased fraction of mobile polymer chains at the film surface can lead to suppression of the glass transition temperature and can increase ductility [87].

5.4.3 Electrodes and Interlayers

Organic electronic devices typically consist of several thin films stacked on top of each other including the organic semiconductor layer electron transport layer, hole transport layer,and electrode. In three-terminal devices, there is also an insulating dielectric layer. Careful selection and design of these functional layers are necessary to mitigate device failure. In the case of the charge transport layers, metal oxides such as ZnO and MoO_3 are often employed and are brittle in nature. It is often desirable to remove oxide-based transport layers to improve flexibility. It has been shown that employing polymer-based charge transport and dielectric layers improves mechanical resilience of flexible devices [97, 109]. For optoelectronic applications, indium tin oxide (ITO) is one the most common transparent conducting electrodes (TCE), but it is brittle and prone to failure. There have been a wide variety of alternative TCEs demonstrated over the years to replace the ITO that includes PEDOT:PSS, graphene, carbon nanotubes (CNTs), silver nanowires (Ag NWs), ultrathin metal films, metal grids, and gallium-based liquid metal electrodes [110–113]. It has been shown that ultrathin metal films coupled with dielectric layers can be used to manage optoelectronic performance while enabling flexible devices [114]. Yao et al. developed highly efficient flexible OLED through an embedded metal composite electrode strategy that retained ~83% of the original luminance even after bending the device 2000 cycles at a radius of ~0.5 mm [115].Embedding nanomaterials such as Ag NWs in a polymer matrix has been one effective approach to gaining improved mechanical resilience and surface uniformity of these electrodes [116, 117].

5.4.4 Interfaces

Durability of the devices rely on strong adhesion and chemical compatibility between the various functional layers. There are a number of factors that will influence adhesion that include physical interlocking (roughness, diffusion), chemical bonding, dispersive forces (van der Waals), and electrostatic forces. Van der Waal forces are the most common adhesion mechanism between the functional layers of the materials, with secondary contributions from electrostatic interactions. Improving surface energy to promote stronger interaction energy between the constituent materials is used to improve wetting during ink processing and subsequent adhesion of a device. For example, UV/O_3 treatment of ZnO has been shown to increase its surface energy and adhesion with organic semiconductor layer [118]. UV/O_3 treatment of PDMS has also been shown to increase adhesion of a polymer semiconductor increasing its COS [2]. Further increase in adhesion is possible with strong bonding mechanisms, such as hydrogen bonding. Song et al. demonstrated the use of D-PEDOT:PSS to achieve hydrogen bonding with a PET substrate as shown in Figure 5.7e [119]. Approaches to improve adhesion through physical mechanisms include surface structures that increase contact area and surface grafting [120]. The surface of the commercial substrates is often intentionally kept rough for better adhesion during roll-to-roll printing. However, excessive roughness of the substrate can cause delamination and premature cracking during mechanical loading [102]. Lastly, the modulus mismatch between different interfaces can also raise interfacial stresses and alter the strain distribution throughout the device structure [19]. Thus, materials with similar elastic modulus are desired to minimize interface stress that increases the driving force for delamination.

5.4.5 Stretchable Device Architecture

There are two primary methods of achieving stretchable devices, employing intrinsically stretchable materials and structural designs that enable stretchability while minimizing the local strain on the functional materials. We covered stretchable materials above and consider structural designs in this section. One effective approach is to exploit film wrinkling where a wrinkled structure flattens upon strain and retracts to a wrinkled state when the strain is removed [121]. This wrinkling allows for the functional components to remain in the elastic regime when large strains are applied to the device. The device is typically placed on an elastomer substrate to drive this wrinkling behavior. Stiff material can also be stretched when employing kirigami cut geometric modification [122]. Shyu et al. has shown that employing the kirigami approach can enhance stretchability of rigid nanocomposites from 4% to 370% without impacting the conductivity over the entire strain range [123]. Figure 5.7f shows two Kirigami-patterned PI films at 0% and 50% strain and associated finite element analysis (FEA) data. The FEA shows only 2% local strain on the strained film while the actual stain of the overall film is 50% indicating the negligible effect on fabricated device in strained position. Fractal designs that employ coil, spring, wavy, and serpentine structures enable stretchability while minimizing strain on the functional components [124]. IIn these designs, the functional structure adjusts its out of plane position to accommodate the imposed in-plane strain when, preventing stress concentration in the material itself.

5.5 Applications

Flexible and stretchable organic electronic devices have been applied in a wide variety of applications. In this section, we highlight demonstrations that employ some of the design considerations outlined above. We organize this section by device type, considering OLEDs, OPVs, OPDs, OECTs, and OFETs.

Flexible and stretchable OLEDs have been applied in numerous applications that include displays, wearable electronics, and bioelectronics. Figure 5.8a shows one such example of OLEDS in a recently launched Samsung Galaxy flip and fold mobile display and Figure 5.8b shows a prototype flexible OLED based photonic

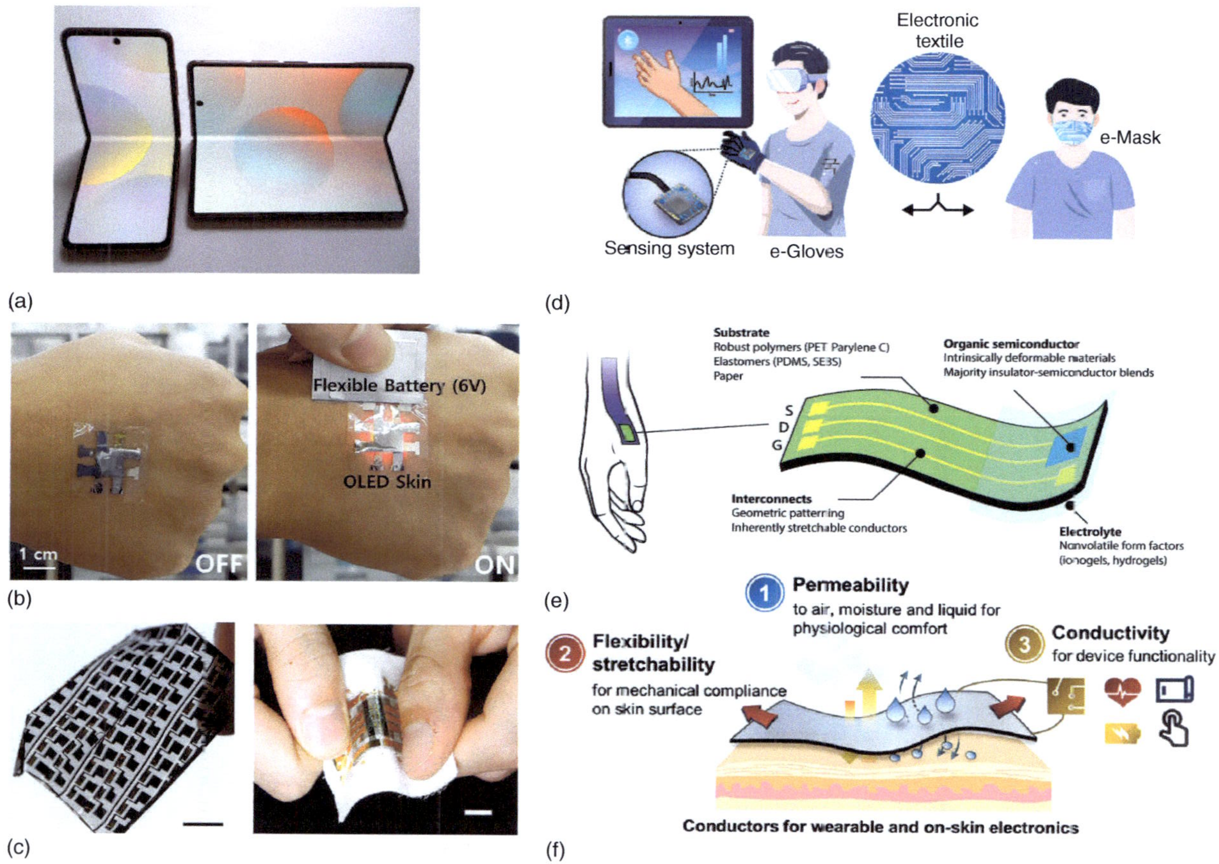

Figure 5.8 (a) Flexible organic light-emitting diode employed in the Samsung Galaxy Flip and Fold mobile display. Source: Ref. [125]/with permission of Vox Media. (b) A picture of flexible OLED-based photonic skin device. Source: Ref. [126]/John Wiley & Sons. (c) Example of a flexible organic solar cell (left). Another flexible OPV cell is employed over a cloth in the right image. The scale bar is 10 mm in both images. Source: Ref. [138]/with permission of National Academy of Sciences. (d) Potential applications of wearable textiles developed from organic field effect transistors. Source: Ref. [141]/with permission of Elsevier. (e) A typical structure of an organic transistor with source, gate, and drain separated by the electrolyte [139]. (f) Design considerations of organic electrochemical device for electronic skin. Source: Ref. [140]/John Wiley & Sons.

skin device integrated on the human body [125, 126]. Flexible OLEDs have been used for lab-on-chip devices, wearable sensors, therapeutic treatment, and optogenetics [127]. For example, flexible OLEDs have been used as light sources for photodynamic therapy to treat bacterial infections [128]. Flexible OPV devices have large power density (W/kg), making them attractive for deployable power sources for expeditions, disaster relief, and other off-grid needs with two examples of flexible OPVs shown in Figure 5.8c. At the other end of the application spectrum, OPV devices also perform well under indoor lighting with opportunities to power small electronic devices, sensors, and various Internet of Things devices [129]. The local power source will eliminate the need for frequent charging or battery replacement and provide a constant power supply [130]. Lastly, flexible and stretchable OPVs provide opportunities to realize architecturally integrated solar power. The devices can be seamlessly integrated into non-planar structures, windows, facades, and roofs [103, 131].

OPDs have been used in healthcare, sporting goods, and imaging equipment [132]. In biomedical applications, OPDs have been used to monitor heart rate, blood oxygen levels, and body temperature [133]. The detectors can be attached to medical devices for biological interrogation or imaging, as demonstrated in endoscopic devices [134]. Researchers have developed flexible X-ray imagers, and fluorescence sensors based on OPDs [135, 136]. Flexible OLEDs and OPDs have been combined to make flexible pulse oximeters [137].

OFETs can be used as an electronic switch in applications such as displays or as sensors, either as the sensing element or as part of a readout array [146]. OFETs have been applied as the backplane array in active-matrix foldable and rollable displays. It has enabled superior durability during flexing with opportunities to advance next-generation televisions, smartphones, smartwatches, and wristbands. Various OFET sensors have been demonstrated that include chemical, pressure, touch, and temperature [146]. A diverse range of flexible and lightweight OFET sensors have been reported and integrated into human skin or cloths to monitor vital signs of individuals. OFETs have also enabled smart electronic textiles to understand muscle activity, body posture or to measure blood pressure, body temperature, and other vital signs [147] Figure 5.8d shows some potential applications of smart electronic textiles developed from flexible and stretchable OFETs.One particularly attractive application of organic electronic materials is OECTs, with mixed ionic and electronic conduction. Mixed conduction provides transduction from ionic signals common in the human body to electrical signals to interface with electronic devices [142]. OECTs can be embedded in wearable or implantable devices for chemical and biological sensing applications to continually monitor a wide range of physiological data [143]. A detailed review of the sensing applications of OECTs such as neural recording, ion sensors, biomolecule detectors, and cell activity monitors has been reported by Marks et al. [144]. The main advantage of using OECT is its capability to amplify weak signals in ion concentration to large changes in electrical current. OECTs as implanted devices, or on human skin, require the same mechanical demands as the other devices, as covered in several recent reviews [139, 143, 145]. Figure 5.8e shows the schematic of a flexible OECT device with important structural features while Figure 5.8f shows the design considerations of an OECT device for electronic skin.

5.6 Conclusion

Organic electronics have numerous attributes that make them attractive for appli cation in flexible and stretchable electronics. While the materials are very thin and with Young's modulus often lower than their inorganic counterpart, there remains a need to design the materials and devices to meet the necessary mechanical demands of these applications. The design and selection of the active layer materials is a major consideration. We reviewed several strategies that have been used to achieve target mechanical responses that can be broadly grouped as molecular design and multi-component films. One of the largest impacts on mechanical response is the MW of the polymers. Using high MW polymers is essential for tough and stretchable films. The use of elastomers in the form of a block copolymer or as a blend has also been shown to be effective, particularly in stretchable device applications. The elastomer supports large elasticity and provides a driving force for shape recovery after strain removal. In addition to the active layer's mechanical properties, the device design must be designed to meet the mechanical demands of a particular target application. There have been a large number of demonstrations of various devices for applications that range from flexible displays to wearable sensors. The most successful demonstration has been flexible active matrix OLED-based displays that are now commercially available. Wearable sensors and electronics for soft robotics may be the most demanding and can require stretchable functionality. In response to this need, there have been impressive material and device design strategies that meet mechanical demands. However, further research and development is needed to commercialize these products. Some of the limitations have been associated with performance, stable operation over long periods of time, and mechanical resilience outside of well-defined experimental conditions. For example, devices may be stretchable but lack the necessary encapsulation of long-term operational stability, or resistance to various unattended mechanical insults beyond the well-defined tensile strain, such as scratch and tear resistance. As these challenges are solved, a plethora of exciting applications await.

Acknowledgments

The authors acknowledge support from the Office of Naval Research Award N6833523C0036 and Office of Naval Research Award N00014-20-1-2155 for enabling the writing of this chapter.

References

1 Kim, T., Kim, J.-H., Kang, T.E. et al. (2015). Flexible, highly efficient all-polymer solar cells. *Nature Communications* 6 (1): 8547.

2 Juillard, S., Planes, E., Matheron, M. et al. (2018). Mechanical reliability of flexible encapsulated organic solar cells: characterization and improvement. *ACS Applied Materials & Interfaces* 10 (35): 29805–29813.

3 Yan, T., Song, W., Huang, J. et al. (2019). 16.67% Rigid and 14.06% flexible organic solar cells enabled by ternary heterojunction strategy. *Advanced Materials* 31 (39): 1902210.

4 Noh, J., Kim, G.-U., Han, S. et al. (2021). Intrinsically stretchable organic solar cells with efficiencies of over 11%. *ACS Energy Letters* 6 (7): 2512–2518.

5 Lipomi, D.J., Chong, H., Vosgueritchian, M. et al. (2012). Toward mechanically robust and intrinsically stretchable organic solar cells: evolution of photovoltaic properties with tensile strain. *Solar Energy Materials and Solar Cells* 107: 355–365.

6 Forrest, S.R. (2020). *Organic Electronics: Foundations to Applications*. USA: Oxford University Press.

7 Someya, T., Bauer, S., and Kaltenbrunner, M. (2017). Imperceptible organic electronics. *MRS Bulletin* 42 (02): 124–130.

8 Yu, K., Rich, S., Lee, S. et al. (2019). Organic photovoltaics: toward self-powered wearable electronics. *Proceedings of the IEEE* 107 (10): 2137–2154.

9 Ren, H., Chen, J., Li, Y., and Tang, J. (2021). Recent progress in organic photodetectors and their applications. *Advanced Science* 8 (1).

10 Wang, Y., Lee, J., Hou, X. et al. (2021). Recent progress and challenges toward highly stable nonfullerene acceptor-based organic solar cells. *Advanced Energy Materials* 11 (5): 2003002.

11 Blau, R., Chen, A.X., Polat, B. et al. (2022). Intrinsically stretchable block copolymer based on PEDOT:PSS for improved performance in bioelectronic applications. *ACS Applied Materials & Interfaces* 14 (4): 4823–4835.

12 Tan, S.T.M., Gumyusenge, A., Quill, T.J. et al. (2022). Mixed ionic–electronic conduction, a multifunctional property in organic conductors. *Advanced Materials* 34 (21): 2110406.

13 Savagatrup, S., Printz, A.D., O'Connor, T.F. et al. (2014). Molecularly Stretchable Electronics. *Chemistry of Materials* 26: 3028–3041.

14 Kim, J.-H., Nizami, A., Hwangbo, Y. et al. (2013). Tensile testing of ultra-thin films on water surface. *Nature Communications* 4 (1): 2520.

15 Stafford, C.M., Harrison, C., Beers, K.L. et al. (2004). A buckling-based metrology for measuring the elastic moduli of polymeric thin films. *Nature Materials* 3 (8): 545–550.

16 Balar, N., Xiong, Y., Ye, L. et al. (2017). Role of polymer segregation on the mechanical behavior of all-polymer solar cell active layers. *ACS Applied Materials & Interfaces* 9 (50): 43886–43892.

17 Printz, A.D., Zaretski, A.V., Savagatrup, S. et al. (2015). Yield point of semiconducting polymer films on stretchable substrates determined by onset of buckling. *ACS Applied Materials & Interfaces* 7 (41): 23257–23264.

18 Chung, J.Y., Douglas, J.F., and Stafford, C.M. (2017). A wrinkling-based method for investigating glassy polymer film relaxation as a function of film thickness and temperature. *The Journal of Chemical Physics* 147 (15): 154902.

19 Balar, N. and Connor, B.T.O. (2017, 2017). Correlating crack onset strain and cohesive fracture energy in polymer semiconductor films. *Macromolecules* 50 (21): 8611–8618.

20 Park, H., Ma, B.S., Kim, Y. et al. (2023). Direct measurement of the thermo-mechanical properties of poly(3-hexylthiophene) thin films on ionic liquid surfaces. *Macromolecules* 56 (4): 1633–1642.

21 Galuska, L.A., Muckley, E.S., Cao, Z. et al. (2021). SMART transfer method to directly compare the mechanical response of water-supported and free-standing ultrathin polymeric films. *Nature Communications* 12 (1): 2347.

22 O'Connell, P.A. and McKenna, G.B. (2007). Novel nanobubble inflation method for determining the viscoelastic properties of ultrathin polymer films. *Review of Scientific Instruments* 78 (1).

23 Wang, B., Luo, D., Li, Z. et al. (2018). Camphor-enabled transfer and mechanical testing of centimeter-scale ultrathin films. *Advanced Materials* 30 (28): 1800888.

24 Zhang, H., Fujiki, K., Ido, K. et al. (2020). Tensile testing of freestanding polymer thin films with non-standard geometries. *Polymer Testing* 91: 106825.

25 O'Connor, B., Chan, E.P., Chan, C. et al. (2010). Correlations between mechanical and electrical properties of polythiophenes. *ACS Nano* 4 (12): 7538–7544.

26 Dupont, S.R., Oliver, M., Krebs, F.C., and Dauskardt, R.H. (2012). Interlayer adhesion in roll-to-roll processed flexible inverted polymer solar cells. *Solar Energy Materials and Solar Cells* 97: 171–175.

27 Brand, V., Levi, K., McGehee, M.D., and Dauskardt, R.H. (2012). Film stresses and electrode buckling in organic solar cells. *Solar Energy Materials and Solar Cells* 103: 80–85.

28 Dauskardt, R.H., Lane, M., Ma, Q., and Krishna, N. (1998). Adhesion and debonding of multi-layer thin film structures. *Engineering Fracture Mechanics* 61 (1): 141–162.

29 Bruner, C. and Dauskardt, R. (2014). Role of molecular weight on the mechanical device properties of organic polymer solar cells. *Macromolecules* 47 (3): 1117–1121.

30 Teoh, S. (2000). Fatigue of biomaterials: a review. *International Journal of Fatigue* 22 (10): 825–837.

31 Borrero-López, O., Hoffman, M., Bendavid, A., and Martin, P.J. (2010). The use of the scratch test to measure the fracture strength of brittle thin films. *Thin Solid Films* 518 (17): 4911–4917.

32 Steinmann, P.A. and Hintermann, H.E. (1989). A review of the mechanical tests for assessment of thin-film adhesion. *Journal of Vacuum Science & Technology A* 7: 2267–2272.

33 Sharma, A., Pan, X., Campbell, J.A. et al. (2017). Unravelling the thermo-mechanical properties of bulk heterojunction blends in polymer solar cells. *Macromolecules* 50 (8): 3347–3354.

34 Akabori, K., Tanaka, K., Nagamura, T. et al. (2005). Molecular motion in ultrathin polystyrene films: dynamic mechanical analysis of surface and interfacial effects. *Macromolecules* 38 (23): 9735–9741.

35 Balar, N., Rech, J.J., Siddika, S. et al. (2022). Resolving the molecular origin of mechanical relaxations in donor–acceptor polymer semiconductors. *Advanced Functional Materials* 32 (4): 2105597.

36 DeLongchamp, D.M., Kline, R.J., Jung, Y. et al. (2008). Molecular basis of mesophase ordering in a thiophene-based copolymer. *Macromolecules* 41 (15): 5709–5715.

37 Balar, N., Siddika, S., Kashani, S. et al. (2020). Role of secondary thermal relaxations in conjugated polymer film toughness. *Chemistry of Materials* 32 (15): 6540–6549.

38 Siddika, S., Peng, Z., Balar, N. et al. (2023). Molecular interactions that drive morphological and mechanical stabilities in organic solar cells. *Joule* 7 (7): 1593–1608.

39 O'Connor, B.T., Awartani, O.M., and Balar, N. (2017). Morphological considerations of organic electronic films for flexible and stretchable devices. *MRS Bulletin* 42 (02): 108–114.

40 Balar, N., Rech, J.J., Henry, R. et al. (2019). The importance of entanglements in optimizing the mechanical and electrical performance of all-polymer solar cells. *Chemistry of Materials* 31 (14): 5124–5132.

41 Bruner, C., Miller, N.C., McGehee, M.D., and Dauskardt, R.H. (2013). Molecular intercalation and cohesion of organic bulk heterojunction photovoltaic devices. *Advanced Functional Materials* 23 (22): 2863–2871.

42 Zhao, B., Pei, D., Jiang, Y. et al. (2021). Simultaneous enhancement of stretchability, strength, and mobility in ultrahigh-molecular-weight poly-(indacenodithiophene-*co*-benzothiadiazole). *Macromolecules* 54 (21): 9896–9905.

43 Choi, J., Kim, W., Kim, D. et al. (2019). Importance of critical molecular weight of semicrystalline n-type polymers for mechanically robust, efficient electroactive thin films. *Chemistry of Materials* 31 (9): 3163–3173.

44 Siddika, S., Balar, N., Booth, R.E., and O'Connor, B.T. (2021). Dynamic mechanical analysis of polymer thin films using a kirigami-inspired support. *ACS Macro Lett* 10 (9): 1107–1112.

45 Sommerville, P.J.W., Li, Y., Dong, B.X. et al. (2020). Elucidating the influence of side-chain circular distribution on the crack onset strain and hole mobility of near-amorphous indacenodithiophene copolymers. *Macromolecules* 53 (17): 7511–7518.

46 Kim, J.-S., Kim, J.-H., Lee, W. et al. (2015). Tuning mechanical and optoelectrical properties of poly(3-hexylthiophene) through systematic regioregularity control. *Macromolecules* 48 (13): 4339–4346.

47 Yuan, Y., Zhu, H., Wang, X. et al. (2023). Enhancing the elasticity of conjugated polymers through precise control of the spacing between the backbone and siloxane side-chains. *ACS Applied Materials & Interfaces* 15 (18): 22341–22350.

48 Savagatrup, S., Printz, A.D., Rodriquez, D., and Lipomi, D.J. (2014). Best of both worlds: conjugated polymers exhibiting good photovoltaic behavior and high tensile elasticity. *Macromolecules* 47 (6): 1981–1992.

49 Sugiyama, F., Kleinschmidt, A.T., Kayser, L.V. et al. (2018). Effects of flexibility and branching of side chains on the mechanical properties of low-bandgap conjugated polymers. *Polymer Chemistry* 9 (33): 4354–4363.

50 You, H., Kim, D., Cho, H. et al. (2018). Shift of the branching point of the side-chain in naphthalenediimide (NDI)-based polymer for enhanced electron mobility and all-polymer solar cell performance. *Advanced Functional Materials* 28 (39): 1803613.

51 Roth, B., Savagatrup, S.V., de los Santos, N. et al. (2016). Mechanical Properties of a Library of Low-Band-Gap Polymers. *Chemistry of Materials* 28 (7): 2363–2373.

52 Sommerville, P.J.W., Balzer, A.H., Lecroy, G. et al. (2023). Influence of side chain interdigitation on strain and charge mobility of planar indacenodithiophene copolymers. *ACS Polymers Au* 3 (1): 59–69.

53 Kim, H.J., Lee, M.Y., Kim, J.-S. et al. (2017). Solution-assembled blends of regioregularity-controlled polythiophenes for coexistence of mechanical resilience and electronic performance. *ACS Applied Materials & Interfaces* 9 (16): 14120–14128.

54 Park, H., Ma, B.S., Kim, J.-S. et al. (2019). Regioregular-*block*-regiorandom poly(3-hexylthiophene) copolymers for mechanically robust and high-performance thin-film transistors. *Macromolecules* 52 (20): 7721–7730.

55 Kim, Y., Park, H., Park, J.S. et al. (2022). Regioregularity-control of conjugated polymers: from synthesis and properties, to photovoltaic device applications. *Journal of Materials Chemistry A* 10 (6): 2672–2696.

56 Virkar, A.A., Mannsfeld, S., Bao, Z., and Stingelin, N. (2010). Organic semiconductor growth and morphology considerations for organic thin-film transistors. *Advanced Materials* 22 (34): 3857–3875.

57 Wang, P., Yang, J., Zhang, Y. et al. (2022). Near-amorphous conjugated polymers: an emerging class of semiconductors for flexible electronics. *ACS Materials Letters* 4 (6): 1112–1123.

58 Zhao, Y., Zhao, X., Zang, Y. et al. (2015). Conjugation-break spacers in semiconducting polymers: impact on polymer processability and charge transport properties. *Macromolecules* 48 (7): 2048–2053.

59 Lu, C., Lee, W., Gu, X. et al. (2017). Effects of molecular structure and packing order on the stretchability of semicrystalline conjugated poly(tetrathienoacene-diketopyrrolopyrrole) polymers. *Advanced Electronic Materials* 3 (2): 1600311.

60 Mun, J., Wang, G.N., Oh, J.Y. et al. (2018). Effect of nonconjugated spacers on mechanical properties of semiconducting polymers for stretchable transistors. *Advanced Functional Materials* 28 (43): 1804222.

61 Seo, S., Lee, J., Kim, D.J. et al. (2023). Poly(dimethylsiloxane)-*block*-PM6 polymer donors for high-performance and mechanically robust polymer solar cells. *Advanced Materials* 35 (24): 2300230.

62 Printz, A.D., Savagatrup, S., Burke, D.J. et al. (2014). Increased elasticity of a low-bandgap conjugated copolymer by random segmentation for mechanically robust solar cells. *RSC Advances* 4 (26): 13635–13643.

63 Yu, X., Li, C., Gao, C. et al. (2021). Incorporation of hydrogen-bonding units into polymeric semiconductors toward boosting charge mobility, intrinsic stretchability, and self-healing ability. *SmartMat* 2 (3): 347–366.

64 Kwon, N.Y., Park, S.H., Kang, H. et al. (2021). Improved stability of all-polymer solar cells using crosslinkable donor and acceptor polymers bearing vinyl moieties in the side-chains. *ACS Applied Materials & Interfaces* 13 (14): 16754–16765.

65 Kim, S.H., Chung, S., Kim, M. et al. (2023). Designing a length-modulated azide photocrosslinker to improve the stretchability of semiconducting polymers. *Advanced Functional Materials* 33 (23): 2212127.

66 Wang, G.-J.N., Shaw, L., Xu, J. et al. (2016). Inducing elasticity through oligo-siloxane crosslinks for intrinsically stretchable semiconducting polymers. *Advanced Functional Materials* 26 (40): 7254–7262.

67 Wang, S., Xu, J., Wang, W. et al. (2018). Skin electronics from scalable fabrication of an intrinsically stretchable transistor array. *Nature* 555 (7694): 83–88.

68 Wang, G.-J.N., Zheng, Y., Zhang, S. et al. (2019). Tuning the cross-linker crystallinity of a stretchable polymer semiconductor. *Chemistry of Materials* 31 (17): 6465–6475.

69 Liu, J., Wang, J., Zhang, Z. et al. (2020). Fully stretchable active-matrix organic light-emitting electrochemical cell array. *Nature Communications* 11 (1): 3362.

70 Zhang, Q., Huang, J., Wang, K., and Huang, W. (2022). Recent structural engineering of polymer semiconductors incorporating hydrogen bonds. *Advanced Materials* 34 (26): 2110639.

71 Zheng, Y., Ashizawa, M., Zhang, S. et al. (2020). Tuning the mechanical properties of a polymer semiconductor by modulating hydrogen bonding interactions. *Chemistry of Materials* 32 (13): 5700–5714.

72 Ocheje, M.U., Selivanova, M., Zhang, S. et al. (2018). Influence of amide-containing side chains on the mechanical properties of diketopyrrolopyrrole-based polymers. *Polymer Chemistry* 9 (46): 5531–5542.

73 Oh, J.Y., Rondeau-Gagné, S., Chiu, Y.-C. et al. (2016). Intrinsically stretchable and healable semiconducting polymer for organic transistors. *Nature* 539 (7629): 411–415.

74 Lee, J., Seo, S., Lee, S. et al. (2022). Intrinsically stretchable, highly efficient organic solar cells enabled by polymer donors featuring hydrogen-bonding spacers. *Advanced Materials* 34 (50): 2207544.

75 Savagatrup, S., Chan, E., Renteria-Garcia, S.M. et al. (2015). Plasticization of PEDOT:PSS by common additives for mechanically robust organic solar cells and wearable sensors. *Advanced Functional Materials* 25 (3): 427–436.

76 Lipomi, D.J., Lee, J.A., Vosgueritchian, M. et al. (2012). Electronic properties of transparent conductive films of PEDOT:PSS on stretchable substrates. *Chemistry of Materials* 24 (2): 373–382.

77 Kim, J.-H., Noh, J., Choi, H. et al. (2017). Mechanical properties of polymer–fullerene bulk heterojunction films: role of nanomorphology of composite films. *Chemistry of Materials* 29 (9): 3954–3961.

78 Müller, C., Goffri, S., Breiby, D.W. et al. (2007). Tough, semiconducting polyethylene-poly(3-hexylthiophene) diblock copolymers. *Advanced Functional Materials* 17 (15): 2674–2679.

79 Sun, T., Scott, J.I., Wang, M. et al. (2017). Plastic deformation of polymer blends as a means to achieve stretchable organic transistors. *Advanced Electronic Materials* 3 (1): 1600388.

80 Ren, H., Tong, Y., Ouyang, M. et al. (2022). Balancing the trade-off between the mechanical and electrical properties of conjugated polymer blend films for organic field-effect transistors. *Journal of Materials Chemistry C* 10 (40): 14921–14928.

81 Peng, Z., Xian, K., Cui, Y. et al. (2021). Thermoplastic elastomer tunes phase structure and promotes stretchability of high-efficiency organic solar cells. *Advanced Materials* 33 (49): 1–14.

82 Dauzon, E., Sallenave, X., Plesse, C. et al. (2022). Versatile methods for improving the mechanical properties of fullerene and non-fullerene bulk heterojunction layers to enable stretchable organic solar cells. *Journal of Materials Chemistry C* 10 (9): 3375–3386.

83 Prete, M., Ogliani, E., Bregnhøj, M. et al. (2021). Synergistic effect of carotenoid and silicone-based additives for photooxidatively stable organic solar cells with enhanced elasticity. *Journal of Materials Chemistry C* 9 (35): 11838–11850.

84 Al Shafe, A., Schrickx, H.M., Ding, K. et al. (2023). Rubber-toughened organic solar cells: miscibility–morphology–performance relations. *ACS Energy Letters* 8 (9): 3720–3726.

85 Jeon, K.-H. and Park, J.-W. (2022). Light-emitting polymer blended with elastomers for stretchable polymer light-emitting diodes. *Macromolecules* 55 (18): 8311–8320.

86 Janasz, L., Borkowski, M., Blom, P.W.M. et al. (2022). Organic semiconductor/insulator blends for elastic field-effect transistors and sensors. *Advanced Functional Materials* 32 (7): 2105456.

87 Xu, J., Wang, S., Wang, G.-J.N. et al. (1979). (2017) Highly stretchable polymer semiconductor films through the nanoconfinement effect. *Science* 355 (6320): 59–64.

88 Li, S., Gao, M., Zhou, K. et al. (2023). Achieving record-high stretchability and mechanical stability in organic photovoltaic blends with a dilute-absorber strategy. *Advanced Materials* 36 (8): 2307278.

89 Choi, J., Kim, W., Kim, S. et al. (2019). Influence of acceptor type and polymer molecular weight on the mechanical properties of polymer solar cells. *Chemistry of Materials* 31 (21): 9057–9069.

90 Kim, W., Choi, J., Kim, J.H. et al. (2018). Comparative study of the mechanical properties of all-polymer and fullerene-polymer solar cells: the importance of polymer acceptors for high fracture resistance. *Chemistry of Materials* 30 (6): 2102–2111.

91 Lee, J.W., Ma, B.S., Choi, J. et al. (2020). Origin of the high donor-acceptor composition tolerance in device performance and mechanical robustness of all-polymer solar cells. *Chemistry of Materials* 32 (1): 582–594.

92 Lee, J.W., Jeong, D., Kim, D.J. et al. (2021). Flexible-spacer incorporated polymer donors enable superior blend miscibility for high-performance and mechanically-robust polymer solar cells. *Energy & Environmental Science* 14 (7): 4067–4076.

93 Fan, Q., Su, W., Chen, S. et al. (2020). Mechanically robust all-polymer solar cells from narrow band gap acceptors with hetero-bridging atoms. *Joule* 4 (3): 658–672.

94 Yang, T. and Zhan, C. (2023). Polymerized A-DA'D-A type small-molecule acceptors for high performance all-polymer solar cells: progress and perspective. *Science China. Chemistry* 66 (9): 2513–2531.

95 Lee, J.-W., Lee, S.-W., Kim, J. et al. (2022). Mechanically robust all-polymer solar cells enabled by polymerized small molecule acceptors featuring flexible siloxane-spacers. *Journal of Materials Chemistry A* 10 (38): 20312–20322.

96 Wang, G., Melkonyan, F.S., Facchetti, A., and Marks, T.J. (2019). All-polymer solar cells: recent progress, challenges, and prospects. *Angewandte Chemie International Edition* 58 (13): 4129–4142.

97 Chen, T., Li, S., Li, Y. et al. (2023). Compromising charge generation and recombination of organic photovoltaics with mixed diluent strategy for certified 19.4% efficiency. *Advanced Materials* 35 (21): 2300400.

98 Zhang, K., Jiang, Z., Qiao, J. et al. (2023). Dredging photocarrier trapping pathways *via* "charge bridge" driven exciton–phonon decoupling enables efficient and photothermal stable quaternary organic solar cells. *Energy & Environmental Science* 16 (8): 3350–3362.

99 Lee, S., Kwon, J.-Y., Yoon, D. et al. (2012). Bendability optimization of flexible optical nanoelectronics via neutral axis engineering. *Nanoscale Research Letters* 7 (1): 256.

100 Sekitani, T., Iba, S., Kato, Y. et al. (2005). Ultraflexible organic field-effect transistors embedded at a neutral strain position. *Applied Physics Letters* 87 (17): 173502.

101 Kaltenbrunner, M., White, M.S., Głowacki, E.D. et al. (2012). Ultrathin and lightweight organic solar cells with high flexibility. *Nature Communications* 3 (1): 770.

102 Someya, T., Kaltenbrunner, M., and Yokota, T. (2015). Ultraflexible organic electronics. *MRS Bulletin* 40 (12): 1130–1137.

103 Saravanapavanantham, M., Mwaura, J., and Bulović, V. (2023). Printed organic photovoltaic modules on transferable ultra-thin substrates as additive power sources. *Small Methods* 7 (1): 2200940.

104 Sekitani, T., Zschieschang, U., Klauk, H., and Someya, T. (2010). Flexible organic transistors and circuits with extreme bending stability. *Nature Materials* 9 (12): 1015–1022.

105 Heng, W., Weihua, L., and Bachagha, K. (2023). Recent progress in flexible electrodes and textile shaped devices for organic solar cells. *Journal of Materials Chemistry A* 11 (3): 1039–1060.

106 Song, R., Yao, S., Liu, Y. et al. (2020). Facile approach to fabricating stretchable organic transistors with laser-patterned Ag nanowire electrodes. *ACS Applied Materials & Interfaces* 12 (45): 50675–50683.

107 Lee, J.-H., Chung, J.Y., and Stafford, C.M. (2012). Effect of confinement on stiffness and fracture of thin amorphous polymer films. *ACS Macro Letters* 1 (1): 122–126.

108 Wang, G., Najafi, F., Ho, K. et al. (2022). Mechanical size effect of freestanding nanoconfined polymer films. *Macromolecules* 55 (4): 1248–1259.

109 Zhang, X., Zhang, H., Li, Y. et al. (2022). Recent progress in hole-transporting layers of conventional organic solar cells with p–i–n structure. *Advanced Functional Materials* 32 (44): 2205398.

110 Kayser, L.V. and Lipomi, D.J. (2019). Stretchable conductive polymers and composites based on PEDOT and PEDOT:PSS. *Advanced Materials* 31 (10): 1806133.

111 Kim, C., An, K., Kang, M. et al. (2022). Facile fabrication of flexible metal grid transparent electrode using inkjet-printed dot array as sacrificial layer. *Scientific Reports* 12 (1): 1572.

112 So, J.-H., Koo, H.-J., Dickey, M.D., and Velev, O.D. (2012). Ionic current rectification in soft-matter diodes with liquid-metal electrodes. *Advanced Functional Materials* 22 (3): 625–631.

113 Pham, V.T.H., Trinh, T.K., Truong, N.T.N., and Park, C. (2017). Liquid eutectic GaIn as an alternative electrode for PTB7:PCBM organic solar cells. *Japanese Journal of Applied Physics* 56 (4): 046501.

114 Liu, C., Zhang, Q., Wang, D. et al. (2018). High performance, biocompatible dielectric thin-film optical filters integrated with flexible substrates and microscale optoelectronic devices. *Advanced Optical Materials* 6 (15): 1800146.

115 Nam, S., Song, M., Kim, D.-H. et al. (2014). Ultrasmooth, extremely deformable and shape recoverable Ag nanowire embedded transparent electrode. *Scientific Reports* 4 (1): 4788.

116 Booth, R.E., Schrickx, H.M., Hanby, G. et al. (2022). Silver nanowire composite electrode enabling highly flexible, robust organic photovoltaics. *Solar RRL* 6 (9): 2200264.

117 Yao, L., Qin, Y., Li, X. et al. (2023). High-efficiency stretchable organic light-emitting diodes based on ultra-flexible printed embedded metal composite electrodes. *InfoMat* 5 (5): e12410.

118 Walzer, K., Maennig, B., Pfeiffer, M., and Leo, K. (2007). Highly efficient organic devices based on electrically doped transport layers. *Chemical Reviews* 107 (4): 1233–1271.

119 Song, W., Fanady, B., Peng, R. et al. (2020). Foldable semitransparent organic solar cells for photovoltaic and photosynthesis. *Advanced Energy Materials* 10 (15): 2000136.

120 Wang, S., Wang, Z., Li, J. et al. (2020). Surface-grafting polymers: from chemistry to organic electronics. *Materials Chemistry Frontiers* 4 (3): 692–714.

121 Wu, H., Kustra, S., Gates, E.M., and Bettinger, C.J. (2013). Topographic substrates as strain relief features in stretchable organic thin film transistors. *Organic Electronics* 14 (6): 1636–1642.

122 Huang, S., Liu, Y., Zhao, Y. et al. (2019). Flexible electronics: stretchable electrodes and their future. *Advanced Functional Materials* 29 (6): 1805924.

123 Shyu, T.C., Damasceno, P.F., Dodd, P.M. et al. (2015). A kirigami approach to engineering elasticity in nanocomposites through patterned defects. *Nature Materials* 14 (8): 785–789.

124 Fan, J.A., Yeo, W.-H., Su, Y. et al. (2014). Fractal design concepts for stretchable electronics. *Nature Communications* 5 (1): 3266.

125 Bohn, D. (2020). *Galaxy Z Flip Review: Temper Your Expectations. The Verge.*

126 Jeon, Y., Choi, H., Park, K., and Choi, K.C. (2020). Flexible organic light-emitting-diode-based photonic skin for attachable phototherapeutics. *Journal of the Society for Information Display* 28 (4): 324–332.

127 Murawski, C. and Gather, M.C. (2021). Emerging biomedical applications of organic light-emitting diodes. *Advanced Optical Materials* 9 (14): 2100269.

128 Lian, C., Piksa, M., Yoshida, K. et al. (2019). Flexible organic light-emitting diodes for antimicrobial photodynamic therapy. *npj Flexible Electronics* 3 (1): 18.

129 Lee, H.K.H., Li, Z., Durrant, J.R., and Tsoi, W.C. (2016). Is organic photovoltaics promising for indoor applications? *Applied Physics Letters* 108 (25): 253301.

130 Hashemi, S.A., Ramakrishna, S., and Aberle, A.G. (2020). Recent progress in flexible–wearable solar cells for self-powered electronic devices. *Energy & Environmental Science* 13 (3): 685–743.

131 Wang, D., Liu, H., Li, Y. et al. (2021). High-performance and eco-friendly semitransparent organic solar cells for greenhouse applications. *Joule* 5 (4): 945–957.

132 Shan, T., Hou, X., Yin, X., and Guo, X. (2022). Organic photodiodes: device engineering and applications. *Frontiers of Optoelectronics* 15 (1): 49.

133 Park, S., Fukuda, K., Wang, M. et al. (2018). Ultraflexible near-infrared organic photodetectors for conformal photoplethysmogram sensors. *Advanced Materials* 30 (34): 1802359.

134 Kielar, M., Gooch, H., Xu, L. et al. (2021). Direct detection of neuronal activity using organic photodetectors. *ACS Photonics* 8 (1): 228–237.

135 Posar, J.A., Davis, J., Alnaghy, S. et al. (2021). Polymer photodetectors for printable, flexible, and fully tissue equivalent X-ray detection with zero-bias operation and ultrafast temporal responses. *Advanced Materials Technologies* 6 (9): 2001298.

136 Gelinck, G.H., Kumar, A., Moet, D. et al. (2013). X-ray imager using solution processed organic transistor arrays and bulk heterojunction photodiodes on thin, flexible plastic substrate. *Organic Electronics* 14 (10): 2602–2609.

137 Lee, H., Kim, E., Lee, Y. et al. (2018). Toward all-day wearable health monitoring: an ultralow-power, reflective organic pulse oximetry sensing patch. *Science Advances* 4 (11): eaas9530.

138 Xu, X., Fukuda, K., Karki, A. et al. (2018). Thermally stable, highly efficient, ultraflexible organic photovoltaics. *Proceedings of the National Academy of Sciences* 115 (18): 4589–4594.

139 Khau, B.V., Scholz, A.D., and Reichmanis, E. (2020). Advances and opportunities in development of deformable organic electrochemical transistors. *Journal of Materials Chemistry C* 8 (43): 15067–15078.

140 Chen, F., Huang, Q., and Zheng, Z. (2022). Permeable conductors for wearable and on-skin electronics. *Small Structures* 3 (1): 2100135.

141 Meena, J.S., Choi, S.B., Jung, S.-B., and Kim, J.-W. (2023). Electronic textiles: New age of wearable technology for healthcare and fitness solutions. *Materials Today Bio* 19: 100565.

142 Yao, Y., Huang, W., Chen, J. et al. (2023). Flexible and stretchable organic electrochemical transistors for physiological sensing devices. *Advanced Materials* 35 (35): 2209906.

143 Rashid, R.B., Ji, X., and Rivnay, J. (2021). Organic electrochemical transistors in bioelectronic circuits. *Biosensors & Bioelectronics* 190: 113461.

144 Marks, A., Griggs, S., Gasparini, N., and Moser, M. (2022). Organic electrochemical transistors: an emerging technology for biosensing. *Advanced Materials Interfaces* 9 (6): 2102039.

145 He, Y., Kukhta, N.A., Marks, A., and Luscombe, C.K. (2022). The effect of side chain engineering on conjugated polymers in organic electrochemical transistors for bioelectronic applications. *Journal of Materials Chemistry C* 10 (7): 2314–2332.

146 Yuvaraja, S., Nawaz, A., Liu, Q. et al. (2020). Organic field-effect transistor-based flexible sensors. *Chemical Society Reviews* 49 (11): 3423–3460.

147 Arab Hassani, F., Shi, Q., Wen, F. et al. (2020). Smart materials for smart healthcare– moving from sensors and actuators to self-sustained nanoenergy nanosystems. *Smart Materials in Medicine* 1: 92–124.

148 Wei, Y., Jang, G.-W., Hsueh, K.F. et al. (1992). Thermal transitions and mechanical properties of films of chemically prepared polyaniline. *Polymer* 33 (2): 314–322.

149 Kalogeras, I.M. and Neagu, E.R. (2004). Interplay of surface and confinement effects on the molecular relaxation dynamics of nanoconfined poly(methyl methacrylate) chains. *The European Physical Journal E* 14 (2): 193–204.

150 Lin, Y.-C., Chen, F.-H., Chiang, Y.-C. et al. (2019). Asymmetric side-chain engineering of isoindigo-based polymers for improved stretchability and applications in field-effect transistors. *ACS Appl Mater Interfaces* 11 (37): 34158–34170.

Part II

Design and Manufacturing

6

Structural Design of Flexible and Stretchable Electronics

Zhaoqian Xie[1,2], *Zichen Zhao*[1,2], *and Raudel Avila*[3]

[1] *Dalian University of Technology, State Key Laboratory of Structural Analysis, Optimization and CAE Software for Industrial Equipment, No.2 Linggong Road, Dalian, Liaoning 116024, PR China*
[2] *Dalian University of Technology, Department of Engineering Mechanics, No.2 Linggong Road, Dalian, Liaoning 116024, PR China*
[3] *Rice University, Department of Mechanical Engineering, Houston, TX 77005, USA*

6.1 Introduction

In recent years, the accelerated development of flexible electronics has influenced applications in multiple fields including bioelectronics, the Internet of Things technologies, and artificial intelligence. These emerging applications are expected to have a long-term impact on society as researchers reimagine the role of electronics and engineer superior device capabilities to seamlessly interact with intrinsically soft or flexible and stretchable systems.

The field of flexible electronics for health applications holds significant importance and impact, as it enables real-time monitoring and output of physiological activity generated by vital organs like the heart, lungs, and brain, providing insights into one's health status [1–6]. Additionally, these flexible devices can, in some cases, stimulate biological tissues via external fields to achieve a desired medical effect. This emerging direction poses new design and structural challenges in terms of achieving large stretchability, robust functionality, reliable stability, and wearability and comfort of flexible electronic devices.

To meet the material and structural performance in flexible electronic applications, researchers typically explore two methods: (i) developing new materials, which involves inventing new formulations for functional materials to achieve high tensile properties and applying them directly to the device; (ii) designing novel structures through mechanics theories, which involves design and analysis of two- or three-dimensional (3D) structures that allow essentially rigid materials to accommodate large deformations without damage, thereby improving the device's comfort when in contact with people.

This chapter will focus on the structural design method to achieve highly flexible device architectures in different dimensional spaces. Sections 6.2–6.4 will review a range of special mechanical structures applied in the field of flexible electronics, with

Mechanics of Flexible and Stretchable Electronics, First Edition. Edited by Yong Zhu and Nanshu Lu.

the hope of inspiring readers to integrate structural design into the future development of flexible electronic devices and discover novel structures that can outperform existing devices to push the limit of flexible electronics. The chapter is organized into three areas: (i) planar designs for stretchable and flexible devices; (ii) 3D structure designs for flexible electronics; and (iii) protective structure designs for flexible electronics.

6.2 Design of Planar Stretchable and Flexible Structures

6.2.1 Wave/Wrinkle Structure Design

The wave/wrinkle structure design is the first method used for designing stretchable inorganic electronics [7]. This design is achieved by introducing buckling into the film-substrate system. Specifically, a hard thin film ribbon is fully bonded to a pre-strained soft substrate, induced by mechanical pre-stretching [7, 8] or eigenstrain [9, 10]. When the pre-strain is released, the hard thin film will buckle and form a highly periodic wave-like structure (Figure 6.1). Under tension or compression, the wavelength and amplitude of the wave/wrinkle structure will change, allowing the device to adapt and withstand a greater strain [7].

To better understand and apply the wave/wrinkle structures in practice, a series of mechanical models have been developed to predict their deformed configurations and stretchability. Assuming that the deformation is small (when the pre-strain $\varepsilon_{pre} < 5\%$) and using the principle of minimum potential energy, the wavelength λ_0, amplitude A_0, and peak strain ε_{peak} can be given [7]:

$$\lambda_0 = 2\pi h_f \left(\frac{\overline{E}_f}{3\overline{E}_s} \right)^{\frac{1}{3}} \tag{6.1}$$

$$A_0 = h_f \sqrt{\frac{\varepsilon_{pre} - \varepsilon_{appl}}{\varepsilon_c} - 1} \tag{6.2}$$

$$\varepsilon_{peak} = 2\varepsilon_c \sqrt{\frac{\varepsilon_{pre} - \varepsilon_{appl}}{\varepsilon_c} - 1} \tag{6.3}$$

where the subscripts "f" and "s" denote the ribbon film and substrate, respectively; $\overline{E} = E/(1 - \nu^2)$ is the plane strain modulus, ν is the Poisson's ratios; h_f is the film thickness; ε_{appl} is the strain applied to the substrate after buckling; and $\varepsilon_c = (3\overline{E}_s/\overline{E}_f)^{\frac{2}{3}}$ is the critical buckling strain, i.e. the lowest strain at which the film will buckle. ε_c is usually very small (for a typical system consisting of Silicon [Si] and polydimethylsiloxane [PDMS] is about 0.034%), combined with Eq. (6.3), the maximum strain in the film after buckling is much smaller than the pre-strain and the tensile strain to which it is subjected, which enables the system to withstand a large deformation.

It is worth noting that the wavelength is independent of the pre-strain, as shown in Eq. (6.1). However, subsequent experimental studies revealed that the

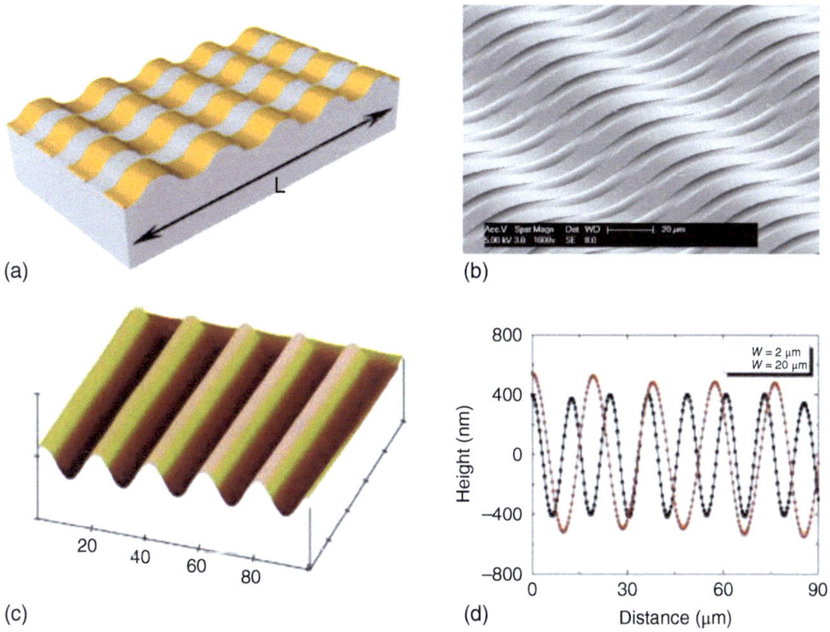

Figure 6.1 (a) Schematic illustration of wave/wrinkle structure (hard thin film ribbon with yellow and substrate with gray); Source: Ref. [11]. Reproduced with permission of AIP Publishing, © 2008. (b) SEM images of the wave structure of wavy diamond ribbons on PDMS substrate system). (c) Atomic force micrographs of the wave structure; (d) Wavelength/amplitude diagram of the wave structure with different ribbon width. Source: Ref. [12]. Reproduced with permission of Elsevier, © 2008.

wavelength after buckling decreases with increasing pre-strain for larger pre-strains (Figure 6.2a). This is mainly because the small deformation assumption no longer holds at this pre strain range. To address this issue, Jiang et al. [13] used finite deformation theory to establish a mapping relationship between the unbuckling and buckling configurations by considering the nonlinear deformation of the substrate. They proposed an analytical solution for the wavelength, amplitude, and maximum strain of film buckling under large substrate deformation expressed as:

$$\lambda = \frac{\lambda_0 \ (1+\varepsilon_{\text{appl}})}{(1+\varepsilon_{\text{pre}}) \ (1+\varepsilon_{\text{appl}}+\zeta)^{\frac{1}{3}}} \tag{6.4}$$

$$A = \frac{A_0}{\sqrt{1+\varepsilon_{\text{pre}}}(1+\varepsilon_{\text{appl}}+\zeta)^{\frac{1}{3}}} \tag{6.5}$$

$$\varepsilon_{\text{peak}} = 2\sqrt{(\varepsilon_{\text{pre}}-\varepsilon_{\text{appl}})\varepsilon_{\text{c}}\frac{(1+\varepsilon_{\text{appl}}+\zeta)^{\frac{1}{3}}}{\sqrt{1+\varepsilon_{\text{pre}}}}} \tag{6.6}$$

where λ_0 and A_0 are the results in Eqs. (6.1) and (6.2), respectively, and ζ can be obtained by the following equation:

$$\zeta = \frac{5(\varepsilon_{\text{pre}}-\varepsilon_{\text{appl}})(1+\varepsilon_{\text{pre}})}{32} \tag{6.7}$$

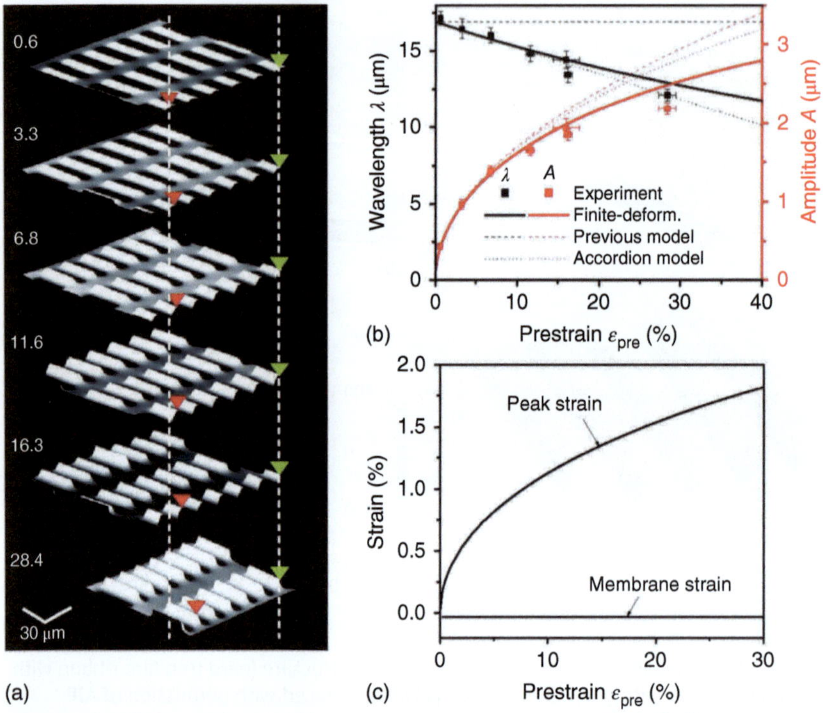

Figure 6.2 (a) Atomic force micrograph images of wave structure with different pre-strains (left, in percent); the arrows and dotted lines define the relative positions on the material, which indicate the wavelength decreases as the pre-strain increases. (b) The relationship between the wavelength/amplitude and pre-strain under the initial model, modified model and experiment (Si-PDMS system, the thickness of Si is 100 nm). (c) The membrane strain and peak strain of ribbon under pre-strain (Si-PDMS system, the thickness of Si is 100 nm). Source: Ref. [13]. Reproduced with permission of National Academy of Sciences, © 2007.

Figure 6.2b shows the wavelength and amplitude of buckled structures of 100 nm thickness Silicon (Si) on PDMS as a function of the pre-strain. When compared to the initial model, the new model strongly agrees with the actual experimental results, providing an effective analytical method for predicting wavelengths and amplitudes at large pre-strains. Figure 6.2c shows that for a 100 nm Si ribbon the strain is very small, which can be neglected, and does not vary with pre-strain. Using Eq. (6.6), the stretchability obtained by pre-straining the Si-PDMS system (~29.2%) is much larger than the fracture strain of a single-crystal Si film (~1.8%), indicating that the system can significantly improve the capacity to withstand tensile deformation in brittle materials.

Figure 6.3 illustrates the detailed processes for preparing the wave single-crystal Si ribbon structure on a soft substrate (e.g. PDMS). The process involves the following steps:

(1) Etching the single-crystal Si plate from the bottom up, leaving a thin layer of single-crystal Si on top and using photolithography to form a Si ribbon pattern on

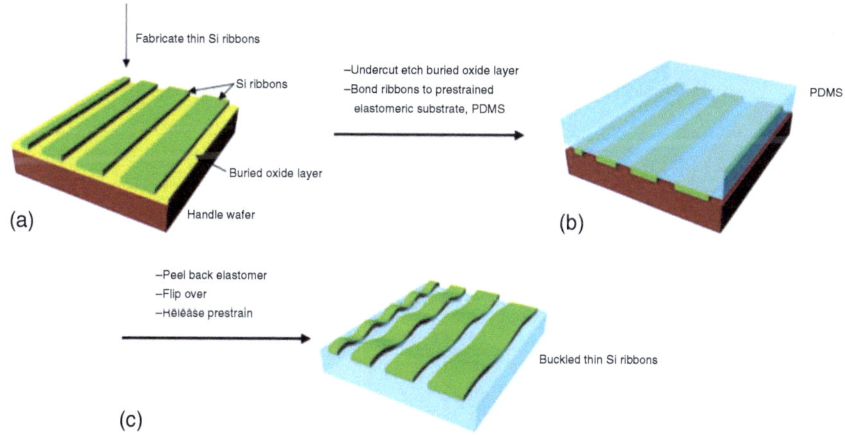

Figure 6.3 A typical wave structure preparation process.(a) Fabricate the Si ribbons on the buried oxide layer. (b) Remove the buried oxide layer by undercut etch and bond Si ribbons to the prestrained elastromic substrate. (c) Peel back the elastomer, flip, and release prestrain to form wave structure. Source: Ref. [12]. Reproduced with permission of Elsevier, © 2008.

the film. The oxide is then etched using anisotropic etching technology, resulting in a Si ribbon that is supported by the wafer but not bonded to it.

(2) The pre-strained substrate (e.g. PDMS) is then placed on the Si ribbon film, creating a bond between the two layers.
(3) The pre-strained substrate (e.g. PDMS) is then peeled back, and the Si ribbon film remains on the substrate. By releasing the pre-strain, the wave/wrinkle structure can be obtained.

Furthermore, a novel approach to forming wave structures involves partially bonding the film to the substrate (Figure 6.4) [15]. This method results in a structure where the wavelength λ and amplitude A are independent of the material's properties and only depend on the bond length and pre-strain as [14]:

$$\lambda = \frac{W_{\text{in}}}{1 + \varepsilon_{\text{pre}}} \tag{6.8}$$

$$A \approx \frac{2}{\pi(1 + \varepsilon_{\text{pre}})} \sqrt{\varepsilon_{\text{pre}} W_{\text{in}} [W_{\text{in}} + W_{\text{act}}(1 + \varepsilon_{\text{pre}})]} \tag{6.9}$$

where W_{in} and W_{act} are the length of activated (bonded) and inactivated (unbonded) sites.

Due to their ease of preparation and high stretchability, wave/wrinkle structures have attracted many researchers and have been used in numerous applications such as micro- and nano-graphic generations [16–18], epidermal sensors [19, 20], and flexible screens [21].

6.2.2 Island–Bridge Structure Design

The island–bridge structure in flexible electronics, which was partially inspired by the wave/wrinkle structure, has undergone significant development that spans to

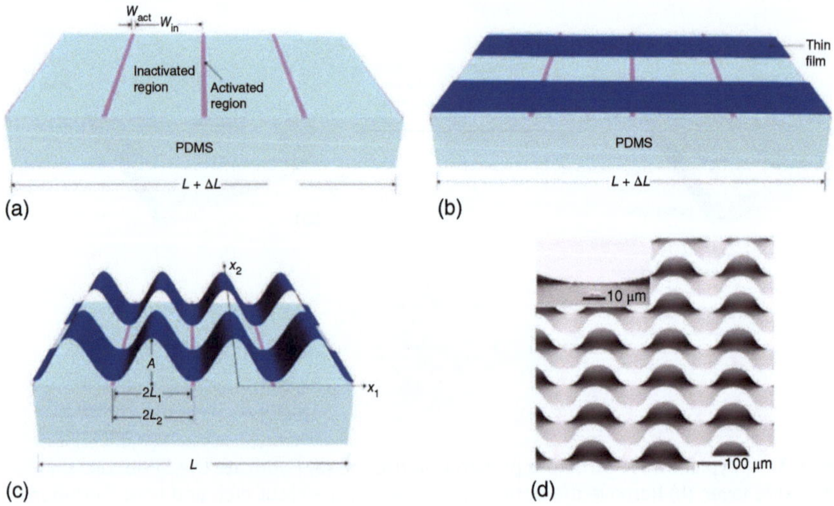

Figure 6.4 Partially bonded wave/wrinkle structure (a) Pre-strained substrate and bonding areas. (b) A thin film parallel to the prestrain direction is attached to the prestrained and patterned PDMS substrate. (c) Schematic images and (d) SEM images. Source: Ref. [14]. Reproduced with permission of AIP Publishing, © 2007.

full areal coverage of body parts using this technique to enhance the system-level stretchability. This structure separates the rigid electronic components into two parts: the island section, which carries the functional components, and the bridge section, which houses the interconnecting wires for the functional components (either coplanar or noncoplanar connecting). When strain is applied, the bridge will deform in and out of the plane to release some of the strain and provide stretchability, which ensures that the rigid island is almost free from stress, and maintains high electrical performance. Consequently, the device's stretchability primarily depends on the bridge design. In this regard, three different island–bridge structures used in flexible electronics will be discussed further.

6.2.2.1 Straight Interconnecting Island–Bridge Structure

The first proposed island–bridge structure is the straight interconnecting structure, which consists of islands strongly bonded to the substrate and straight interconnecting wires between them that may or may not be bonded on the pre-strained substrate. As previously discussed in Section 6.2.1, the bonded straight wires will form the wave/wrinkle structures depending on the bonded and unbonded lengths. Here, the focus of the discussion is on unbonded straight wires. After the pre-strain is released, the unbonded wire buckles out-of-plane to form an arch bridge shape (as shown in Figure 6.5a), which enhances the stretchability of flexible electronic devices. Experiments and finite element analysis (FEA) have confirmed that when the structure is stretched and twisted in various directions, the output voltage remains essentially the same as that of the undeformed structure (as illustrated in Figure 6.5b). Consequently, the device can achieve complex deformations with various levels of curvature (as demonstrated in Figure 6.5c).

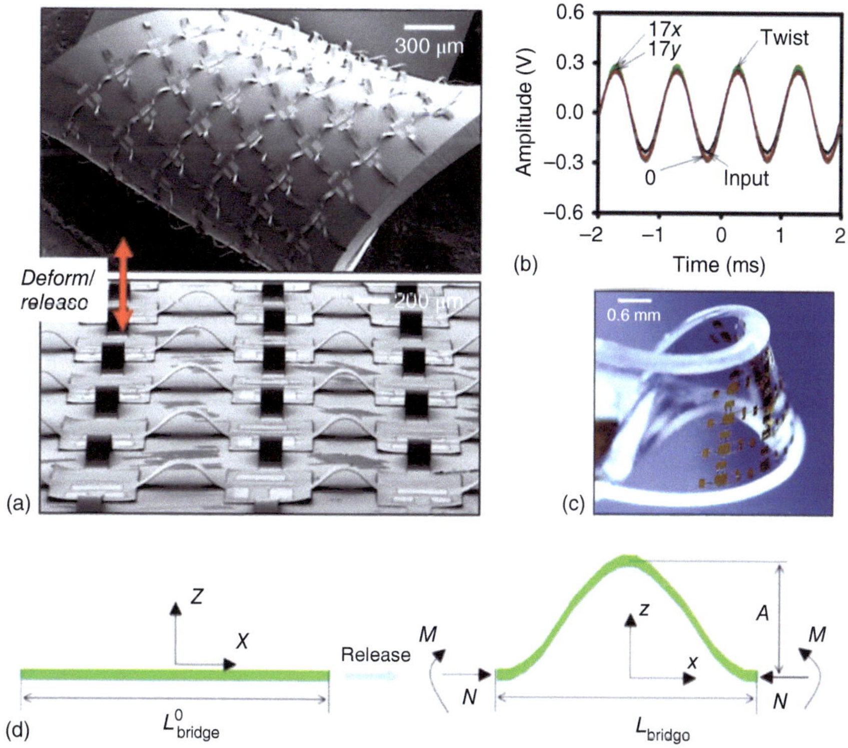

Figure 6.5 (a) SEM images of an array of CMOS inverters under complex deformation (top) and undeformed state (bottom, ≈20% pre-strain). (b) The corresponding electrical output as a function of time for a sinusoidal input under different states (undeformed, 17% tensile strains along the x direction, 17% tensile strains along the y direction, twist). (c) The optical image of a complex deformed device. Source: Ref. [22]. Reproduced with permission of National Academy of Sciences, © 2008. (d) Schematic illustration of straight-interconnecting island-bridge structure. Source: Ref. [23]. Reproduced with permission of AIP Publishing, © 2009.

Song et al. [23] assumed that the wires can be appropriately modeled as the fixed beam supported at both ends (Figure 6.5d). Using the minimization of total energy, they derived the amplitude A of the bridge, and the maximum strains of the island $\left(\varepsilon^{\max}_{\text{bridge}}\right)$ and bridge $\left(\varepsilon^{\max}_{\text{island}}\right)$ as well as the total stretchability $(\varepsilon_{\text{stretchability}})$ as:

$$A = \frac{2L^0_{\text{bridge}}}{\pi}\sqrt{\frac{L^0_{\text{bridge}} - L_{\text{bridge}}}{L^0_{\text{bridge}}} - \varepsilon_{\text{c}}} \tag{6.10}$$

$$\varepsilon^{\max}_{\text{bridge}} = 2\pi\frac{h_{\text{bridge}}}{L^0_{\text{bridge}}}\sqrt{\frac{\varepsilon_{\text{pre}}}{1+\varepsilon_{\text{pre}}}} \tag{6.11}$$

$$\varepsilon^{\max}_{\text{island}} = \frac{\left(1-v^2_{\text{island}}\right)E_{\text{bridge}}h^2_{\text{bridge}}}{E_{\text{island}}h^2_{\text{island}}}\varepsilon^{\max}_{\text{bridge}} \tag{6.12}$$

$$\varepsilon_{\text{stretchability}} = \frac{\varepsilon_{\text{pre}}}{1 + (1 + \varepsilon_{\text{pre}})\frac{L^0_{\text{island}}}{L^0_{\text{bridge}}}} \tag{6.13}$$

where h_{island}, L^0_{island}, E_{island}, and v_{island} are the thickness, initial length, elastic modulus, and Poisson's ratio of the island respectively; h_{bridge}, L^0_{bridge}, L_{bridge}, and E_{bridge} are the thickness, initial length, length after buckling, and elastic modulus of the bridge; $\varepsilon_c = \frac{\pi^2 h^2_{\text{bridge}}}{\left[3\left(L^0_{\text{bridge}}\right)^2\right]}$ is the critical buckling strain for Euler buckling of the beam. From Eq. (6.10), only the condition $\left(L^0_{\text{bridge}} - L_{\text{bridge}}\right)/L^0_{\text{bridge}} \geq \varepsilon_c$ is satisfied, and buckling occurs. Since the ratio $h_{\text{bridge}}/L^0_{\text{bridge}} \ll 1$, Eq. (6.11) reveals that the maximum strain of the wire is much smaller than the pre-strain of the substrate. Further reducing the strain on the island can be achieved by increasing the thickness of the island (according to Eq. (6.12)). To improve the overall stretchability of the device, according to the terms in Eq. (6.13), the following suggestions can be adopted: lengthening the wire, reducing the length of the island, and increasing the pre-strain.

The preparation steps of the straight interconnecting island–bridge structure design are illustrated in Figure 6.6, which is like the wave/wrinkle structure, but only islands are bonded to the substrate whereas the bridges are nonbonded. After the gradual release of pre-strain, the arch bridge is formed as a result.

Due to the excellent mechanical and electrical properties of the resulting structure, the straight wire island–bridge construction can increase the stretchability of electronic devices by up to 50% approximately without compromising the electrical performance. This makes them a desirable design strategy that has been widely adopted in stretchable electronic systems including eyeball-like digital cameras [24, 25], flexible screens [26], and photovoltaic devices [27].

6.2.2.2 Serpentine Interconnecting Island–Bridge Structure

Based on the mechanics analysis above, the straight interconnecting island–bridge structure is an efficient way to improve the stretchability of the electronics.

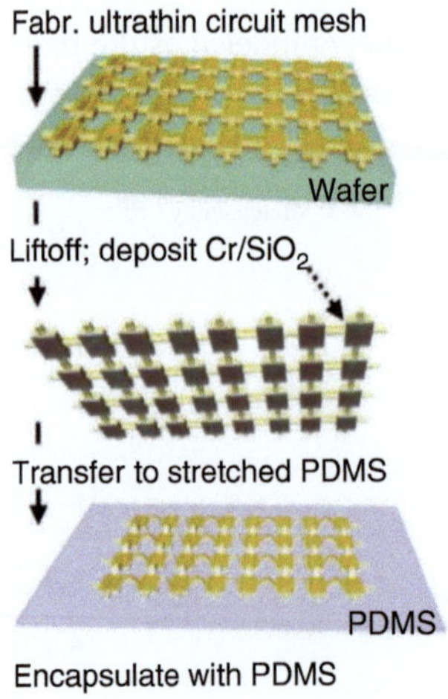

Figure 6.6 The preparation process of straight interconnecting island–bridge structure design. Source: Ref. [22]. Reproduced with permission of National Academy of Sciences, © 2008.

However, the straight interconnecting structures naturally lead to conflicts between the goal of creating a system with a high stretchability but reliably incorporating enough functional components for it to work properly. This conflict is captured as a filling ratio, i.e. high stretchability requires more space between the islands (from Eq. (6.13)), whereas the ratio of the area of functional components to the area of electronic devices will have to decrease. The quantitative relationship between the system stretchability and functional component filling ratio is proposed by Zhang et al. [28]:

$$\text{Stretchability of the system} = (1 - \sqrt{\text{Filling ratio}}) \times (\text{Stretchability of the interconnect}) \quad (6.14)$$

To address the problem of balancing stretchability and functional component coverage, a new configuration was developed by combining the serpentine wire structure design with the island–bridge structure design, known as the serpentine interconnecting island–bridge structure. In this structure, the straight wires are replaced with serpentine wires, which exhibit better stretchability at smaller spacing between islands. The structure does not require pre-straining, and when stretched, the wires bend and twist out of the plane through out-of-plane buckling, reducing the accumulation of strain energy in the serpentine interconnecting wires. Similar to the straight wire structure, the islands are nearly free from strain, but the total stretchability can reach approximately 100% [22]. Figure 6.7a,b illustrate the diagram and optical photos of the serpentine interconnecting island–bridge structures used in flexible electronics.

Zhang et al. [29] proposed a scaling law and derived the analytical expressions for the critical buckling strain ε_{cr} and the maximum local strain ε_{max} for a multiunit serpentine guide (Figure 6.7c) in tension:

$$\varepsilon_{cr} = \frac{\alpha(\alpha^2 + 6) + \frac{3\pi}{4}(2\alpha^2 + 1) - \frac{9(\alpha^2 + \pi\alpha + 2)^2}{8\alpha+\pi+8(2\alpha+\pi)m^2}}{f_1(m) + f_2(m)\alpha + f_3(m)\alpha^2}\sqrt{\frac{G}{E}}\frac{t^2}{w^2} \quad (6.15)$$

$$\varepsilon_{max} = g_1(m, \alpha)\sqrt{\varepsilon_{appl}}\frac{t}{l_1} + g_2(m, \alpha)\varepsilon_{appl}^2\frac{w}{l_1} \quad (6.16)$$

where $\alpha = l_2/l_1$ is the height-span ratio of the wire, m is the unit number of the serpentine wire, w is the section width, t is the thickness, G is the torsional modulus, E is the Young's modulus, f_1, f_2, and f_3 are the contributions of each second-order term of length scale to the lateral buckling, ε_{appl} is the strain applied at both ends of the wire, and g_1, g_2 are the coefficients depending on the number of unit cells m and the length/spacing ratio α. The relationship between ε_{cr} and the cross-sectional aspect ratio is shown in Figure 6.8.

To avoid damage caused by contact between the wires and adjacent materials, the fully bonded serpentine interconnect structures are also proposed. These structures still remain highly stretchable, experiments show that elastic stretchability can generally reach about 54% (non-prestrained substrate) and about 120% (40% pre-strained substrate) [28].

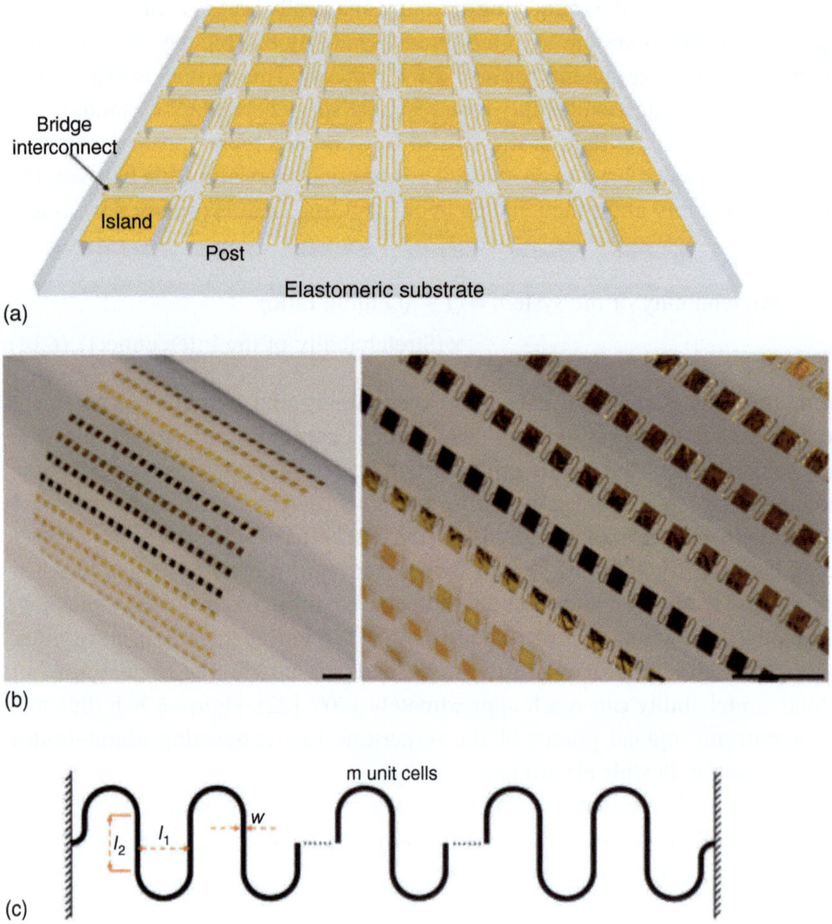

Figure 6.7 (a) Schematic images of serpentine interconnecting island–bridge structure. (b) Optical images of serpentine interconnecting island–bridge structure. (c) The geometric schematic of serpentine wire. Source: Ref. [29]. Reproduced with permission of Royal Society of Chemistry, © 2013.

To achieve stretchability in fully bonded serpentine interconnect structures with pre-strained substrates (Figure 6.9), the thickness of the serpentine wires is a major factor affecting its stretchability. Increasing the metal thickness leads to a significant decrease in elastic stretchability, while thinner metals result in greater stretchability (Figure 6.10). Some experiments have shown a significant nonlinear decrease in elastic stretchability, which is due to the transition from local buckling to global buckling. Further, the buckling wavelength λ is given by [28]:

$$\lambda = 2\pi\left[\frac{\overline{E}_{\mathrm{PI}}}{3\overline{E}_{\mathrm{s}}}(2t_{\mathrm{PI}} + t_{\mathrm{mental}})^3 + \frac{\overline{E}_{\mathrm{metal}} - \overline{E}_{\mathrm{PI}}}{3\overline{E}_{\mathrm{s}}}t_{\mathrm{metal}}^3\right]^{\frac{1}{3}} g(\varepsilon_{\mathrm{appl}}) \tag{6.17}$$

where $g(\varepsilon_{\mathrm{appl}}) = \left[1 + \frac{5}{32}(1 + \varepsilon_{\mathrm{appl}} - \sqrt{1 + \varepsilon_{\mathrm{appl}}})\right]^{-1/3} / (\sqrt{1 + \varepsilon_{\mathrm{appl}}})$; $\overline{E} = E/(1 - \nu^2)$ is the plane strain modulus, ν is the Poisson's ratios, t is the thickness, and $\varepsilon_{\mathrm{appl}}$ is the

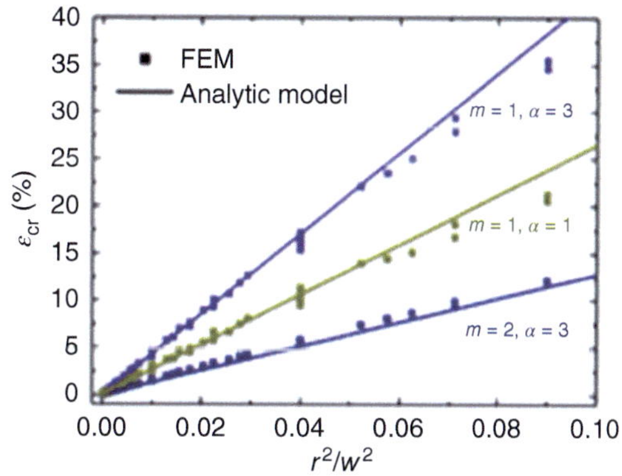

Figure 6.8 The dependences of critical buckling strain on the square of the cross-sectional aspect ratio. Source: Ref. [29]. Reproduced with permission of Royal Society of Chemistry, © 2013.

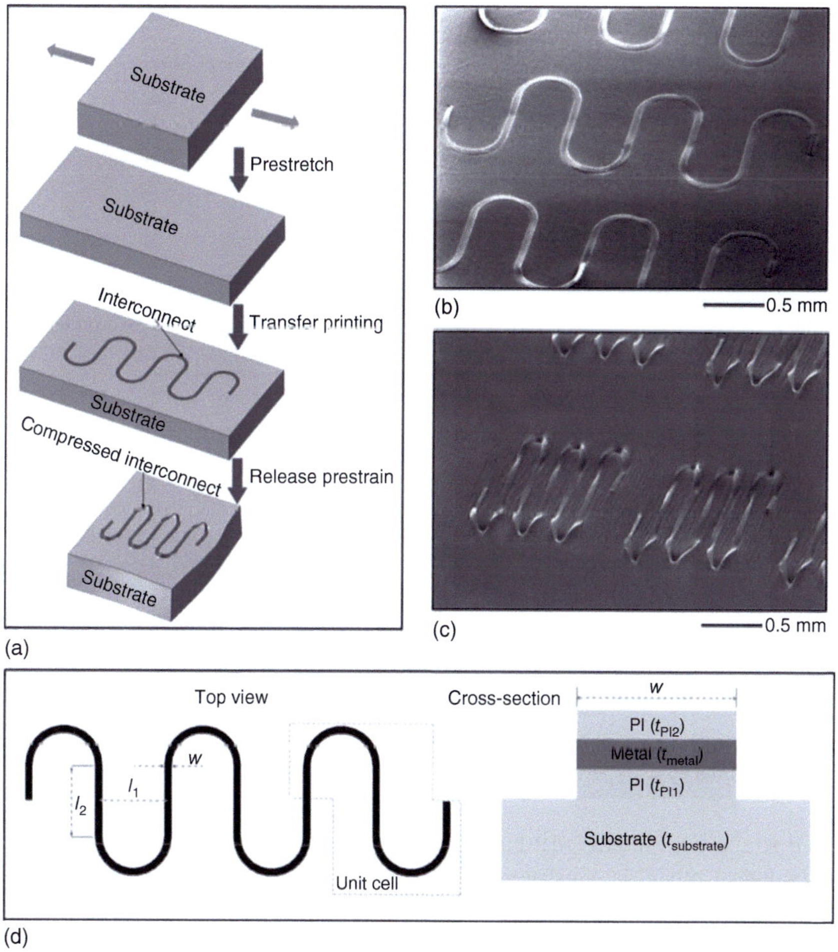

Figure 6.9 (a) Preparation flow of the fully bonded serpentine interconnects. The SEM images of (b) before pre-strain release and (c) after pre-strain released fully bonded serpentine wires. (d) Top view and cross-section view of the structure. Source: Ref. [28]. Reproduced with permission of John Wiley & Sons, © 2014.

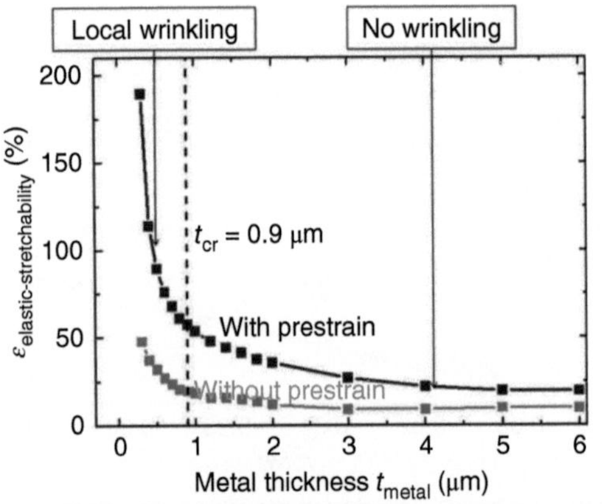

Figure 6.10 The elastic stretchability as a function of metal thickness. Source: Ref. [28]. Reproduced with permission of John Wiley & Sons, © 2014.

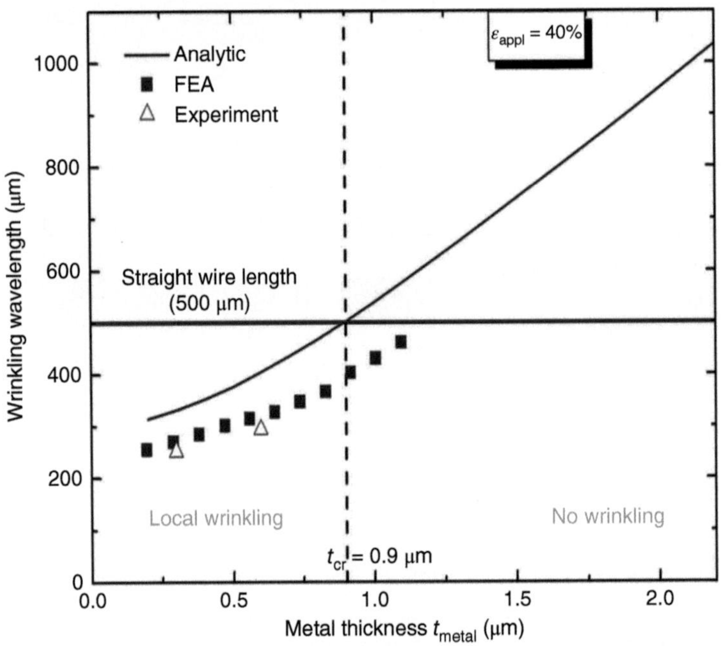

Figure 6.11 The relationships between wrinkling wavelength and metal thickness. Source: Ref. [28]. Reproduced with permission of John Wiley & Sons, © 2014.

applied strain. Global and local buckling occurs for $\lambda \geq l_2$ and $\lambda < l_2$, respectively. The analytical model is consistent with both the FEA and experimental results (Figure 6.11).

Further, Pan et al. [30] developed experiments and theory to understand the influence of the elastomer substrate thickness on the buckling mode and the device's stretchability. The schematic diagram of the device used is shown in Figure 6.12. They discovered that increasing the substrate thickness results in insensitivity to the serpentine interconnect if it is thick enough to induce the local wrinkling mode.

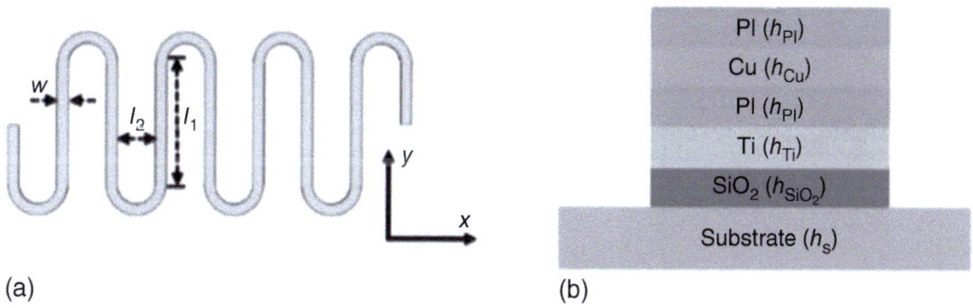

Figure 6.12 Schematic diagram of serpentine wire in (a) top view and (b) cross-section view. Source: Ref. [30]. Reproduced with permission of John Wiley & Sons, © 2017.

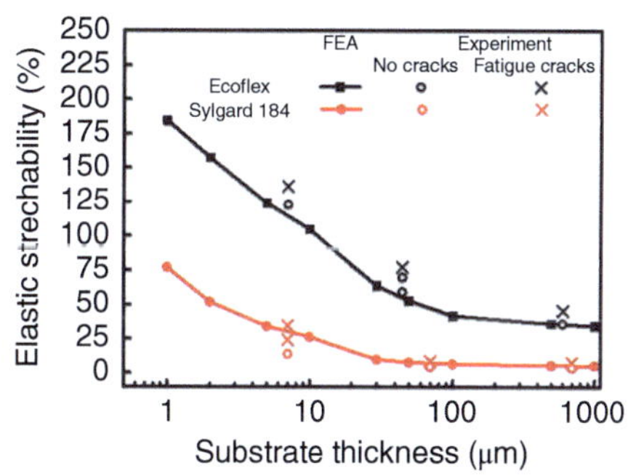

Figure 6.13 The elastic stretchability as a function of substrate thickness. Source: Ref. [30]. Reproduced with permission of John Wiley & Sons, © 2017.

However, when global buckling occurs, decreasing the substrate thickness leads to a significant increase in elastic stretchability (Figure 6.13). The buckling mode can be judged by the critical length l_{cr} [30]:

$$l_{cr} = 4\pi\sqrt{\overline{EI}\left\{\frac{[\overline{E}_f/(3\overline{E}_s)]^{2/3}}{\overline{E}_f h_f + \overline{E}_s h_s} - \frac{0.3}{G_s(h_f + h_s)}\right\}} \tag{6.18}$$

where $h_f = \sqrt{12EI/\sum_{i=1}^{5}\overline{E}_i h_i}$ is the efficient thickness of multilayer wires, $\overline{E}_f = \sum_{i=1}^{5}\overline{E}_i h_i/h_f$ is the effective plane-strain modulus of multilayer wires. h_s, $\overline{E}_s$, and ν_s are the height, plane-strain modulus, and Poisson's ratio of the substrate. $G_s = \overline{E}_s/[2(1+\nu_s)]$ is the shear modulus of the substrate. The effective bending stiffness $\overline{EI}$ of the devices is given as follows:

$$EI = \frac{\left(\overline{E}_f h_f^2 - \overline{E}_s h_s^2\right)^2 + 4\overline{E}_s h_s \overline{E}_f h_f (h_s + h_f)^2}{12(\overline{E}_s h_s + \overline{E}_f h_f)} \tag{6.19}$$

Global and local buckling occurs for $l > l_{cr}$ and $l < l_{cr}$, respectively, where l is the length of the film. FEA and scanning electron microscope (SEM) characterization results show that the experimental results are consistent with the analytical model results (Figure 6.14).

Due to its superior performance and space-saving abilities, the serpentine structure is widely used in flexible electronic devices. Xie et al. [31] designed a stretchable near-field communication antenna based on a serpentine microstructure, which

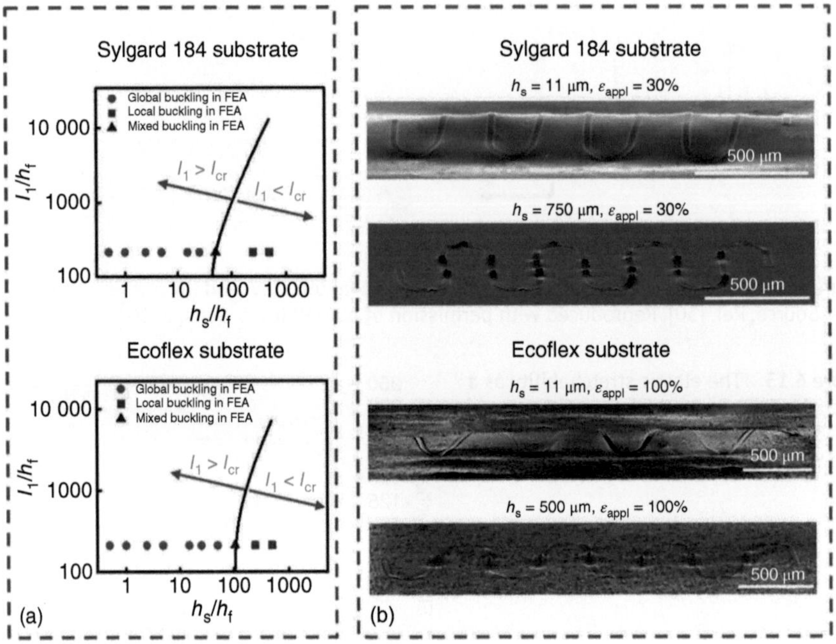

Figure 6.14 (a) Buckling modes obtained by FEA. (b) SEM images of deformed serpentine interconnects with different substrate materials and thicknesses. Source: Ref. [30]. Reproduced with permission of John Wiley & Sons, © 2017.

enables wireless, battery-free power and data transmission, works stably under large deformation, and seemly integrate with human skin. Chung et al. [32] used the serpentine structure to design a wireless epidermal electronic for neonatal intensive care monitoring.

6.2.2.3 Fractal-Inspired Interconnecting Island–Bridge Structure

Another effective way to make full use of the limited space between the functional islands is by using fractal-inspired interconnect design, also known as self-similar design. In this design method, geometry can be divided into several parts (or self-similar units) and each part is the reduced shape of the whole. Xu et al. [33] were the first to adapt fractal designs for flexible electronic applications. Further, the method has been extended to other flexible electronic designs, especially in island–bridge design. Figure 6.15 shows several representative fractal interconnect designs with their corresponding FEA and MicroXCT images.

Zhang et al. [35] used the recursive method to solve the equivalent stiffness of all the levels of a structure with thick fractal wires (wires of similar thickness and width that do not buckle under tension) and established an analytical model. The illustration of the self-similar fractal rectangular and serpentine wire from order 1–4 is detailed in Figure 6.16. They derived the analytical solution expression for the elastic elongation of any thick fractal wire, where the stretchability of the rectangular structure is given by:

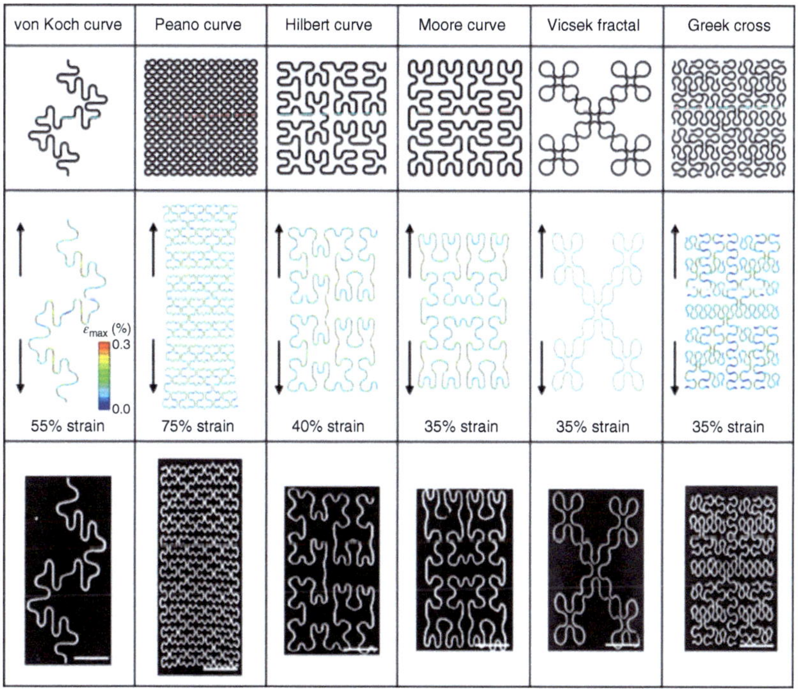

Figure 6.15 The representative fractal interconnects design with FEA and MicroXCT images. Source: Ref. [34]. Reproduced with permission of Springer Nature, © 2014.

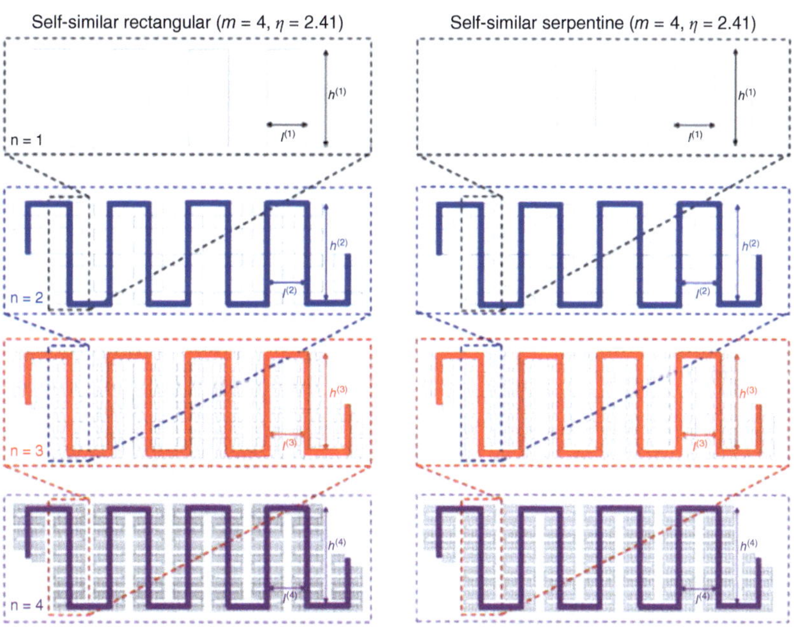

Figure 6.16 Schematic illustration of the fractal rectangular and serpentine wire from order one to four. Source: Ref. [35]. Reproduced with permission of Elsevier, © 2013.

$$\varepsilon^{(1)}_{\text{stretchability}} = \frac{\varepsilon_{\text{yield}} l^{(1)}}{\omega} \frac{\eta^{(1)}}{12} \times \frac{16[m^{(1)}]^2[\eta^{(1)}+1][\eta^{(1)}+3] - [\eta^{(1)}+6]^2}{4[m^{(1)}]^2[\eta^{(1)}+1] + 3m^{(1)}[\eta^{(1)}+2] - \eta^{(1)} - 6}$$
$$\text{(fractal series } n = 1)$$
$$\varepsilon^{(n)}_{\text{stretchability}} = \frac{\varepsilon_{\text{yield}} l^{(n)}}{\omega} \frac{2}{m^{(n)}} \times \left| \frac{\overline{T}^{(n)}_{11}\left[\overline{T}^{(n)}_{23}\right]^2 - \overline{T}^{(n)}_{11}\overline{T}^{(n)}_{22}\overline{T}^{(n)}_{33} + \overline{T}^{(n)}_{33}\left[\overline{T}^{(n)}_{12}\right]^2}{2\overline{T}^{(n)}_{12}\left[\overline{T}^{(n)}_{33} - \overline{T}^{(n)}_{23}\right] + \left\{\left[\overline{T}^{(n)}_{23}\right]^2 - \overline{T}^{(n)}_{22}\overline{T}^{(n)}_{33}\right\}\eta^{(n)} - \frac{\eta^{(n-1)}\eta^{(n)}}{2m^{(n-1)}}\overline{T}^{(n)}_{12}\overline{T}^{(n)}_{33}} \right|$$
$$\text{(fractal series } n \geq 2) \tag{6.20}$$

The stretchability of the serpentine structure is

$$\varepsilon_{\text{stretchability}} = \frac{2\varepsilon_{\text{yield}} l^{(1)}}{m^{(1)}\omega} \frac{\left(\overline{T}^{(1)}_{11}\overline{T}^{(1)}_{22}\overline{T}^{(1)}_{33} - \overline{T}^{(1)}_{11}\left[\overline{T}^{(1)}_{23}\right]^2 - \overline{T}^{(1)}_{33}\left[\overline{T}^{(1)}_{12}\right]^2\right)}{2\overline{T}^{(1)}_{12}\overline{T}^{(1)}_{23} - 3\overline{T}^{(1)}_{12}\overline{T}^{(1)}_{33} + \left\{\overline{T}^{(1)}_{22}\overline{T}^{(1)}_{33} - \left[\overline{T}^{(1)}_{23}\right]^2\right\}(\eta^{(1)} - 1) + \sqrt{\left\{\overline{T}^{(1)}_{22}\overline{T}^{(1)}_{33} - \left[\overline{T}^{(1)}_{23}\right]^2\right\}^2 + \left[\overline{T}^{(1)}_{12}\right]^2\left[\overline{T}^{(1)}_{33}\right]^2}}$$
$$\text{(fractal series } n = 1) \tag{6.21}$$

where $\varepsilon_{\text{yield}}$ is the yield strain of the material, $l^{(n)}$, $\eta^{(n)}$, $m^{(n)}$, and ω are respectively the span, span-height ratio, number of elements, and width of the first-order fractal structure, $\overline{T}^{(n)}_{ij}$ $(i, j = 1,2,3)$ is the flexibility matrix component of the nth-order structure.

Figure 6.17 shows the relationship between the dimensionless elastic stretchability of strictly self-similar serpentine wire $\overline{\varepsilon}_{\text{stretchability}} = \varepsilon_{\text{stretchability}} w/[\varepsilon l^{(n)}]$ and the self-similar series n of the wire. An increase in stretchability can be observed with an increase in self-similar order. Each time the self-similar order is increased by one, the stretchability increases by more than a factor of one, demonstrating that the fractal design is an effective method for enhancing the elastic stretchability of the structure.

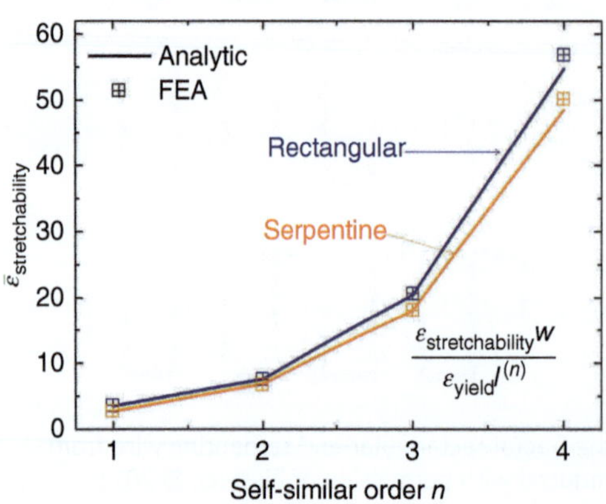

Figure 6.17 Relationship between dimensionless elastic stretchability and the number of self-similarity steps of a conductor. Source: Ref. [35]. Reproduced with permission of Elsevier, © 2013.

6.3 Design of Three-Dimensional Flexible Electronic Structures

6.3.1 Helical Design

Helical structures (Figure 6.18) consist of thin and slender (thickness much less than width, width much less than arc length) ribbons, which ensures that the deformation is mainly controlled by out-of-plane bending in the width direction. These structures can deform like a spring and absorb large strains. Theoretical research has focused on the buckling and post-buckling behavior of helical designs. The theoretical displacement field model for the buckling-guided 3D helical structure was established by Liu et al. [36]:

$$U_1 = bR_{\sin}\left(\frac{\pi\theta}{2\theta_0}\right)\left(2-\frac{\theta}{\theta_0}\right)^2\left(\frac{\theta}{\theta_0}\right)^2\sqrt{\varepsilon_{\mathrm{appl}}} \tag{6.22}$$

$$U_2 = -R\varepsilon_{\mathrm{appl}}\cos\left(\frac{\pi\theta}{\theta_0}\right)\left(1-\frac{\theta}{\theta_0}\right)\left(\frac{\theta}{\theta_0}\right)\times\left[c_1\frac{\theta}{\theta_0}\left(2-\frac{\theta}{\theta_0}\right)+\frac{\theta_0}{2}\sin\theta_0\right] \tag{6.23}$$

$$U_3 = R\varepsilon_{\mathrm{appl}}\times\left[\cos\left(\frac{\theta_0}{2}\right)^2\cos\left(\frac{\pi\theta}{\theta_0}\right)\left(1-\frac{\theta}{\theta_0}\right)\theta + c_2\sin\left(\frac{\pi\theta}{\theta_0}\right)\left(1-\frac{\theta}{\theta_0}\right)^3\left(\frac{\theta}{\theta_0}\right)\right] \tag{6.24}$$

where R is the arc radius of the serpentine wire, θ_0 is the arc top angle, θ is the introduced parametric coordinate indicating the position of the arc along the central axis, corresponding to the arc length $S = R\theta$; $\varepsilon_{\mathrm{appl}}$ is the externally applied strain, b, c_1, and c_2 are dimensionless parameters to be determined based on the principle of minimization of energy.

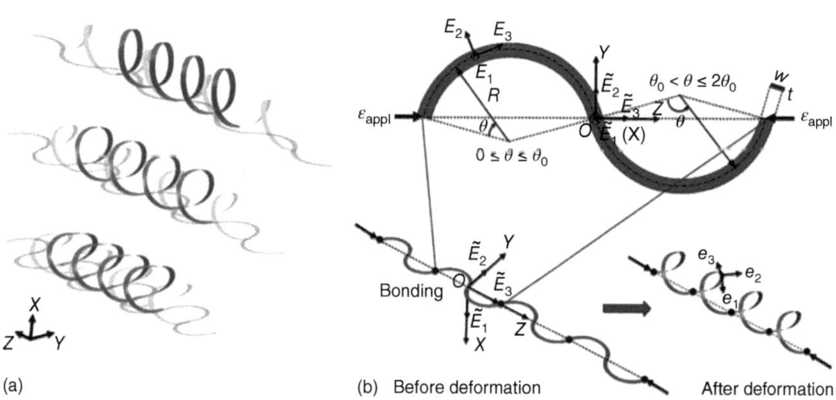

Figure 6.18 (a) Schematic of helical structure and (b) the mechanics model and coordinate system. Source: Ref. [36]. Reproduced with permission of John Wiley & Sons, © 2016.

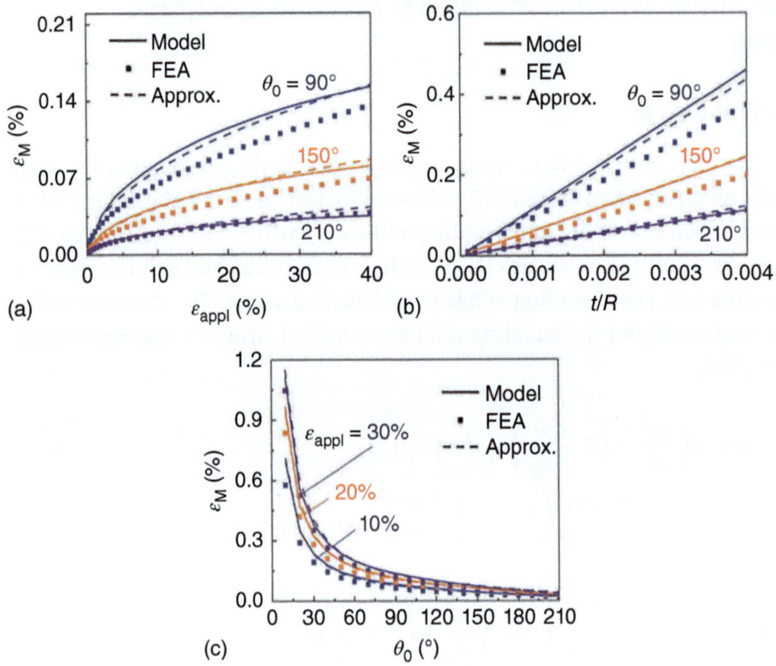

Figure 6.19 Comparation between the precise model calculation, approximate solution, and FEA for a maximum principal strain of material under different (a) applied strain; (b) normalized thickness; and (c) arc angle. Source: Ref. [36]. Reproduced with permission of John Wiley & Sons, © 2016.

The maximum principal strain in the helical structure can be expressed as [36]:

$$\varepsilon_M = F_2(\theta_0)\frac{t}{R}\sqrt{\varepsilon_{appl}} \tag{6.25}$$

where F_2 is the function of θ_0, t is the thickness of wire, which indicates that the maximum strain is only dependent on the applied strain, arc angle, and normalized thickness. Figure 6.19 shows the strong agreement between the FEA and precise model calculations using Eq. (6.25) highlighting that the equation can accurately predict the maximum strain.

Figure 6.20 illustrates the schematic assembly procedures for controlling the buckling of helical structures. The deformation mode of the structure changes significantly with different slender ribbon patterns or bonded points. Various 3D structures with different mode ratios have been studied by Xu et al. [38].

Additionally, Luan et al. [39] creatively introduced microfluidic structures into a mechanically guided 3D assembly method; thus, proposing a new approach to create artificially complex vascular networks.

6.3.2 Origami-Inspired Design

Origami is an ancient paper-folding art that transforms 2D patterns into complex 3D structures using crease patterns (as shown in Figure 6.21a). The crease

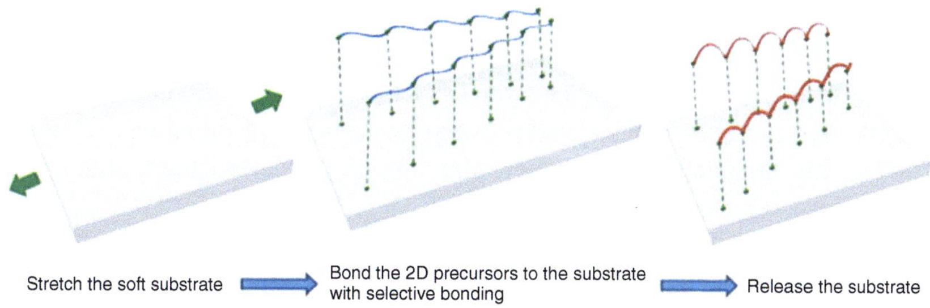

Figure 6.20 Assembly procedures for controlled buckling of the helical structures. Source: Ref. [37]. Reproduced with permission of Elsevier, © 2018.

Figure 6.21 Schematics of (a) Origami structure. Source: Ref. [40]. Reproduced with permission of John Wiley & Sons, © 2016. (b) Miura folding. Source: Ref. [41]. Reproduced with permission of Springer Nature, © 2014.

patterns enable the structure to be repeatedly folded and unfolded, protecting the functional surfaces from large strains. Cheng et al. [42] and Song et al. [41] introduced this concept to flexible electronics, achieving unprecedented deformability while maintaining high throughput and performance comparable to rigid electronics. This approach provides a new direction for the development of flexible electronics.

The structure of origami is mainly controlled by the local crease formed on the 2D pattern. Miura folding (Figure 6.21b) is one of the most well-known origami designs and can be adopted to engineer foldable electronic platforms. To achieve significant folding deformation, the strips should have a small thickness ratio and a small length ratio. As the compression deformation of the entire structure mostly occurs at the crease, it is important to avoid crease fracture during the design process. Generally, reducing the crease thickness or increasing the crease length can effectively reduce the maximum strain on the material. Moreover, some typical origami configurations are studied to find the kinematic and dynamical mechanisms, which have been summarized recently by Lu et al. [43] and Zhu et al. [44].

6.3.3 Kirigami-Inspired Design

Kirigami is an ancient art that involves graphic design and is realized by strategically cutting geometries to realize a pattern. This approach has also been adopted in flexible electronics design, providing the advantage of greatly reducing stress concentration during the buckling stage [45]. Figure 6.22a presents a comparison of silicon structures with different cutting percentages, where the high cutting percentage structure undergoes significant folding and experiences a lower maximum strain compared to the structure with a lower cutting percentage. This paper-cut structure also offers good design flexibility, which allows for better control of geometric shapes during buckling assembly. The structure is highly correlated with the cutting path. For instance, cutting along the radial and circumferential directions tends to produce symmetrical structures, while cutting along the serpentine path results in an antisymmetric configuration, as shown in the circular structure in Figure 6.22b.

There are several other 3D design methods for flexible electronics, too. For instance, Fu et al. [46] achieved different three-dimensional structures by adjusting the loading road strength of the same two-dimensional configuration. Cheng et al. [47] used microlattice design to regulate the deformation of structural 3D molding. And inspired by the DNA and proteins fold in three dimensions, Zhang et al. [48] used microfolding to design 3D morphable systems (Figure 6.23), which can be easily accessed and modulated *in situ*, providing a new strategy apart from the method based on substrate. Figure 6.24 shows that an epicardial bioelectronic robot constructed by the microfolding method can enable holistic monitoring of cardiac contractility. It is expected that 3D design methods for flexible electronic devices will continue to expand and be one of the key directions for future research development.

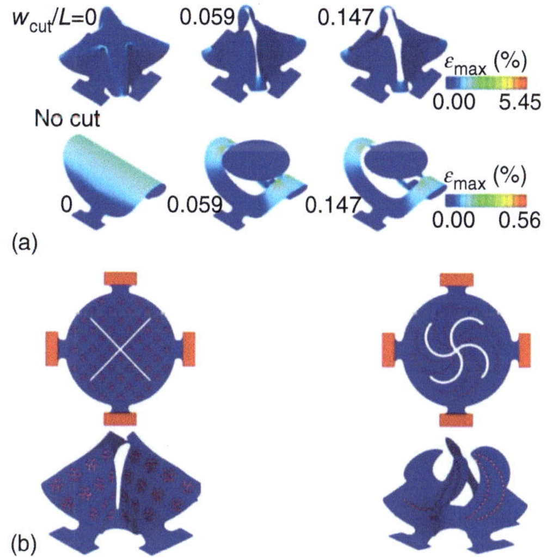

Figure 6.22 (a) Structures with different cutting percentages. (b) Path-related configuration changes. Source: Ref. [45]. Reproduced with permission of National Academy of Sciences, © 2015.

6.4 Design of Protective Structures for Flexible Electronic Devices

6.4.1 Strain Limited Structure Design

While flexible electronics can accommodate a wide range of deformations, including those caused by large joint movements, these deformations can still accumulate over many cycles and lead to damage or device failure. Therefore, there is a need for protective structures to be integrated into flexible electronic designs.

Jiang et al. [49] first proposed a strategy to protect flexible devices through a method called strain limitation. They designed a structure with a "J" shaped stress-strain curve, which has a low modulus at small deformations, allowing for flexibility without limiting the deformation of flexible electronics or skin. Then, the modulus increases rapidly at larger deformations to provide protection to the electronics. This design inspired the development of strain-limited structures, which have rapidly evolved to fully deployable platforms (Figure 6.25).

Ma et al. [50] proposed a method for preparing structures with a bilinear stress–strain behavior (Figure 6.26). Here, the rigid film buckles and has a low modulus before the strain transition point (initial stretch state). However, once the rigid film is straightened and the strain reaches the transition point, the modulus increases sharply. The transition strain $\varepsilon_{\text{transition}}$ is calculated as:

$$\varepsilon_{\text{transiton}} = \frac{\varepsilon_{\text{pre}} - \frac{1}{4}\left(\frac{9\overline{E}_f h^3}{\overline{E}_s H^3}\right)^{\frac{1}{3}}}{1 + \frac{1}{4}\left(\frac{9\overline{E}_f h^3}{\overline{E}_s H^3}\right)^{\frac{1}{3}}} \tag{6.26}$$

As mentioned in Section 6.2, there are two types of buckling models: global buckling and local buckling. Under global buckling, the strain-limiting effect will not be

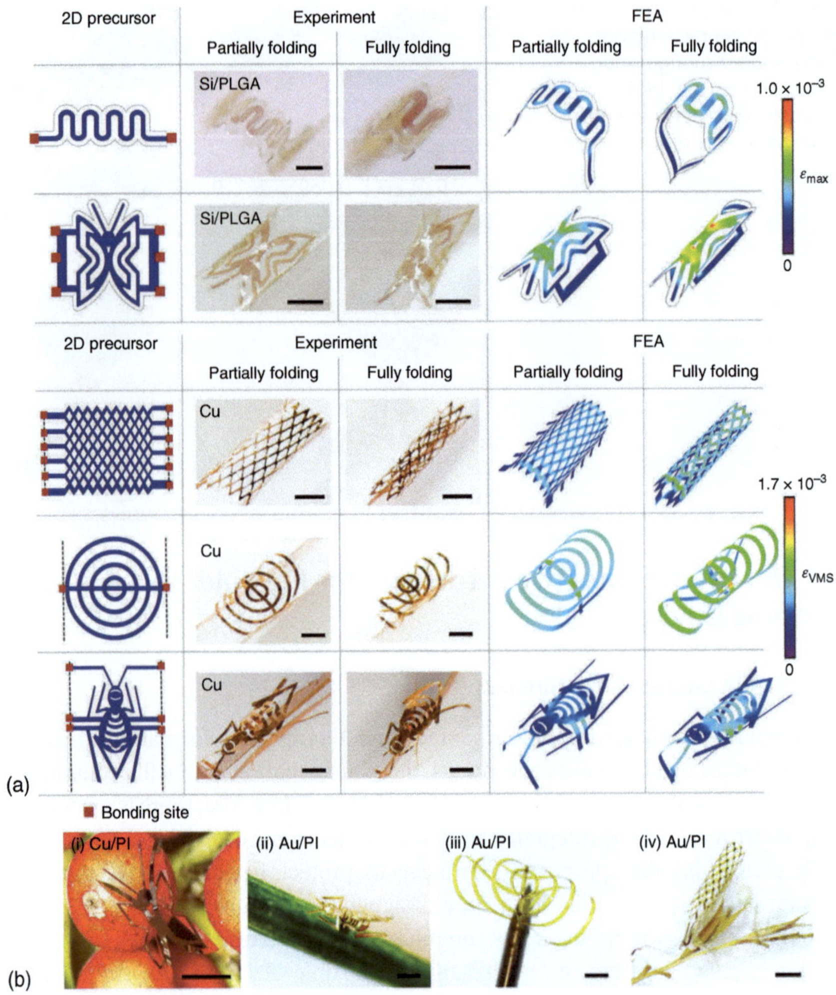

Figure 6.23 Structures realized by deterministic microfolding (a) Examples of microfolding structures. (b) Optical images of freestanding 3D mesostructures on different objects. Source: Ref. [48]/American Association for the Advancement of Science/CC BY 4.0.

effective. The criterion of two buckling models is given in reference [11]:

$$L_{cr} = 4\pi\sqrt{EI\left[\frac{(\overline{E}_f/3\overline{E}_s)^{2/3}}{\overline{E}_s H + \overline{E}_f h} - \frac{0.3}{G_s(H+h)}\right]} \tag{6.27}$$

where $\overline{E} = E/(1 - \nu^2)$, E_s, and ν_s are the elastic modulus and Poisson's ratio of the soft substrate, respectively, while E_f, and ν_f are the elastic modulus and Poisson's ratio of the hard film, respectively. H is the thickness of the soft substrate, and h is the thickness of the hard thin film. G_s is the shear modulus of the soft substrate,

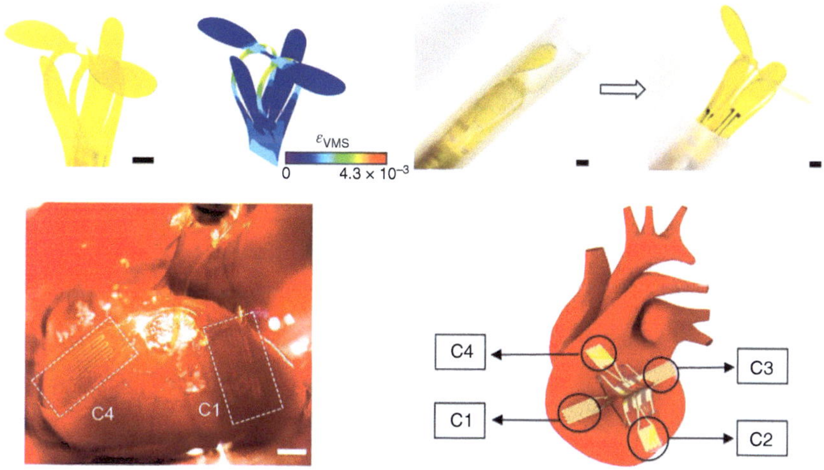

Figure 6.24 Epicardial bioelectronic robot formed by microfolding.
Source: Ref. [48]/American Association for the Advancement of Science/CC BY 4.0.

$G_s = \overline{E}_s/[2(1 + \nu_s)]$. The bending stiffness of the overall structure is:

$$EI = \frac{(\overline{E}_f h^2 - \overline{E}_s H^2)^2 + 4\overline{E}_s H \overline{E}_f h (H + h)^2}{12(\overline{E}_s H + \overline{E}_f h)} \tag{6.28}$$

The figure of critical film length that separates local wrinkling from global buckling is shown in Figure 6.27. Global and local buckling occur for $L > L_{cr}$ and $L < l_{cr}$, respectively.

They have also extended this approach from unidirectional to multidirectional strain limitation. A bi-directional pre-strained stiff thin mesh film combined with structures can achieve a multidirectional strain limitation (Figure 6.28). The transition strain is:

$$\varepsilon^i_{\text{transition}} = \frac{\varepsilon^i_{\text{pre}} - \frac{W}{6(W+S)}\left(\frac{9\overline{E}_f h^3}{\overline{E}_s H^3}\right)^{\frac{1}{3}}}{1 + \frac{W}{6(W+S)}\left(\frac{9\overline{E}_f h^3}{\overline{E}_s H^3}\right)^{\frac{1}{3}}} \tag{6.29}$$

where i denotes to x- or y-direction; W is the width of the mesh, and S is the space in the mesh (see Figure 6.28a).

6.4.2 Strain Isolation Structure Design

Inorganic electronic devices have superior electronic properties than organic electronics but are mechanically fragile and susceptible to failure due to their individual material layouts. To safeguard active electronic components, strain isolation designs have been developed that can be employed in the production

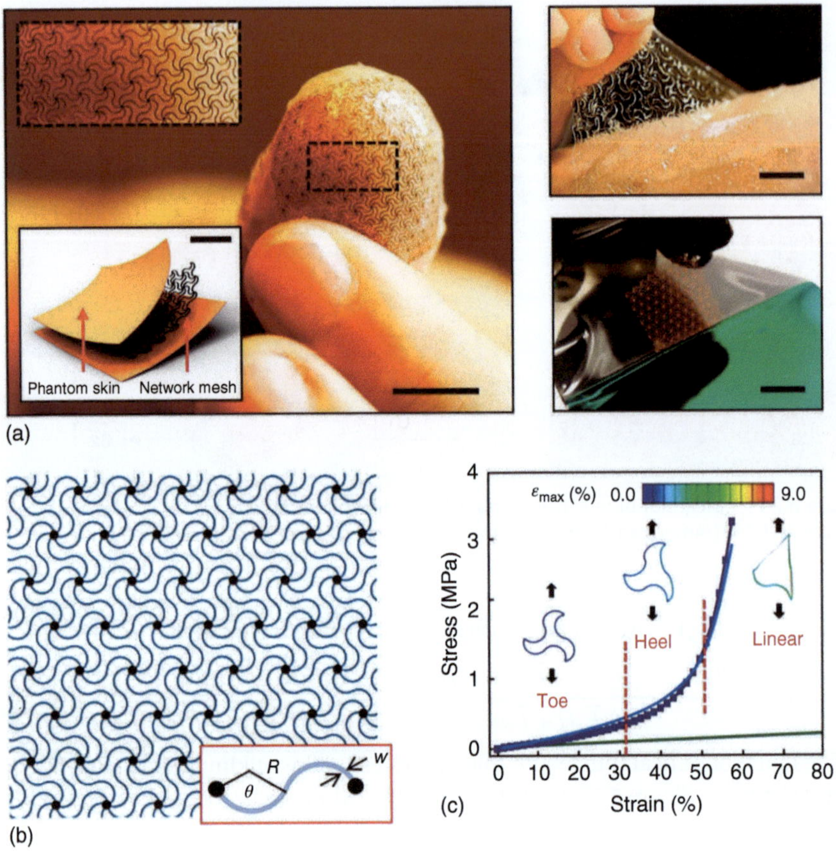

Figure 6.25 (a) The optical images of the J-shaped strain limited structure. (b) Schematic diagram of the strain limiting structure. (c) The stress-strain curve of the J-shaped strain limited structure. Source: Ref. [49]/Springer Nature/CC BY 4.0.

and preparation of stretchable inorganic devices. Furthermore, the strain isolation approach is also suitable for organic devices when the necessary stretchability surpasses the stretchability of a single component.

The strain isolation method can vary depending on the specific application. One approach is the core/shell strain isolation structure [51], as illustrated in Figure 6.29. This structure consists of three layers: an outer shell made of Ecoflex, an internal shell structure wrapped around a core structure composed of ultrasoft Silbione, and flexible electronics enclosed in the innermost layer. This design helps to protect the active electronic components from mechanical strain while maintaining their electrical functionality. Moreover, it can effectively reduce the stimulation of flexible electronic devices on human skin and improve wearability and comfort. The structure can be optimized in terms of thickness and edge gradient to improve wearing comfort.

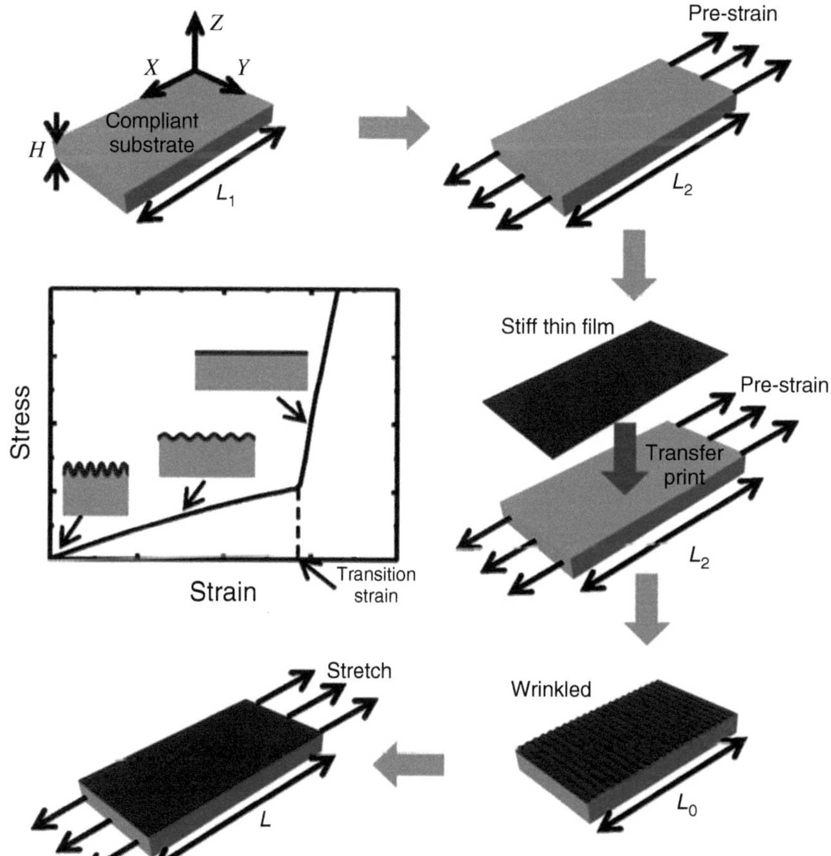

Figure 6.26 Schematic illustrations for fabricating bilinear strain-limiting structures. Source: Ref. [50]. Reproduced with permission of John Wiley & Sons, © 2016.

Ma et al. [52] proposed a new way to isolate the strain. They use an ultrasoft film with a cavity filled with a fluid (water or gas) to achieve strain isolation (Figure 6.30). But for this configuration, the roof is prone to collapse and the adhesion between the layers of elastomer above and below the cavity must be taken into account in the design. The normalized total potential energy can be expressed as the function of normalized collapse length:

$$\frac{a^3 U_{\text{total}}}{Dh^2} = \frac{12}{\left(1 - \left(\frac{b}{a}\right)\right)^3} - 2\bar{\gamma}\frac{b}{a} \tag{6.30}$$

$$\frac{a^3 U_{\text{total}}}{Dh^2} = \frac{48\left(4 + 7\left(\frac{b}{a}\right) + 4\left(\frac{b}{a}\right)^2\right)}{\left(1 - \left(\frac{b}{a}\right)\right)^5} - 2\bar{\gamma}\frac{b}{a} \tag{6.31}$$

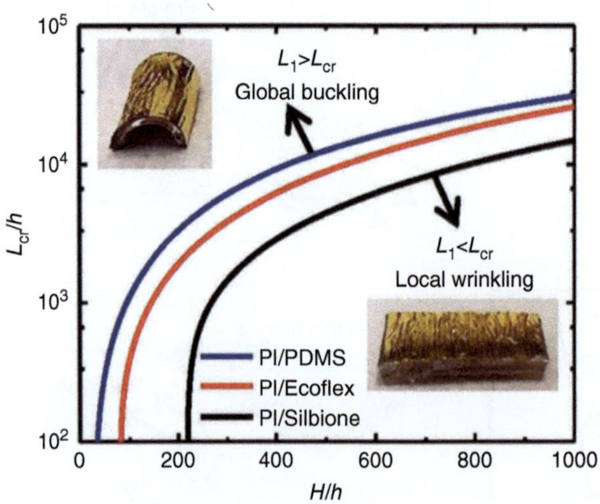

Figure 6.27 The figure of critical film length that separates local wrinkling from global buckling. Source: Ref. [50]. Reproduced with permission of John Wiley & Sons, © 2016.

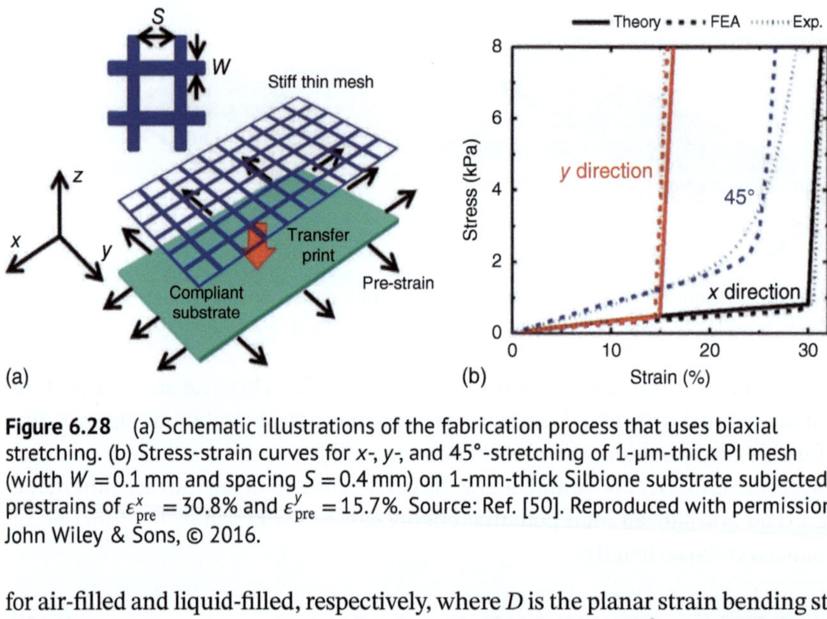

Figure 6.28 (a) Schematic illustrations of the fabrication process that uses biaxial stretching. (b) Stress-strain curves for x-, y-, and 45°-stretching of 1-μm-thick PI mesh (width $W = 0.1$ mm and spacing $S = 0.4$ mm) on 1-mm-thick Silbione substrate subjected to prestrains of $\varepsilon_{\text{pre}}^{x} = 30.8\%$ and $\varepsilon_{\text{pre}}^{y} = 15.7\%$. Source: Ref. [50]. Reproduced with permission of John Wiley & Sons, © 2016.

for air-filled and liquid-filled, respectively, where D is the planar strain bending stiffness of the top layer, $D = E_{\text{Eco}}t^3/\left[12\left(1 - v_{\text{Eco}}^2\right)\right]$; $\overline{\gamma} = \alpha^4\gamma/(Dh^2)$ is the normalized adhesion energy, E_{Eco} and v_{Eco} are the elastic modulus and Poisson ratio of the top layer material, respectively.

The images of the relationship between normalized total potential energy and normalized collapse length with different normalized values of the work of adhesion are shown in Figure 6.31.

Another approach for strain isolation is microstructural design. Li et al. [53] proposed a microisland design to isolate the strain caused by the thermal mismatch

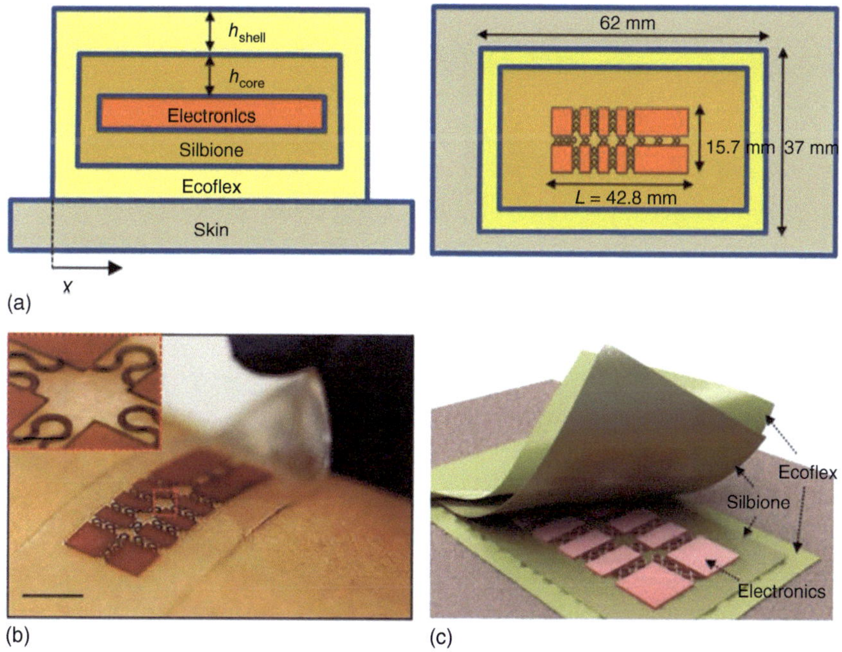

Figure 6.29 Core/shell strain isolation structure diagram: (a) Cross-section (Left) and top view (Right). (b) Physical image. (c) Schematic diagram. Source: Ref. [51]. Reproduced with permission of John Wiley & Sons, © 2015.

Figure 6.30 Schematic images of cavity-type substrate strain isolation design. Source: Ref. [52]. Reproduced with permission of John Wiley & Sons, © 2017.

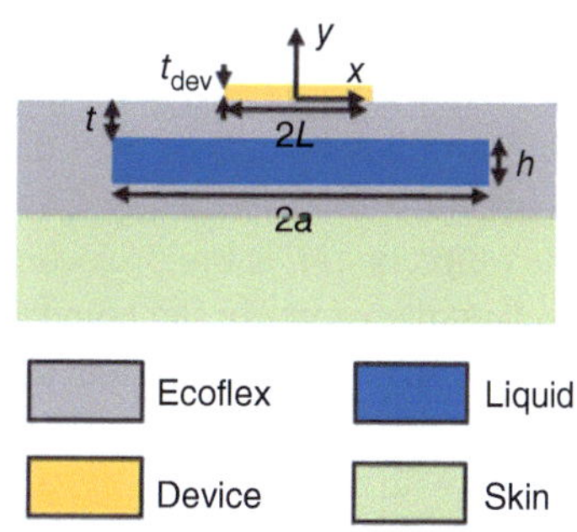

when the hard films are deposited into the substrate. Figure 6.32a gives the illustration of microisland structure of the stiff film deposited onto a soft substrate. The relationship between the strain isolation ratio $\varepsilon_i^f/\varepsilon_i^s$ and the geometric and material parameters of the microisland are deduced:

$$\frac{\varepsilon_i^f}{\varepsilon_i^s} = \left(1 - \frac{\cosh\left(\frac{l_i}{\sqrt{\left(h_s{}^2\overline{E}_s + 2h_s h_f \overline{E}_f\right)/2G_s}} \right)}{\cosh\left(\frac{L_i/2}{\sqrt{\left(h_s{}^2\overline{E}_s + 2h_s h_f \overline{E}_f\right)/2G_s}} \right)} \right) \tag{6.32}$$

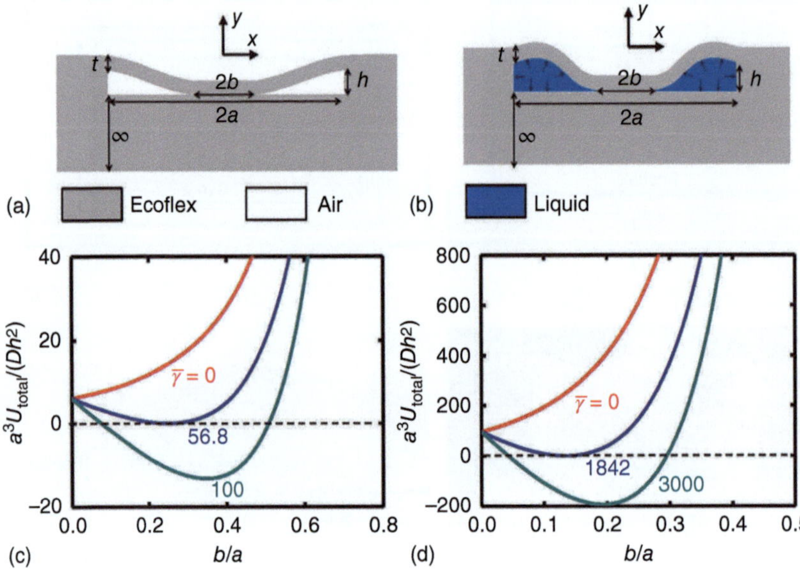

Figure 6.31 Comparison of cavity collapse between inflation and filling: Schematic of cavity collapse filled with (a) air and (b) liquid. Relationship between normalized total potential energy ($a^3U_{total}(Dh^2)$) and normalized collapse length (b/a) with several normalized values of the work of adhesion (γ) in (c) gas-filled chamber and (d) liquid-filled chamber. Source: Ref. [52]. Reproduced with permission of John Wiley & Sons, © 2017.

where $\varepsilon_i^{\mathrm{f}}$ is the thermal strain mismatch with microislands for strain isolation in the i direction, $\varepsilon_i^{\mathrm{s}}$ is the thermal strain mismatch between the stiff film and soft substrate without microislands in the i direction; h_{s} and h_{f} are the thickness of the microislands and film, respectively. L_i is the characteristic dimension of the microislands in the i direction. l_i is the projection distance from any point on a microisland to the rectangular center in the i direction.

The worst strain isolation ratio is also analyzed and appears at the center of microisland, i.e. the point where $l_x = l_y = 0$. The analytic model for a round island design with characteristics D is given by:

$$\frac{\hat{\varepsilon}_i^{\mathrm{f}}}{\varepsilon_i^{\mathrm{s}}} = 1 - \frac{1}{\cosh(M)} \tag{6.33}$$

where $\hat{\varepsilon}_i^{\mathrm{f}}$ represents the maximum thermal strain mismatch in the center of the microisland. M is defined as $M = D/\sqrt{2\left(h_{\mathrm{s}}^2\overline{E}_{\mathrm{s}} + 2h_{\mathrm{s}}h_{\mathrm{f}}\overline{E}_{\mathrm{f}}\right)/G_{\mathrm{s}}}$. The relationship between wrinkle amplitude and pre-strain satisfies:

$$A = h_{\mathrm{f}}\sqrt{\frac{\varepsilon_i^{\mathrm{s}}}{\varepsilon_{\mathrm{IC}}} - 1} \tag{6.34}$$

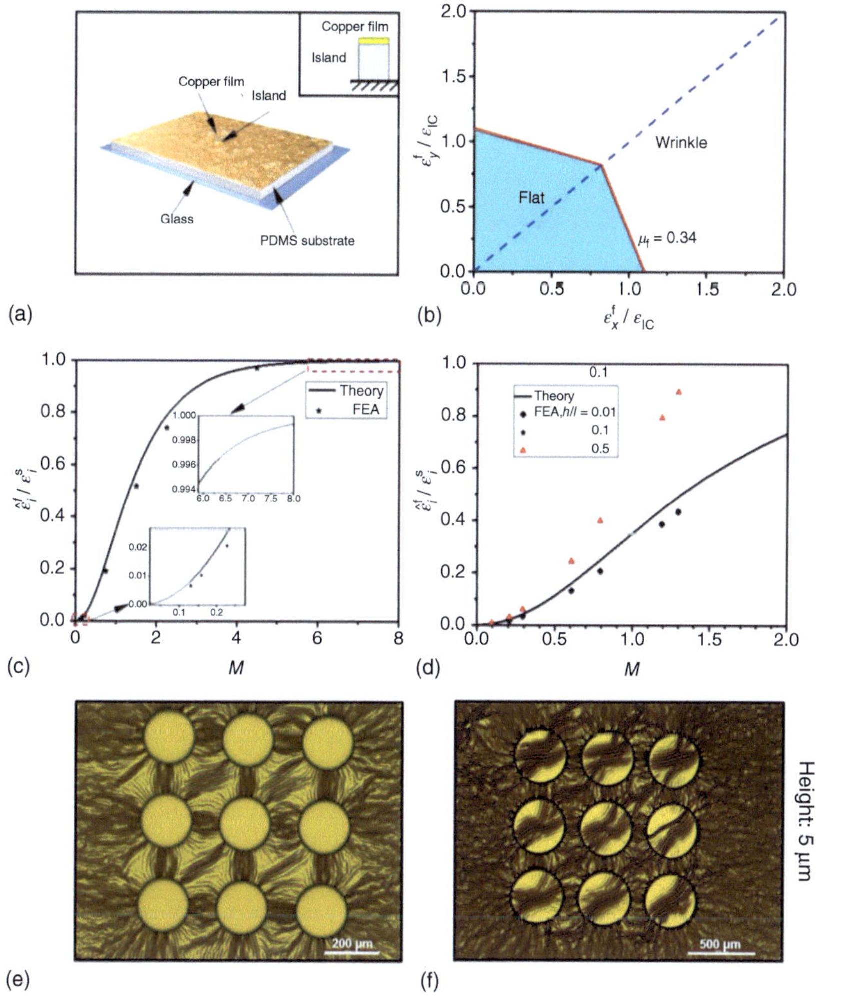

Figure 6.32 (a) Simplified simulation model. (b) Microstructure leveling/fold phase diagram during bidirectional stretching. (c) Relationship between microstructure size and strain isolation ratio. (d) Theory and simulation error analysis. (e) The surface morphology after coating with the diameter of microstructure smaller than the critical size. (f) Surface morphology of the microstructure is larger than the critical size. Source: Ref. [53]. Reproduced with permission of John Wiley & Sons, © 2019.

where $\varepsilon_{IC} = (3\overline{E}_s/\overline{E}_f)^{2/3}/[4(1+\mu_f)]$. Theoretical, numerical, and experimental results for microisland strain isolation structures are shown in Figure 6.32. The preparation flow of a micro island structure is shown in Figure 6.33.

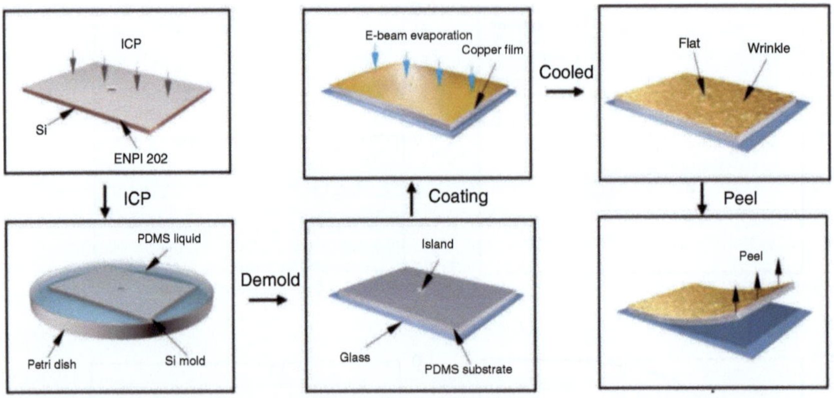

Figure 6.33 The preparation process of microisland structure. Source: Ref. [53]. Reproduced with permission of John Wiley & Sons, © 2019.

References

1 Han, M., Chen, L., Aras, K. et al. (2020). Catheter-integrated soft multilayer electronic arrays for multiplexed sensing and actuation during cardiac surgery. *Nature Biomedical Engineering* 4 (10): 997–1009.

2 Jeong, H., Lee, J.Y., Lee, K. et al. (2021). Differential cardiopulmonary monitoring system for artifact-canceled physiological tracking of athletes, workers, and COVID-19 patients. *Science Advances* 7 (20): eabg3092.

3 Rwei, A.Y., Lu, W., Wu, C. et al. (2020). A wireless, skin-interfaced biosensor for cerebral hemodynamic monitoring in pediatric care. *Proceedings of the National Academy of Sciences* 117 (50): 31674–31684.

4 Yang, Y., Wu, M., Wegener, A.J. et al. (2022). Preparation and use of wireless reprogrammable multilateral optogenetic devices for behavioral neuroscience. *Nature Protocols* 17 (4): 1073–1096.

5 Ausra, J., Wu, M., Zhang, X. et al. (2021). Wireless, battery-free, subdermally implantable platforms for transcranial and long-range optogenetics in freely moving animals. *Proceedings of the National Academy of Sciences* 118 (30): e2025775118.

6 Lu, W., Bai, W., Zhang, H. et al. (2021). Wireless, implantable catheter-type oximeter designed for cardiac oxygen saturation. *Science Advances* 7 (7): eabe0579.

7 Khang, D.Y., Jiang, H., Huang, Y. et al. (2006). A stretchable form of single-crystal silicon for high-performance electronics on rubber substrates. *Science* 311 (5758): 208–212.

8 Jones, J., Lacour, S.P., Wagner, S. et al. (2004). Stretchable wavy metal interconnects. *Journal of Vacuum Science and Technology A* 22 (4): 1723–1725.

9 Cavallo, F., Huang, Y., Dent, E.W. et al. (2014). Neurite guidance and three-dimensional confinement via compliant semiconductor scaffolds. *ACS Nano* 8 (12): 12219–12227.

10 Bowden, N., Brittain, S., Evans, A.G. et al. (1998). Spontaneous formation of ordered structures in thin films of metals supported on an elastomeric polymer. *Nature* 393 (6681): 146–149.

11 Wang, S., Song, J., Kim, D.H. et al. (2008). Local versus global buckling of thin films on elastomeric substrates. *Applied Physics Letters* 93 (2): 023126.

12 Jiang, H., Khang, D.Y., Fei, H. et al. (2008). Finite width effect of thin-films buckling on compliant substrate: experimental and theoretical studies. *Journal of the Mechanics and Physics of Solids* 56 (8): 2585–2598.

13 Jiang, H., Khang, D.Y., Song, J. et al. (2007). Finite deformation mechanics in buckled thin films on compliant supports. *Proceedings of the National Academy of Sciences* 104 (40): 15607–15612.

14 Jiang, H., Sun, Y., Rogers, J.A. et al. (2007). Mechanics of precisely controlled thin film buckling on elastomeric substrate. *Applied Physics Letters* 90 (13): 133119.

15 Sun, Y., Choi, W.M., Jiang, H. et al. (2006). Controlled buckling of semiconductor nanoribbons for stretchable electronics. *Nature Nanotechnology* 1 (3): 201–207.

16 Sharp, J.S. and Jones, R.A.L. (2002). Micro-buckling as a route towards surface patterning. *Advanced Materials* 14 (11): 799.

17 Yoo, P.J., Suh, K.Y., Park, S.Y. et al. (2002). Physical self-assembly of microstructures by anisotropic buckling. *Advanced Materials* 14 (19): 1383–1387.

18 Schmid, H., Wolf, H., Allenspach, R. et al. (2003). Preparation of metallic films on elastomeric stamps and their application for contact processing and contact printing. *Advanced Functional Materials* 13 (2): 145–153.

19 Cantarella, G., Vogt, C., Hopf, R. et al. (2017). Buckled thin-film transistors and circuits on soft elastomers for stretchable electronics. *ACS Applied Materials & Interfaces* 9 (34): 28750–28757.

20 Qi, D., Liu, Z., Yu, M. et al. (2015). Highly stretchable gold nanobelts with sinusoidal structures for recording electrocorticograms. *Advanced Materials* 27 (20): 3145–3151.

21 Jang, J. (2006). Displays develop a new flexibility. *Materials Today* 9 (4): 46–52.

22 Kim, D.H., Song, J., Choi, W.M. et al. (2008). Materials and noncoplanar mesh designs for integrated circuits with linear elastic responses to extreme mechanical deformations. *Proceedings of the National Academy of Sciences* 105 (48): 18675–18680.

23 Song, J., Huang, Y., Xiao, J. et al. (2009). Mechanics of noncoplanar mesh design for stretchable electronic circuits. *Journal of Applied Physics* 105 (12): 123516.

24 Ko, H.C., Stoykovich, M.P., Song, J. et al. (2008). A hemispherical electronic eye camera based on compressible silicon optoelectronics. *Nature* 454 (7205): 748–753.

25 Ko, H.C., Shin, G., Wang, S. et al. (2009). Curvilinear electronics formed using silicon membrane circuits and elastomeric transfer elements. *Small* 5 (23): 2703–2709.

26 Park, S.I., Xiong, Y., Kim, R.H. et al. (2009). Printed assemblies of inorganic light-emitting diodes for deformable and semitransparent displays. *Science* 325 (5943): 977–981.

27 Lee, J., Wu, J., Ryu, J.H. et al. (2012). Stretchable semiconductor technologies with high areal coverages and strain-limiting behavior: demonstration in high-efficiency dual-junction GaInP/GaAs photovoltaics. *Small* 8 (12): 1851–1856.

28 Zhang, Y., Wang, S., Li, X. et al. (2014). Experimental and theoretical studies of serpentine microstructures bonded to prestrained elastomers for stretchable electronics. *Advanced Functional Materials* 24 (14): 2028–2037.

29 Zhang, Y., Xu, S., Fu, H. et al. (2013). Buckling in serpentine microstructures and applications in elastomer-supported ultra-stretchable electronics with high areal coverage. *Soft Matter* 9 (33): 8062–8070.

30 Pan, T., Pharr, M., Ma, Y. et al. (2017). Experimental and theoretical studies of serpentine interconnects on ultrathin elastomers for stretchable electronics. *Advanced Functional Materials* 27 (37): 1702589.

31 Xie, Z., Ji, B., and Huo, Q. (2018). Mechanics design of stretchable near field communication antenna with serpentine wires. *Journal of Applied Mechanics* 85 (4): 045001.

32 Chung, H.U., Kim, B.H., Lee, J.Y. et al. (2019). Binodal, wireless epidermal electronic systems with in-sensor analytics for neonatal intensive care. *Science* 363 (6430): eaau0780.

33 Xu, S., Zhang, Y., Cho, J. et al. (2013). Stretchable batteries with self-similar serpentine interconnects and integrated wireless recharging systems. *Nature Communications* 4 (1): 1543.

34 Fan, J.A., Yeo, W.H., Su, Y. et al. (2014). Fractal design concepts for stretchable electronics. *Nature Communications* 5 (1): 3266.

35 Zhang, Y., Fu, H., Su, Y. et al. (2013). Mechanics of ultra-stretchable self-similar serpentine interconnects. *Acta Materialia* 61 (20): 7816–7827.

36 Liu, Y., Yan, Z., Lin, Q. et al. (2016). Guided formation of 3D helical mesostructures by mechanical buckling: analytical modeling and experimental validation. *Advanced Functional Materials* 26 (17): 2909–2918.

37 Fan, Z., Hwang, K.C., Rogers, J.A. et al. (2018). A double perturbation method of postbuckling analysis in 2D curved beams for assembly of 3D ribbon-shaped structures. *Journal of the Mechanics and Physics of Solids* 111: 215–238.

38 Xu, S., Yan, Z., Jang, K.I. et al. (2015). Assembly of micro/nanomaterials into complex, three-dimensional architectures by compressive buckling. *Science* 347 (6218): 154–159.

39 Luan, H., Zhang, Q., Liu, T.L. et al. (2021). Complex 3D microfluidic architectures formed by mechanically guided compressive buckling. *Science Advances* 7 (43): eabj3686.

40 Yan, Z., Zhang, F., Wang, J. et al. (2016). Controlled mechanical buckling for origami-inspired construction of 3D microstructures in advanced materials. *Advanced Functional Materials* 26 (16): 2629–2639.

41 Song, Z., Ma, T., Tang, R. et al. (2014). Origami lithium-ion batteries. *Nature Communications* 5 (1): 3140.

42 Cheng, Q., Song, Z., Ma, T. et al. (2013). Folding paper-based lithium-ion batteries for higher areal energy densities. *Nano Letters* 13 (10): 4969–4974.

43 Lu, L., Leanza, S., and Zhao, R.R. (2023). Origami with rotational symmetry: a review on their mechanics and design. *Applied Mechanics Reviews* 75 (5): 050801.

44 Zhu, Y., Schenk, M., and Filipov, E.T. (2022). A review on origami simulations: from kinematics, to mechanics, toward multiphysics. *Applied Mechanics Reviews* 74 (3): 030801.

45 Zhang, Y., Yan, Z., Nan, K. et al. (2015). A mechanically driven form of Kirigami as a route to 3D mesostructures in micro/nanomembranes. *Proceedings of the National Academy of Sciences* 112 (38): 11757–11764.

46 Fu, H., Nan, K., Bai, W. et al. (2018). Morphable 3D mesostructures and microelectronic devices by multistable buckling mechanics. *Nature Materials* 17 (3): 268–276.

47 Cheng, X., Fan, Z., Yao, S. et al. (2023). Programming 3D curved mesosurfaces using microlattice designs. *Science* 379 (6638): 1225–1232.

48 Zhang, L., Zhang, Z., Weisbecker, H. et al. (2022). 3D morphable systems via deterministic microfolding for vibrational sensing, robotic implants, and reconfigurable telecommunication. *Science Advances* 8 (51): eade0838.

49 Jang, K.I., Chung, H.U., Xu, S. et al. (2015). Soft network composite materials with deterministic and bio-inspired designs. *Nature Communications* 6 (1): 6566.

50 Ma, Y., Jang, K.I., Wang, L. et al. (2016). Design of strain-limiting substrate materials for stretchable and flexible electronics. *Advanced Functional Materials* 26 (29): 5345–5351.

51 Lee, C.H., Ma, Y., Jang, K.I. et al. (2015). Soft core/shell packages for stretchable electronics. *Advanced Functional Materials* 25 (24): 3698–3704.

52 Ma, Y., Pharr, M., Wang, L. et al. (2017). Soft elastomers with ionic liquid-filled cavities as strain isolating substrates for wearable electronics. *Small* 13 (9): 1602954.

53 Li, H., Wang, Z., Lu, S. et al. (2019). Elastomers with microislands as strain isolating substrates for stretchable electronics. *Advanced Materials Technologies* 4 (2): 1800365.

7

Laser-Based Fabrication Process Development for Flexible and Stretchable Electronics

Jung Jae Park, Minwoo Kim, and Seung Hwan Ko

Seoul National University, Applied Nano and Thermal Science Lab, Department of Mechanical Engineering, Gwanak-gu, Seoul 08826, South Korea

7.1 Introduction

In recent years, soft electronics have gained considerable attention due to their potential for versatile applications such as soft robotics [1–4], wearable electronics [5–10], and healthcare monitoring devices [11–15]. In order for these soft electronic devices to be mounted on a curved surface and operate reliably under mechanical deformation such as human movement, electronic components of the devices, including electronic circuits, sensors, actuators, and heaters, should be flexible and stretchable. However, the conventional processes, including photolithography patterning and vacuum deposition method, are incompatible with the fabrication of flexible and stretchable electronic devices, which are called next-generation electronics. This is because diverse soft materials such as elastomers, commonly utilized as substrates in flexible and stretchable electronics, are usually vulnerable to the etching process and heat treatment at high temperatures used in the conventional manufacturing [16–19]. Furthermore, these conventional manufacturing methods consist of considerable steps and require expensive and complex mechanical apparatus for vacuum forming, etching, exposure on a mask, and heat treatment.

So, many researchers need an alternative fabrication process for soft electronics instead of the conventional processes, although the development of the photolithographic fabrication with vacuum deposition has facilitated the emergence of various high-performance integrated electronic devices [20–24]. In such needs, laser-based fabrication techniques have emerged as a promising approach, possessing suitable features for the implementation of soft electronics [25–29].

First, the manufacturing process based on a laser is a facile and fast process that can be utilized at room temperature and atmospheric pressure. The process does not need sophisticated apparatus such as vacuum pump, chamber, heat generator, and patterning mask and facilitates rapid patterning with high-resolution features in a simple procedure. These characteristics of laser fabrication, such as simplicity

Jung Jae Park and Minwoo Kim contributed equally to this work.

Mechanics of Flexible and Stretchable Electronics, First Edition. Edited by Yong Zhu and Nanshu Lu.

and immediacy, allow for facile prototyping of flexible and stretchable electronics. This is important for research and development, where various and repetitive attempts are indispensable. Second, the laser process approves enormous freedom of material selection and process conditions. Because the procedure operates basically at ambient room temperature, a wide range of materials including metals, polymers, semiconductors, and composites can be applied without specific constraints. Besides, if other process conditions, such as specific temperature or in a certain solution, are required, the process can be scaled up by optionally adding proper equipment to the basic laser setup. Finally, the laser-based fabrication is highly customizable by controlling versatile parameters of the process consisting of wavelength, power, scanning speed, repetition rate, and type of wave: continuous or pulsed. This customizability allows for a wide assortment of laser fabrication including laser sintering, ablation, and cutting in a single step. For example, one can form a conductive layer with a desired pattern on a substrate by sintering and erasing the conductor partly by ablation and then cutting whole layers into a specific structure by cutting in a single laser with a step process [30–32]. Additionally, a laser can be utilized to synthesize micro- and nanomaterials with unique properties and functionalities for soft electronics by inducing a photothermal effect, hydrothermal reaction, and pyrolysis process.

In this review, we present an overview of a laser-based process for the fabrication of flexible and stretchable electronics. First, we focus on the representative manufacturing processes based on laser techniques including sintering, ablation, cutting, and hydrothermal growth. Then, to gain a comprehensive understanding of the adaptability of laser-based methods, we introduce a wide range of applications in versatile areas such as skin electronics, soft robotics, virtual reality, and functional electronic components. Finally, we discuss the remarkable prospects of laser-based fabrication methods and expect that the features introduced in this review will pave the way for the development of advanced soft electronics.

7.2 Representative Laser-Based Fabrication Process

Among manufacturing processes with laser, laser sintering is a method for achieving a desired pattern by instantaneously melting materials locally [33–37]. When materials are irradiated with a laser, photothermal effect induced by laser locally generates thermal energy by which the materials melt and coalesce instantly [38–42]. Since the line width of the sintered material pattern is usually on a similar scale of laser beam spot size, a complicated pattern with high resolution can be accomplished by the laser sintering. In addition, this method is highly customizable because various parameters including laser power, scanning speed, and wavelength can be finely adjusted for assorted materials. For example, Ko et al. presented laser direct writing process facilitating high-resolution metal patterning in ambient room temperature via selective laser sintering of metal nanoparticles. Firstly, they prepared gold nanoparticle solution and printed the solution on a polymer substrate using an inkjet printing system. Next, the group laser-sintered selectively the printed

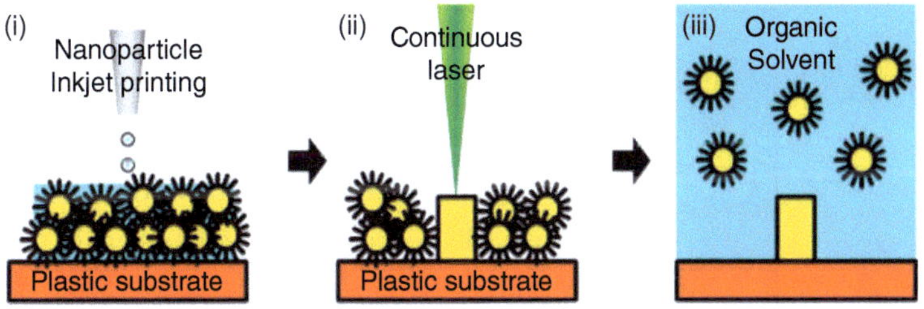

Figure 7.1 Laser sintering fabrication. The circles indicate gold nanoparticles and the rectangle represents a sintered gold conductor with a desired pattern. Source: Reproduced with permission from Ref. [43], IOP Publishing. Copyright 2007.

metal nanoparticles through a focused Ar ion laser and then a sintered conductor was formed with the desired shape. After laser scanning was completed with the desired pattern, they rinsed the residues of the laser-unexposed nanoparticles in an organic solvent. Consequently, the deposition of patterned conductor with high resolution was achieved. As a proof of concept, the group successfully fabricated organic field effect transistors with submicron critical features using the laser direct writing process with inkjet printing (Figure 7.1) [43].

Unlike laser sintering manner, laser ablation is a technique for accomplishing a specific shape by eliminating materials locally [44–48]. When materials are exposed by laser, laser energy is absorbed at the surface of materials, causing the superficial materials to vaporize or sublimate and leaving well-defined vacant space. Furthermore, laser with sufficiently high power enables it to cut through materials in a certain pattern, which is called laser cutting [49–52]. By laser ablation and cutting, removing partial layers of materials and cutting through whole layers of materials are simply realized, which facilitates achievement of high-resolution complex patterns. For instance, Lu et al. showed a technique for micropatterning liquid metal via laser ablation process. They coated liquid metal on an elastomer substrate and laser-ablated the liquid metal film into a desired feature with microscale dimension. For direct ablation of liquid metal, the group utilized pulsed UV laser with 355 nm wavelength. As a result, they demonstrated a stretchable photoplethysmogram (PPG) sensing device where conventional electronic components such as integrated circuit (IC) chip, resistor, and capacitor were interconnected by deformable liquid metal circuits. The implemented device, worn as a wristband, clearly detected the PPG waveform and successfully measured heart rate and arterial blood oxygenation (Figure 7.2) [53].

On the other hand, for selective and on-demand synthesis of various functional materials such as nanoparticles, nanowires, and nanotubes, laser-induced hydrothermal growth is a decent and facile method that exploits the advantages of laser process. [54–59] In this process, laser beam is focused on a certain spot in desired precursor solution typically containing metal salts or organic reactants. Then, a localized temperature field by laser irradiation initiates nucleation of target material and induces growth of the material with high precision, leading to the formation of diverse functional materials with unique structures and

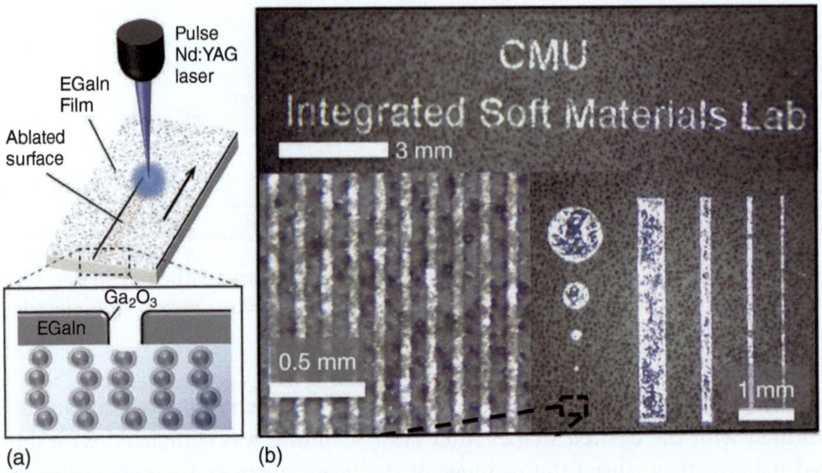

Figure 7.2 Laser ablation fabrication. (a) Schematic of liquid metal patterning using laser ablation. (b) Optical images of laser-fabricated liquid metal patterns. Source: Reproduced with permission from Ref. [53], American Chemical Society. Copyright 2017.

properties. For example, Lee et al. exhibited a method for fabricating hierarchical heterogeneous nanowires by well-controlled synthesis of nanowires on nanowires via laser-induced hydrothermal growth. First, they transferred silver nanowires on a glass substrate and deposited zinc oxide quantum dot seeds onto the nanowires by wetting them with the seed solution. Then, they soaked the sample in precursor solution for zinc oxide nanowire and irradiated the sample with 532 nm wavelength laser to activate hydrothermal growth of zinc oxide nanowire on silver nanowire. As a result, the group successfully demonstrated zinc oxide nanowire branches grown on silver nanowire backbone and showed that diameters of a bunch of the branches can be controlled in accordance with laser polarization. Moreover, they created a nanowire-based UV sensor with heterogeneous hierarchy of zinc oxide–silver nanowires, demonstrating the potential of this method for fabricating flexible and stretchable electronics with compact size and high performance (Figure 7.3) [60].

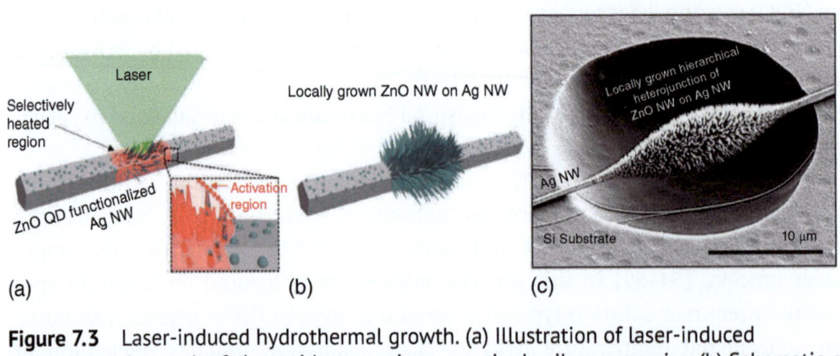

Figure 7.3 Laser-induced hydrothermal growth. (a) Illustration of laser-induced hydrothermal growth of zinc oxide nanowires on a single silver nanowire. (b) Schematic of synthesized zinc oxide nanowire branches on a silver nanowire. (c) SEM image of a grown heterogeneous hierarchy of zinc oxide–silver nanowire. Source: Reproduced with permission from Ref. [60], American Chemical Society. Copyright 2017.

7.3 Applications Based on Laser Fabrication

For development of flexible and stretchable electronics, it is essential to fabricate various configurations of circuits on thin film-type or elastomer-based substrates, or without substrates. Laser is utilized for this purpose, providing fast and digital maskless patterning methods [61–66]. For instance, Kim et al. demonstrated evolvable skin electronics of which circuit design can be modified and new functionalities can be added by *in situ* and *in operando* adaptation using laser-based fabrication. During the fabrication process, they utilized visible laser with 532 nm wavelength and UV laser with 355 nm wavelength for different purposes. First of all, they spin-coated silver nanoparticle (AgNP) ink on ultrathin substrate of colorless polyimide (CPI) and selectively formed metal pattern by visible laser sintering of metal nanoparticles. To insulate the patterned circuit, they deposited additional CPI solution after the removal of residual AgNPs. Next, they precisely formed vertical interconnect accesses (VIAs), allowing separate wires on different layers to be connected, by UV laser ablation of the insulation layer. After spin-coating of AgNP ink, the VIAs were filled with AgNPs and then second circuit connected to the first circuit on the bottom layer was fabricated through successive sintering and cleaning processes. Furthermore, they introduced the erasing process by selective ablation of the upper circuit using low-powered UV laser without damaging the substrate and the formerly fabricated circuit. As a proof of the concept, they demonstrated extending the originally integrated electronic device containing temperature sensors by adding new features such as skin condition tracking, physiology sensing, motion tracking, and radio frequency (RF) power delivery (Figure 7.4a) [30].

By controlling penetration depth of laser, a freestanding conductor based on liquid metal composite can be accomplished without substrates or encapsulated microchannels. For example, Kim et al. developed freestanding stretchable liquid metal thin-film conductor with high electrical conductivity. The freestanding conductor was fabricated by continuous visible laser with 532 nm wavelength. First, they prepared liquid metal–silver nanowire thin film through vacuum filtration process. Then, the film was patterned into desired shapes by direct laser patterning with computer-aided design (CAD). In this process, the depth-controlled laser process by fast laser scanning speed allowed localized multiphase sintering to occur only on the surface of the film. Since the multiphase sintering, which is concurrence of nanowelding of silver nanowires and aggregation of liquid metal, induced etch-resistivity, the film was distinguished in-depth as well as in-plane. After etching process of sodium hydroxide solution, the non-sintered region was etched away and the sintered region became freestanding thin-film conductor. The freestanding conductor had remarkable attributes such as stand-alone feature, customizability, and shape-adaptability stemming from the laser process as well as high electrical conductivity originating in liquid metal and silver nanowire. As a proof, the group introduced representative examples of the freestanding conductor applied to environments with various constraints where conventional circuits cannot be applied, which include folding 90° horizontally and vertically,

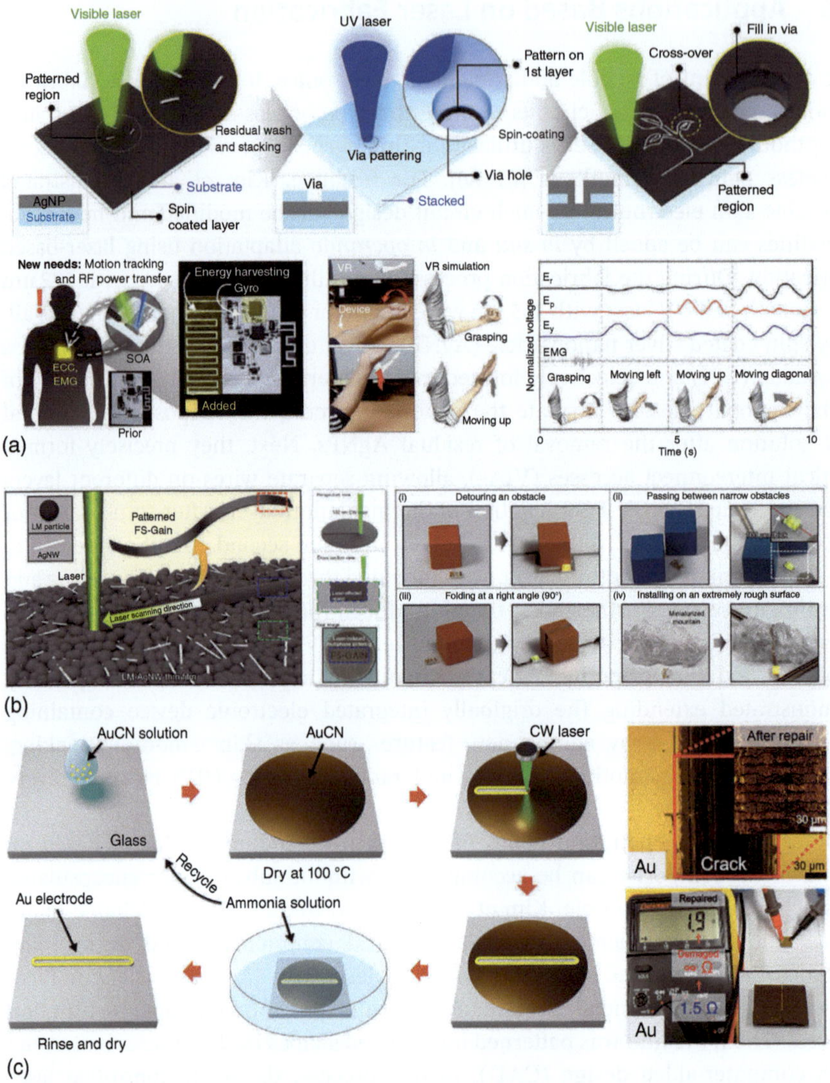

Figure 7.4 Laser-based fabrication of electronic circuits. (a) Skin electronics integrated with laser-fabricated electronic circuits. Source: Reproduced with permission from Ref. [30], John Wiley & Sons. Copyright 2022. (b) Freestanding liquid metal–silver nanowire composite conductor. Source: From Ref. [67]/Springer Nature/CC BY 4.0. (c) Formation of gold electrode using laser refining. Source: From Ref. [68]/MDPI/CC BY 4.0.

passing narrow gaps, and installing on a surface with high curvature and roughness (Figure 7.4b) [67].

A laser is a suitable tool for facile and on-demand fabrication of micropatterned conductors. Especially, as reported by Lim et al., selective laser refining of gold cyanide (AuCN) is a cost- and time-effective process for gold patterning. The reason is that selective laser refining enables the extraction of straight gold from

AuCN, which is usually the first product from primitive gold ore and writes directly desired patterns using *in situ* refinement and alignment without synthesis of gold nanoparticles. To demonstrate the process, the group drop-casted AuCN dissolved in ammonia solution on target substrate and dried the solution. Then, they utilized continuous wave (CW) laser with 532 nm wavelength for refining of AuCN and selective patterning of gold. In this process, thermal decomposition of AuCN and partial sintering of refined gold occurred simultaneously through laser-induced photothermal reaction. Also, the sintered gold agglomerated together parallel to laser scanning direction by thermocapillary force, reducing defects of the sintered gold conductor. Next, the sample was rinsed with ammonia solution to eliminate residual AuCN and then the patterned gold conductor was achieved. As a practical way to utilize the selective laser refining process, the group demonstrated *in situ* facile restoration of damaged conductors (Figure 7.4c) [68].

Laser cutting can be utilized to fabricate conductors with complex configurations such as serpentine patterns. Since the serpentine pattern endows stretchability to a metal film, which is flexible but not stretchable, it can be applied to a conductor for stretchable electronics [69–73]. For example, Lee et al. exhibited a stretchable skin-like thermo-haptic device using serpentine-patterned copper film as an interconnection conductor. To find the optimal design according to trade-off relationship between stretchability and conductivity, they performed parametric study with serpentine design parameters. Then, based on the calculated results, they patterned copper film to the serpentine design using direct laser cutting of 355 nm wavelength laser. Finally, they integrated thermoelectric pellets with the serpentine copper conductor and then fabricated the stretchable skin-like thermo-haptic glove. The thermo-haptic device operated stably under the dynamic motion of a hand and deformation of skin. In addition, they reconstructed the artificial thermal feeling of touching assorted hot and cold objects with the device in virtual reality (VR) space (Figure 7.5a) [74].

Similarly, the group also presented thermally controlled imperceptible artificial skin using laser cutting and ablation processes. They utilized laser cutting to fabricate thermally conductive elastomer and electrically conductive copper film into desired shape. Then, they laser-ablated and peeled off insulating elastomer coated on the copper conductor to be selectively exposed for interconnection between thermoelectric pellets. Lastly, they achieved the imperceptible artificial skin with active camouflage ability in visible to infrared range by adding thermochromic liquid crystal layer to the thermoelectric device (Figure 7.5b) [75].

A laser can also be utilized to realize stretchable electronic circuits with high conductivity by activating liquid metal [53, 77–80]. For example, Liu et al. showed stretchable elastomer-liquid metal circuits with high electrical conductivity and high-resolution patterns through direct laser writing process. First, they prepared a double emulsion composed of ethanol, polydimethylsiloxane (PDMS) precursor, and liquid metal using a two-step ultrasonication process. In the emulsion, liquid metal microparticles coated with PDMS precursors were dispersed stably in ethanol as the solvent. Then, they covered the emulsion on a PDMS substrate using a rod coating method. After ethanol evaporated and PDMS precursors polymerized, a

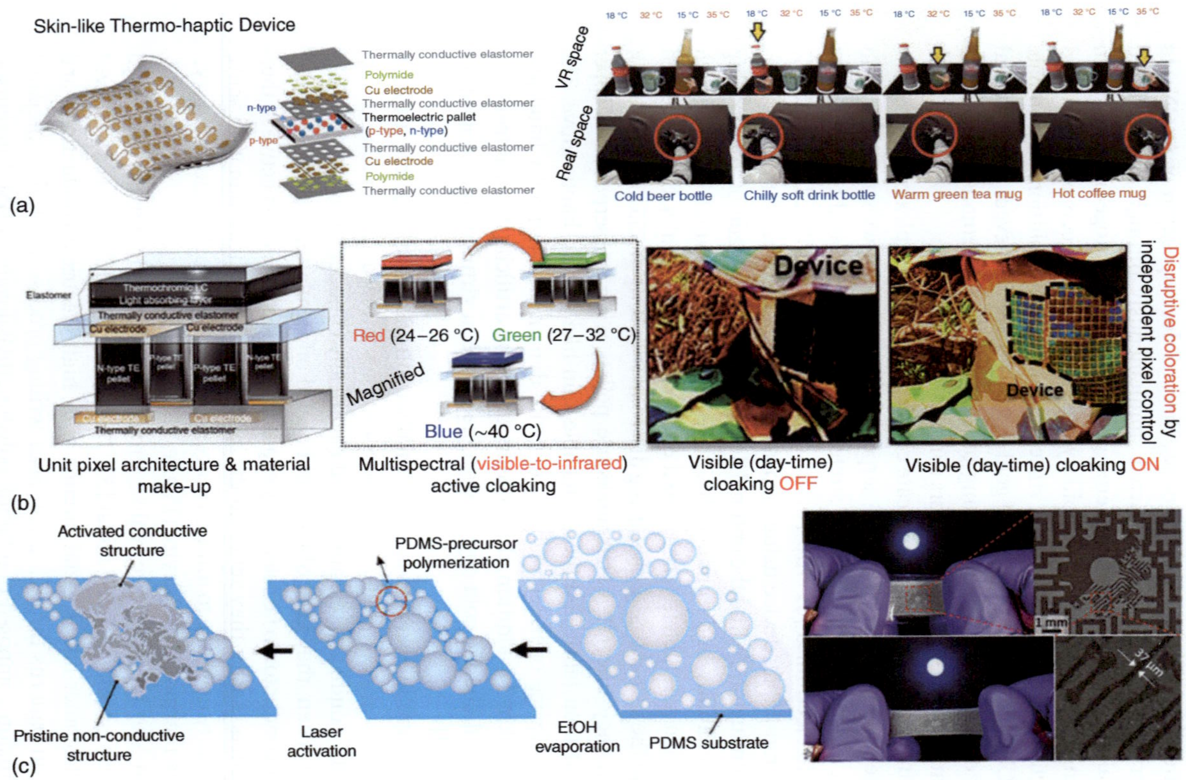

Figure 7.5 Laser-based fabrication of patterned electrodes. (a) Copper-based serpentine electrode for stretchable skin-like thermohaptic device. Source: Reproduced with permission from Ref. [74], John Wiley & Sons. Copyright 2020. (b) Thermally controlled imperceptible artificial skin integrated with laser-patterned serpentine electrodes. Source: Reproduced with permission from Ref. [75], John Wiley & Sons. Copyright 2020. (c) Laser-printed liquid metal-elastomer conductor. Source: Reproduced with permission from Ref. [76], American Chemical Society. Copyright 2021.

nonconductive polymer-liquid metal composite film was formed with high adhesion to each liquid metal particle and the substrate. Next, they activated the composite film selectively with a desired pattern via 355 nm wavelength laser, causing the patterned region of the film to convert into conductive state. The laser-activated PDMS-liquid metal composite was highly conductive and strain-invariant as well as stretchable. As a proof, the group fabricated a stretchable electronic circuit with a complicated pattern connecting an LED. Then, they demonstrated that the LED shone brightly without degradation in brightness while twisting, bending, and stretching the circuit (Figure 7.5c) [76].

Laser irradiation can transform phases of certain polymers such as poly(3,4-ethylenedioxythophene):poly(styrene sulfonate) (PEDOT : PSS) and polyvinylidene fluoride (PVDF) [81, 82]. Among them, PEDOT : PSS as a conductive polymer has multiple phases where PEDOT is a conductive core and PSS is an insulative shell. Laser treatment can lead to separation of the phases, which enhances electrical properties and aqueous stability of PEDOT : PSS. For instance, Won et al. fabricated water-stable PEDOT : PSS hydrogels with high electrical conductivity based on laser-induced phase separation of PEDOT : PSS. To induce the phase separation, they utilized an intense electric field and photothermal reaction instantaneously activated by a continuous 532 nm laser.

Furthermore, they added plasmonic gold nanoparticles for enhanced interaction between PEDOT : PSS and the laser. The addition of gold nanoparticles and the laser treatment also transformed the structure of PEDOT crystallite from a benzoid structure to a quinoid structure, which means enhancement of electrical conductivity in separated PEDOT domains. Finally, they established the selective patterning of highly conductive PEDOT : PSS hydrogels by the laser-induced phase separation process and successfully developed the bioelectronic device for neural signal recording and neural stimulation (Figure 7.6a) [82].

Kim et al. demonstrated biocompatible cost-effective electrophysiological sensors with laser-treated copper–gold core–shell nanowires (CuAuNWs). In this research, they utilized nanowelding process to enhance electrical performance of CuAuNWs as conductor and cutting process to realize desired patterns of a sensor. First, the group synthesized copper nanowires through a hydrothermal method. Then, they performed subsequent ligand exchange of copper nanowires and gold deposition on the ligand-exchanged nanowires, which resulted in a synthesis of CuAuNWs. Next, after transfer of the CuAuNW film onto a sacrificial layer by vacuum transfer process, they nanowelded the film using 532 nm wavelength laser and spin-coated polyurethane acrylate (PUA) solution on the film. Then, they laser-patterned the PUA-coated CuAuNW film to the desired pattern of electrophysiological sensor using 355 nm wavelength laser. The fabricated electrophysiological sensor operated stably on forearm and chest, which showed clear signal of electromyogram according to muscle contracts and apparently detected the QRS complex of electrocardiogram (Figure 7.6b) [83].

In flexible and stretchable electronic devices, various electronic components including functional semiconductors and metal electrodes are indispensable but the connection of different materials brings about defective interfaces between them.

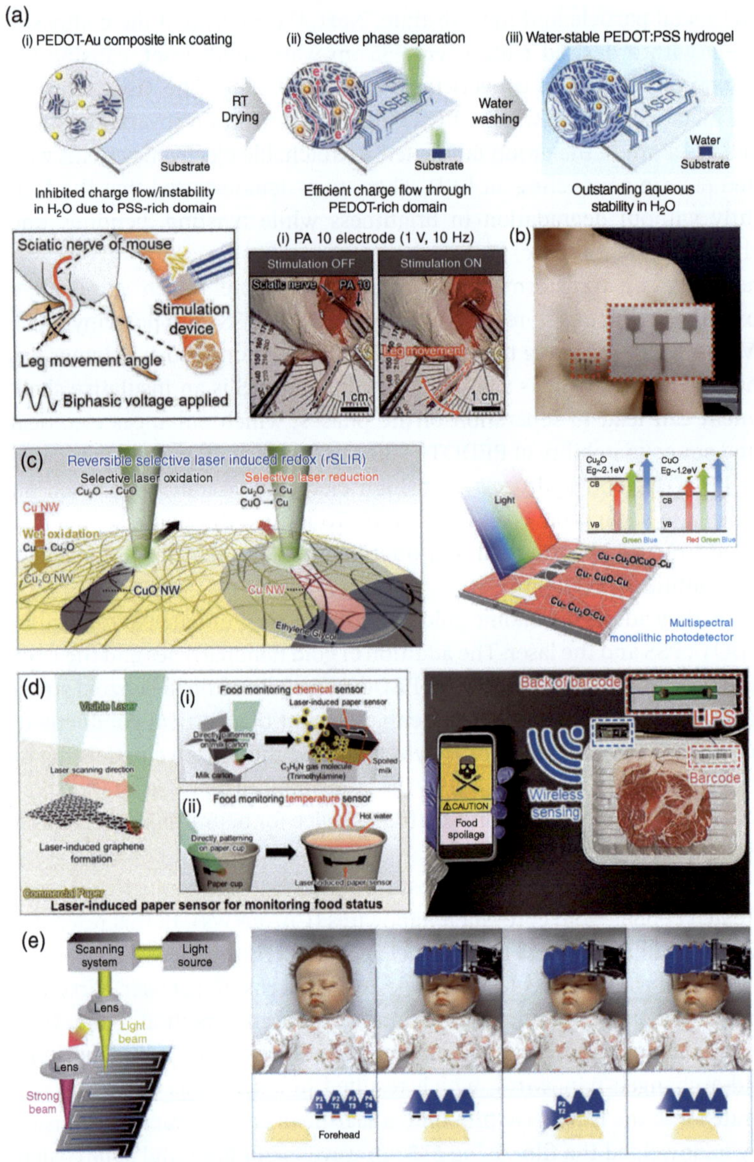

Figure 7.6 Laser-based fabrication of sensors. (a) Digital selective transformation of conductive polymer for bio-signal sensing. Source: From Ref. [82]/American Association for Advancement of Science/CC BY 4.0. (b) Laser-assisted fabrication of CuAuNW-based electrophysiological sensor. Source: Reproduced with permission from Ref. [83], John Wiley & Sons. Copyright 2020. (c) Multispectral photodetector integrated by reversible selective laser-induced redox. Source: From Ref. [84]/Springer Nature/CC BY 4.0. (d) Laser-induced paper sensor for real-time food spoilage monitoring. Source: Reproduced with permission from Ref. [85], Elsevier. Copyright 2022. (e) Stretchable multimodal sensor array with kirigami structure for human-robot interaction. Source: From Ref. [32]/Springer Nature/CC BY 4.0.

However, laser process offers the possibility of monolithic fabrication of electronic components with seamless interfaces between semiconductors and metal electrodes [86–90]. For example, Bang et al. established a reversible selective laser-induced redox process controlling transition between oxidation states of transition metals especially such as copper. For the process, they adopted a CW laser with 532 nm wavelength delivering photothermal energy. Interestingly, the photothermal energy causes selectively either oxidation or reduction depending on environmental conditions. Laser irradiation under atmospheric conditions induces the oxidation from copper(I) oxide (Cu_2O) to copper(II) oxide (CuO) and laser treatment with reducing agents leads to the reduction from Cu_2O and CuO to copper. Exploiting reversible and selective conversion between oxidation states of copper through laser-induced redox, the group fabricated flexible multispectral photodetectors with sensitivity for multiple wavelengths. First, in the fabrication steps, they transferred copper nanowires onto a substrate through vacuum filtration method. Then, they transformed the copper nanowires into Cu_2O nanowires by wet oxidation process. The Cu_2O nanowires were selectively laser-oxidated and laser-reduced and then encapsulated by PUA, which resulted in seamlessly integrated photodetectors including three configurations of Cu–Cu_2O–Cu, Cu–CuO–Cu, and Cu–Cu_2O&CuO–Cu. The manufactured multispectral photodetectors successfully showed multispectral response generating photocurrents depending on wavelength and intensity of visible light (Figure 7.6c) [84].

Laser treatment can produce functional nanomaterials directly from natural substances such as wood and leaves [91–93]. This enables us to fabricate biodegradable functional devices with common materials around us. For example, Jung et al. presented a method to develop graphene with porous structure, named laser-induced graphene (LIG), on a paper that originates from nature and is common in daily life. They utilized continuous visible laser activating a photothermal pyrolysis reaction by which irradiated surface of a paper turns into LIG. Also, they conducted a parametric study with varying laser power and scanning speed, and successfully fabricated mechanically and electrically stable LIG without damaging substrate. As a proof, the group demonstrated laser-induced paper sensor for wireless, real-time monitoring of food freshness with chemical and temperature sensing. They fabricated the sensor on the paper of commercial food packages such as milk cartons, paper cups, and barcode paper and the sensor clearly detected spoilage of the food (Figure 7.6d) [85].

By manipulating process parameters of laser such as power, frequency, duty cycle, and scanning speed, laser ablation and cutting can be realized in a single process. This process enables to fabrication of components of a multimodal soft sensor array, including various sensors and conductive wires, at the same time. For instance, Ham et al. manufactured a laser-patterned kirigami sensor array for soft robotic skin using the process. First, the group prepared a polyethylene terephthalate (PET) film coated with reflective aluminum. Then, they adopted a diode UV laser with 354.7 nm wavelength for ablation of conductive layer and cutting of both conductive and substrate layer. In weak irradiation mode with low power and high scanning speed, the metal layer was laser-ablated without damaging the substrate into meandering shape and interdigitated comb pattern for temperature sensor and proximity sensor, respectively. On the other hand, both layers were laser-cut entirely in strong irradiation mode with high power and low scanning speed, resulting in the film being patterned as kirigami structure for stretchable conductive wires connecting sensors. Finally, the fabricated kirigami sensor array with four temperature sensors and four proximity sensors was integrated on the soft pneumatic hand. Since the kirigami structure granted stretchability to the sensor array, it operated reliably during the flexing of the robotic fingers. As a proof, they demonstrated human–robot interaction by measuring proximity and temperature while the robotic hand approached the forehead of a doll with gentle contact (Figure 7.6e) [32].

Laser cutting can endow a kirigami structure to an electrode [94–96]. Since the kirigami patterned electrode possesses tunable stretchability, it can be used as a sensing probe for bio-signal. For instance, Won et al. presented an ultra-stretchable and transparent kirigami electrode fabricated by 365 nm wavelength UV laser cutting. The group prepared a thin and transparent film with embedded silver nanowire percolation network, which was flexible but not stretchable. Then, they laser cut the film with desired patterns with the aid of CAD program. The cut granted stretchability to the film, and its stretchability could be easily tunable with various kirigami patterns such as uniaxial, biaxial, and square spiral. The laser-cut film endured over 400% tensile strain with negligible resistance change. Since the film was stretchable, strain-invariant, and transparent, it was appropriate for acquiring bio-signals such as electrocardiogram, electrooculogram, and electromyogram. As a proof, they measured electromyogram signals from forearms with the film and controlled a drone with varying electromyogram signals by changing the posture of the forearms (Figure 7.7a) [97].

By controlling the power of laser, sintering and ablation can be achieved in a single laser with one step process. This process realizes microscopic stiffness tuning and macroscopic structural patterning of the exposed target film at the same time. For example, Kim et al. fabricated a highly sensitive crack sensor with the process. They first prepared silver nanoparticle-coated CPI film on a glass substrate. Then, they utilized 365 nm UV laser to ablate the film into the desired serpentine pattern and to anneal, or sinter, the other area for stiffness enhancement to induce crack for the sensor. They theoretically analyzed the crack formation mechanism and figured out the optimal power for crack formation, which was obviously lower than ablation power. Moreover, they implemented deep-learning analysis for the crack sensor signal in order to identify full hand motion with only one sensor on a wrist. They successfully realized the system by generating the same hand motion on display (Figure 7.7b) [98].

Laser sintering can be utilized to fabricate a microscopic structure such as wavy pattern on the exposed area [100–102]. This wavy pattern can identify pressure since higher pressure will penetrate more into the valley of the wave. In another research, Kim et al. fabricated the deep-learned crack sensor devised a transparent and wearable three-dimensional touch sensor with a wavy pattern created by laser exposure. In this research, they prepared an elastomer with spin-coated silver nanoparticles. Then, at this time, they exposed a 532 nm wavelength visible laser

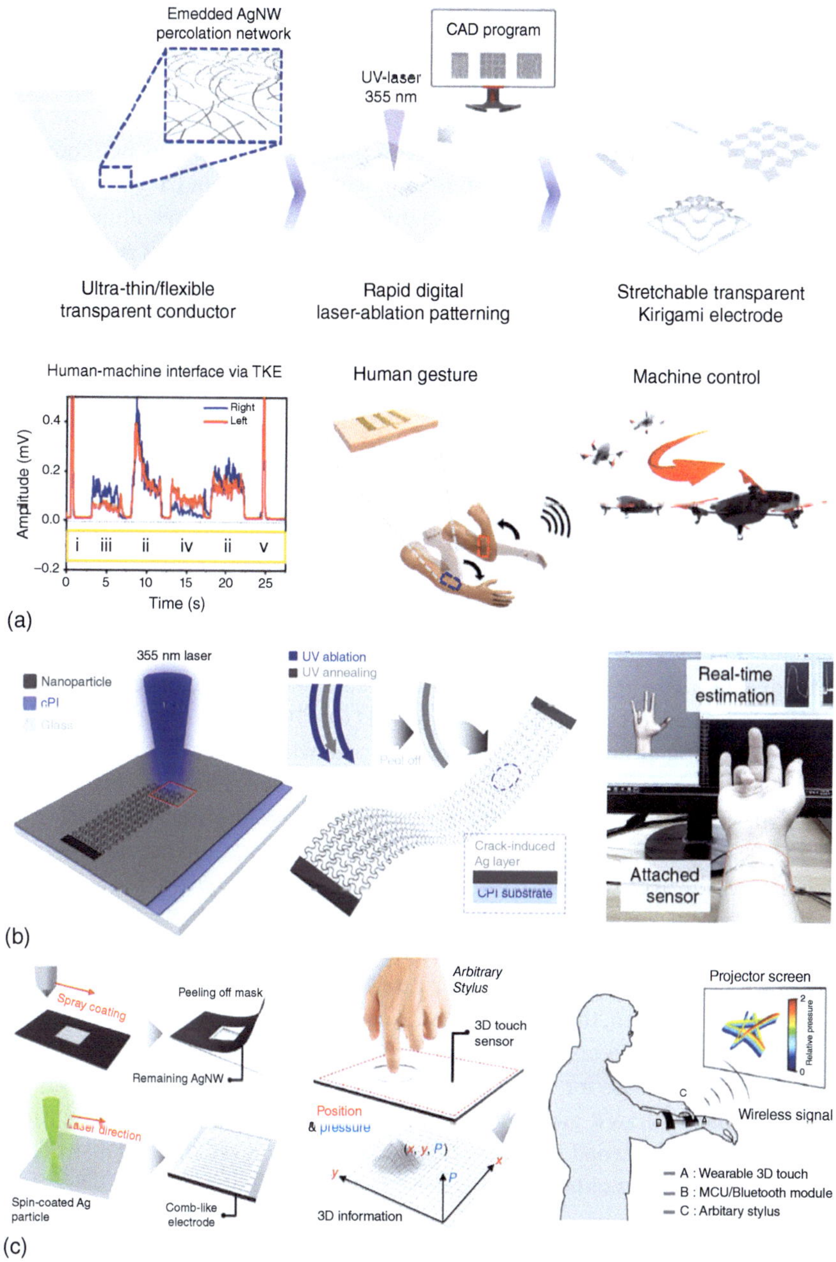

Figure 7.7 Laser-based fabrication of wearable sensors. (a) Stretchable transparent Kirigami electrode for bio-signal sensing. Source: From Ref. [97], American Chemical Society. Copyright 2019. (b) Laser-fabricated crack sensor for the epicentral human motions. Source: From Ref. [98]/Springer Nature/CC BY 4.0. (c). Transparent wearable three-dimensional touch sensor. Source: From Ref. [99]/Springer Nature/CC BY 4.0.

to the silver nanoparticles. The heat generated by the laser locally melted the nanoparticles and induced thermal capillary flow by temperature gradient between the adjacent nonexposed portion. The flow permanently left a wavy structure in the film after the laser was removed and the molten region re-solidified. The group, which is similar to the aforementioned research, found the optimal laser power for the creation of the regular wave structure. Too much low power was not enough to create a wavy structure, whereas too high power generated an irregular balling structure. They utilized this wavy structure thin film to create a three-dimensional touch sensor, which could sense the x position, y position, and pressure at the same time. They introduced another layer of transparent silver nanowire-coated thin film on top of the wavy structure. They utilized the conventional potential reading method for position sensing. For pressure sensing, they exploited the feature that as the wavy structure is compressed more, the covered portion of the structure gets larger. In this way, they devised a transparent and wearable three-dimensional sensing device for the first time (Figure 7.7c) [99].

Soft heating device is one of the major components in next-generation electronics. Unlike conventional rigid and bulk heaters, recently developed heaters are small and soft for application in various emerging fields [103–105]. In the process of fabricating a soft heater, patterning the heating component, which is usually an electrical conductor, is necessary for acquiring desired areal resistance for optimizing the operating temperature range and power. Since the laser is an adequate tool to pattern a conductor without damaging the underlying soft substrates, various research on fabricating soft heaters by laser has been performed [106–108]. For instance, Kim et al. replicated a feeling of heat in a virtual reality space with a laser-fabricated highly stretchable copper nanowire heater. In the fabrication process, they utilized 532 nm wavelength laser and 355 nm wavelength laser for different purposes. First, they prepared a copper nanowire network on a sacrificial substrate and exposed 532 nm wavelength visible laser for nano-welding, or annealing, the network for high conductivity. Then, they coated the nano-welded network with UV-curable PUA for suppressing copper oxidation. Next, they cured the PUA with 355 nm wavelength UV laser with serpentine pattern to grant stretchability to the composite of PUA and nanowire network. Then, they transferred the composite to the thermal release tape and to the water-soluble tape sequentially. Finally, the composite was transferred to the adhesive substrate for fabricating a thermal haptic glove. With an array of heaters attached to the VR glove, they successfully replicated the different thermal sensations of touching wood and copper, respectively. Moreover, they reconstructed the thermal feeling of holding a hot water cup that was being filled in VR space (Figure 7.8a) [31].

Since a heater's performance depends on the areal resistance of the conductor, a careful manipulation of the conductivity in a single device can realize a heater with various heating abilities. Laser is an appropriate apparatus for the manipulation due to its facile power and speed control. For example, Cho et al. created a heater with a stretchable, seamless, and variable resistance conductor. The conductor was fabricated by exposing 532 nm wavelength visible laser with varying laser power for resistance control. At first, they created a liquid metal–silver nanowire thin film

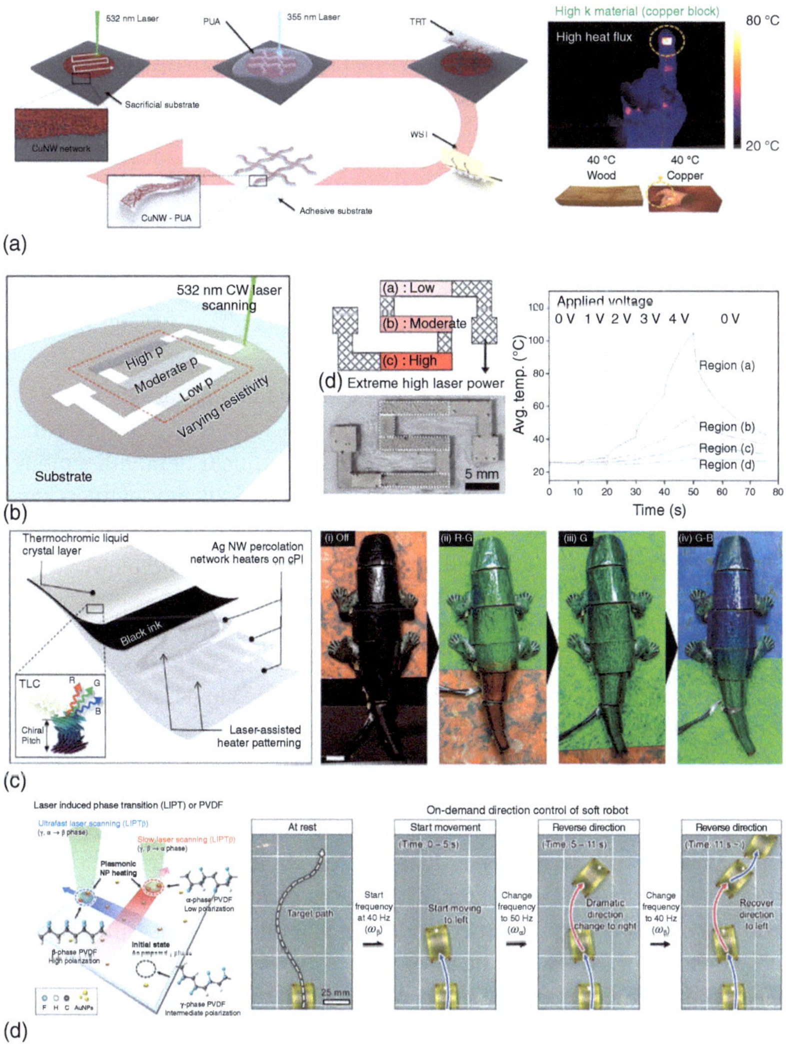

Figure 7.8 Laser-based fabrication of heaters and actuators. (a) Stretchable and oxidation-free Cu nanowire-based heater for reconstruction of thermal sensation. Source: Reproduced with permission from Ref. [31], The Royal Society of Chemistry. Copyright 2020. (b) Monolithically programmed stretchable conductor. Source: Reproduced with permission from Ref. [109], John Wiley & Sons. Copyright 2022. (c) Biomimetic chameleon soft robot with artificial camouflage skin. Source: Reproduced with permission from Ref. [110]/Springer Nature/CC BY 4.0. (d) Digital selective reversible phase control for piezoelectric polymer. Source: Reproduced with permission from Ref. [81], John Wiley & Sons. Copyright 2022.

on an elastomer by vacuum filtration. Then, they exposed the film with varying laser power in order to achieve diverse resistivities in a single film. Exposing the laser to the film induced entanglement between the liquid metal particles and the nanowires. This phenomenon is analogous to laser sintering of nanomaterials, which causes conductivity enhancement by merging two individual nanomaterials.

The high-power laser induced vigorous entanglement between the liquid metal particles and the nanowires, which finally attained high conductivity. On the other hand, the low-power laser induced weak entanglement between the two, which attained low conductivity. Interestingly, regardless of the power of the laser, the laser-exposed portion of the film anchored to the elastomeric substrate which facilitated the portion to endure basic solution treatment. This effect enabled cleaning the laser-unexposed portion to wash away and leaving the laser-exposed portion only. As a proof of a concept, the group fabricated a heater with four different conductivities in a single conductor by varying the laser power. When a certain voltage was imposed to the conductor, low-power treated area became hotter, and vice versa, as expected (Figure 7.8b) [109].

In fabricating multiple layers of heaters with color actuators, various features, such as achieving conductivity, heater patterning, via formation, and external power connection, need to be considered. Unlike conventional photolithography processes, the laser-based fabrication process can facilitate all the features performed in non-vacuum, ambient room temperature conditions with high quality. For instance, Kim et al. invented a biomimetic chameleon soft robot based on a laser fabrication. The robot consists of multilayered silver nanowire heaters connected with multiple via holes filled with laser-sintered silver nanoparticles. The fabrication process for multilayered heaters is as follows. A CPI is spin-coated on a glass substrate. Then, silver nanoparticles are spin-coated and laser sintered by 532 nm wavelength visible laser for contact area creation. Next, silver nanowires are spray-coated and laser-ablated to be patterned. This process creates the first heating layer. Then, a CPI layer is spin-coated above to form a second layer. To create via hole for electrical connection between different layers, 352 nm wavelength UV laser is exposed to the second layer. The created via hole is filled with silver nanoparticles by spin-coating and sintered by 532 nm wavelength visible laser. Then again, silver nanowires are spray-coated and laser-patterned to constitute second heating layer. The third layer is fabricated in the same way as the second layer. Finally, thermochromic liquid crystal ink is deposited on the top layer, completing the fabrication of color-changeable skin of the biomimetic chameleon soft robot. The multilayered heater with different conductor designs in each layer successfully duplicated the chameleon, as shown in Figure 7.8c. As explained above, the whole process of patterning, via hole forming and sintering was conducted by laser in ambient conditions without damaging underlying layers [110].

Laser treatment can transform phases of certain materials such as PVDF and PEDOT : PSS [81, 82]. Especially for materials that have multiple phases such as PVDF, a careful control of laser parameters enables reversible phase transition between them. For example, Won et al. established a laser-induced phase transition process for PVDF. They utilized a plasmonic nanomaterial of gold nanoparticles for enhanced laser absorption. They successfully realized reversible phase transition of PVDF from pristine gamma phase to either alpha phase or beta phase with different laser parameters. Exploiting different resonant frequencies of different phases, they manufactured a direction-controllable soft robot with seamlessly integrated phases of PVDF. First, they prepared a pristine gamma phase PVDF film. Then, they deposited gold nanoparticles on top of the film. Finally, they exposed laser to the film with different parameters to create a seamless multiphase PVDF film. The transition between alpha phase and beta phase was reversible with careful laser parameter control. They analyzed that the alpha phase was induced by slow laser scan with a thermodynamics dominant process, whereas the beta phase was induced by fast laser scan with an electrostatics dominant process. The fabricated multiphase-embedded thin PVDF film was successfully actuated with varying frequencies of an external voltage to control the direction (Figure 7.8d) [81].

Laser treatment is suitable for fabricating microfluidic channels [111–113]. Since the laser beam power follows Gaussian distribution, the cross-section of laser-affected portion forms a hemispheric structure. By taking advantage of this feature, one can readily create a desired pattern of microfluidic channel with CAD on an elastomer or other functional substrate. As this process is based on pyrolysis of a substrate by localized laser exposure, it is called as successive laser pyrolysis. As an example of successive laser pyrolysis, Shin et al. presented a seamless patterning of PDMS. By multiple laser treatments with varying laser parameters, they successfully controlled the shape of the PDMS. The group fabricated not only a microfluidic channel but also diverse three-dimensional structures that could not be achieved previously such as seamless hollow cubes and pillar arrays. They also manufactured a pneumatic gripper with laser-pyrolyzed PDMS. The fabrication process starts with drawing an initiator point with an ink. Since 532 nm wavelength visible laser cannot be absorbed to a transparent PDMS, a black initiator point that can absorb the laser is necessary. After drawing one, the laser is exposed from the point with a desired pattern. The heat generated from the laser exposure is strong enough to pyrolyze the surface of the elastomer, and the pyrolyzed surface, which is dark, becomes another heat absorption spot realizing successive laser pyrolysis. This process is possible for both top and bottom surfaces of the elastomer because it is transparent for the laser beam to pass through facilitating the fabrication of various unique structures and channels (Figure 7.9a) [114].

Lim et al. demonstrated another research about laser pyrolysis. In this research, the group laser pyrolyzed polyimide film. They figured out that highly porous polyimide fiber was explosively fabricated by fast and high-power laser. They named the phenomenon as pyrolytic jetting of polyimide. Similar to the PDMS laser pyrolysis mentioned above, the polyimide laser pyrolysis created a similar hemispherical cross-section on the polyimide film with various laser parameters. They also exposed the laser multiple times to achieve deeper channels or even punch a hole. As a proof of concept, the group fabricated a patterned silver electrode by using the fabricated channel as a master mold (Figure 7.9b) [115].

Laser-induced hydrothermal growth is an effective and facile way to fabricate diverse functional materials on demand. Recently, Song et al. presented a method to create apatite with various inorganic ions attached by nanosecond laser pulse-induced hydrothermal growth. First, a thermal reservoir that consists of titanium in this article is filled with desired precursor solution. The precursor solution comprised inorganic ions such as calcium, magnesium, strontium, and zinc with phosphate ions. Then, the laser is exposed on the titanium surface

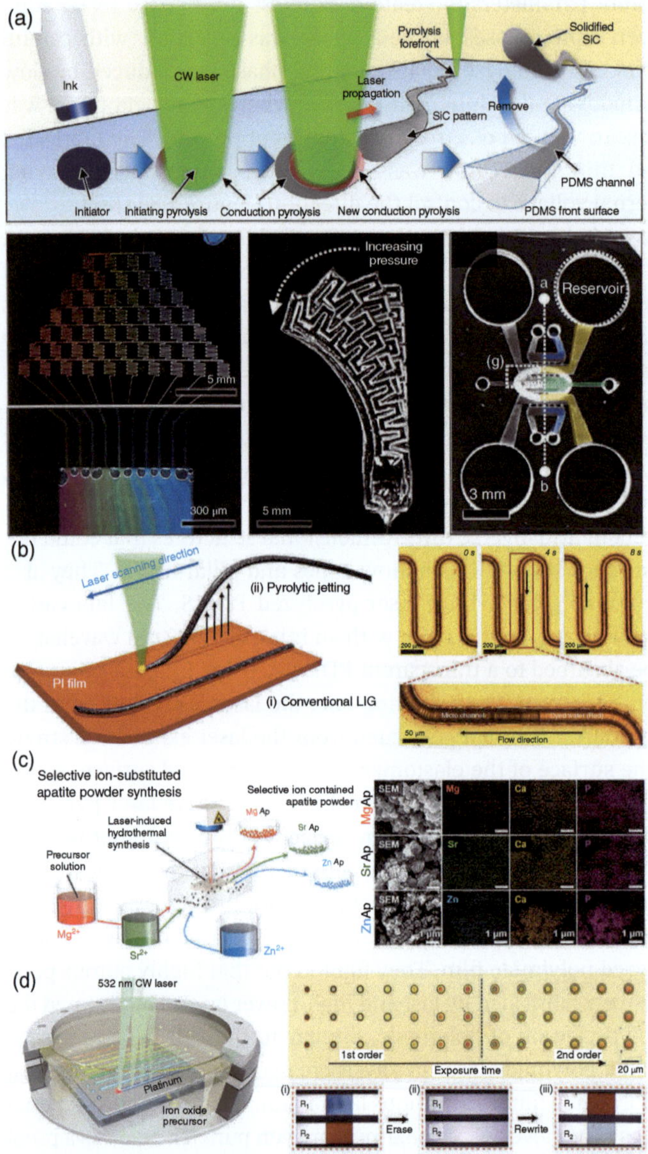

Figure 7.9 Laser-based fabrication of micro-channels and optoelectronic devices. (a) Monolithic digital patterning of PDMS with laser pyrolysis. Source: Reproduced with permission from Ref. [114], Springer Nature. Copyright 2021. (b) Monolithic digital patterning of polyimide film by laser-induced pyrolytic jetting. Source: Reproduced with permission from Ref. [115], Elsevier. Copyright 2022. (c) Synthesis of various apatites by laser-induced hydrothermal growth. Source: Reproduced with permission from Ref. [116], American Chemical Society. Copyright 2022. (d) Digital laser micropainting for reprogrammable optoelectronic device. Source: Reproduced with permission from Ref. [117], John Wiley & Sons. Copyright 2020.

creating apatite powders. The group successfully manipulated the powder size by laser parameters. They claimed that the formation of the powders stems from high heat generated on the surface of the laser-exposed titanium (Figure 7.9c) [116].

Similarly, Lee et al. exhibited a laser-based micropainting with structural coloration. The main phenomenon that they utilized was laser-induced hydrothermal growth of the metal oxide layer. They prepared a platinum-coated glass substrate and an iron oxide precursor solution was filled. Then, they exposed 532 nm wavelength visible laser to induce hydrothermal growth of an iron oxide layer. In this study, they controlled the laser scanning speed to modify the thickness of the layer, which enabled achieving the coloration by thin-film interference. Besides, by substituting the precursor solution to ethylene glycol, they could selectively erase the iron oxide with the same laser exposure. This granted the process with the reprogrammability (Figure 7.9d) [117].

7.4 Perspectives and Conclusion

We reviewed the laser-based fabrication techniques, as alternative fabrication methods to replace conventional manufacturing processes such as lithographic patterning and vacuum deposition for implementing soft electronic devices. Laser has become a suitable tool for realizing flexible and stretchable electronics due to its exclusive properties, competent abilities, and unrestrained scalability.

First, the laser manufacturing process can be utilized simply and immediately at room temperature and atmospheric environment. This implies that intricate mechanical equipment, including chamber, vacuum pump, and heat generator for deposition, and etching tank containing toxic developer, are unnecessary for the laser process. Also, unlike the conventional lithographic process, patterning mask is not required to create micro- and nanostructures by laser. Therefore, the laser process is an accessible approach for many researchers to prototype soft electronic devices because it is a facile and fast method to form functional micro- and nanostructures on a scale similar to or smaller than the scale of laser beam.

Second, the laser fabrication process does not restrict material selection and process conditions. Various types of materials such as inorganic materials, elastomers, ceramics, and composites can be treated by laser because the process is performed at room temperature if other specific conditions are not required. In addition, laser can deliver its energy to desired spots only during the process. This means that laser-induced reactions can be locally and selectively generated, allowing laser treatment to be carried out without undesirable damage to surrounding area composed of materials that may be vulnerable to heat. Some laser-induced reactions even differ not only in *xy* plane but also in *z*-axis, that is the depth direction, which causes to convert upper conductive region into freestanding electrode. Also, the laser process is expandable, and one can reconfigure the process by adding appropriate equipment if different conditions are needed such as in an aqueous solution.

Third, the fabrication process based on laser can be easily controlled and freely customized by varying process parameters including intensity, wavelength, scanning speed, and duty cycle. Well-controlled process parameters allow desired processes such as laser sintering, ablation, and cutting to occur at a desired region, which means that complicated micro- and nanostructures like hierarchical and heterogeneous constitutions are easily realized by laser. Macroscale patterning can also be constructed by laser-like conventional UV laser machining, so it is possible to achieve structure with various scales from nano- and microscale to macroscale in a single-step laser process. Moreover, the laser process covers synthesis of micro- and nanomaterials with uncommon characteristics and valuable functionalities via versatile laser-induced processes such as hydrothermal growth, pyrolysis, phase transition, and redox between metal and metal oxide.

Based on the abovementioned advantages, the laser-based fabrication process has been used to implement soft electronics in diverse areas. In soft robotics and wearable electronics, electronic components of devices should be flexible and stretchable to endure and conform to mechanical deformation of mounted surfaces. For this, soft circuits are fabricated by laser-patterned serpentine or kirigami structure, wearable touch panel is achieved through three-dimensional wavy structure formed by laser sintering, and flexible actuator is realized via phase transition of polymer induced by laser. In various sensors, selective laser-induced redox process implements monolithic fabrication of flexible photodetector with seamless interfaces between semiconductors and metal electrodes, and laser-induced pyrolysis of paper allows fabrication of biodegradable sensor. Above this, further applications in other fields such as energy devices [118–120], bioelectronics [121–123], and green technology are also achieved by laser-based fabrication and the approach using laser could pave the way for the advanced development of flexible and stretchable electronic devices.

Author Contributions

J.J.P. and M.K. contributed equally to this work. All the authors wrote and edited the manuscript.

References

1 Deng, H., Sattari, K., Xie, Y. et al. (2020). Laser reprogramming magnetic anisotropy in soft composites for reconfigurable 3D shaping. *Nature Communications* 11: 6325.

2 Amiri Moghadam, A.A., Alaie, S., Deb Nath, S. et al. (2018). Laser cutting as a rapid method for fabricating thin soft pneumatic actuators and robots. *Soft Robotics* 5: 443–451.

3 Du, T., Sun, L., and Wan, J. (2022). A worm-like crawling soft robot with pneumatic actuators based on selective laser sintering of TPU powder. *Biomimetics (Basel)* 7: 205.

4 Dong, X., Luo, X., Zhao, H. et al. (2022). Recent advances in biomimetic soft robotics: fabrication approaches, driven strategies and applications. *Soft Matter* 18: 7699–7734.

5 Pinheiro, T., Correia, R., Morais, M. et al. (2022). Water peel-off transfer of electronically enhanced, paper-based laser-induced graphene for wearable electronics. *ACS Nano* 16: 20633–20646.

6 Liang, S., Chen, X., Li, F., and Song, N. (2022). Laser-engraved liquid metal circuit for wearable electronics. *Bioengineering (Basel)* 9: 59.

7 Raza, T., Tufail, M.K., Ali, A. et al. (2022). Wearable and flexible multifunctional sensor based on laser-induced graphene for the sports monitoring system. *ACS Applied Materials & Interfaces* 14: 54170–54181.

8 Wang, H., Zhao, Z., Liu, P., and Guo, X. (2022). A soft and stretchable electronics using laser-induced graphene on polyimide/PDMS composite substrate. *npj Flexible Electronics* 6: 26.

9 Fu, X.Y., Chen, Z.D., Zhang, Y.L. et al. (2019). Direct laser writing of flexible planar supercapacitors based on GO and black phosphorus quantum dot nanocomposites. *Nanoscale* 11: 9133–9140.

10 Liu, S., Yuen, M.C., White, E.L. et al. (2018). Laser sintering of liquid metal nanoparticles for scalable manufacturing of soft and flexible electronics. *ACS Applied Materials & Interfaces* 10: 28232–28241.

11 Chen, X., Hou, Z., Li, G. et al. (2022). A laser-scribed wearable strain sensing system powered by an integrated rechargeable thin-film zinc-air battery for a long-time continuous healthcare monitoring. *Nano Energy* 101.

12 Huang, Y., Tao, L.Q., Yu, J. et al. (2020). Health monitoring and automatic notification device based on laser-induced graphene. *IEEE Transactions on Electron Devices* 67: 4488–4492.

13 Kang, S., Naqi, M., Shin, J. et al. (2021). Laser-processed stretchable-gradient interconnection-based temperature sensor for a real-time monitoring system. *ACS Applied Electronic Materials* 3: 5601–5607.

14 Pang, Y., Chen, B.K., Yu, S.F., and Lingamanaik, S.N. (2020). Enhanced laser speckle optical sensor for in situ strain sensing and structural health monitoring. *Optics Letters* 45: 2331–2334.

15 Fan, Z., Diao, X., Hu, K. et al. (2020). Structural health monitoring of metal-to-glass-ceramics penetration during thermal cycling aging using femto-laser inscribed FBG sensors. *Scientific Reports* 10: 12330.

16 Wang, J., Zhang, C., Deng, Y., and Zhang, P. (2022). A review of research on the effect of temperature on the properties of polyurethane foams. *Polymers (Basel)* 14: 4586.

17 Geng, P., Zhao, J., Wu, W. et al. (2018). Effect of thermal processing and heat treatment condition on 3D printing PPS properties. *Polymers (Basel)* 10: 875.

18 Nikolaidis, A.K. and Achilias, D.S. (2018). Thermal degradation kinetics and viscoelastic behavior of poly(methyl methacrylate)/organomodified montmorillonite nanocomposites prepared via in situ bulk radical polymerization. *Polymers (Basel)* 10: 491.

19 Dattilo, S., Puglisi, C., Mirabella, E.F. et al. (2020). Thermal degradation processes of aromatic poly(ether sulfone) random copolymers bearing pendant carboxyl groups. *Polymers (Basel)* 12: 1810.

20 Chen, R., Wang, X., Li, X. et al. (2021). A comprehensive nano-interpenetrating semiconducting photoresist toward all-photolithography organic electronics. *Science Advances* 7: eabg0659.

21 Yao, Y., Zhang, L., Leydecker, T., and Samori, P. (2018). Direct photolithography on molecular crystals for high performance organic optoelectronic devices. *Journal of the American Chemical Society* 140: 6984–6990.

22 Hwang, H.S., Zakhidov, A.A., Lee, J.K. et al. (2008). Dry photolithographic patterning process for organic electronic devices using supercritical carbon dioxide as a solvent. *Journal of Materials Chemistry* 18.

23 Jang, J., Song, Y., Yoo, D. et al. (2016). The development of fluorous photolithographic materials and their applications to achieve flexible organic electronic devices. *Flexible and Printed Electronics* 1: 023001.

24 Lee, J.-K., Taylor, P.G., Zakhidov, A.A. et al. (2009). Orthogonal processing: a novel photolithographic patterning method for organic electronics. *Journal of Photopolymer Science and Technology* 22: 565–569.

25 Dong, L., Fan, K., Feng, Y. et al. (2022). Preparation technology of stretchable electrode based on laser cutting. *Machines* 10: 854.

26 Dallinger, A., Keller, K., Fitzek, H., and Greco, F. (2020). Stretchable and skin-conformable conductors based on polyurethane/laser-induced graphene. *ACS Applied Materials & Interfaces* 12: 19855–19865.

27 Yeo, J., Hong, S., Lee, D. et al. (2012). Next generation non-vacuum, maskless, low temperature nanoparticle ink laser digital direct metal patterning for a large area flexible electronics. *PLoS One* 7: e42315.

28 Gillan, L., Hiltunen, J., Behfar, M.H., and Rönkä, K. (2022). Advances in design and manufacture of stretchable electronics. *Japanese Journal of Applied Physics* 61: SE0804.

29 Green Marques, D., Alhais Lopes, P., de Almeida, A.T. et al. (2019). Reliable interfaces for EGaIn multi-layer stretchable circuits and microelectronics. *Lab on a Chip* 19: 897–906.

30 Kim, K.K., Choi, J., Kim, J.H. et al. (2021). Evolvable skin electronics by in situ and in operando adaptation. *Advanced Functional Materials* 32: 2106329.

31 Kim, D., Bang, J., Lee, W. et al. (2020). Highly stretchable and oxidation-resistive Cu nanowire heater for replication of the feeling of heat in a virtual world. *Journal of Materials Chemistry A* 8: 8281–8291.

32 Ham, J., Han, A.K., Cutkosky, M.R., and Bao, Z. (2022). UV-laser-machined stretchable multi-modal sensor network for soft robot interaction. *npj Flexible Electronics* 6: 94.

33 DiNoro, J.N., Paxton, N.C., Skewes, J. et al. (2022). Laser sintering approaches for bone tissue engineering. *Polymers (Basel)* 14: 2336.

34 Kumar, S. (2003). Selective laser sintering: a qualitative and objective approach. *JOM* 55: 43–47.

35 Roy, N.K., Behera, D., Dibua, O.G. et al. (2019). A novel microscale selective laser sintering (mu-SLS) process for the fabrication of microelectronic parts. *Microsystems & Nanoengineering* 5: 64.

36 Liu, S., Reed, S.N., Higgins, M.J. et al. (2019). Oxide rupture-induced conductivity in liquid metal nanoparticles by laser and thermal sintering. *Nanoscale* 11: 17615–17629.

37 Liu, S., Yuen, M.C., and Kramer-Bottiglio, R. (2019). Reconfigurable electronic devices enabled by laser-sintered liquid metal nanoparticles. *Flexible and Printed Electronics* 4: 015004.

38 Colombo, L.L., Vanzulli, S.I., Blazquez-Castro, A. et al. (2019). Photothermal effect by 808-nm laser irradiation of melanin: a proof-of-concept study of photothermal therapy using B16-F10 melanotic melanoma growing in BALB/c mice. *Biomedical Optics Express* 10: 2932–2941.

39 Noroozi, M., Mohammadi, B., Radiman, S. et al. (2019). Photothermal effect of modulating laser irradiation on the thermal diffusivity of Al(2)O(3) nanofluids. *Nanoscale Research Letters* 14: 37.

40 Ishibashi, Y. et al. (2020). Nanosecond laser photothermal effect-triggered amplification of photochromic reactions in diarylethene nanoparticles. *Chemical Communications* 56: 7088–7091.

41 Werner, D., Hashimoto, S., and Uwada, T. (2010). Remarkable photothermal effect of interband excitation on nanosecond laser-induced reshaping and size reduction of pseudospherical gold nanoparticles in aqueous solution. *Langmuir* 26: 9956–9963.

42 Chen, X., Chen, Y., Yan, M., and Qiu, M. (2012). Nanosecond photothermal effects in plasmonic nanostructures. *ACS Nano* 6: 2550–2557.

43 Ko, S.H. et al. (2007). All-inkjet-printed flexible electronics fabrication on a polymer substrate by low-temperature high-resolution selective laser sintering of metal nanoparticles. *Nanotechnology* 18: 345202.

44 Russo, P., Hu, A., Compagnini, G. et al. (2014). Femtosecond laser ablation of highly oriented pyrolytic graphite: a green route for large-scale production of porous graphene and graphene quantum dots. *Nanoscale* 6: 2381–2389.

45 Huang, H., Lai, J., Lu, J., and Li, Z. (2019). Pulsed laser ablation of bulk target and particle products in liquid for nanomaterial fabrication. *AIP Advances* 9: 015307.

46 Solomon, J.M., Ahmad, S.I., Dave, A. et al. (2022). Ultrafast laser ablation, intrinsic threshold, and nanopatterning of monolayer molybdenum disulfide. *Scientific Reports* 12: 6910.

47 Jordan, R., Cole, D., Lunney, J. et al. (1995). Pulsed laser ablation of copper. *Applied Surface Science* 86: 24–28.

48 Dawood, A., Ahmed, N., Bashir, S. et al. (2022). Laser ablation of copper alloy under varying environmental conditions to achieve purpose-built surface structures. *Coatings* 12: 1972.

49 Bartlett, M.D., Markvicka, E.J., and Majidi, C. (2016). Rapid fabrication of soft, multilayered electronics for wearable biomonitoring. *Advanced Functional Materials* 26: 8496–8504.

50 Lim, C., Shin, Y., Jung, J. et al. (2019). Stretchable conductive nanocomposite based on alginate hydrogel and silver nanowires for wearable electronics. *APL Materials* 7: 031502.

51 Park, C., Jung, H., Lee, H. et al. (2018). One-step laser encapsulation of nano-cracking strain sensors. *Sensors (Basel)* 18: 2673.

52 Keller, K., Grafinger, D., and Greco, F. (2021). Printed and laser-scribed stretchable conductors on thin elastomers for soft and wearable electronics. *Frontiers in Materials* 8: 688133.

53 Lu, T., Markvicka, E.J., Jin, Y., and Majidi, C. (2017). Soft-matter printed circuit board with UV laser micropatterning. *ACS Applied Materials & Interfaces* 9: 22055–22062.

54 Yeo, J., Hong, S., Kim, G. et al. (2015). Laser-induced hydrothermal growth of heterogeneous metal-oxide nanowire on flexible substrate by laser absorption layer design. *ACS Nano* 9: 6059–6068.

55 In, J.B., Kwon, H.J., Lee, D. et al. (2014). In situ monitoring of laser-assisted hydrothermal growth of ZnO nanowires: thermally deactivating growth kinetics. *Small* 10: 741–749.

56 Yeo, J., Hong, S., Wanit, M. et al. (2013). Rapid, one-step, digital selective growth of ZnO nanowires on 3D structures using laser induced hydrothermal growth. *Advanced Functional Materials* 23: 3316–3323.

57 Hong, S. et al. (2013). Low-temperature rapid fabrication of ZnO nanowire UV sensor array by laser-induced local hydrothermal growth. *Journal of Nanomaterials* 2013: 1–7.

58 Kimiagar, S. and Abrinaei, F. (2023). Laser-assisted hydrothermal synthesis of MoS_2 nanosheets under different laser energies and potential application in nonlinear optics. *Optik* 272: 170305.

59 Kong, H., Kim, H., Hwang, S. et al. (2022). Laser-induced hydrothermal growth of iron oxide nanoparticles on diverse substrates for flexible micro-supercapacitors. *ACS Applied Nano Materials* 5: 4102–4111.

60 Lee, H., Manorotkul, W., Lee, J. et al. (2017). Nanowire-on-nanowire: all-nanowire electronics by on-demand selective integration of hierarchical heterogeneous nanowires. *ACS Nano* 11: 12311–12317.

61 Zheng, B., Zhao, G., Yan, Z. et al. (2022). Direct freeform laser fabrication of 3D conformable electronics. *Advanced Functional Materials* 33: 2210084.

62 Lu, Z., Wang, M., Zhang, P., and Li, C. (2022). Femtosecond laser machining of flexible printed circuit boards. *Journal of Physics: Conference Series* 2185: 012082.

63 Piqué, A., Mathews, S.A., Pratap, B. et al. (2006). Embedding electronic circuits by laser direct-write. *Microelectronic Engineering* 83: 2527–2533.

64 Park, S., Lee, H., Kim, Y.-J., and Lee, P.S. (2018). Fully laser-patterned stretchable microsupercapacitors integrated with soft electronic circuit components. *NPG Asia Materials* 10: 959–969.

65 Zhao, G., Ling, Y., Su, Y. et al. (2022). Laser-scribed conductive, photoactive transition metal oxide on soft elastomers for Janus on-skin electronics and soft actuators. *Science Advances* 8: eabp9734.

66 Ji, S.Y., Choi, W., Kim, H.Y. et al. (2018). Fully solution-processable fabrication of multi-layered circuits on a flexible substrate using laser processing. *Materials (Basel)* 11: 268.

67 Kim, M., Cho, C., Shin, W. et al. (2022). Nanowire-assisted freestanding liquid metal thin-film patterns for highly stretchable electrodes on 3D surfaces. *npj Flexible Electronics* 6: 99.

68 Lim, J., Ham, J., Lee, W. et al. (2021). A transformative gold patterning through selective laser refining of cyanide. *Nanomaterials (Basel)* 11: 1921.

69 Nasreldin, M., Delattre, R., Marchiori, B. et al. (2019). Microstructured electrodes supported on serpentine interconnects for stretchable electronics. *APL Materials* 7: 031507.

70 Xu, S., Zhang, Y., Cho, J. et al. (2013). Stretchable batteries with self-similar serpentine interconnects and integrated wireless recharging systems. *Nature Communications* 4: 1543.

71 Kim, D.H., Lu, N., Ma, R. et al. (2011). Epidermal electronics. *Science* 333: 838–843.

72 Huang, Z., Hao, Y., Li, Y. et al. (2018). Three-dimensional integrated stretchable electronics. *Nature Electronics* 1: 473–480.

73 Ji, B., Xie, Z., Hong, W. et al. (2020). Stretchable Parylene-C electrodes enabled by serpentine structures on arbitrary elastomers by silicone rubber adhesive. *Journal of Materiomics* 6: 330–338.

74 Lee, J., Sul, H., Lee, W. et al. (2020). Stretchable skin-like cooling/heating device for reconstruction of artificial thermal sensation in virtual reality. *Advanced Functional Materials* 30: 1909171.

75 Lee, J., Sul, H., Jung, Y. et al. (2020). Thermally controlled, active imperceptible artificial skin in visible-to-infrared range. *Advanced Functional Materials* 30: 2003328.

76 Liu, S., Kim, S.Y., Henry, K.E. et al. (2021). Printed and laser-activated liquid metal-elastomer conductors enabled by ethanol/PDMS/liquid metal double emulsions. *ACS Applied Materials & Interfaces* 13: 28729–28736.

77 Vallabh, C.K.P., Xiong, Y., and Zhao, X. (2021). 3D Printing of liquid metal EGaIn through laser induced forward transfer: a proof-of-concept study. *Manufacturing Letters* 28: 42–45.

78 Chambel, A., Sanati, A.L., Lopes, P.A. et al. (2022). Laser writing of eutectic gallium-indium alloy graphene-oxide electrodes and semitransparent conductors. *Advanced Materials Technologies* 7: 2101238.

79 Zhang, J., Zhang, C., Li, H. et al. (2022). Controlling the oxidation and wettability of liquid metal via femtosecond laser for high-resolution flexible electronics. *Frontiers in Chemistry* 10: 965891.

80 Pan, C., Kumar, K., Li, J. et al. (2018). Visually imperceptible liquid-metal circuits for transparent, stretchable electronics with direct laser writing. *Advanced Materials* 30: e1706937.

81 Won, D., Cho, H., Kim, H. et al. (2022). Digital selective reversible phase control of monolithically integrated heterogeneous piezoelectric polymer for frequency dependent unimorph. *Advanced Optical Materials* 10: 2201206.

82 Won, D., Kim, J., Choi, J. et al. (2022). Digital selective transformation and patterning of highly conductive hydrogel bioelectronics by laser-induced phase separation. *Science Advances* 8: eabo3209.

83 Kim, D., Bang, J., Won, P. et al. (2020). Biocompatible cost-effective electrophysiological monitoring with oxidation-free Cu–Au core–shell nanowire. *Advanced Materials Technologies* 5: 2000661.

84 Bang, J., Jung, Y., Kim, H. et al. (2022). Multi-bandgap monolithic metal nanowire percolation network sensor integration by reversible selective laser-induced redox. *Nanomicro Letters* 14: 49.

85 Jung, Y., Min, J., Choi, J. et al. (2022). Smart paper electronics by laser-induced graphene for biodegradable real-time food spoilage monitoring. *Applied Materials Today* 29: 101589.

86 Yang, Y., Zhao, H., Ren, X., and Huang, Y. (2021). Monolithic integration of laser onto multilayer silicon nitride photonic integrated circuits with high efficiency at telecom wavelength. *Optics Express* 29: 28912–28923.

87 Liang, G., Zeng, Y., Hu, X. et al. (2017). Monolithic semiconductor lasers with dynamically tunable linear-to-circular polarization. *ACS Photonics* 4: 517–524.

88 Li, N., Liu, K., Sorger, V.J., and Sadana, D.K. (2015). Monolithic III–V on silicon plasmonic nanolaser structure for optical interconnects. *Scientific Reports* 5: 14067.

89 Park, C.J., Kim, J.W., Namgung, S.D. et al. (2020). Selective photo-thermal modulation of ZnO/Pt interface for monolithic 3D integration of oxide-based resistive random access memory. *Applied Surface Science* 520: 146380.

90 Rauter, P., Lin, J., Genevet, P. et al. (2014). Electrically pumped semiconductor laser with monolithic control of circular polarization. *Proceedings of the National Academy of Sciences of the United States of America* 111: E5623–E5632.

91 Ye, R., Chyan, Y., Zhang, J. et al. (2017). Laser-induced graphene formation on wood. *Advanced Materials* 29: 1702211.

92 Le, T.S.D., Park, S., An, J. et al. (2019). Ultrafast laser pulses enable one-step graphene patterning on woods and leaves for green electronics. *Advanced Functional Materials* 29: 1902771.

93 Liu, W., Huang, Y., Peng, Y. et al. (2020). Stable wearable strain sensors on textiles by direct laser writing of graphene. *ACS Applied Nano Materials* 3: 283–293.

94 Bachmann, A.L., Hanrahan, B., Dickey, M.D., and Lazarus, N. (2022). Self-folding PCB kirigami: rapid prototyping of 3D electronics via laser cutting and forming. *ACS Applied Materials & Interfaces* 14: 14774–14782.

95 Xu, R., Zverev, A., Hung, A. et al. (2018). Kirigami-inspired, highly stretchable micro-supercapacitor patches fabricated by laser conversion and cutting. *Microsystems & Nanoengineering* 4: 36.

96 Wang, Z., Wang, Z., Ko, W.H. et al. (2021). An autonomous laser kirigami method with low-cost real-time vision-based surface deformation feedback system. *The International Journal of Advanced Manufacturing Technology* 118: 1873–1883.

97 Won, P., Park, J.J., Lee, T. et al. (2019). Stretchable and transparent kirigami conductor of nanowire percolation network for electronic skin applications. *Nano Letters* 19: 6087–6096.

98 Kim, K.K., Ha, I., Kim, M. et al. (2020). A deep-learned skin sensor decoding the epicentral human motions. *Nature Communications* 11: 2149.

99 Kim, K.K., Ha, I., Won, P. et al. (2019). Transparent wearable three-dimensional touch by self-generated multiscale structure. *Nature Communications* 10: 2582.

100 Pham, M.S., Dovgyy, B., Hooper, P.A. et al. (2020). The role of side-branching in microstructure development in laser powder-bed fusion. *Nature Communications* 11: 749.

101 Tang, J., Geng, X., Li, D. et al. (2021). Machine learning-based microstructure prediction during laser sintering of alumina. *Scientific Reports* 11: 10724.

102 Kempen, K., Yasa, E., Thijs, L. et al. (2011). Microstructure and mechanical properties of selective laser melted 18Ni-300 steel. *Physics Procedia* 12: 255–263.

103 Lin, S.-Y., Zhang, T.Y., Lu, Q. et al. (2017). High-performance graphene-based flexible heater for wearable applications. *RSC Advances* 7: 27001–27006.

104 Lee, D., Bang, G., Byun, M., and Choi, D. (2020). Highly flexible, transparent and conductive ultrathin silver film heaters for wearable electronics applications. *Thin Solid Films* 697: 137835.

105 Claypole, A., Claypole, J., Leeder, J. et al. (2022). Large area, stretchable, wearable, screen-printed carbon heaters for use in elite sport. *Journal of Coatings Technology and Research* 20: 261–273.

106 Huang, Y., Chen, X., Yu, J., and Tao, L.-Q. (2021). Graphene-based film heater fabricated by laser writing. *Materials Letters* 284: 128869.

107 Zhang, J., Xu, R., Feng, J. et al. (2021). Laser direct writing of flexible heaters on polymer substrates. *Industrial & Engineering Chemistry Research* 60: 11161–11170.

108 Koo, S. (2022). Flexible heater fabrication using amino acid-based ink and laser-direct writing. *Micromachines (Basel)* 13: 2209.

109 Cho, C., Shin, W., Kim, M. et al. (2022). Monolithically programmed stretchable conductor by laser-induced entanglement of liquid metal and metallic nanowire backbone. *Small* 18: e2202841.

110 Kim, H., Choi, J., Kim, K.K. et al. (2021). Biomimetic chameleon soft robot with artificial crypsis and disruptive coloration skin. *Nature Communications* 12: 4658.

111 Ghosh, R., Gopalakrishnan, S., Savitha, R. et al. (2019). Fabrication of laser printed microfluidic paper-based analytical devices (LP-microPADs) for point-of-care applications. *Scientific Reports* 9: 7896.

112 Wlodarczyk, K.L., Hand, D.P., and Maroto-Valer, M.M. (2019). Maskless, rapid manufacturing of glass microfluidic devices using a picosecond pulsed laser. *Scientific Reports* 9: 20215.

113 Yong, J., Zhan, Z., Singh, S.C. et al. (2019). Microfluidic channels fabrication based on underwater superpolymphobic microgrooves produced by femtosecond laser direct writing. *ACS Applied Polymer Materials* 1: 2819–2825.

114 Shin, J., Ko, J., Jeong, S. et al. (2021). Monolithic digital patterning of polydimethylsiloxane with successive laser pyrolysis. *Nature Materials* 20: 100–107.

115 Lim, J., Park, S., Cho, H. et al. (2022). Monolithic digital patterning of polyimide by laser-induced pyrolytic jetting. *Chemical Engineering Journal* 428: 131050.

116 Song, S., Um, S.H., Park, J. et al. (2022). Rapid synthesis of multifunctional apatite via the laser-induced hydrothermal process. *ACS Nano* 16: 12840–12851.
117 Lee, Y., Kwon, J., Lim, J. et al. (2020). Digital laser micropainting for reprogrammable optoelectronic applications. *Advanced Functional Materials* 31: 2006854.
118 Wang, Y., Zhao, Y., and Qu, L. (2021). Laser fabrication of functional micro-supercapacitors. *Journal of Energy Chemistry* 59: 642–665.
119 Fu, X.-Y., Chen, Z.D., Han, D.D. et al. (2020). Laser fabrication of graphene-based supercapacitors. *Photonics Research* 8: 577–588.
120 Guo, H., Yan, J., Jiang, L. et al. (2022). Femtosecond laser Bessel beam fabrication of a supercapacitor with a nanoscale electrode gap for high specific volumetric capacitance. *ACS Applied Materials & Interfaces* 14: 39220–39229.
121 Zhu, J., Liu, S., Hu, Z. et al. (2021). Laser-induced graphene non-enzymatic glucose sensors for on-body measurements. *Biosensors & Bioelectronics* 193: 113606.
122 Xu, Y., Fei, Q., Page, M. et al. (2021). Laser-induced graphene for bioelectronics and soft actuators. *Nano Research* 14: 3033–3050.
123 Gerasimenko, A.Y., Kurilova, U.E., Suetina, I.A. et al. (2021). Laser technology for the formation of bioelectronic nanocomposites based on single-walled carbon nanotubes and proteins with different structures, electrical conductivity and biocompatibility. *Applied Sciences* 11: 8036.

8

Electrospinning Manufacturing of Stretchable Electronics

Yinhui Li[1], *Kan Li*[2], *Yunlei Zhou*[3], *and YongAn Huang*[4]

[1] *Taiyuan University of Technology, Micro-Nano System Research Center, College of Information and Optical Engineering, Taiyuan, Shanxi 030024, China*
[2] *Huazhong University of Science and Technology, State Key Laboratory of Digital Manufacturing Equipment and Technology, Flexible Electronics Research Center, Wuhan 430074, China*
[3] *Xidian University, Institute of Technology, Hangzhou, Zhejiang 311231, China*
[4] *Huazhong University of Science and Technology, State Key Laboratory of Digital Manufacturing Equipment and Technology, Flexible Electronics Research Center, Wuhan 430074, China*

8.1 Background

Stretchable electronics have captured widespread attention due to their potential applications in robotics, wearables, and bio-compatible implantable devices [1, 2]. Notably, artificial electronic skin is a flexible sensor that emulates the sensory functions of human skin, converting various stimuli—pressure, strain, shear force, temperature, and humidity—into electrical signals [3–8]. The device holds promising prospects in robotics automation, flexible electronics, wearable technology, the biomedical field, and human–computer interactions [9–16]. To mimic human skin, artificial electronic skin must conform to irregular and moving surfaces, detect multiple mechanical stimuli, and meet requirements such as high stretchability, sensitivity, broad sensing range, and rapid response/recovery times.

Research has highlighted flexible and highly sensitive pressure sensors made from conductive polymers, gold nanowires, graphene, and carbon nanotubes (CNTs) as sensing elements. These incorporate micro- and nanostructures (nanofiber, fabrics, and 3D architectures) based on various sensing mechanisms (e.g. resistive, piezoelectric) [17–28]. Electrospinning emerges as an ideal method for the cost-effective and large-area fabrication of highly stretchable and sensitive sensors. The evolution of electrospinning, from fiber fabrication to digital electrohydrodynamic (EHD) direct-writing, simplifies processes, enhances device performance, and reduces costs, marking a promising path for the development of high-performance organic electronics economically and innovatively.

The fiber mat and fabric become electrically conductive with relatively low filler content via surface coating, largely preserving elasticity [29, 30]. However, achieving high electrical conductivity in conductive polymer composites often requires high filler loading, which can lead to a loss of elasticity due to the modulus

Mechanics of Flexible and Stretchable Electronics, First Edition. Edited by Yong Zhu and Nanshu Lu.

mismatch between polymers and fillers. The next generation of electronic devices combines exceptional sensitivity, flexibility, and stretchability, finding applications in smart electronics, medical diagnostics, wearable devices, artificial organs, and human–computer interactions.

8.2 High-Precision Manufacturing

8.2.1 Inkjet Printing

Inkjet printing is a method that directly deposits functional materials onto specific substrates, showcasing its potential with excellent material utilization, maskless pattern manufacturing, and cost-effectiveness. This technique is particularly appealing for creating electronic devices due to its ability to avoid the expensive and complex multistep lithography processes that could potentially damage organic materials and flexible substrates. Various inkjet printers, based on different actuation mechanisms such as thermal [31], piezoelectric [32], and aerosol [33], have been commercialized and are widely used in the electronics industry. The thermal and piezoelectric mechanisms are primarily developed for the actuation of the inkjet nozzle head, where droplets of very small diameters are ejected at each pulse, generated by either thermal or piezoelectric actuators within the inkjet nozzle head. Figure 8.1a shows the droplet actuation mechanism of an inkjet materials printer, where ink is supplied from a reservoir and a driving signal is applied to the actuator, generating droplets according to the designed geometry.

Inkjet printers are categorized into two types: continuous inkjet and drop-on-demand (Figure 8.1b). In the continuous system, ink flows from a reservoir to the nozzles. A charging electrode draws ink from the nozzle, forming a droplet that is then directed to the substrate by deflection plates. Conversely, the drop-on-demand system operates based on a digital input signal that reflects the pattern design, generating each droplet as needed. The substrate is precisely aligned perpendicular to the nozzle and moves in synchronization with droplet generation to form patterns or thin films. Commercial inkjet printheads may feature up to 128 nozzles, with droplet sizes as small as 1.0 pl. The inkjet printing process involves three key steps: droplet ejection and formation, liquid–solid interaction upon droplet contact with the substrate, and ink droplet drying and solidification to form the final deposit [36, 37]. In comparison, inkjet printing offers advantages over traditional patterning technologies such as photolithography [38], micro-contact printing [39], and shadow masking [40].

For a wide range of applications, inkjet printing can utilize a variety of inks based on [41–44] and conductive fillers [45–48]. However, stretchable applications pose unique challenges. Unlike conductive pastes that can incorporate a high percentage of conductive particles, inkjet inks must maintain a low viscosity (10–20 cP) to ensure proper printing, limiting the concentration of particles in the ink. Nonetheless, successful stretchable applications have been achieved with custom inks. For example, commercial inks containing 30–35% silver nanoparticles

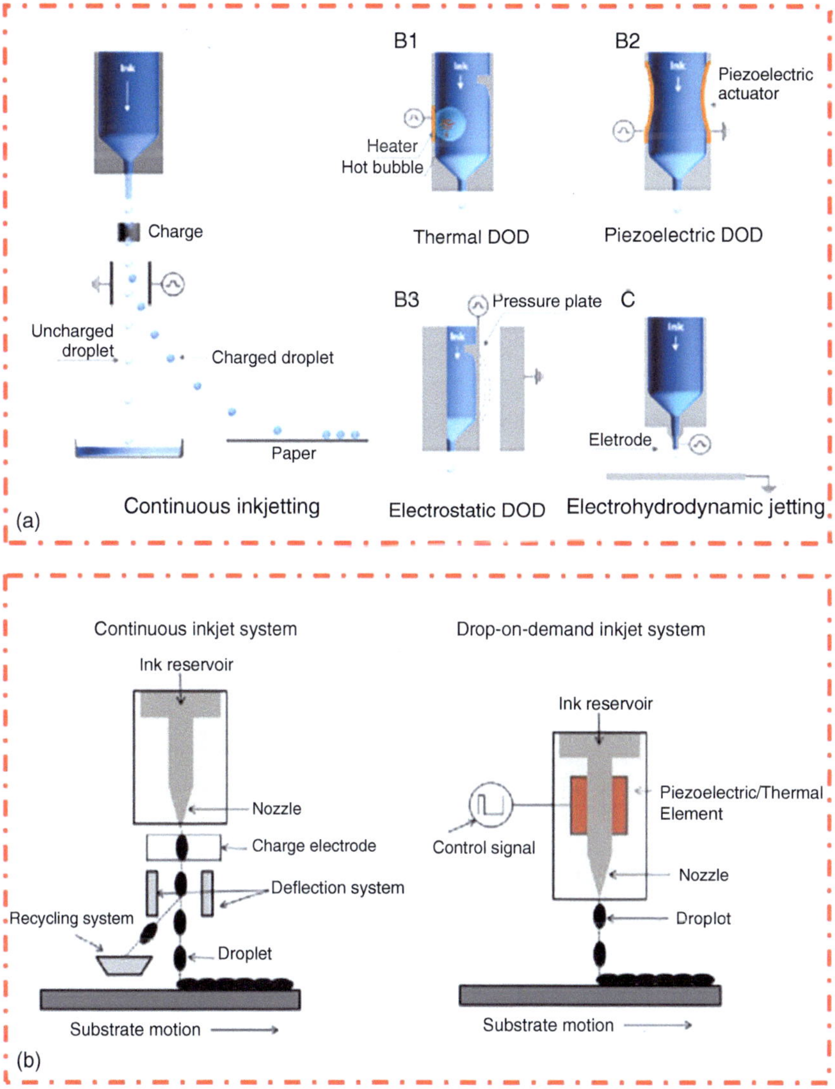

Figure 8.1 (a) Illustrates the actuation mechanism of a droplet in an inkjet material printer. Source: Ref. [34]/with permission of American Chemical Society; (b) Provides a schematic diagram of continuous and drop-on-demand inkjet systems. Source: Ref. [35]/IntechOpen/CC BY 3.0.

(Ag NPs) mixed with polydimethylsiloxane (PDMS) achieve a low viscosity of 10–17 cP, making them suitable for printing stretchable devices [49]. Additionally, materials such as poly(methyl methacrylate) (PMMA), propylene glycol methyl ether acetate (PGMEA), poly(vinyl alcohol) (PVA), and PDMS [50] have been utilized either purely or as composites in stretchable matrices, with the printed systems demonstrating the ability to withstand strains up to 50% without electrical breakdown and maintaining stability for 1000 stretching cycles at over 40% strain.

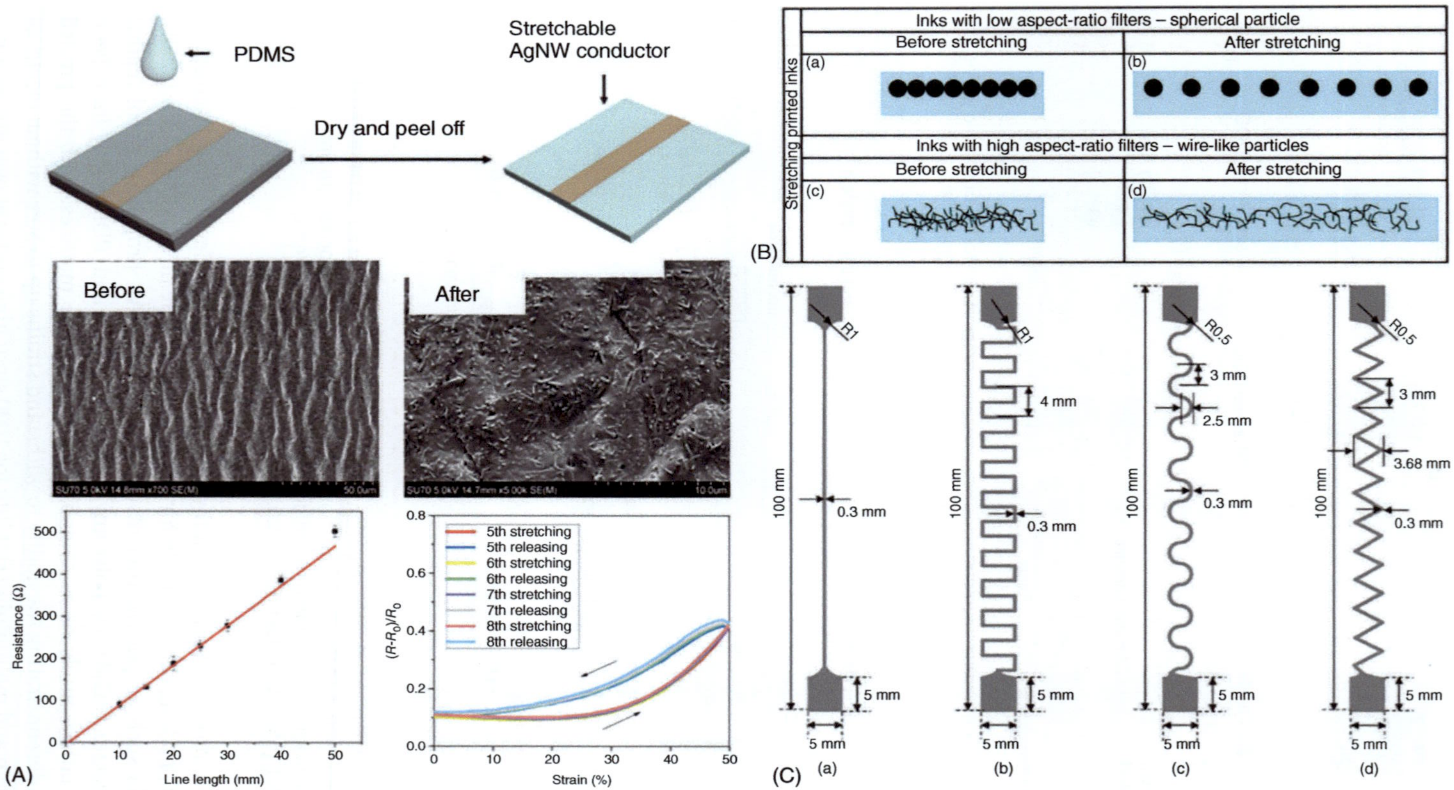

Figure 8.2 (A) Schematic illustration of the fabrication process for a stretchable Ag nanowire conductor, including SEM images depicting the conductor before and after stretching. Source: Ref. [51]/with permission of American Chemical Society; (B) Mechanism of spherical particle-based inks before and after stretching, showcasing spherical particles like NPs (a and b) and with high aspect-ratio fillers (C and D). Source: Ref. [52]/The Royal Society of Chemistry; (C) Explores various meandering geometries in circuit design for stretchable circuits. Source: Ref. [53]/with permission of Elsevier.

Two critical factors significantly impact the performance of stretchable devices: (i) The use of conductive nanoparticles (NPs) presents challenges in achieving ink formulations for stretchable electronics. The presence of multiple connection points between NPs can enhance the electromechanical behavior of the ink under strain. However, the filler size is a pivotal consideration due to the potential for nozzle clogging during the printing process. Huang et al. introduced a method of transferring inkjet-printed patterns to circumvent the clogging issue [51], achieving a conductivity of approximately 8.9×10^3 S/m with a nanowire concentration of 10 mg/ml dispersed in an ethanol solution as illustrated in Figure 8.2A. (ii) The geometry of filler particles and the circuit design are both crucial in determining the electrical and stretchable performance of the device. Even slight stretching of a pattern containing these particles can lead to electrical changes due to the disconnection at the particle junctions. This characteristic makes particle-based inks challenging for applications that must endure significant mechanical stress but ideal for strain sensors. Fernandes et al. proposed utilizing round-like, high-aspect-ratio nanotubes or nanowires as filler particles [52], as depicted in Figure 8.2B. Such high-aspect-ratio fillers like nanotubes or nanowires can minimize abrupt electrical fluctuations and achieve a lower percolation threshold compared to spherical particles due to the establishment of electrical connections between fillers at stretchable points, thereby enhancing the stretchability of the application.

For stretchable electronic systems to function effectively, the design of stretchable circuits is paramount. Kim et al. [53] demonstrated that the extensibility of circuits increases in the order of single line, zigzag, rectangular pulse, and horseshoe meandering geometries as illustrated in Figure 8.2C. This progression highlights the importance of choosing the right circuit geometry to optimize the stretchability and performance of electronic systems, ensuring they can accommodate the mechanical demands of various applications.

8.2.2 EHD Printing

To enhance resolution and ensure compatibility, the adaptability of flexible printing modes—including electrospray, electrospinning, and EHD jet printing—has been highlighted as a focal point in the current research within printed electronics and biomedical fields. EHD printing, which employs a high electric field force to draw fluids into thin jets or small droplets rather than pushing them, offers distinct advantages such as high resolution (<200 nm) and compatibility with a broad viscosity range of inks (1–100,000 cP). This technique is particularly well-suited for creating high-resolution stretchable electronics, representing a significant leap forward in the field. Considerable progress has been made in the direct-writing of isolated electrospun fibers, markedly broadening the scope of electrospinning for the fabrication of micro-/nanodevices. These applications include electronic components and structures used for nanofabrication, either as sacrificial elements or as masks.

The development of electrospinning from simple fiber fabrication to digital EHD direct-writing is illustrated in Figure 8.3. This evolution is categorized into four stages: traditional electrospinning for the fabrication of nonwoven nanofibers,

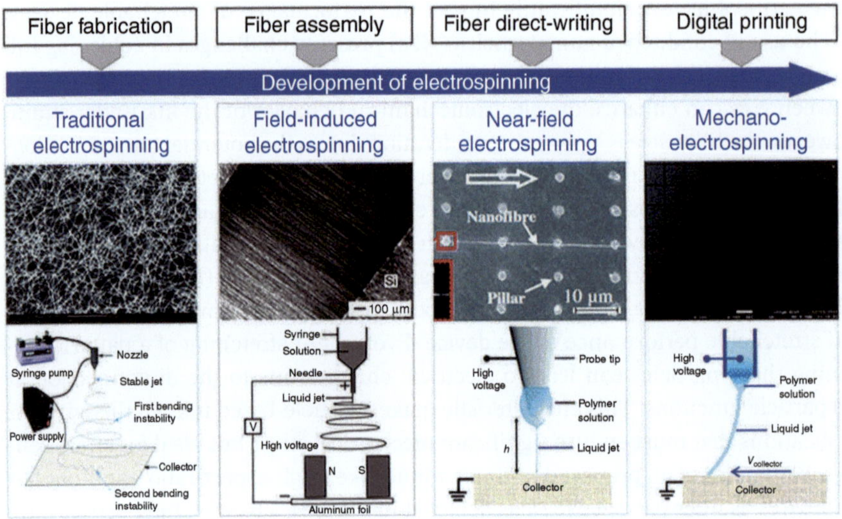

Figure 8.3 Evolution of electrospinning technology, from fiber fabrication to digital EHD direct-writing. Source: Ref. [54]/with permission of American Chemical Society.

electrospinning with assisted-electrode or dynamic collector for fiber assembly [55], near-field electrospinning (NFES) with the ability to position electrospun nanofibers [56], and mechano-electrospinning (MES) for positioning the electrospun nanofibers and tuning the morphology of the nanofibers in one step [57]. The traditional electrospinning process results in the deposition of fibers in a random, nonwoven mat, which makes it impossible to produce single-fiber devices. Techniques such as using auxiliary electrodes [58] or a dynamic mechanical collector [59, 60] can align fibers, but they do not offer control over individual fibers. Controlling a single fiber is crucial for the fabrication of structures and devices at the micro-/nanoscale. This need for precision has led to advancements in electrospinning, transforming it from a method that produces disordered fibers to a highly controllable direct-writing technology.

EHD direct-writing is a straightforward, cost-effective, and highly efficient method for printing fibers with diameters ranging from several micrometers to tens of nanometers, as shown in Figure 8.4. This method overlaps with and advances contemporary textile fiber technology [62–65]. A groundbreaking improvement in the control of single fibers has been achieved by reducing the needle-to-collector distance h from more than 10 cm to just 1 cm. This reduction enables precise control over the position, shape, and morphology of individual electrospun fibers, with h playing a critical role in the dynamic behavior and morphology of the fibers [66]. Yongxian An et al. have classified electrospinning into three types based on the needle-to-collector distance: far-field ($h > h_{\text{straight}} + h_{\text{whipping\_1}}$), medium-field ($h_{\text{straight}} < h < h_{\text{straight}} + h_{\text{whipping\_1}}$), and near-field electrospinning ($h < h_{\text{straight}}$), with the differences summarized in Table 1. In far-field electrospinning, an applied voltage of 10–50 kV is typically used, while NFES requires only 0.6–3 kV. By adjusting the needle-to-collector distance, the morphologies of electrospun fibers can

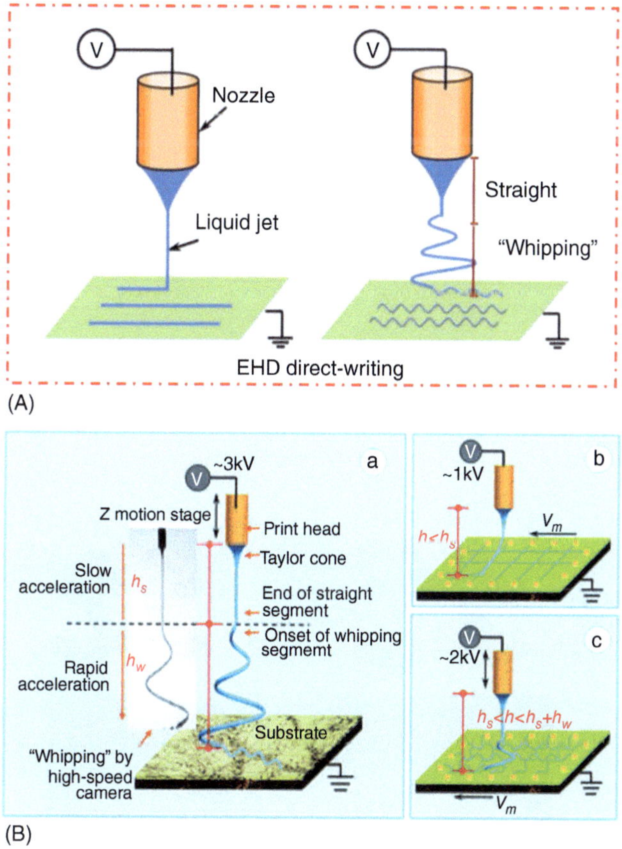

Figure 8.4 (A) Schematic EHD direct-writing. Source: Adapted from Ref. [61]; (B) Standard electrospinning (far-field electrospinning) used to fabricate a nonwoven mat. Source: Ref. [54] with permission of The Royal Society of Chemistry.

vary from fiber to ribbon, depending on the degree of solidification. This technique combines dip-pen, inkjet, and electrospinning technologies to offer controllable electrospinning for sub-100 nm nanofabrication.

The EHD direct-writing technique's ability to precisely deposit single fibers makes it an excellent choice for creating sacrificial/masking structures or templates for micro/nanofabrication [69–71]. It shows significant potential for developing flexible micro/nano devices and expands the application range of traditional electrospinning, especially in the emerging field of flexible and stretchable electronics. The formation of these devices is influenced by three types of operating parameters: material factors (molecular weight [72, 73], viscosity [74], conductivity [75], permittivity surface tension [76], and substrate surface [77]), process factors (including needle-to-collector distance [78, 79], nozzle layout [80], auxiliary electrodes [81], applied voltage [82, 83], flow rate, and collector velocity [84, 85]), and circumstance factors (humidity [86], temperature [87], and liquid viscosity [88]). Previous research has shown EHD printing's compatibility with a wide variety of

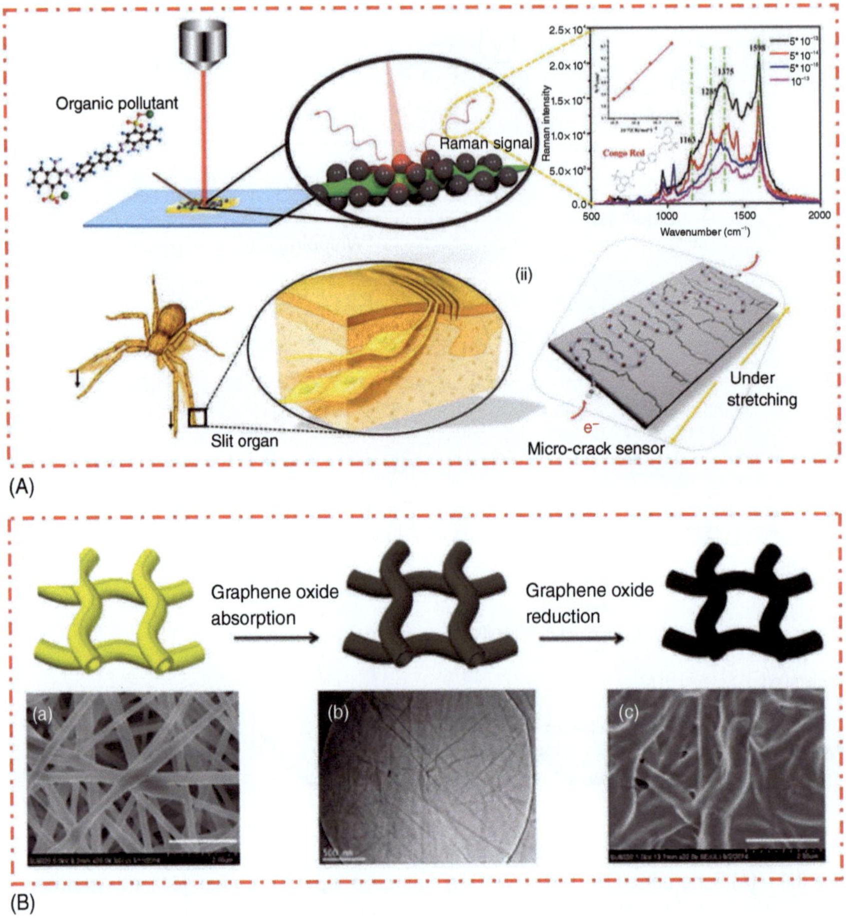

Figure 8.5 (A) Fabrication process [67] and (B) quantitative analysis of a composite nonwoven mat made of elastomer fibers [68].

inks, including quantum dots [89], metal NPs [90], and hybrid perovskites[91], allowing for the creation of various high-resolution 2D and 3D structures [92, 93] smaller than 1 μm.

Conductive and piezoelectric polymers are particularly prominent among inks used in EHD printing because they can exploit deformations induced by small forces through pressure, mechanical vibration, elongation/compression, bending, or twisting [94] for functionality. For instance, Lu et al. [95] have developed stretchable strain sensors using aligned microfibrous arrays of poly(3,4-ethylenedioxythiophene): poly(styrene sulfonate)–poly(vinyl pyrrolidone) (PEDOT:PSS-PVP) with unique curled architectures. This was achieved using a novel electrospinning setup that features a reciprocating-type spinneret moving in simple harmonic motion. Additionally, Persano et al. [96] used a rotating collector to electrospin aligned nanofibrous arrays of poly(vinylidene fluoride-*co*-trifluoroethylene) (PVDF-TrFe), notable for their attractive piezoelectric properties, with average diameters of

260 nm. EHD direct-writing streamlines the fabrication processes, enhances device performance, and reduces costs, promising to revolutionize the production of high-performance organic electronics cost-effectively and innovatively.

8.3 Electrospinning Stretchable Structure

The mechanical and sensing properties of conductive polymer composites are largely determined by the flexibility of their raw materials and their structural design. Unlike traditional sensing materials such as semiconductors and metals, which are rigid and brittle, conductive polymer composites require innovative structural designs to improve their stretchability and flexibility. This is crucial for their application in stretchable and flexible electronic devices. Techniques like electrospinning are often employed to create complex topological structures that enhance these materials' flexibility.

Intrinsically flexible organics, including PDMS [97], polyurethane (PU) [98], thermoplastic polyurethane (TPU) [99, 100], polyimide (PI) [101], polyethylene terephthalate (PET) [102], and hydrogels based on PVA [19], are preferred as the soft matrix in flexible sensor design. These materials are directly combined with active materials to create sensors that are not only flexible but also comfortable and skin-friendly. Additionally, the flexibility and skin affinity of these composites can be further improved by fabricating fibers or fibrous polymer mats through melt or solution spinning [103–105].

For optimal electrical conductivity, materials such as CNTs [106], carbon nanostructures (CNSs) [107], graphene [108], silver nanowires (AgNWs) [109, 110], Mxene [111, 112], and conductive polymers (polyaniline [113] and PEDOT:PSS [114]) are mixed with the elastomer before spinning or deposited on the elastomer mats. However, achieving high electrical conductivity often requires a high loading of conductive fillers, which can compromise the material's elasticity due to the mismatch in modulus between the polymers and fillers. A solution to this issue is the application of conductive fillers via surface coating on fiber mats and fabrics. This approach maintains the elasticity of the materials, which makes them electrically conductive, even at low filler content [113, 114]. Such strategies are particularly beneficial for creating fiber-based, comfortable, breathable, and flexible electronics capable of sensing mechanical force, temperature, humidity, and gas.

8.3.1 Stretchable Nanofiber Mats

Elastomer fiber mat sensors, capable of detecting mechanical stimuli, temperature, and vapor from chemicals, are increasingly sought after for their potential in wearable electronics and human–robot interfacing devices. Their appeal lies in their excellent wearability, skin compatibility, durability, etc. Generally, the creation of multifunctional sensors hinges on two crucial aspects: the selection of specific materials and the meticulous design of their structure. It often requires integrating various functional materials that are responsive to different stimuli, such as strain, humidity, and temperature, in the fabrication of elastomer-based flexible sensors that can react to multiple forms of stimuli, including mechanical deformation, humidity, and temperature. For example, Chen et al. [67] showcased a multifunctional sensing platform utilizing cellulose nanocrystals and Ag NPs for detecting strain and humidity (Figure 8.5a). The hygroscopic nature of cellulose nanocrystals allows water molecules to easily diffuse into the microcrack conductive network, thereby reconnecting the microcrack junctions, which leads to changes in the electrical signals related to stretch and strain. Similarly, Hu et al. [68] developed flexible and stretchable strain sensors by reducing graphene oxide (GO) on electrospun PU nanofiber mats.

Enhancing the sensitivity and responsiveness of flexible and stretchable wearable sensors has led researchers to explore sensor structures with multi-dimensional architecture. Yan et al. [115] introduced an all-encompassing, multi-responsive TPU nanofiber mat sensor. This sensor consists of crimped elastomer fibers with platinum NPs deposited on their surface. The 1D TPU nanofibers can transform into various 3D nanofiber architectures through the controlled relaxation of aligned elastomer polymers when exposed to a poor solvent annealing process. These 3D architectures enable the strain sensor to detect a broader range of tensile strain and pressure with heightened sensitivity, due to the precise control over fiber morphology and structure, as illustrated in Figure 8.6a. Kun et al. [116] outlined an effective, cost-efficient method for creating a highly sensitive, stretchable electronic skin. This skin detects dynamic and static pressure, strain, and flexion through an elastic GO-doped PU nanofiber membrane, coated with an ultrathin layer of conductive poly(3,4-ethylenedioxythiophene) (PEDOT). The nanofiber-based electronic skin sensor exhibits high-pressure sensitivity (up to 20.6 kPa^{-1}), a broad sensing range (1 Pa to 20 kPa), exceptional cycling stability and repeatability (over 10,000 cycles), and remarkable strain sensitivity across a broad range (up to approximately 550%). This flexible and stretchable nanofiber-based sensor, with its superior multimodal mechanical sensitivity, holds potential for applications in wearable artificial skin for health monitoring, medical diagnostics, tactile sensors, and humanoid robotics (see Figure 8.6b).

8.3.2 Stretchable Yarns and Fabrics

For artificial electronic skin to function effectively, it must be a flexible sensor capable of detecting a variety of mechanical forces, including pressure, strain, and bending. It should also possess soft elasticity to conform to irregular surfaces and linear wearability for seamless integration into textiles. Research has explored flexible pressure sensors utilizing materials such as nanowires, CNTs, polymer nanofibers, metal NPs, and graphene. Similarly, stretchable strain sensors employing stretchable materials and innovative architectures have been developed.

Gong et al. [117] developed a pressure sensor by incorporating Au nanowire into tissue paper through the electrospinning method. The unique porous and rough texture of the tissue paper, combined with the "hairy" Au nanowires, endowed the pressure sensor with high sensitivity, enabling it to detect pressure, bending, and twisting

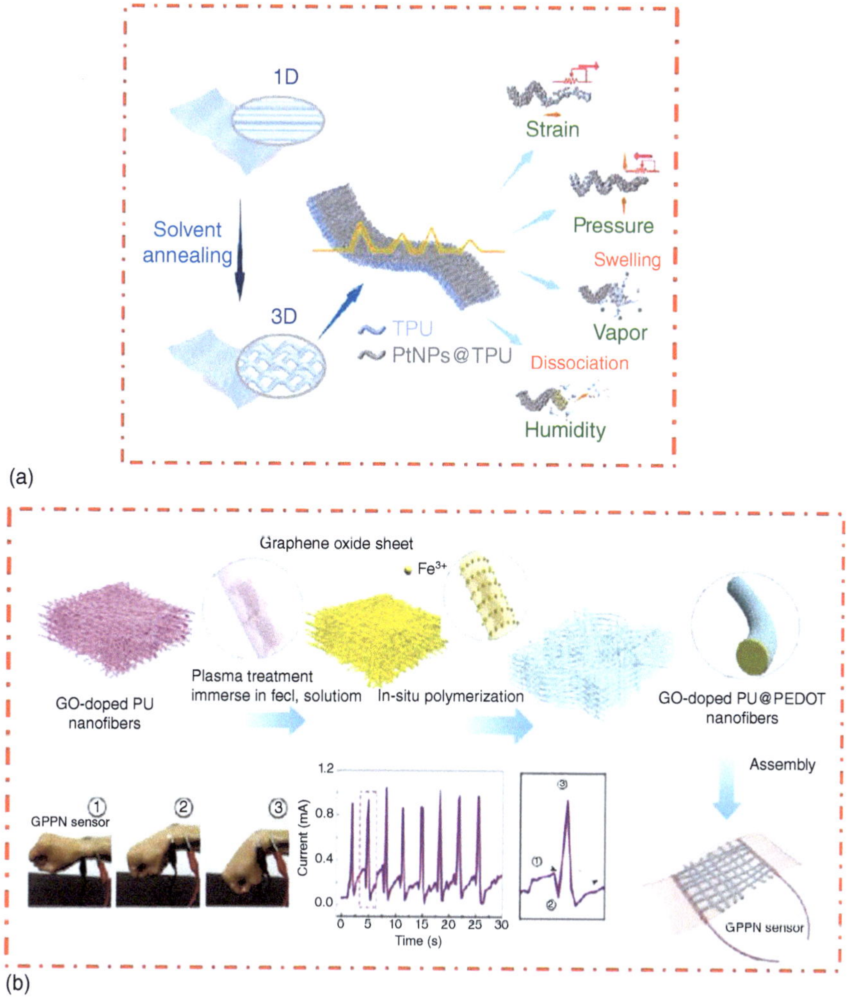

Figure 8.6 (a) Schematic illustration of the fabrication process for a multifunctional sensor [115] and (b) outlines its potential applications in monitoring human activities and the surrounding environment [116].

effectively. Zhang et al. [118] showcased a strain sensor made from carbonized plain weave cotton fabric, notable for its broad operational strain range, which extends up to 140%. Wang et al. [119] designed an ultrasensitive strain sensor by embedding Ag nanowires into a polyolefin elastomer nanofibrous yarn. The sensor's gauge factor surged exponentially with increasing strain, achieving a remarkable peak as high as 1.392×10^4 at a strain of 64%.

Furthermore, the development of a linearly weaved, flexible, and retractable multimodal mechanical sensor capable of detecting mechanical forces from pressure, strain, and bending has also been pursued [120, 121]. Ko and coworkers [122] proposed a multimodal electronic skin based on CNT-composite elastomer nanofiber films. Created via the electrospinning method with an interlocked microstructure,

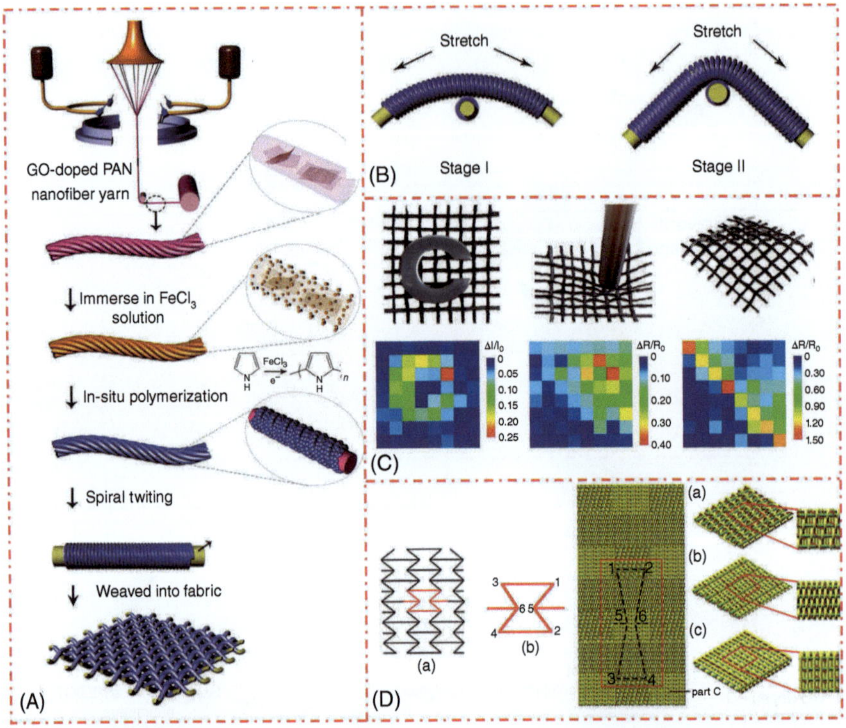

Figure 8.7 (A) Schematic illustration of the electronic fabric fabrication process; (B) Depicts the fiber at Stage I and Stage II; (C) Electronic fabric in a flexible, bent state. Source: Ref. [123]/with permission of John Wiley & Sons; (D) Formation and arrangement of re-entrant hexagonal geometry and different parts. Source: Adapted from Ref. [124].

this skin can detect various mechanical stimuli, including pressure, shear force, strain, bending, and torsion. Nan et al. [123] fabricated a wearable fabric-like electronic skin from continuous GO-doped polyacrylonitrile nanofiber yarns through a straightforward electrospinning process. Conductive polypyrrole was applied to the nanofibers' surface through in situ polymerization to create composite yarns, which were then woven into a fabric sensor. This sensor demonstrated high sensitivity (4.08/N) and the ability to detect mechanical stresses from static and dynamic pressure, strain, and bending. The sensor's high sensitivity is attributed to its hierarchical structure, which spans from macroscopic yarns of ~100% to submicron-scale fibers and nanometer-scale particles of 0.1% (Figure 8.7).

Additionally, stretchable fiber-like sensors based on negative Poisson's ratio (auxetic) materials have gained popularity. Such materials expand laterally under tensile stress, offering numerous advantages [125–129], including improved shear stiffness [130], synclastic curvature in sheets and panels, enhanced fracture toughness, and indentation resistance [1], particularly notable for their excellent energy absorption and damping properties [131]. Recent studies have focused on auxetic yarns and fabrics [124, 132–134], with Zulifqar and Hu reporting on biaxial-stretchable auxetic woven fabrics created using non-auxetic yarns and weaving technology. These helix auxetic fabrics demonstrate a negative Poisson ratio effect in both weft and warp directions across a wide range of tensile strain, an effect that can be amplified by employing all elastic yarns in the weft direction [124].

8.4 Application in Stretchable Electronics

This section provides an overview and critical analysis of recent advancements in flexible and stretchable electronic devices utilizing electrospun ultrathin fibers. We cover a range of applications including strain and pressure sensors, organic field-effect transistors (OFETs), and transparent electrodes.

8.4.1 Strain and Pressure Sensor

The evolving landscape of electronics is increasingly focused on developing flexible and stretchable devices for sensing applications, which are attracting significant attention for their potential in renewable integrated power sources for future wearable electronics [135, 136]. Electrospinning mats and fabrics, in particular, have shown promise in this area, especially for creating flexible and stretchable pressure and strain sensors [137]. Arrays of fibers produced through EHD techniques have demonstrated superior performance in flexible and stretchable devices compared to nonwoven mats or isolated strands, as shown in Figure 8.8A.

Shen et al. [138] have developed a stretchable strain sensor composed of Ag nanowires and a poly(3,4-ethylenedioxythiophene):poly(styrene sulfonate) patterned microstructure. This sensor exhibits high sensitivity (gauge factor ~81), extensive stretchability (up to 150%), and remarkable long-term reliability (maintaining performance over 10^4 cycles at 150% strain), as illustrated in Figure 8.8B. This is attributed to the curled architecture of the spun fibrous polymer arrays, which offers a threefold improvement over traditional electrospun nonwoven mats. Such advancements are critical for the development of stretchable devices, with potential applications in soft robotics, elastic semiconductors, and elastic solar cells.

In another study, Tang et al. [139] prepared highly stretchable composite nanofiber yarns from CNT/TPU via multi-needle liquid-bath electrospinning. By uniformly dispersing CNTs within the flexible TPU matrix, the resulting composite nanofiber yarn achieved a breaking elongation of up to 476% (Figure 8.9a). The integration of a small amount of CNTs changes the color of the filaments from white to black, indicating the incorporation of CNTs (Figure 8.9b). With 0.8 wt% CNT, the conductivity of these composite nanofiber yarns reached 7.09 μS/cm, a sevenfold increase compared to pure nanofiber yarn. Moreover, when these yarns were dip-coated with 4 wt% CNT ink, their conductivity improved to 0.62 S/cm, five times higher than the uncoated CNT/TPU composite yarns, with a resistance of only 4.43 kΩ/cm. These composite yarns have been effectively used in applications such as monitoring leg movements and evaluating bending degrees (Figure 8.9c), showcasing their potential in smart sports bandages and the medical field. Additionally, other conductive and flexible polymers and composites, like

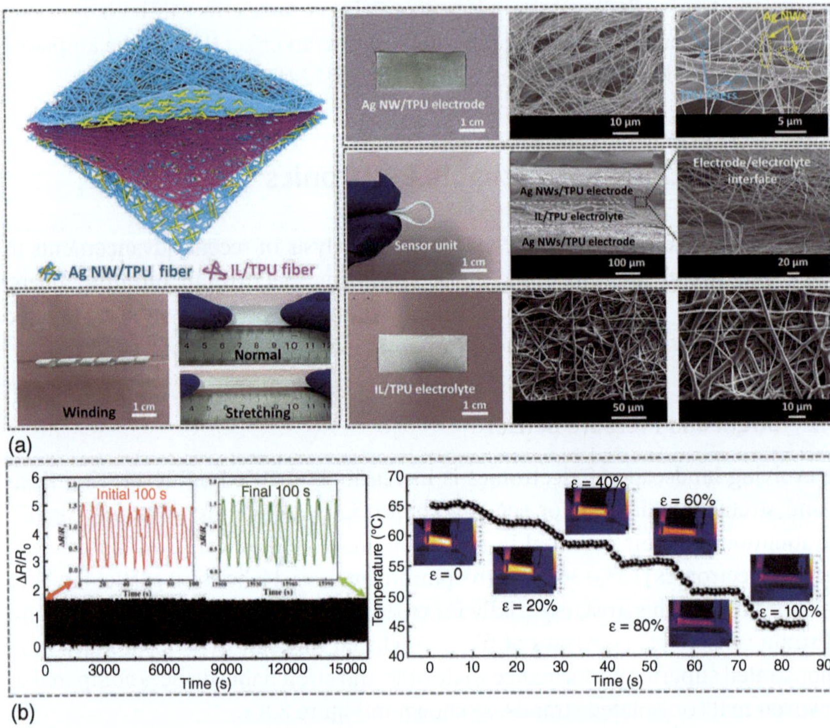

Figure 8.8 (a) Structural design and morphology characterization of the all-nanofiber tactile sensor. Source: Ref. [137]/with permission of Elsevier; (b) Relative resistance changes over time curves for a strain of 20% and the temperature responses to strains for the stretchable heater. Source: Ref. [138]/with permission of John Wiley & Sons.

poly(vinylidene fluoride-*co*-trifluoroethylene) (PVDF-TrFe) [140], can be utilized to fabricate stretchable and flexible strain sensors through various electrospinning methods.

8.4.2 Organic Field-Effect Transistors

OFETs have become increasingly significant in both academic research and industrial applications, particularly for their potential in wearable, implantable, and skin-like electronics [141–143]. One of the key advantages of OFETs is their compatibility with low fabrication temperatures, which broadens the range of substrate materials that can be used, including flexible polymeric substrates. This compatibility is crucial for developing a variety of low-cost electronic applications, such as radio frequency identification (RFID) tags, next-generation displays, and chemical sensors [144–150].

Despite the promise, the application of polymer-based OFETs has been limited by their relatively low charge-carrier mobilities, often falling below 0.01 $cm^2/(V\ s)$ [151]. However, electrospun nanofibers, with their one-dimensional (1D) nanostructures, have been shown to enhance charge transfer and minimize grain

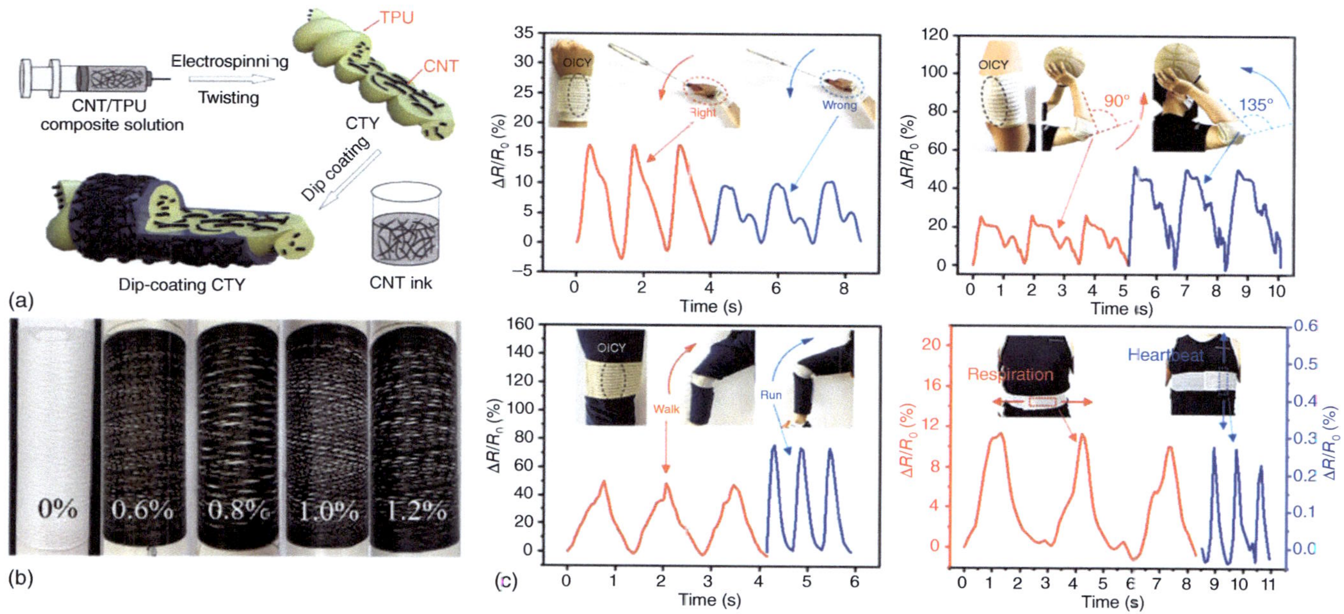

Figure 8.9 (a) Flow chart detailing the construction process of a dual CNT network on dip-coated composite nanofiber yarn; (b) Photograph of filaments with varying contents of CNT; (c) Application of the smart sports bandage. Source: Ref. [139]/with permission of Elsevier.

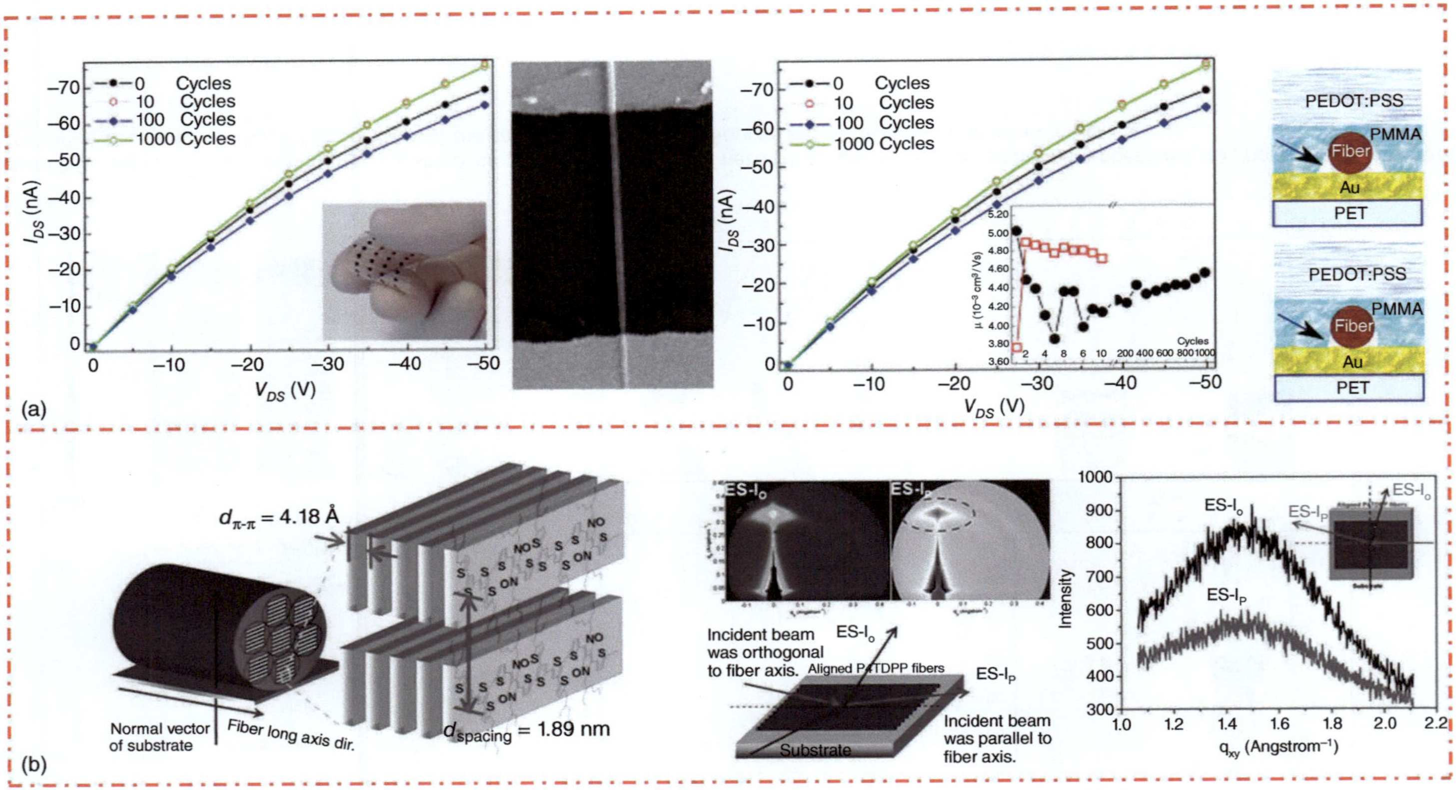

Figure 8.10 (a) Photograph of P3DT/P3HT fiber-based FET arrays. Source: Ref. [156]/with permission of Elsevier; (b) Schematic representation of the molecular packing structure within electrospun nanofibers, accompanied by a grazing incidence X-ray diffraction (GIXD) pattern and in-plane diffraction profiles of the electrospun nanofibers. Source: Ref. [154]/with permission of John Wiley & Sons.

boundary effects, leading to increased field-effect mobility [152–155]. Manuelli et al. [156] reported that the charge-carrier mobility of electrospun nanofibers exceeds that measured in planar conditions, stabilizing after approximately 1000 bending cycles (Figure 8.10a). Similarly, Lin et al. [154] found that the electrospinning process can significantly improve π–π molecular stacking and the orientation of crystallites within nanofibers, achieving a high mobility of 0.305 $cm^2/(V\ s)$ (Figure 8.10b), which substantially surpasses that of spin-coated film.

The capability for continuous nanofiber production via electrospinning also facilitates efficient and cost-effective assembly of device components for flexible electronics. Crucially, electrospinning can produce nanofibers with uniaxial alignment, an essential feature for OFETs where randomly oriented structures can complicate microscale device fabrication and channel area refinement [157]. Lee et al. [158] developed organic thin-film transistors using electrospun-aligned poly(3-hexylthiophene) (P3HT) nanofibers (Figure 8.11a). The devices are highly flexible and transparent, with ion gel achieving perfect contact with the P3HT fibers

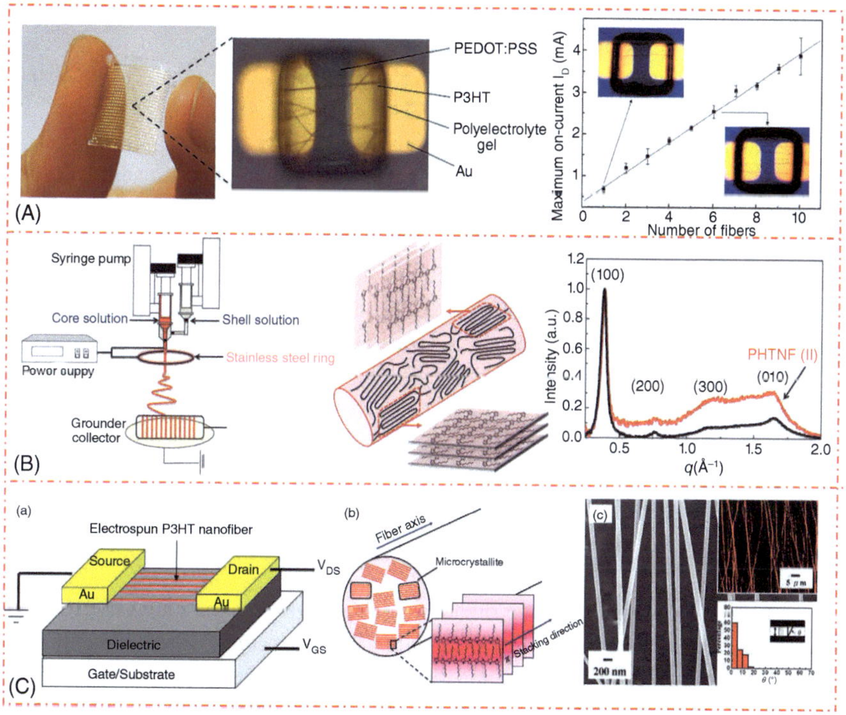

Figure 8.11 (A) Device structure of arrayed organic field-effect transistors (OFETs) and a graph showing the on-current as a function of the number of fibers bridging the source and drain electrodes. Source: Ref. [158]/with permission of American Chemical Society; (B) Schematic of the coaxial electrospinning setup with a secondary electric field, alongside microstructure details and XRD patterns. Source: Ref. [159]/with permission of John Wiley & Sons; (C) Offers a schematic representation of the coaxial electrospinning setup and process used to fabricate aligned, inner microstructures of single electrospun P3HT nanofibers, as well as FE-SEM images. Source: Ref. [160]/American Chemical Society.

(Figure 8.11a). Devices incorporating single and multiple (10) P3HT fibers demonstrated on-current values of 0.7 and 3.8 mA, respectively, and operated at voltages below 2 V. The average field-effect mobility and on/off ratio were approximately 2 $cm^2/(V\ s)$ and 10^5, respectively. Importantly, these devices showed a linear relationship between the number of fibers and the maximum on-current, indicating that their electrical properties can be finely controlled by adjusting the fiber count.

To enhance the performance of P3HT OFETs, research has identified two primary strategies: (i) the introduction of a secondary electric field during the electrospinning process, and (ii) the alignment of electrospun nanofibers. These approaches are designed to improve charge-carrier mobility, a critical factor for the efficient operation of OFETs.

Chen et al. [159] demonstrated the first approach by incorporating an extra electric field below the spinneret in their electrospinning setup to produce P3HT nanofibers. This additional electric field exerts an extensional force that elongates the liquid jet, resulting in thinner fibers. The effect of this technique is significant; the charge-carrier mobilities of the nanofibers fabricated with and without secondary electric field were found to be 1.54×10^{-4} and 1.62×10^{-1} $cm^2/(V\ s)$, respectively (Figure 8.11b). This represents a dramatic enhancement in mobility of over 1000 times attributed to effective charge transport through the delocalized electrons along the highly extended and oriented P3HT backbones, as opposed to conventional π–π stacking.

Another approach to improving OFET performance involves aligning electrospun nanofibers. Chen et al. [160] employed a two-fluid coaxial electrospinning technique, using P3HT as the core material and PMMA as the shell, to produce aligned P3HT nanofibers. Subsequent removal of the PMMA shell (Figure 8.11c) revealed that the carrier mobility of the OFETs based on aligned P3HT nanofibers could be enhanced by up to three orders of magnitude (1.92×10^{-1} $cm^2/(V\ s)$) with an on/off ratio of 4.45×10^4. Especially when using a low shell flow rate. However, it was noted that thermal annealing temperatures above 100 °C could cause relaxation of the P3HT orientation within the nanofibers, leading to a decrease in field-effect mobility.

8.4.3 Optoelectronic Devices

In the realm of flexible electronic devices, such as sensors, the electrode plays a critical role due to its need for favorable mechanical strength, large capacitance, and higher energy density. Significant research efforts have been focused on the development of flexible electrodes, with electrospinning and electropolymerization being key techniques in their fabrication [160, 161]. Liu et al. [162] have pioneered the fabrication of a flexible TiO_2-based battery electrode using a novel approach. They utilized nonwoven activated carbon fabric composed of electrospun carbon nanofibers, approximately 300 nm in diameter, as a structural support. These carbon nanofibers are interconnected, forming a conductive 3D network. This network facilitates easy ion access and endows the electrode with good conductivity, which is essential for high-performance flexible electronic devices.

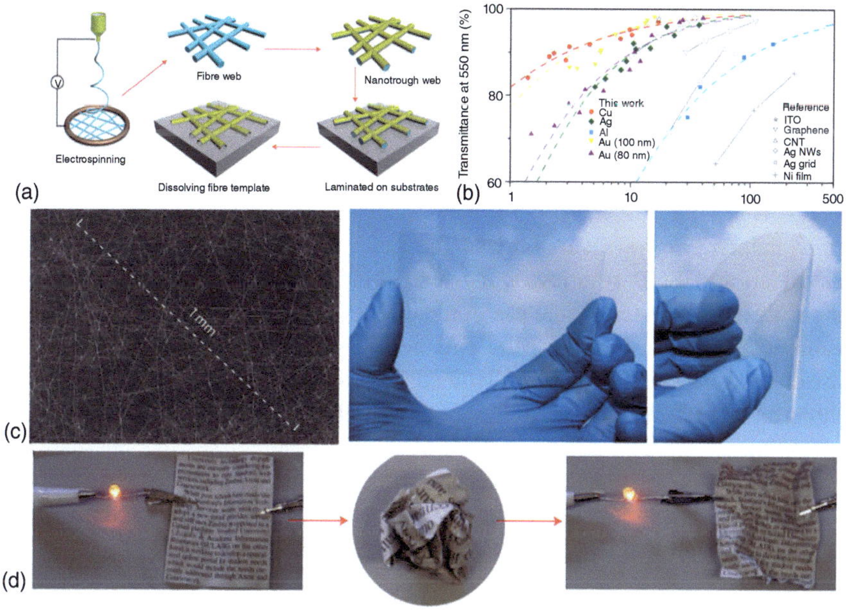

Figure 8.12 (a) Schematic of the polymer-nanofiber templating process used for fabricating nanotroughs; (b) Sheet resistance versus optical transmission (at 550 nm) made of copper, gold, silver, and aluminum. (c) SEM images of gold nanotrough networks, highlighting their easy transferability onto various substrates; (d) Photographs of conducting paper created by transferring gold nanotrough networks onto paper. Source: Ref. [163]/with permission of Springer Nature.

In the domain of optoelectronics, particularly for applications requiring transparent electrodes, Cui's group [163] has made significant advancements. They developed a transparent electrode based on a metal nanotrough network, with electrospun nanofibers serving as the template. This method produces nanotroughs that are 400 nm wide and over 1 mm long, preferentially coated on one side of the polymer fibers (Figure 8.12). These fibers are not only mechanically flexible and robust but can also be easily transferred onto various substrates, enabling integration with macroscopic systems.

The durability and adaptability of these transparent electrodes were demonstrated by transferring the nanotrough network onto paper. Even after the electrode was crushed and then unfolded, it remained conductive, with only minimal changes in resistance (Figure 8.6e,f). The resilience is attributed to the continuous nature of the nanotroughs, which can undergo nanoscale deformations to alleviate stress without breaking continuity.

However, a notable challenge with electrospun transparent electrodes is their relatively high surface roughness, which can be problematic for practical applications. To address this issue, an annealing process is sometimes employed after electrospinning [164, 165]. This process melts the nanofibers, transforming them into a continuous film and fusing adjacent fibers at their junctions. Such treatment not only reduces the likelihood of electrical shorting in devices built atop the electrode [164] but also enhances optical transmittance [165], making these electrodes more suitable for a wide range of optoelectronic applications.

8.5 Conclusions

In this chapter, we have summarized the main approaches of electrospinning, fiber inkjet, and EHD printing, along with their mechanisms, for the development of flexible and stretchable electronic devices or sensors. EHD direct-writing streamlines the fabrication process, enhances device performance, and reduces fabrication costs. This approach undoubtedly paves a new path for the creation of high-performance organic electronics, both cost-effectively and innovatively. The electrospinning strategy is particularly advantageous for producing fiber-based, comfortable, and breathable flexible electronics. These are capable of sensing mechanical force, temperature, humidity, and gas, making them ideal for a wide range of stretchable and flexible electronic applications.

References

1 Haghayegh, M., Cao, R., Zabihi, F. et al. (2022). Recent advances in stretchable, wearable and bio-compatible triboelectric nanogenerators. *Journal of Materials Chemistry C* 10 (32): 11439–11471.

2 Rajan, K., Garofalo, E., and Chiolerio, A. (2022). Wearable intrinsically soft, stretchable, flexible devices for memories and computing. *Sensors* 18 (2): 367.

3 Potyrailo, R.A., Ghiradella, H., Vertiatchikh, A. et al. (2007). Morpho butterfly wing scales demonstrate highly selective vapour response. *Nature Photonics* 1: 123–128.

4 Vignolini, S., Moyroud, E., Hingant, T. et al. (2015). The flower of *Hibiscus trionum* is both visibly and measurably iridescent. *The New Phytologist* 205: 97–101.

5 Liu, F., Dong, B.Q., Liu, X.H. et al. (2009). Structural color change in longhorn beetles Tmesisternus isabellae. *Optics Express* 17: 16183.

6 Sun, J., Bhushan, B., and Tong, J. (2013). Structural coloration in nature. *RSC Advances* 3: 14862–14889.

7 Shen, H., Wang, Z., Wu, Y., and Yang, B. (2016). One-dimensional photonic crystals: fabrication, responsiveness and emerging applications in 3D construction. *RSC Advances* 6: 4505–4520.

8 Mulder, D.J., Schenning, A.P.H.J., and Bastiaansen, C.W.M. (2014). Chiral-nematic liquid crystals as one dimensional photonic materials in optical sensors. *Journal of Materials Chemistry C* 2: 6695–6705.

9 Moirangthem, M., Arts, R., Merkx, M., and Schenning, A.P.H.J. (2016). An optical sensor based on a photonic polymer film to detect calcium in serum. *Advanced Functional Materials* 26: 1154–1160.

10 Lova, P., Bastianini, C., Giusto, P. et al. (2016). Label-free vapor selectivity in poly(*p*-phenylene oxide) photonic crystal sensors. *ACS Applied Materials & Interfaces* 8: 31941–31950.

11 Khan, M.K., Bsoul, A., Walus, K. et al. (2015). Photonic patterns printed in chiral nematic mesoporous resins. *Angewandte Chemie, International Edition* 54: 4304–4308.

12 Nam, H., Song, K., Ha, D., and Kim, T. (2016). Inkjet printing based mono-layered photonic crystal patterning for anti-counterfeiting structural colors. *Scientific Reports* 6 (1): 30855.

13 Moirangthem, M. and Schenning, A.P.H.J. (2018). Full color camouflage in a printable photonic blue-colored polymer. *ACS Applied Materials & Interfaces* 10: 4168–4172.

14 Fang, Y., Ni, Y., Leo, S.Y. et al. (2015). Direct writing of three-dimensional macroporous photonic crystals on pressure- responsive shape memory polymers. *ACS Applied Materials & Interfaces* 7: 23650–23659.

15 Ge, J., Goebl, J., He, L. et al. (2009). Rewritable photonic paper with hygroscopic salt solution as ink. *Advanced Materials* 21: 4259–4264.

16 Moirangthem, M., Scheers, A.F., and Schenning, A.P.H.J. (2018). A full color photonic polymer, rewritable with a liquid crystal ink. *Chemical Communications* 54: 4425–4428.

17 Bai, L., Xie, Z., Wang, W. et al. (2014). Bio-inspired vapor-responsive colloidal photonic crystal patterns by inkjet printing. *ACS Nano* 8: 11094–11100.

18 van der Zande, B.M.I., Steenbakkers, J., Lub, J. et al. (2005). Mass transport phenomena during lithographic polymerization of nematic monomers monitored with interferometry. *Journal of Applied Physics* 97 (12): 123519.

19 Zhang, P., Kragt, A.J.J., Schenning, A.P.H.J. et al. (2018). An easily coatable temperature responsive cholesteric liquid crystal oligomer for making structural colour patterns. *Journal of Materials Chemistry C* 6: 7184–7187.

20 Burgess, I.B., Mishchenko, L., Hatton, B.D. et al. (2011). Encoding complex wettability patterns in chemically functionalized 3D photonic crystals. *Journal of the American Chemical Society* 133: 12430–12432.

21 Lee, H.S., Kim, J.H., Lee, J.S. et al. (2014). Magnetoresponsive discoidal photonic crystals toward active color pigments. *Advanced Materials* 26: 5801–5807.

22 Mitov, M. (2012). Cholesteric liquid crystals with a broad light reflection band. *Advanced Materials* 24: 6260–6276.

23 Fan, J., Li, Y., Bisoyi, H.K. et al. (2015). Light-directing omnidirectional circularly polarized reflection from liquid-crystal droplets. *Angewandte Chemie, International Edition* 54: 2160–2164.

24 Bisoyi, H.K. and Li, Q. (2016). Light-driven liquid crystalline materials: from photo-induced phase transitions and property modulations to applications. *Chemical Reviews* 116: 15089–15166.

25 Bisoyi, H.K. and Li, Q. (2016). Light-directed dynamic chirality inversion in functional self-organized helical superstructures. *Angewandte Chemie, International Edition* 55: 2994–3010.

26 van Heeswijk, E.P.A., Kloos, J.J.H., de Heer, J. et al. (2018). Well-adhering, easily producible photonic reflective coatings for plastic substrates. *ACS Applied Materials & Interfaces* 10: 30008–30013.

27 Kragt, A.J.J., Broer, D.J., and Schenning, A.P.H.J. (2018). Easily processable and programmable responsive semi-interpenetrating liquid crystalline polymer network coatings with changing reflectivities and surface topographies. *Advanced Functional Materials* 28 (6): 1704756.

28 Zhang, W., Kragt, S., Schenning, A.P.H.J. et al. (2017). Easily processable temperature-responsive infrared reflective polymer coatings. *ACS Omega* 2: 3475–3482.

29 Jia, J., Pu, J.-H., Liu, J.-H. et al. (2020). Surface structure engineering for a bionic fiber-based sensor toward linear, tunable, and multifunctional sensing. *Materials Horizons* 7 (9): 2450–2459.

30 Wu, L., Luo, J., Li, Y. et al. (2021). Emulsion dipping based superhydrophobic, temperature tolerant, and multifunctional coatings for smart strain sensing applications. *Composites Science and Technology* 216: 109045.

31 Cui, X., Boland, T., and DD'Lima, D., K Lotz, M. (2012). Thermal inkjet printing in tissue engineering and regenerative medicine. *Recent Patents on Drug Delivery & Formulation* 6 (2): 149–155.

32 Saunders, R.E., Gough, J.E., and Derby, B. (2008). Delivery of human fibroblast cells by piezoelectric drop-on-demand inkjet printing. *Biomaterials* 29 (2): 193–203.

33 Deiner, L.J. and Reitz, T.L. (2017). Inkjet and aerosol jet printing of electrochemical devices for energy conversion and storage. *Advanced Engineering Materials* 19 (7): 1600878.

34 Li, X., Liu, B., Pei, B. et al. (2020). Inkjet bioprinting of biomaterials. *Chemical Reviews* 120: 10793–10833.

35 Khan, S., Ali, S., and Bermak, A. (2019). Smart manufacturing technologies for printed electronics. *Hybrid Nanomaterials-Flexible Electronics Materials.*

36 Machekposhti, S.A., Movahed, S., and Narayan, R.J. (2020). Physicochemical parameters that underlie inkjet printing for medical applications. *Biophysical Reviews* 1: 011301.

37 Sohrabi, S. and Liu, Y. (2018). Modeling thermal inkjet and cell printing process using modified pseudopotential and thermal lattice Boltzmann methods. *Physical Review E* 97: 033105.

38 Sushko, O., Pigeon, M., Donnan, R.S. et al. (2017). Comparative study of sub-THz FSS filters fabricated by inkjet printing, microprecision material printing, and photolithography. *IEEE Transactions on Terahertz Science and Technology* 7 (2): 184–190.

39 Fan, Z., Ho, J.C., Jacobson, Z.A. et al. (2008). Wafer-scale assembly of highly ordered semiconductor nanowire arrays by contact printing. *Nano Letters* 8 (1): 20–25.

40 Yang, Y., Chang, S.C., Bharathan, J., and Liu, J. (2000). Organic/polymeric electroluminescent devices processed by hybrid ink-jet printing. *Journal of Materials Science: Materials in Electronics* 11: 89–96.

41 Wang, S., Xu, J., Wang, W. et al. (2018). Skin electronics from scalable fabrication of an intrinsically stretchable transistor array. *Nature* 555: 83–88.

42 Tekin, E., de Gans, B.-J., and Schubert, U.S. (2004). Ink-jet printing of polymers–from single dots to thin film libraries. *Journal of Materials Chemistry* 14: 2627–2632.

43 Samusjew, A., Kratzer, M., Moser, A. et al. (2017). Inkjet printing of soft, stretchable optical waveguides through the photo-polymerization of high-profile linear patterns. *ACS Applied Materials & Interfaces* 9: 4941–4947.

44 Yoon, J., Joo, Y., Lee, B, et al. (2018). Stretchable active-matrix light-emitting diode array using printed electric components on plastic and elastomer hybrid substrate. *SID International Symposium Digest of Technical Paper*, Vol. 49, 1925–1927.

45 Schlatter, S., Rosset, S., and Shea, H. (2017). *Inkjet Printing of Carbon Black Electrodes for Dielectric Elastomer Actuators* (ed. Y. Bar-Cohen), 1016311. SPIE.

46 Vuorinen, T., Niittynen, J., Kankkunen, T. et al. (2016). Inkjet-printed graphene/PEDOT: PSS temperature sensors on a skin-conformable polyurethane substrate. *Scientific Reports* 6: 1–8.

47 Wan, H., Li, X., Zhang, L. et al. (2018). Rapidly responsive and flexible chiral nematic cellulose nanocrystal composites as multifunctional rewritable photonic papers with eco-friendly inks. *ACS Applied Materials & Interfaces* 10: 5918–5925.

48 Li, P., Zhang, Y., and Zheng, Z. (2019). Polymer-assisted metal deposition (PAMD) for flexible and wearable electronics: principle, materials, printing, and devices. *Advanced Materials* 1902987.

49 Abu-Khalaf, J.M., Al-Ghussain, L., and Ala'aldeen Al-Halhouli. (2018). Fabrication of stretchable circuits on polydimethylsiloxane (PDMS) pre-stretched substrates by inkjet printing silver nanoparticles. *Materials* 11 (12): 2377.

50 Al-Halhouli, A., Qitouqa, H., Alashqar, A., and Abu-Khalaf, J. (2018). Inkjet printing for the fabrication of flexible/stretchable wearable electronic devices and sensors. *Sensor Review* 38 (4): 438–452.

51 Huang, Q., Al-Milaji, K.N., and Zhao, H. (2018). Inkjet printing of silver nanowires for stretchable heaters. *ACS Applied Nano Materials* 1: 4528–4536.

52 Fernandes, D.F., Majidi, C., and Tavakoli, M. (2019). Digitally printed stretchable electronics: a review. *Journal of Materials Chemistry C* 7: 14035–14068.

53 Kim, K.-S., Jung, K.-H., and Jung, S.-B. (2014). Design and fabrication of screen-printed silver circuits for stretchable electronics. *Microelectronic Engineering* 120: 216–220.

54 Gutbrod, S.R., Sulkin, M.S., Rogers, J.A., and Efimov, I.R. (2014). Patient-specific flexible and stretchable devices for cardiac diagnostics and therapy. *Progress in Biophysics and Molecular Biology* 115 (2–3): 244–251.

55 Li, D., Wang, Y.L., and Xia, Y.N. (2003). Electrospinning of polymeric and ceramic nanofibers as uniaxially aligned arrays. *Nano Letters* 3: 1167–1171.

56 Sun, D., Chang, C., Li, S., and Lin, L. (2006). Near-field electrospinning. *Nano Letters* 6 (4): 839–842.

57 Huang, Y.A., Wang, X.M., Duan, Y.Q. et al. (2012). Controllable self-organization of colloid microarrays based on finite length effects of electrospun ribbons. *Soft Matter* 8: 8302–8311.

58 Teo, W.E. and Ramakrishna, S. (2006). A review on electrospinning design and nanofibre assemblies. *Nanotechnology* 17: R89–R106.

59 Bu, N.B., Huang, Y.A., Wang, X.M., and Yin, Z.P. (2012). Continuously tunable and oriented nanofiber direct-written by mechano-electrospinning. *Materials and Manufacturing Processes* 27: 1318–1323.

60 Huang, Y.A., Bu, N.B., Duan, Y.Q. et al. (2013). *Electrohydrodynamic direct-writing. Nanoscale* 5: 12007–12017.

61 Zhu, M., Duan, Y., Liu, N. et al. (2019). Electrohydrodynamically printed high-resolution full-color hybrid perovskites. *Advanced Functional Materials* 29: 1903294.

62 Bu, N., Huang, Y., Deng, H., and Yin, Z. (2012). Tunable bead-on-string microstructures fabricated by mechano-electrospinning. *Journal of Physics D: Applied Physics* 45: 405301.

63 Ramakrishna, S., Fujihara, K., Teo, W.E. et al. (2006). Electrospun nanofibers: solving global issues. *Materials Today* 9: 40.

64 Inagaki, M., Yang, Y., and Kang, F.Y. (2012). Carbon nanofibers prepared via electrospinning. *Advanced Materials* 24: 2547.

65 Huang, Y., Ding, Y., Bian, J. et al. (2017). Hyper-stretchable self-powered sensors based on electrohydrodynamically printed, self-similar piezoelectric nano/microfibers. *Nano Energy* 40: 432.

66 Duan, Y., Ding, Y., Xu, Z. et al. (2017). Helix electrohydrodynamic printing of highly aligned serpentine micro/nanofibers. *Polymers* 9: 434.

67 Chen, Z., Wang, Y., Yang, Y. et al. (2021). Multifunctional sensing platform based on green-synthesized silver nanostructure and microcrack architecture. *Chemical Engineering Journal* 403 (1): 126388.

68 Hu, D.Q., Wang, Q.H., Yu, J.X. et al. (2016). Highly stretchable strain sensors using an electrospun polyurethane nanofiber/graphene composite. *Journal of Nanoscience and Nanotechnology* 16: 5839–5842.

69 Zou, W., Haibo, Y., Lv, X. et al. (2022). Electrohydrodynamic direct-writing fabrication of microstructure- enhanced microelectrode arrays for customized and curved physiological electronics. *Advanced Materials Interfaces* 9: 2201197.

70 Zhang, Z., He, H., Wanlin, F. et al. (2020). Electro-hydrodynamic direct-writing technology toward patterned ultra-thin fibers: advances. *Materials and Applications* 35: 100942.

71 Wang, D., Abbas, Z., Liangkun, L. et al. (2022). Simulation and printing of microdroplets using straight electrode-based electrohydrodynamic jet for flexible substrate. *Micromachines* 13 (10): 1727.

72 Wang, K., Paine, M.D., and Stark, J.P. (2009). Fully voltage-controlled electrohydrodynamic jet printing of conductive silver tracks with a sub-100 μm linewidth. *Journal of Applied Physics* 106: 024907.

73 Kameoka, J., Verbridge, S.S., Liu, H.Q. et al. (2004). Fabrication of suspended silica glass nanofibers from polymeric materials using a scanned electrospinning source. *Nano Letters* 4: 2105.

74 Braly, I.L., Qilettes, D.W., Pazos-Outon, L.M. et al. (2018). Hybrid perovskite films approaching the radiative limit with over 90% photoluminescence quantum efficiency. *Nature Photonics* 12: 355.

75 Zhang, Z., Yao, J., Liang, L. et al. (2018). High-performance photocoupler based on perovskite light emitting diode and photodetector. *ACS Applied Materials & Interfaces* 10: 39441.

76 Shigeta, K., He, Y., Sutanto, E. et al. (2012). Functional protein microarrays by electrohydrodynamic jet printing. *Journal of the American Chemical Society* 84: 10012.

77 Wang, X., Xu, L., Zheng, G. et al. (2012). Pulsed electrohydrodynamic printing of conductive silver patterns on demand. *Science China Technological Sciences* 55: 1603.

78 Coppola, S., Nasti, G., Vespini, V. et al. (2018). From electrohydrodynamic instabilities of liquids to the high-resolution ink-jet printing through pyroelectric driving power. *Journal of Micro/Nanolithography, MEMS, and MOEMS* 17: 031206.

79 Bi, C., Wang, S., Wen, W. et al. (2018). Room-temperature construction of mixed-halide perovskite quantum dots with high photoluminescence quantum yield. *Journal of Physical Chemistry C* 122: 5151.

80 Mao, J., Lin, H., Ye, F. et al. (2018). All-perovskite emission architecture for white light-emitting diodes. *ACS Nano* 12: 10486.

81 Zou, W., Yu, H., Zhou, P., and Liu, L. (2019). Electrohydrodynamically printed high-resolution full-color hybrid perovskites. *Materials and Design* 166: 274158.

82 An, B.W., Kim, K., Lee, H. et al. (2015). High-resolution printing of 3D structures using an electrohydrodynamic inkjet with multiple functional inks. *Advanced Materials* 27: 4322.

83 Kroupa, D.M., Roh, J.Y., Milstein, T.J. et al. (2018). Quantum-cutting ytterbium-doped CsPb $(Cl_{1-x}Br_x)_3$ perovskite thin films with photoluminescence quantum yields over 190%. *ACS Energy Letters* 3: 2390.

84 Mondal, N., De, A., and Samanta, A. (2019). Achieving near-unity photoluminescence efficiency for blue-violet-emitting perovskite nanocrystals. *ACS Energy Letters* 4: 32.

85 Zhong, J.W., Zhong, Q.Z., Hu, Q.Y. et al. (2015). Stretchable self-powered fiber-based strain sensor. *Advanced Functional Materials* 25: 1798–1803. 2012, 84, 10012.

86 Li, K., Wang, D., Wang, Q. et al. (2018). Thermally assisted electrohydrodynamic jet high-resolution printing of high-molecular weight biopolymer 3D structures. *Macromolecular Materials and Engineering* 303: 1800345.

87 Dutta, A., Behera, R.K., Pal, P. et al. (2019). Near-unity photoluminescence quantum efficiency for all $CsPbX_3$ (X = Cl, Br, and I) perovskite nanocrystals: a generic synthesis approach. *Angewandte Chemie* 131: 5608.

88 Song, J., Li, J., Li, X. et al. (2015). Quantum dot light-emitting diodes based on inorganic perovskite cesium lead halides ($CsPbX_3$). *Advanced Materials* 27: 7162023.

89 Kim, B.H., Onses, M.S., Lim, J.B. et al. (2015). High-resolution patterns of quantum dots formed by electrohydrodynamic jet printing for light-emitting diodes. *Nano Letters* 15: 969.

90 Kang, K., Yang, D., Park, J. et al. (2017). Micropatterning of metal oxide nanofibers by electrohydrodynamic (EHD) printing towards highly integrated and multiplexed gas sensor applications. *Sensors and Actuators, B: Chemical* 250: 574–583.

91 Li, H., Liu, N., Shao, Z. et al. (2019). Coffee ring elimination and crystalline control of electrohydrodynamically printed high-viscosity perovskites. *Journal of Materials Chemistry C* 7: 14867.

92 Paulina, R.F., Correr, W., Mendonça, C.R., and Almeida, J.M.P. (2020). Single-step printing of metallic nanoparticles in 2D micropatterns. *Journal of Nanoparticle Research* 22: 260.

93 Gao, D. and Zhou, J.G. (2019). Designs and applications of electrohydrodynamic 3D printing. *International Journal of Bioprinting* 5 (11): PMC7415867172.

94 Lin, S., Liu, X., Liu, J. et al. (2019). Anti-fatigue-fracture hydrogels. *Science Advances* 5: eaau8528.

95 Lu, C., Mohwald, H., and Fery, A. (2008). Large-scale regioselective formation of well-defined stable wrinkles of multilayered films via embossing. *Chemistry of Materials* 20: 7052–7059.

96 Lacour, S.P., Wagner, S., Huang, Z., and Suo, Z. (2003). Stretchable gold conductors on elastomeric substrates. *Applied Physics Letters* 82: 2404–2406.

97 Zhang, F.T., Xu, L., Chen, J.H. et al. (2018). Electroless deposition metals on poly-(dimethylsiloxane) with strong adhesion as flexible and stretchable conductive materials. *ACS Applied Materials & Interfaces* 10 (2): 2075–2082.

98 Wang, Y., Chao, M., Wan, P., and Zhang, L. (2020). A wearable breathable pressure sensor from metal-organic framework derived nano composites for highly sensitive broad- range healthcare monitoring. *Nano Energy* 70: 104560.

99 Li, B., Luo, J., Huang, X. et al. (2020). A highly stretchable, super-hydrophobic strain sensor based on poly dopamine and graphene reinforced nanofiber composite for human motion monitoring. *Composites, Part B: Engineering* 181: 107580.

100 Chen, J., Wang, F., Zhu, G. et al. (2021). Breathable strain/temperature sensor based on fibrous networks of ionogels capable of monitoring human motion, respiration, and proximity. *ACS Applied Materials & Interfaces* 13 (43): 51567–51577.

101 Jiang, Y., He, Q., Cai, J. et al. (2020). Flexible strain sensor with tunable sensitivity via microscale electrical breakdown in graphene/polyimide thin films. *ACS Applied Materials & Interfaces* 12 (52): 58317–58325.

102 Wang, D., Li, D., Zhao, M. et al. (2018). Multifunctional wearable smart device based on conductive reduced graphene oxide/polyester fabric. *Applied Surface Science* 454: 218–226.

103 Xiong, J., Chen, J., and Lee, P.S. (2021). Functional fibers and fabrics for soft robotics, wearables, and human-robot interface. *Advanced Materials* 33 (19): 2002640.

104 Wang, X., Liu, X., and Schubert, D.W. (2021). Highly sensitive ultrathin flexible thermoplastic polyurethane/carbon black fibrous film strain sensor with adjustable scaffold networks. *Nano Micro Letters* 13 (1): 64.

105 Wang, H., Zhang, Y., Liang, X., and Zhang, Y. (2021). Smart fibers and textiles for personal health management. *ACS Nano* 15: 12497–12508.

106 Zhang, H., Liu, D., Lee, J.H. et al. (2021). Anisotropic, wrinkled, and crack-bridging structure for ultrasensitive, highly selective multidirec-tional strain sensors. *Nano Micro Letters* 13 (1): 122.

107 Ke, K., Solouki Bonab, V., Yuan, D., and Manas Zloczower, I. (2018). Piezoresistive thermoplastic polyurethane nanocomposites with carbon nanostructures. *Carbon* 139: 52–58.

108 Xiao, W., Wang, L., Li, B. et al. (2022). Interface-engineered reduced graphene oxide assembly on nanofiber surface for high performance strain and temperature sensing. *Journal of Colloid and Interface Science* 608 (1): 931–941.

109 Li, Y., Wang, S., Xiao, Z.-C. et al. (2020). Flexible TPU strain sensors with tunable sensitivity and stretchability by coupling AgNWs with rGO. *Journal of Materials Chemistry C* 8 (12): 4040–4048.

110 Zhou, K., Xu, W., Yu, Y. et al. (2021). Tunable and nacre-mimetic multifunctional electronic skins for highly stretchable contact-noncontact sensing. *Small* 17 (31): 2100542.

111 Pu, J.H., Zhao, X., Zha, X.J. et al. (2019). Multilayer structured AgNW/WPU-MXene fiber strain sensors with ultrahigh sensitivity and a wide operating range for wearable monitoring and healthcare. *Journal of Materials Chemistry A* 7 (26): 15913–15923.

112 Xu, W., Hu, S., Zhao, Y. et al. (2021). Nacre-inspired tunable strain sensor with synergistic interfacial interaction for sign language interpretation. *Nano Energy* 90: 106606.

113 Yu, G.F., Yan, X., Yu, M. et al. (2016). Patterned, highly stretchable and conductive nanofibrous PANI/PVDF strain sensors based on electrospinning and in situ polymerization. *Nanoscale* 8 (5): 2944–2950.

114 Farrokhnia, M., Amoabediny, G., Ebrahimi, M. et al. (2022). Ultrasensitive early detection of insulin antibody employing novel electrochemical nano-biosensor based on control-lable electro-fabrication process. *Talanta* 238 (1): 122947.

115 Li, Y., Jia, J., Hua, Y. et al. (2022). Macromolecule relaxation directed 3D nanofifiber architecture in stretchable fibrous mats for wearable multifunctional sensors. *ACS Applied Materials & Interfaces* 14 (13): 15678–15686.

116 Qi, K., He, J., Wang, H. et al. (2017). A highly stretchable nanofiber-based electronic skin with pressure-, strain-, and flexion-sensitive properties for health and motion monitoring. *ACS Applied Materials & Interfaces* 9 (49): 42951–42960.

117 Gong, S., Schwalb, W., Wang, Y. et al. (2014). A wearable and highly sensitive pressure sensor with ultrathin gold nanowires. *Nature Communications* 5 (1): 3132.

118 Zhang, M., Wang, C., Wang, H. et al. (2016). Carbonized cotton fabric for high-performance wearable strain sensors. *Advanced Functional Materials* 27: 1604765.

119 Zhong, W., Liu, C., Xiang, C. et al. (2017). Continuously producible ultrasensitive wearable strain sensor assembled with three-dimensional interpenetrating Ag nanowires/polyolefin elastomer nanofibrous composite yarn. *ACS Applied Materials & Interfaces* 9 (48): 42058–42066.

120 Pang, C., Lee, G.Y., Kim, T.I. et al. (2012). A flexible and highly sensitive strain-gauge sensor using reversible interlocking of nanofibres. *Nature Materials* 11: 795.

121 Park, S., Kim, H., Vosgueritchian, M. et al. (2014). Stretchable energy-harvesting tactile electronic skin capable of differentiating multiple mechanical stimuli modes. *Advanced Materials* 26: 7324.

122 Park, J., Lee, Y., Hong, J. et al. (2014). Tactile-direction-sensitive and stretchable electronic skins based on human-skin-inspired interlocked microstructures. *ACS Nano* 8: 12020.

123 Nan, N., He, J., You, X. et al. (2019). A stretchable, highly sensitive, and multimodal mechanical fabric sensor based on electrospun conductive nanofiber yarn for wearable electronics. *Advanced Materials Technologies* 4: 1800338.

124 Zulifqar, A. and Hu, H. (2019). Development of bi-stretch auxetic woven fabrics based on re-entrant hexagonal geometry. *Physica Status Solidi B* 256: 1800172.

125 Miller, W., Ren, Z., Smith, C.W., and Evans, K.E. (2012). A negative Poisson's ratio carbon fibre composite using a negative Poisson's ratio yarn reinforcement. *Composites Science and Technology* 72: 761–766.

126 Papadopoulou, A., Laucks, J., and Tibbits, S. (2017). Auxetic materials in design and architecture. *Nature Reviews Materials* 2: 170–178.

127 Sloan, M.R., Wright, J.R., and Evans, K.E. (2011). The helical auxetic yarn a novel structure for composites and textiles; geometry, manufacture and mechanical properties. *Mechanics of Materials* 43: 476–486.

128 Steffens, F., Rana, S., and Fangueiro, R. (2016). Development of novel auxetic textile structures using high performance fibres. *Materials and Design* 106: 81–89.

129 Rabbi, M.F., Chalivendra, V., and Kim, Y. (2018). Dynamic constitutive response of novel auxetic Kevlar®/epoxy composites. *Composite Structures* 195: 1–13.

130 Miller, W., Hook, P.B., Smith, C.W. et al. (2009). The manufacture and characterisation of a novel, low modulus, negative Poisson's ratio composite. *Composites Science and Technology* 69: 651–655.

131 Chang, Y., Ma, P., and Jiang, G. (2017). Energy absorption property of warp-knitted spacer fabrics with negative Possion's ratio under low velocity impact. *Composite Structures* 182: 471–477.

132 McAfee, J. and Faisal, N.H. (2017). Parametric sensitivity analysis to maximise auxetic effect of polymeric fibre based helical yarn. *Composite Structures* 162: 1–12.

133 Zhang, G., Ghita, O.R., and Evans, K.E. (2016). Dynamic thermo-mechanical and impact properties of helical auxetic yarns. *Composites Part B: Engineering* 99: 494–505.

134 Ng, W.S. and Hu, H. (2018). Woven fabrics made of auxetic plied yarns. *Polymers* 10: 226.

135 Htwe, Y.Z.N. and Mariatti, M. (2022). Printed graphene and hybrid conductive inks for flexible, stretchable, and wearable electronics: progress, opportunities, and challenges. *Journal of Science: Advanced Materials and Devices* 7 (2): 100435.

136 Dong, K., Peng, X., and Wang, Z.L. (2020). Fiber/fabric-based piezoelectric and triboelectric nanogenerators for flexible/stretchable and wearable electronics and artificial intelligence. *Advanced Materials* 32: 1902549.

137 Wang, P., Liu, J., Wei, Y. et al. (2022). Flexible, stretchable, breathable and sweatproof all-nanofiber iontronic tactile sensor for continuous and comfortable knee joint motion monitoring. *Nano Energy* 103 (1): 107768.

138 Shen, G., Chen, B., Liang, T. et al. (2020). Transparent and stretchable strain sensors with improved sensitivity and reliability based on Ag NWs and PEDOT: PSS patterned microstructures. *Advanced Electronic Materials* 6: 1901360.

139 Tang, J., Yuting, W., Ma, S. et al. (2022). Flexible strain sensor based on CNT/TPU composite nanofiber yarn for smart sports bandage. *Composites Part B: Engineering*. 232: 109605.

140 Persano, L., Dagdeviren, C., Su, Y.W. et al. (2013). High performance piezoelectric devices based on aligned arrays of nanofibers of poly(vinylidenefluoride-*co*-trifluoroethylene). *Nature Communications* 4: 1633.

141 Liang, J., Li, L., Chen, D. et al. (2015). Intrinsically stretchable and transparent thin-film transistors based on printable silver nanowires, carbon nanotubes and an elastomeric dielectric. *Nature Communications* 6: 7647.

142 Pang, C., Koo, J.H., Nguyen, A. et al. (2015). Highly skin-conformal microhairy sensor for pulse signal amplification. *Advanced Materials* 27: 634–640.

143 Kaltenbrunner, M., Sekitani, T., Reeder, J. et al. (2013). An ultra-lightweight design for imperceptible plastic electronics. *Nature* 499: 458–463.

144 Klauk, H. (2006). *Organic Electronics, Materials, Manufacturing, and Applications*. Weinheim: Wiley-VCH.

145 Bao, Z. and Locklin, J. (2007). *Organic Field-Effect Transistors*. New York: CRG Press.

146 Subramanian, V. (2007). *Organic Field-Effect Transistors*, Optical Science and Engineering Series, vol. 128, 489–505. Boca Raton, FL: CRC Press.

147 Forrest, S.R. (2004). The path to ubiquitous and low-cost organic electronic appliances on plastic. *Nature* 428: 911.

148 Mabeck, J.T. and Malliaras, G.G. (2006). Chemical and biological sensors based on organic thin-film transistors. *Bioanalytical Chemistry* 384: 343.

149 Brabec, C.J., Sariciftci, N.S., and Hummelen, J.C. (2001). Plastic solar cells. *Advanced Functional Materials* 11: 15.

150 Lim, B., Baeg, K.J., Jeong, H.G. et al. (2009). A new poly (thienylenevinylene) derivative with high mobility and oxidative stability for organic thin-film transistors and solar cells. *Advanced Materials* 21: 2808.

151 Wang, W., Li, Z., Xu, X. et al. (2011). Au-doped polyacrylonitrile–polyaniline core–shell electrospun nanofibers having high field-effect mobilities. *Small* 7: 597.

152 Wang, W., Lu, X., Li, Z. et al. (2011). One-dimensional polyelectrolyte/polymeric semiconductor core/shell structure: sulfonated poly(arylene ether ketone)/polyaniline nanofibers for organic field-effect transistors. *Advanced Materials* 23: 5109.

153 Briseno, A.L., Mannsfeld, S.C.B., Lu, X.M. et al. (2007). Fabrication of field-effect transistors from hexathiapentacene single-crystal nanowires. *Nano Letters* 7: 668.

154 Lin, C.J., Hsu, J.C., Tsai, J.H. et al. (2011). High-performance FETs prepared from electrospun aligned P4TDPP nanofibers. *Macromolecular Chemistry and Physics* 212: 2452.

155 Cho, H., Min, S.-Y., and Lee, T.-W. (2013). Electrospun organic nanofiber electronics and photonics. *Macromolecular Materials and Engineering* 298: 475.

156 Manuelli, A., Persano, L., and Pisignano, D. (2014). Flexible organic field-effect transistors based on electrospun conjugated polymer nanofibers with high bending stability. *Organic Electronics* 15 (5): 1056–1061.

157 Wang, X.M., Liu, Y.Q., Chen, Q.Z. et al. (2021). Recent advances in stretchable field-effect transistors. *Journal of Materials Chemistry C* 9: 7796–7828.

158 Lee, S.W., Lee, H.J., Choi, J.H. et al. (2010). Periodic array of polyelectrolyte-gated organic transistors from electrospun poly(3-hexylthiophene) nanofibers. *Nano Letters* 10: 347.

159 Chen, J.-Y., Hung-Chin, W., Chiu, Y.-C. et al. (2015). Electrospun poly(3-hexylthiophene) nanofibers with highly extended and oriented chains through secondary electric field for high-performance field-effect transistors. *Advanced Electronic Materials* 1: 1400028.

160 (a) Chen, J.-Y., Kuo, C.-C., Lai, C.-S. et al. (2011). Manipulation on the morphology and electrical properties of aligned electrospun nanofibers of poly(3-hexylthiophene) for field-effect transistor applications. *Macromolecules* 44 (8): 2883–2892; (b) Wang, D.W., Li, F., Zhao, J. et al. (2009). Fabrication of graphene/polyaniline composite paper via in situ anodic electropolymerization for high-performance flexible electrode. *ACS Nano* 3: 1745.

161 Wang, Z.L., Guo, R., Li, G.R. et al. (2012). Polyaniline nanotube arrays as high-performance flexible electrodes for electrochemical energy storage devices. *Journal of Materials Chemistry* 22: 2401.

162 Liu, S., Wang, Z., Yu, C. et al. (2013). A flexible TiO_2 (B)-based battery electrode with superior power rate and ultralong cycle life. *Advances Materials* 25: 3462.

163 Wu, H., Kong, D., Ruan, Z.C. et al. (2013). A transparent electrode based on a metal nanotrough network. *Nature Nanotechnology* 8: 421.

164 Wu, H., Hu, L., Rowell, M.W. et al. (2010). Electrospun metal nanofiber webs as high-performance transparent electrode. *Nano Letters* 10: 4242.

165 Huang, Y.L., Baji, A., Tien, H.W. et al. (2012). Self-assembly of silver–graphene hybrid on electrospun polyurethane nanofibers as flexible transparent conductive thin films. *Carbon* 50: 3473.

9

Mechanics-Guided 3D Assembly of Flexible Electronics

Guoquan Luo, Jianzhong Zhao, Xu Cheng, and Yihui Zhang

Tsinghua University, AML, Department of Engineering Mechanics, Laboratory of Flexible Electronics Technology, No. 1, Qinghua Yuan, Haidian District, Beijing 100084, PR China

9.1 Introduction

3D flexible electronics represent an area of focusing interest in recent years. Compared with classical 2D flexible electronics, 3D flexible electronics can offer better device performances or advantages in applications. For example, 3D flexible electronics can provide better conformability with nondevelopable curvy surfaces of biological tissues or organs, without inducing wrinkles or additional stresses, which is useful in bio-integrated applications [1, 2]. Due to the specific 3D design, the flexible electronics can also offer enhanced electromagnetic performances [3, 4], and enable the tunability of the spatial orientation of electromagnetic or optical fields [5, 6].

Currently, the fabrication of 3D flexible electronics mainly rely on stress-induced rolling/folding [7, 8], transfer printing on curve surfaces [9–12], multi-material 3D printing [13, 14], and mechanics-guided 3D assembly [4, 15]. Among the various fabrication methods of 3D flexible electronics, the mechanics-guided 3D assembly represents an emerging type of fabrication method developed in recent years [2, 4, 16–21]. This assembly method mainly relies on a prestrained elastomer substrate to provide forces that drive the transformation of predefined 2D precursor structures into controlled 3D configurations (as shown in Figure 9.1). This method is applicable to a wide range of advanced materials (such as device-grade silicon, GaAs, 2D materials (graphene and MoS_2), and other electronic materials) over a variety of length scales (from sub-microscale to millimeter-scale) because it is compatible with state-of-the-art planar fabrication approaches and processing techniques.

In this chapter, we introduce the fundamental aspects of the mechanics-guided 3D assembly and discuss the representative applications of 3D flexible electronic devices formed using this method. The chapter begins with the discussions on design strategies of mechanics-guided assembly in Section 9.2, covering the latest progress of

Mechanics of Flexible and Stretchable Electronics, First Edition. Edited by Yong Zhu and Nanshu Lu.

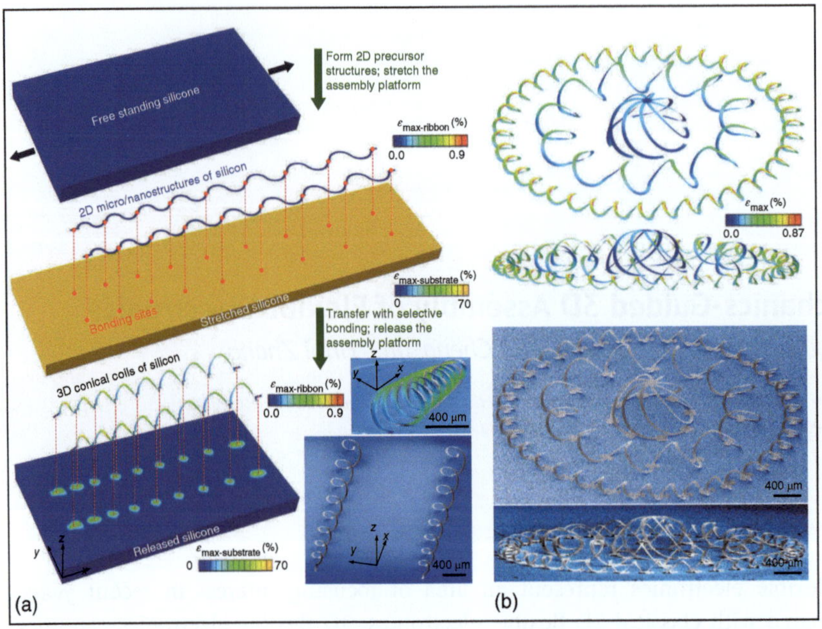

Figure 9.1 Process for mechanics-guided assembly of 3D mesostructures in monocrystalline silicon. (a) FEA results illustrate the assembly process that begins with transferring the 2D precursor structures (fabricated by planar processing technologies) onto a prestretched substrate at selective bonding sites. Release of the prestrain in the substrate induces compressive forces that transform the 2D precursor structure into 3D configurations. (b) A complex 3D mesostructure formed from a 2D precursor that consists of closed-loop circular filamentary serpentines and radially oriented ribbons selectively bonded to a biaxially stretched elastomer substrate. Source: Adapted from Ref. [4]/with permission of American Association for the Advancement of Science.

2D precursor designs, elastomer substrate designs, and loading conditions used in this method. The underlying mechanics of such assembly methods are summarized in Section 9.3, involving the finite deformation models of postbuckling analyses, the effect of interface delamination, as well as the stability analyses of morphable 3D structures. Section 9.4 summarizes the representative applications of 3D flexible electronics fabricated using the mechanics-guided assembly. Finally, an outlook on the existing challenges and future opportunities is provided.

9.2 Design Strategies of Mechanics-Guided Assembly

In the mechanics-guided assembly, the structural design plays a crucial role in expanding the accessible range of 3D geometric configurations. According to the nature of the assembly process, the structural designs mainly focus on three aspects, including 2D precursor designs, elastomer substrate designs, and design strategy of loading conditions.

9.2.1 2D Precursor Designs

9.2.1.1 Origami/Kirigami Design Strategy

The Origami design, an ancient art of China and Japan, mainly relies on the strategic design of folding creases and folding sequences to form 3D architectures. In the mechanics-guided 3D assembly, the creases can be realized by decreasing the bending stiffness of local regions in the 2D precursor, either through reduction of the width/thickness or through introduction of the material plasticity. When the 2D precursor buckles, the bending deformation would concentrate in those weak stiffness regions, thereby resulting in the formation of desired folding deformations. It is noteworthy that the crease region usually undergoes relatively high stress levels, and therefore, the structure design should ensure that no material failures occur in the crease region [22, 23].

Figure 9.2a shows two examples of Origami mesostructures achieved by reducing the thickness of 2D precursor at the crease regions. The 2D precursor contains eight ribbon structures connected with a circle at the center through creases. Assembly of such precursor with a carefully selected level of biaxial prestrain in the substrate yields a nearly closed cylindrical shell configuration. A windmill 3D configuration with well-defined chirality can also be achieved by this strategy [24].

Figure 9.2b illustrates another type of Origami design strategy that utilizes material plasticity. These two precursors share a common crease geometry and are fabricated with copper and polyethylene terephthalate (PET) materials, respectively. For the copper precursor, the bending deformation concentrates at the crease regions, leading to a controlled plastic yielding without failure. The plastic yielding softens the stiffness at local crease regions, thereby resulting in much larger folding deformations than the PET structure, where the crease regions are still in the elastic regime. Finite element analysis (FEA) results agree with the experiments in the above-mentioned cases [25].

For 2D membrane precursors (typically with in-plane aspect ratios of <5 : 1), the 3D assembly process would lead to “kink-induced” stress concentrations that cause material failure. To resolve this issue, Zhang et al. [16] introduced the Kirigami design strategy to relieve the stress concentration and reduce the peak stresses in the structure, thereby enabling the formation of membrane structures with a broad class of 3D geometries. In particular, this work demonstrated the fabrication of a series of 3D silicon membrane mesostructures and hybrid nanomembrane–nanoribbon systems, including heterogeneous combinations with polymers and metals, with critical dimensions that range from 100 nm to 30 mm.

9.2.1.2 Multilayer and Multilevel Design Strategy

The multilayer design strategy mainly utilizes multilayer 2D precursors to allow the formation of 3D structures with denser and more complex topologies, as compared to single-layer 2D precursors.

Figure 9.2c illustrates two examples of multilayer precursor designs, both consisting of several parallel straight ribbons that are bonded with each other at selective sites. Specifically, the first design has three precursor layers, in which the blue,

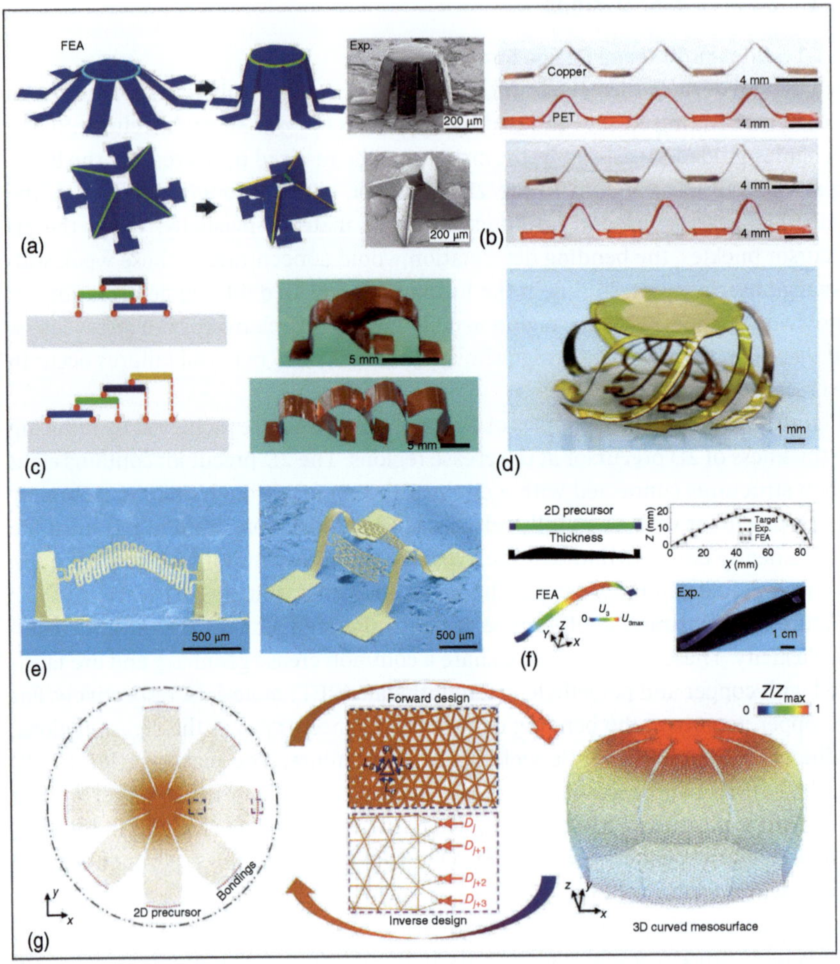

Figure 9.2 Design strategies for 2D precursors. (a) Origami design strategy is based on local thickness reduction, which enables the formation of a cylindrical shell (upper) and a windmill-like mesostructure (lower). (b) Origami design strategy is based on material plasticity. The thickness for copper and PET is 30 and 42 μm, respectively. The loading strain is 18% for the upper panel and 32% for the lower panel. (c) Illustration of the multilayer design strategy, where the various layers are selectively bonded with each other, and red dots in the left panel indicate the bonding area. (d) Optical image of a multilayer cage-shaped mesostructure. (e) Multilevel mesostructure with an ultralow-stiffness second-level substructure. (f) Inverse design of a 2D ribbon precursor with an optimized thickness distribution. (g) Microlattice design to program the curvature of 3D mesosurfaces. Source: (a) Adapted from Ref. [24] © 2016/John Wiley & Sons. (b) Adapted from Ref. [25]. © 2017/Elsevier. (c) Reproduced from Ref. [17]/with permission of American Association for the Advancement of Science. (d) Adapted from Ref. [26]. © 2021/IOP Publishing. (e) Adapted from Ref. [27]. © 2019/Springer Nature. (f) Adapted from Ref. [28]/with permission of John Wiley & Sons. (g) Reproduced from Ref. [29]/with permission of American Association for the Advancement of Science.

green, and dark blue layers have two, one, and zero ends bonded to the substrate, respectively. In the second design, all ribbons bond to the substrate at least at one site, thereby leading to 3D configurations with limited out-of-plane displacements. In this multilayer design, the bonding position between different layers provides more design flexibilities, which further expands the geometric configurations accessible to the mechanics-guided assembly [17]. A similar multilayer design strategy can also be used to form a 3D spiral structure [6, 30]. Inspired by *Enhydra lutris fur*, Cheng et al. [26] designed a multilayer cage-shaped sensor capable of multistage load bearing and collapse prevention under out-of-plane compression. As shown in Figure 9.2d, the bilayer structure consists of two separated layers, including a smaller bottom layer and a larger top layer. When subjected to external compression or impact, the top layer is used to realize the first level of load bearing. Under further compression, the top layer contacts the bottom one, and both come into play to achieve the second level of load bearing, thereby protecting the tender epithelial layer. Additionally, the bottom layer can retain a thin layer of air on the surface of the substrate that prevents the top layer from collapsing and adhering to the substrate.

Figure 9.2c illustrates two examples of multilevel design strategies. Here, the straight ribbon serves as the first-level structure, and the serpentine structure (or fractal serpentine structure) serves as the second-level structure. By controllably releasing the prestrain of the elastomer substrate, the first-level structure is first buckled, which then lifts off the second-level structure from the substrate. Afterward, the 3D assembly of the second-level structure can be realized by introducing a sacrificial layer design and releasing the prestrain of the elastomer substrate in the perpendicular direction [27]. In the mechanics-guided assembly, the bending stiffness of the precursor has to be sufficiently large to enable the assembly process. However, 3D piezoelectric microstructures used for energy harvesting require ultralow bending stiffnesses to offer a relatively low resonant frequency (e.g. below 100 Hz). In this context, the ultralow-stiffness 3D mesostructures realized through the multilevel design strategy could create interesting opportunities in energy harvesters and biomedical devices [31].

9.2.1.3 Design Strategy Based on Spatial Stiffness Control

The design strategy based on spatial stiffness control exploits spatial distributions of bending stiffness in the 2D precursor to achieve the desired curvature distribution, thereby allowing the programming of the assembled 3D shape. The stiffness control can be realized by spatially varying the thickness [28], width [32], or porosity [29] of the 2D precursor.

Figure 9.2f shows a spatial thickness design in the ribbon to yield the desired curved profile through the mechanics-guided assembly. Here, the function of thickness distribution serves as the design parameter. A finite deformation model gives the analytic solutions of the spatial variations of thickness and the prestrain in the substrate. By incorporating the ribbon width as an additional design parameter, this design method can be extended to reproduce centrally symmetric 3D surfaces in a discretized manner [28].

Figure 9.2g illustrates the microlattice design concept inspired by biological 3D porous structures, where the spatial porosity distribution can be realized by adjusting the ribbon widths at different locations. This strategy bypasses the challenges in fine control of gradient thickness distributions and allows the assembly of complex functional mesostructures with significantly reduced lateral sizes. With the assistance of an analytical model and the machine learning method, the porosity distribution of the 2D precursor can be optimized for a broad range of target 3D surfaces [29].

9.2.2 Elastomer Substrate Designs

For the mechanics-guided assembly of 3D structures on a homogeneous substrate with uniform thickness, the uniaxial/biaxial stretching only induces a uniform strain distribution in the substrate. By contrast, for a structurally designed substrate (e.g. nonuniform, inhomogeneous, or curvilinear), more complex strain distributions can be realized, involving either strain gradients or shear strain components, which provides more design space than the case of conventional homogeneous substrates.

9.2.2.1 Engineered Planar Substrate Design Strategy

Figure 9.3a demonstrates the design of a heterogeneous substrate with spatial variation of the material modulus. The heterogeneous substrate consists of three kinds of elastomeric material segments with different moduli, i.e. the soft, medium, and hard material segments denoted by green, blue, and light gray, respectively. Tensile forces applied to the two ends of the substrate induce spatially varying strain patterns correlated to the strategic variation in the modulus of the substrate material. The butterfly-like precursors are bonded to those material segments. Once the prestrained substrate is released, the spatially varying strain in the substrate induces a varying degree of compressive buckling at different locations, accordingly forming an array of spatially nonuniform 3D "butterflies" with wings of different tilt angles [33].

Figure 9.3b illustrates the design strategy of a heterogeneous substrate with spatial variation in thickness. Here, the spatial thickness variation of the substrate was realized through molding and curing. In the assembly process, the compressive force transferred to the precursor varies spatially in a corresponding manner, thereby leading to the formation of nonuniform 3D structures. The thick region on the substrate delivers small compressive forces, while the thin region delivers large compressive forces [34].

Figure 9.3c illustrates the Kirigami substrate design strategy, where a series of cuts through the thickness are fabricated on the elastomer substrate. Biaxially stretching the elastomer to a strain of 100% deforms the elastomer in a manner that causes adjacent units to rotate in opposite directions. During the two stages of deformation, the shapes of the cuts change from rectangular to ellipsoidal to roughly circular. Then, the 2D precursor was selectively bonded onto the prestrained elastomer. During the first stage of relaxation of the substrate prestrain, the 2D precursor first buckled

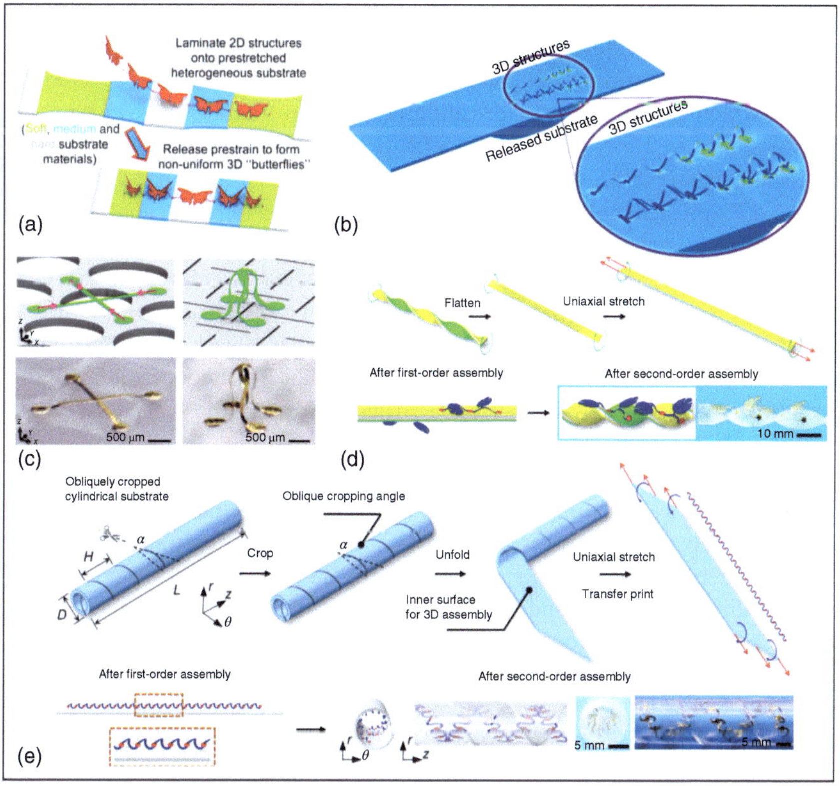

Figure 9.3 Design strategies for elastomer substrates. (a) Substrate design with engineered variations in the material modulus. Owing to the nonuniform strain field in the substrate, the 2D precursors are transformed into a series of 3D butterfly-like structures with different shapes. (b) Substrate design with engineered variations in the thickness. The magnified view highlights spatial variations in loadings delivered to the 2D precursor. (c) Kirigami substrate design that allows coupled buckling and twisting deformations of the 2D precursor. (d) Ordered 3D assembly process on the substrate with curved surfaces. The helicoidal substrate is flattened by torsional and tensile loadings, followed by the integration of leaf-like precursor structures and gradual release of the pre-applied loadings. (e) Conceptual illustration of the assembly of 3D structures on the inner surface of a cylindrical tube, where the substrate is cropped obliquely, flattened by bending deformations, and then prestretched, before the integration with the 2D precursor. Bottom panel presents the ordered assembly process of a hierarchical 3D helical structures inside cylindrical tubes with FEA predictions and an optical image. Source: (a) Adapted from Ref. [33]. (b) Adapted from Ref. [34]. (c) Reproduced from Ref. [35]/with permission of National Academy of Sciences. (d) Adapted from Ref. [20]. © 2022/American Association for the Advancement of Science – AAAS. (e) Adapted from Ref. [20]. © 2022/American Association for the Advancement of Science – AAAS.

into a 3D shape. During the second-stage relaxation, the bonding sites experienced rotational motions, which induced controlled twisting deformations of the precursor structure, thereby enabling the formation of a chiral 3D configuration, which is hard to achieve using a uniform elastomer [35].

9.2.2.2 Curvilinear Substrate Design Strategy

Electronic systems capable of conformal integration with complex curved 3D surfaces are of growing interest. Figure 9.3d illustrates the process of the ordered 3D assembly, which can enable the fabrication of 3D mesostructures on a curvilinear elastomer substrate. Here, a thin elastomer with a helicoidal shape substrate served as the assembly substrate, which was fabricated by casting and curing liquid elastomers in a 3D-printed mold. Torsional and stretch loadings were first applied to flatten the substrate. Then, an additional level of prestretch was applied, which provided elastic energy in the substrate that could be used for the first-assembly process. Then, during the release of the prestretch, the leaf-like precursors were assembled into 3D configurations through the compressive buckling. Further release of the pre-applied torsional loadings recovers the substrate to the helicoidal 3D shape, resulting in the spatial rearrangement of two leaf-like 3D structures during the second assembly. Different from the assembly on the flat substrate, these two leaf-like structures become very close to each other in the final state of second assembly, although they are initially located at two different sides of the flattened substrate [20].

Figure 9.3e illustrates the assembly process of 3D structures on the inner surface of a cylinder, which is quite challenging because it is usually difficult to transfer 2D precursors directly onto a cylindrical inner surface with small radii. Xue et al. [20] proposed a strategy by introducing oblique cropping to the cylindrical substrate, such that the substrate can be flattened by bending deformations. Then, an additional level of uniaxial prestrain was applied to the flattened substrate to drive the first assembly of the 2D precursor structure. The second assembly allowed the formation of the 3D structure on the inner surface of the cylindrical substrate (Figure 9.3e, lower right panel). The final 3D configuration features a hierarchical 3D helical topology, in which the small helices (pitch ~4 mm) form through the compressive buckling of serpentine structures (during first assembly), and the large helices (pitch ~20 mm) arise from the oblique rolling of the cropped substrate (during second assembly) [20].

9.2.3 Strategy of Loading Conditions

Excluding mechanical loadings introduced, other forms of physical loadings, such as electric and magnetic fields, can also be introduced to enable a multi-field loading in the mechanics-guided assembly, such that the ability to control/adjust the assembly process can be strengthened.

9.2.3.1 Mechanical Loading Strategy

During the mechanics-guided assembly process, the release of the uniaxial or biaxial prestrain in the substrate delivers compressive loadings to the 2D precursors. By selecting the time sequences or path of compressive loadings, specially engineered 2D precursors can be assembled into different stable 3D configurations. Figure 9.4a illustrates a loading path strategy that utilizes the release sequences of the biaxial prestrain to control the buckling configuration of 2D precursor. A cross-shaped precursor, with strategic distributions of creases, was fabricated and bonded to

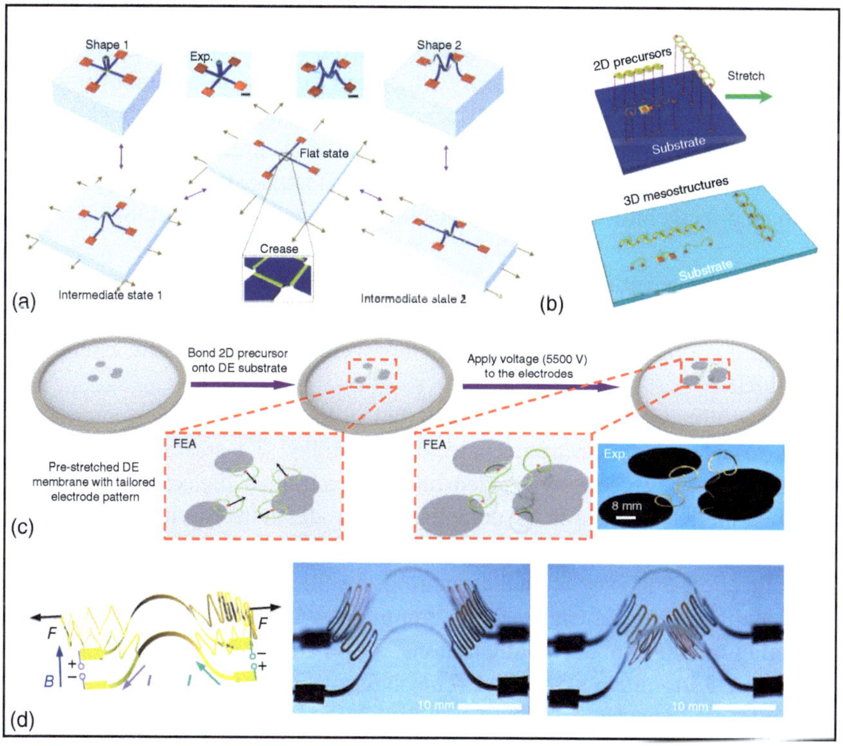

Figure 9.4 Loading strategies for the mechanics-guided assembly. (a) Loading path-controlled strategy to form 3D mesostructures with bistable or multistable configurations. (b) Tensile buckling process to assemble 3D mesostructures. The red regions of the 2D precursors indicate bonding regions with the substrate. (c) Electromechanically coupled loadings based on dielectric elastomer substrates. By applying a high voltage to the electrodes (dark gray), the 2D precursor (green) is transformed into the desired 3D configuration. (d) 3D ribbon mesostructures actuated by Lorentz forces and magnetic forces. Source: (a) Reproduced from Ref. [18]/with permission of Springer. (b) Ref. [36]/Springer Nature/CC BY 4.0. (c) Adapted from Ref. [37] © 2019/Oxford University Press. (d) Adapted from Ref. [31] © 2021/National Academy of Sciences.

a biaxially prestretched substrate. Releasing the prestrain equally along different directions resulted in the formation of Shape 1.

In contrast, by releasing the biaxial prestrain sequentially, the precursor was deformed into Shape 2. As such, the 3D configuration can be switched between the two shapes by stretching the substrate and releasing it with a different loading path. During the aforementioned process, the precursor can be optimized to maintain an elastic deformation region. Then, the entire process is reversible, and the two configurations can be reshaped into each other repetitively [18]. Based on this loading path controlled strategy, a class of very simple ribbon elements (with only a single branch and two ends) that can be reshaped between distinct 3D configurations was identified, which could serve as building-block structures to design complex reconfigurable mesostructures through a bottom-up strategy [38].

Excluding the compressive buckling, the tensile buckling can also be harnessed to achieve the mechanics-guided assembly. Based on the tensile buckling strategy, the prestrain of the substrate is not necessary, which can simplify the assembly process. Figure 9.4b illustrates the process of 3D assembly based on the tensile buckling. It begins with transfer printing and selective bonding of 2D precursors onto an elastomeric substrate. Following the stretching of the substrate, the precursor can be transformed directly into 3D configurations through tensile buckling [36].

9.2.3.2 Electric/Magnetic-Field-Assisted Loading Strategy

During the mechanics-guided assembly process, reconfigurable 3D structures can be achieved if additional electric/magnetic forces can be applied to the assembled 3D structures.

Figure 9.4c illustrates a loading strategy that utilizes dielectric elastomer (DE) actuators to achieve an electromechanically controlled 3D assembly. Here, the 2D precursors were selectively bonded onto a prestrained dielectric elastomers membrane. Applied voltages on the compliant electrodes at the top/bottom surfaces induced Maxwell forces that expanded the dielectric membrane biaxially along the in-plane direction. Such actuation deformations resulted in the inward movements of four bonding sites and triggered the compressive buckling of 2D precursors, transforming them into 3D configurations. Compared with the conventional mechanical loading process, this electromechanical loading method can achieve on-demand local assembly, with a faster loading speed [37].

A strategy that utilizes electroadhesion-mediated interfaces could realize reconfigurable 3D mesostructures [39]. A simple straight-ribbon precursor is taken as an example to illustrate the strategy. The 2D precursor consisting of a patterned top conductive metal layer (e.g. copper) and a bottom supporting layer (e.g. polymer) is transferred and selectively bonded onto a prestrained substrate. Compliant electrodes (e.g. carbon grease) are placed onto the bottom surface of the substrate. Applying a high voltage between the conductive metal layer of 2D precursor and compliant electrodes results in Maxwell forces between the interfaces, thereby forming a strong interface governed by the electroadhesion. After the release of the substrate prestrain, the ribbon structure is buckled into a two-wave mode, which is different from the single-wave mode in the case of no electroadhesion .

Figure 9.4d presents the utility of Lorentz forces to control the buckling shapes of 3D mesostructures. In this example, the 2D precursor consists of two straight ribbons connected with a pair of serpentine copper wires. When the biaxial prestrain is released, a controlled compressive buckling occurs, transforming the 2D precursor into a double-layer, multistable 3D configuration made up of two supporting ribbons and two reconfigurable serpentine ribbons. The assembled 3D mesostructure is then placed in a uniform magnetic field. Connecting the exposed metal layers at the bonding sites to external current sources generates Lorentz forces that can enable the 3D reconfiguration among various stable and/or temporary states. The amplitude and the direction of the current in each serpentine ribbon can be adjusted remotely and rapidly, therefore enabling remote and accurate control of the Lorentz force applied to each serpentine ribbon [31].

9.3 Mechanics Modeling and Analyses of the 3D Assembly

The mechanics-guided 3D assembly methods typically involve the 2D-to-3D geometric transformation of a patterned high-modulus thin film (e.g. 2D precursor) on a low-modulus elastomer substrate much thicker than the film, driven by compressive buckling. The mechanics-guided assembly process, which involves the interface delamination of such thin film/substrate system, usually depends on five main factors: (i) geometric layout of the 2D precursor and bonding sites, (ii) material parameters of the thin film and substrate, (iii) interfacial adhesion, (iv) loading form of the substrate, and (v) loading magnitude. Hence, a mechanics model that can capture the key assembly mechanisms and accurately predict the assembled 3D configurations is essential to the design. In recent years, several representative methods were exploited to develop theoretical models of the mechanics-guided assembly, such as the elastica theory [40, 41], the energy method [42–47], and the perturbation method [48, 49]. The FEA simulations and experimental observations were also widely used to unveil the complex deformation mechanism associated with the interface delamination and to validate the theoretical models.

In this section, we first summarize the theoretical models of representative 2D precursors without interfacial adhesion, including straight ribbons, helical structures, frame structures, and 2D curved ribbons. The content then shifts to the assembly with interfacial adhesion, noting that the interfacial adhesion between the thin film and elastomer substrate is vital to a successful assembly. Finally, the mechanics of loading-path controlled 3D assembly are introduced, which allows the construction of design diagrams to distinguish the monostability and bistability for a variety of geometric parameters.

9.3.1 Buckling Analysis of 2D Precursors

9.3.1.1 Straight Ribbons

The island-bridge design represents a general strategy widely used in stretchable electronics. The coplanar island-bridge design was initially proposed by Lacour et al. [50], where active devices were interconnected by stretchable wavy, metal wires, while the noncoplanar island-bridge design was first introduced by Kim et al. [51], aiming to achieve a substantial increase of the stretchability. Under stretching, the rigid islands (which have high effective stiffness) keep almost undeformed (e.g. with <1% strain) to ensure the mechanical integrity of functional materials, while the interconnects (which have low effective stiffness) deform to render the stretchability. In the island-bridge design, the straight interconnects can be either strongly bonded or weakly bonded to a pre-stretched elastomeric substrate, resulting in coplanar wavy interconnects [50] or noncoplanar arc-shaped interconnects [51], upon release of the prestrain ε_{pre}. Figure 9.5a shows a deformed stretchable array of complementary metal-oxide-semiconductor (CMOS) inverters, and the insets highlight three typical deformations in the structure, i.e. diagonal stretching, twisting, and bending [51].

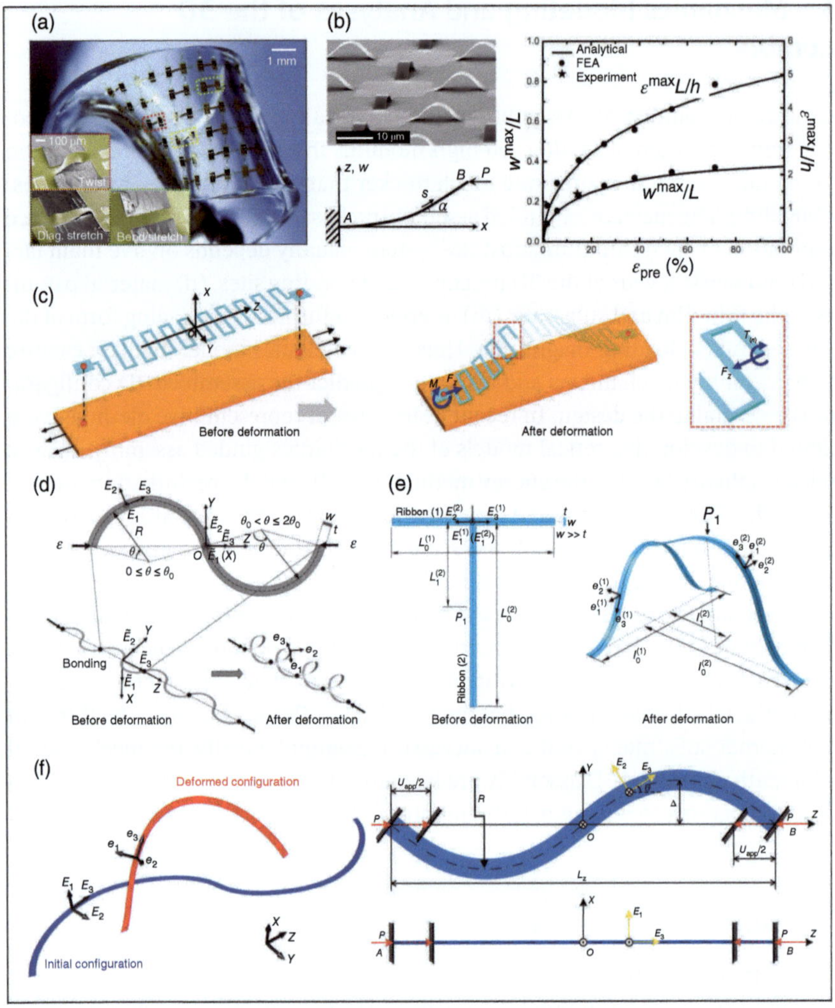

Figure 9.5 Buckling analysis of 2D precursors with various geometric configurations. (a) Optical image of a stretchable array of CMOS inverters under combined bending and twisting deformations, with the insets (colorized) showing three different types of deformation mode: diagonal stretching, twisting, and bending. (b) Scanning electron microscope (SEM) images of the silicon island-bridge structure on a PDMS substrate. Schematic illustration of the mechanics model for the buckled bridge by the release of prestrain in the substrate. Normalized maximum deflection and maximum strain of the bridge versus the prestrain of the substrate. (c) Illustration of the torsion-dominated buckling of serpentine structures to form 3D architectures with ultralow rigidity. (d) Illustration of the mechanics-guided assembly of 3D helical mesostructures and key dimensions associated with the model. (e) Illustration of the mechanics-guided assembly of frame mesostructures, along with the coordinate system and key geometric parameters. (f) Illustration of different coordinate systems associated with the buckling deformations of a planar, curved beam and the theoretical model for the end-to-end compression. Source: (a) Adapted from Ref. [51]. © 2008/National Academy of Sciences. (b) Adapted from Ref. [40] © 2013/Royal Society of Chemistry. (c) Reproduced from Ref. [52]/with permission of Elsevier. (d) Reproduced from Ref. [42]/with permission of John Wiley & Sons. (e) Reproduced from Ref. [43]/with permission of The Royal Society. (f) Reproduced from Ref. [49]/with permission of Elsevier.

Song et al. [46] developed an analytic model to investigate the buckling of island-bridge structures with straight interconnects, in which the sinusoidal function was adopted to characterize the arc-shaped profile. This model applies mainly to the case that the maximum deflection of the interconnect is relatively small as compared to the bridge length [44–47, 51–53]. Since the assumption of sinusoidal function is not accurate enough in the regime of large deflections of the interconnect, Li et al. [40] developed a finite deformation model that discards the assumption of sinusoidal profile for bridge deformation. Based on the elastica theory, this work presented the postbuckling analyses with consideration of the geometrically nonlinear effect. Specifically, as shown in Figure 9.5b, α is the angle which the tangent at a point of the curved bridge AB makes with the x-axis; s is the distance along the axis of the curved bridge from A; w is the deflection in the z-direction and w_B is the deflection at B; and P is the force applied at B by the adjacent part of AB. The governing equation of the bridge can be given by [40]

$$EI\frac{d\alpha}{ds} = P(w_B - w) \tag{9.1}$$

where EI is the flexural rigidity of the bridge, E is Young's modulus, and I is the moment of inertia. The maximum deflection and maximum strain of the bridge can be expressed as

$$w^{max} = \frac{\beta_\alpha L}{K(\beta_\alpha)} \text{ and } \varepsilon^{max} = 4\beta_\alpha K(\beta_\alpha)\frac{h}{L} \tag{9.2}$$

where $\beta_\alpha = \sin(\alpha/2)$, and $K(\beta_\alpha)$ is the complete elliptic integral of the first kind. The results of the analytical mechanics model agree well with those from FEA and the experiments (Figure 9.5b).

Serpentine structures are of growing interest due to their low stiffness and high stretchability. Zhao et al. [52] developed an analytical model for the compressive buckling of serpentine structures (Figure 9.5c), showing that the buckling of serpentine structures obeys similar rules as that of a straight ribbon, with the effective torsional rigidity ($\beta_t Gbh^3\lambda/H$) playing the role of the bending rigidity. The effective torsional rigidity involves not only the cross-sectional profile (ribbon width b and thickness h) but also the geometry parameters unique in the serpentine structures, which significantly enlarges the design space, where $\beta_t Gbh^3$ is the torsional rigidity, G is the shear modulus, λ is the length of short ribbons, and H is the length of long ribbons. β_t is a torsional parameter that depends on h/b, and β_t is approximately 1/3 for $h << b$. The displacements of mechanics-guided serpentine structures have the same forms as those of a straight ribbon with fixed-fixed boundaries.

9.3.1.2 Helical Structures

3D helical structures are attractive for applications in a broad range of microsystem technologies due to their outstanding mechanical and electromagnetic properties. Liu et al. [42] developed an analytic model of compressive buckling in 2D serpentine-shaped ribbons (Figure 9.5d). An approximate scaling between the general displacements (displacements and twist angle) and the applied compressive strain was derived [41], i.e. $\phi \propto \sqrt{\varepsilon}$, $U_1 \propto \sqrt{\varepsilon}$, $U_2 \propto \varepsilon$, and $U_3 \propto \varepsilon$, where ε is the compressive strain $\varepsilon = \frac{\varepsilon_{pre}}{1+\varepsilon_{pre}}$. The approximate relations between the general displacements and the typical geometry, material and loading parameters can be observed in FEA, indicating that the functions of displacement components and twist angle can be assumed, with several unknown coefficients [41]. According to the energy minimization methods, these unknown coefficients can be then solved. The configurations of 3D helical structures are affected by three groups of fabrication-related parameters, including the loading parameter (applied compressive strain ε), the material parameters (Young's modulus E and Poisson's ratio ν), and the geometrics parameters. In the design stage, these fabrication-related parameters should be optimized not only to yield desired 3D configurations but also to ensure that brittle fracture or plastic yield can be avoided in the assembly process.

9.3.1.3 Frame Structures

In various reported 3D structures and functional devices based on mechanics-guided assembly, there are two types of 3D structures that involve ribbon-type components, including straight and curved ribbons without any branches/joints [40, 41, 45–47] and frame ribbon structures with multiple branches and/or joints [19, 43, 54]. The coupling of bending/twisting deformations among different ribbon components of the 3D frame structures makes the analyses more complicated. Liu et al. [43] reported an analytic model of compressive buckling in frame structures consisting of straight ribbons based on the energy methods (Figure 9.5e). Similar to the helical structures, an approximate scaling between the general displacements and the applied compressive strain was adopted [42]. In the developed analytic model, three typical frame geometries were considered, including "+," "T," and "H" shaped designs. Such an energy-based model can provide predictions of the maximum out-of-plane displacements, mode ratios, and maximum strains of the frame structures, which can be used as design references in practical applications.

9.3.1.4 2D Curved Ribbons

To systematically understand the assembly process of general 3D ribbon-shaped structures, Fan et al. [49] established a double perturbation method to shed light on the relationship between the 3D shape and the associated parameters for assembly. By introducing two perturbation parameters related to the initial configuration and the deformation, the highly nonlinear governing equations can be transformed into a series of solvable, linear equations. One of the perturbation parameters a_0 is set as the ratio of maximum out-of-plane displacement U_1 to the total arc-length L_S, and the other parameter, $b_0 = \frac{\Delta}{L_S}$, is associated with the flexing level of initial planar curved ribbons, where Δ is the maximum rise of the axis, as shown in Figure 9.5f. In the special condition of an arc-shaped ribbon with a constant curvature radius, $b = \frac{L_S}{R}$ is more preferably used, due to the simplified analyses, where R is the curvature radius. Based on the double perturbation method [49], the analytic solutions of displacement/curvature components during the postbuckling of a general planar ribbon can be obtained, when neglecting high-order terms. Three representative ribbon shapes (sinusoidal, polynomial, and arc configurations) were used to show the validity of the theoretical method, through comparisons to the results of experiment and FEA.

9.3.2 Interfacial Adhesion in the Mechanics-Guided Assembly

The interfacial adhesion of thin film/substrate systems plays a vital role in the mechanics-guided assembly of 3D structures. For example, global buckling is dominant only in the condition of very weak adhesion, while no buckling or local buckling becomes the prevailing mode for relatively strong adhesions [39, 45, 47, 54–56]. Besides, a spatial interface control of the film/substrate system is essential for a successful assembly.

9.3.2.1 Design Diagrams of Delamination States

As shown in Figure 9.6a, Ko et al. [47] reported the experimental and theoretical studies on the mechanics of island-bridge designs. The 1D array of silicon islands and polyimide (PI) interconnects were transferred to a thin piece of polydimethylsiloxane (PDMS) with a uniaxial prestrain, and then the configurations were monitored during the release of the prestrain (Figure 9.6a, left panel). Without any interfacial adhesion, the interconnect would buckle once the compressive strain exceeds the Euler buckling strain ε_c. With adhesion, the buckling behavior becomes more complicated, since the delamination of the interface between the PDMS substrate and the PI interconnect should be taken into account. For weak adhesion $\gamma \le 8Eh\varepsilon_c^2$, the compressive strain separating global buckling $\left(\varepsilon \ge \varepsilon_c + \sqrt{\frac{2\gamma}{Eh}}\right)$ and no buckling $\left(\varepsilon < \varepsilon_c + \sqrt{\frac{2\gamma}{Eh}}\right)$ was obtained [47], where γ is the work of adhesion (energy per unit area of interface). For relatively strong adhesion $\gamma > 8Eh\varepsilon_c^2$, the buckling modes are controlled by $\varepsilon < 5\left(\frac{\gamma\varepsilon_c}{8Eh}\right)^{2/5}$ (for no buckling), $5\left(\frac{\gamma\varepsilon_c}{8Eh}\right)^{2/5} \le \varepsilon \le \varepsilon_c + \frac{\gamma}{2Eh\varepsilon_c}$ (for local buckling) and $\varepsilon > \varepsilon_c + \frac{\gamma}{2Eh\varepsilon_c}$ (for global buckling). These relations distinguish the various buckling modes in the phase diagram of the normalized strain versus work of adhesion (Figure 9.6a, right panel).

The low bending stiffnesses of the serpentine structures create difficulties in overcoming the interfacial adhesion energy to allow delamination from the underlying elastomeric substrate. Li et al. [56] presented the phase diagram of buckled states of serpentine structures (no buckling, local buckling, and global buckling), as shown in Figure 9.6b. The phase diagram can be explicitly expressed by only two dimensionless combinations of parameters, i.e. the normalized compression and the normalized adhesion.

Nonlinear buckling and delamination behaviors in material systems that combine both weak and strong interfaces are foundational to the assembly process, but they can be difficult to control, especially for complex geometries. Liu et al. [54] presented theoretical and experimental studies of the fundamental aspects of adhesion and delamination. By quantifying the effects of various essential parameters on the 3D assembly process, general design diagrams were established, involving four dominant delamination states (wrinkling, partial delamination of the weak interface, full

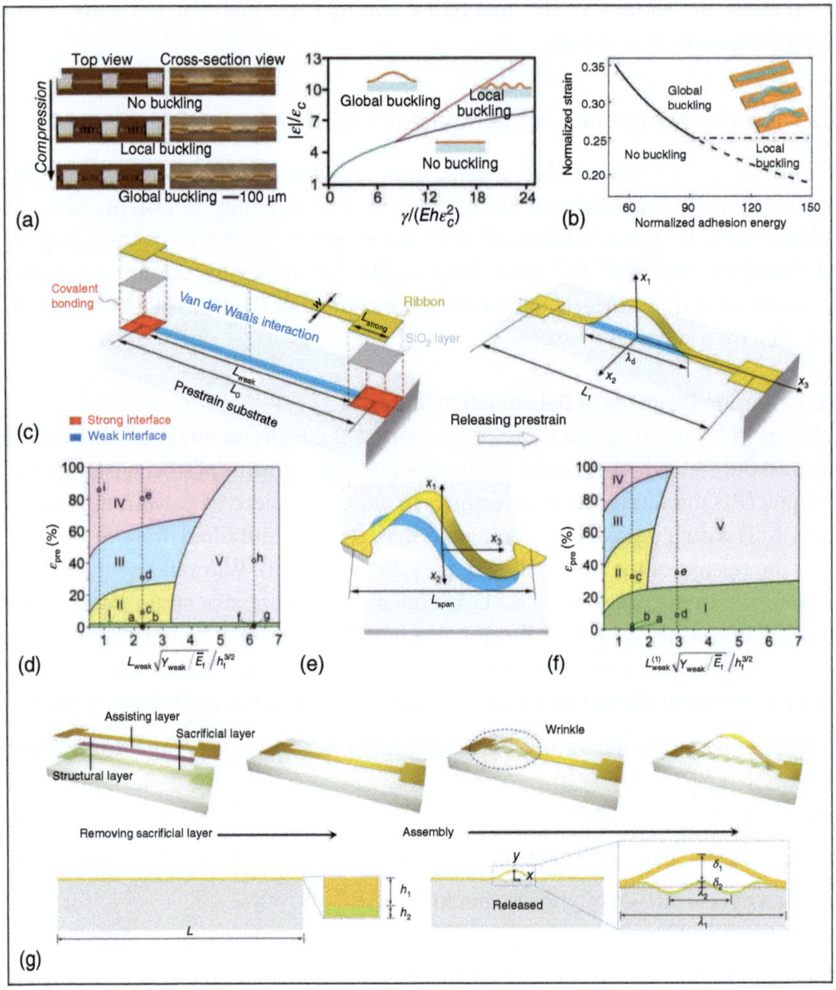

Figure 9.6 Interfacial adhesion in the mechanics-guided assembly. (a) Top and cross-sectional views of a linear array of interconnected silicon islands on a PDMS substrate subjected to different levels of compressive strains. The right panel shows a phase diagram of the normalized strain versus the normalized adhesion energy for different buckling states. (b) Phase diagram for different buckling states of serpentine structures. (c) Delamination processes associated with mechanics-guided assembly of 3D straight ribbons that include both strong and weak interfaces. (d) Design diagrams for straight ribbons with SU-8/D-Skin material systems. (e) Illustration of a helical structure formed through compressive buckling of a serpentine ribbon. (f) Design diagram for the helical ribbon formed with 120° arc angle for the 2D precursor. (g) Schematic illustration of the wrinkling-assisted strategy for controlled delamination in the 3D assembly, and the mechanics model for studying the controlled interface delamination in a bilayer film/substrate system. Source: (a) Reproduced from Ref. [47]/with permission of John Wiley & Sons. (b) Reproduced from Ref. [56]/with permission of Elsevier. (c–f) Reproduced from Ref. [54]/with permission of National Academy of Sciences. (g) Reproduced from Ref. [55]/with permission of Elsevier.

delamination of the weak interface, and partial delamination of the strong interface), as shown in Figure 9.6c. In this work, a dimensionless length of the weak interface was introduced, $\overline{L}_{\text{weak}} = L_{\text{weak}}\sqrt{\gamma/\overline{E}}/h^{3/2}$, with $\overline{E}$ being the plane-strain moduli of the straight ribbon, which indicates that the effects of four parameters on the delamination process can be reflected in the form of a single combined parameter. Design diagrams for straight ribbon structures in SU-8/D-Skin material systems are shown in Figure 9.6d. Figure 9.6e and f further presents the design diagram for the mechanics-guided 3D assembly of helical structures formed using serpentine precursor with 120° arc angles. These diagrams provide guidelines for the selection of engineering parameters that avoid interface-related failure.

9.3.2.2 Controlled Interface Delamination

To realize controlled delamination in the mechanics-guided 3D assembly, Shuai et al. [55] proposed a wrinkling-assisted strategy, in which an additional assisting layer is introduced such that a weaker film/assisting-layer interface replaces the original film/substrate interface, and the wrinkles formed in the assisting layer induce additional driving forces to separate the film/assisting-layer interface. Figure 9.6g illustrates the deformed configuration of such bilayer film/substrate system. The interface crack initiates in the bilayer film/substrate system, when the compressive displacement surpasses a threshold. Then the delaminated region of the film undergoes a buckled deformation, while the assisting layer fully bonded with the substrate experiences wrinkled deformations. In the delaminated region, the wrinkling deformation of the assisting layer can be characterized by $y_2 = \delta_2 \cos\left(\frac{2\pi x}{\lambda_2}\right)$, while the configuration of the structural layer can be approximately assumed as $y_1 = \frac{\delta_1}{2}\left[1 + \cos\left(\frac{2\pi x}{\lambda_1}\right)\right]$, where $\lambda_{1,2}$ and $\delta_{1,2}$ are the wavelength and the amplitude, respectively. Taking into account the geometric relation and the inextensibility of the buckled film, the two amplitudes ($\delta_{1,2}$) and the wavelength (λ_2) can be all expressed as functions of the compressive displacement ΔL and λ_1. The energy minimization can be then used to determine the dimensionless wavelength $\tilde{\lambda}_1 = \lambda_1/\Delta L$ as

$$\tilde{\lambda}_1 = \sqrt[3]{\frac{4\pi^2}{\tilde{\Delta L}^2} + \sqrt{\frac{16\pi^4}{\tilde{\Delta L}^4} - \frac{K_a^3}{27}}} + \sqrt[3]{\frac{4\pi^2}{\tilde{\Delta L}^2} - \sqrt{\frac{16\pi^4}{\tilde{\Delta L}^4} - \frac{K_a^3}{27}}} \tag{9.3}$$

where $\tilde{\Delta L} = \Delta L/l_c$ is the dimensionless compressive displacement, $l_c = \sqrt{B_1/\gamma}$ is a characteristic length that balances bending rigidity (B_1) and adhesion of the structural layer [57]. A combined material/geometry parameter, $K_a = \overline{E}_2 h_2 \varepsilon_c/\gamma$, was then defined, where $\overline{E}_2$ and h_2 denote the plane-strain moduli and thickness of the assisting layer, respectively. The model shows that the combined parameter ($K_a = \overline{E}_2 h_2 \varepsilon_c/\gamma$) governs the process of delamination in the bilayer film/substrate system, offering a useful design reference for the mechanics-guided 3D assembly. Furthermore, the wrinkling-assisted strategy can enable the assembly of highly flexible 3D structures and those with large areas attached to the substrate in the initial configuration, which is particularly challenging to achieve by using previous strategies.

9.3.3 Loading-Path Controlled Assembly

Recent advances in mechanics-guided 3D assembly provided a powerful approach to achieving reconfigurable 3D structures, which relied on certain loading paths to trigger the stabilization of distinct 3D buckling modes in specially engineered precursor structures.

9.3.3.1 Reconfigurable Cross-Shaped Structures

Luo et al. [19] presented a finite-deformation model to analyze the stability of different buckling modes induced during the simultaneous and sequential compressions of precursor structures with cross-shaped ribbon geometries. Based on the perturbation method, a theoretical model was developed to predict the postbuckling deformations in segmented cross-shaped ribbon structures with a prescribed number of creases. By introducing a dimensionless bending moment μ as the small parameter, the governing equation of a uniform beam in the segmented ribbon can be expressed as:

$$\begin{cases} \dfrac{d^2\overline{\theta}}{d\overline{S}^2} + \dfrac{1}{\mu}\sin(\mu\overline{\theta}) = 0 \\ \overline{\theta}\Big|_{\overline{S}=0} = \dfrac{\theta_1 - \theta_p}{\mu}, \quad \dfrac{d\overline{\theta}}{d\overline{S}}\Big|_{\overline{S}=0} = 1 \end{cases} \tag{9.4}$$

The analytical solution $\overline{\theta}(\overline{S})$ of Eq. (9.4) can be obtained based on the perturbation method and boundary conditions. Then, the deformed configuration of the cross-shaped structure can be obtained. Based on the analyses of the strain-energy landscape, a stability coefficient was proposed to evaluate the bistability of cross-shaped ribbon structures. As shown in Figure 9.7a,b, there are two possible buckling modes of the cross-shaped structure, i.e. pop-down and pop-up modes, with strain energy $E_{\text{pop-down}}$ and $E_{\text{pop-up}}$, respectively. The pop-down mode is relatively unstable in view of the strain energy. Based on the energy profile, there is a critical configuration (between pop-down and pop-up modes) that gives a local energy maximal. The stability of the pop-down mode then depends highly on the magnitude of the energy barrier ($E_{\text{barrier}} = E_{\text{critical}} - E_{\text{pop-down}}$) between the critical configuration and the pop-down mode. Then, a dimensionless energy barrier E_r normalized by the strain energy $E_{\text{pop-down}}$ can be introduced to quantify the stability of pop-down mode:

$$E_r = \frac{E_{\text{barrier}}}{E_{\text{pop-down}}} \tag{9.5}$$

A simple criterion to judge the reconfigurability of the cross-shaped patterns can be introduced as $E_r \geq E_{\text{threshold}}$, where $E_{\text{threshold}}$ is the threshold of dimensionless energy barriers to enable bistable configurations. This model was validated by FEA and experimental measurements for a broad range of cross-shaped structures.

9.3.3.2 Reconfigurable Structures Harnessing Interface Mechanics

The film/substrate interface can be actively controlled to alter the buckling mode of the 2D precursor. Figure 9.7c presents a bistable H-shaped 2D precursor whose

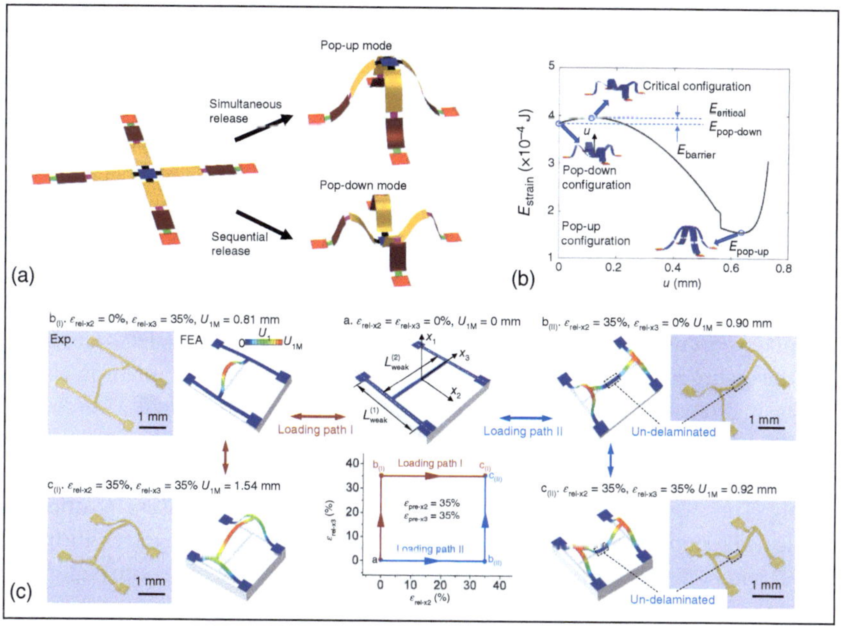

Figure 9.7 Mechanics of loading-path controlled assembly. (a) Illustration of the cross-shaped 2D precursor for the bistability analysis through the loading-path controlled assembly. (b) Energy analysis for a typical cross-shaped structure with three creases in the substructure. The insets show the configurations at three different out-of-plane displacements. (c) Illustration of the design strategy that harnesses the interface mechanics to achieve reconfigurability through loading-path controlled assembly for the H-shaped 2D precursor. Source: (a) and (b) Reproduced from Ref. [19]/with permission of Elsevier. (c) Reproduced from Ref. [54]/with permission of National Academy of Sciences.

reconfiguration can be realized through control of the loading path [54]. The H-shaped structure consists of three straight ribbon components (two of them with length $L^{(1)}_{\text{weak}}$ and the other with length $L^{(2)}_{\text{weak}}$), selectively bonded on a biaxially prestretched substrate. Appropriate choices of geometric parameters ensure full delamination of the ribbon along the x_3 direction after releasing the x_3-directional prestrain. Subsequent release of the prestrain along the x_2 direction then leads to the formation of a pop-up mesostructure (lower left panel; following loading path I). If the prestrain along the x_2 direction is released first, although the two ribbons along the x_2 direction fully delaminate, the central segment of the ribbon along the x_3 direction remains on the substrate (upper right panel), due to the insufficient driving force to separate the interface at this region. So, the structure is finally stabilized with a pop-down mode (lower right panel, following loading path II). The reconfigurability of the H-shaped precursor depends on the prestrain ε_{pre}, the dimensionless length $\overline{L}^{(1)}_{\text{weak}} = L^{(1)}_{\text{weak}}\sqrt{\gamma_{\text{weak}}/\overline{E}_{\text{f}}/h_{\text{f}}^{3/2}}$ and the length ratio $L^{(2)}_{\text{weak}}/L^{(1)}_{\text{weak}}$. A design diagram based on FEA could be constructed to guide the selection of these parameters to enable reconfigurability.

9.4 Applications of 3D Flexible Electronics

9.4.1 Flexible Sensors

Figure 9.8a shows a 3D capacitive force sensor fabricated by the mechanics-guided assembly method. This capacitive sensor consists of two lower electrodes on the PDMS substrate and a laminated 3D seesaw-like mesostructure in PI/Cu/PI serving as the upper electrode. The two lower electrodes and the upper electrode form a differential capacitor. The capacitance of this differential capacitor changes due to the variation of the distance between the lower and upper electrodes when the out-of-plane compressive force is applied on the edge of the upper electrode, from which the compressive force can be measured. The sensor can offer high-precision force sensing (resolution ~5.22 nN) and a broad adjustment range (by ~33 times) of sensitivity [58].

Figure 9.8b shows a 3D piezoresistive force sensor through the mechanics-guided assembly. The 2D precursor consists of several Si-NM piezoresistors positioned in vertical direction and horizontal directions. The fabrication involves forming p-doped (boron) Si-NM piezoresistive elements by impurity diffusion onto patterned PI films and assembling PI precursor into 3D configuration. The 3D configuration enables simultaneous measurements of the normal force and the shear force, according to the signal outputs from the piezoresistive elements [59].

Figure 9.8c shows a mechanically reliable 3D flexible Janus temperature sensor capable of decoupled temperature measurement of the skin temperature and target object temperature. The flexible device consists of the temperature sensor components sitting on the top surface and at the bottom surface of the elastomer substrate, respectively. By adopting an anti-fatigue design strategy, a sacrificial layer of PI is added on the top surface of copper interconnects in the 3D flexible sensor, allowing a significant increase of the fatigue life (>10^5 compressive cycles). When the device is mounted on the robot's skin, the underlying sensor measures the internal temperature while the top sensor can detect the temperature fluctuation of the surrounding environment, facilitating the mimicking of real human feelings (e.g. hot, warm, cool, or cold). Thereby, with the decoupled Janus temperature measurement capability, this flexible sensor can be potentially used to impart the pronunciation of real human feelings to soft intelligent robotics [60].

9.4.2 Tunable Electromagnetic Devices

Electromagnetic devices fabricated by mechanics-guided assembly have attracted growing attention, because the 3D topology can enhance the electromagnetic performances [3] and enable the tunability of the spatial orientation of electromagnetic field [38]. Besides, the electromagnetic properties can be easily tuned through the substrate loading [64].

Figure 9.8d shows a 3D hemispherical electrically small antenna. In this example, the width and spacing of the meander lines on the precursor are 150 and 500 μm, respectively. A well-designed precursor shape ensures a nearly closed 3D hemispher-

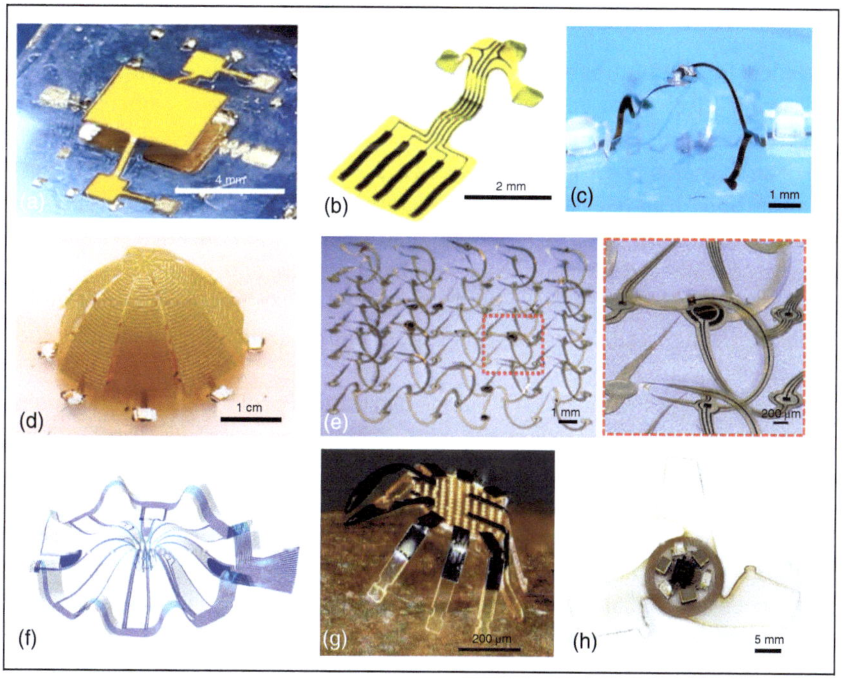

Figure 9.8 Mechanically assembled 3D flexible electronics and their applications. (a) 3D capacitive force sensor. (b) Multimodal 3D piezoresistive force sensor capable of normal force, shear force, bending, and temperature sensing. (c) 3D mechanically reliable Janus temperature sensor. (d) Hemispherical electrically small antenna with tunable frequency. (e) 3D electronic tissue scaffold with integrated electrodes and connection wires. (f) Illustration of a 3D neural interface for cortical spheroids with optical, thermal, and electrochemical capabilities. (g) Submillimeter-scale wireless terrestrial robots actuated by laser scanning. (h) 3D IoT flier is capable of measurement of fine dust pollution through the light dosimetry method. Source: (a) Ref. [58]/Springer Nature/CC BY 4.0. (b) Adapted from Ref. [59]. © 2019/American Chemical Society. (c) Adapted from Ref. [60]. © 2021/John Wiley & Sons. (d) Adapted from Ref. [3]. © 2019/John Wiley & Sons. (e) Adapted from Ref. [61]. © 2020/Elsevier. (f) Adapted from Ref. [1]. © 2021/American Association for the Advancement of Science – AAAS. (g) Adapted from Ref. [62]. © 2022/American Association for the Advancement of Science – AAAS. (h) Adapted from Ref. [63]. © 2021/Springer Nature.

ical configuration, which is beneficial to its electromagnetic performance. According to experimental measurements, the quality factor Q of this electrically small antenna is very close to its fundamental limits Q_{lb} ($Q/Q_{lb} = 1.1$)[3].

Bai et al. [38] developed a reconfigurable 3D antenna whose radiation pattern can be switched among various different operational modes. The antenna is composed of four pairs of bistable whip antenna components that can be reshaped between two distinct configurations following the loading-path controlled strategy. Considering the angle dependence of the sequential release of the biaxial prestrain in the substrate, the antenna could be adjusted to operate in nine distinct modes, suggesting the capability to achieve beam scanning in a discrete mode using a single antenna.

9.4.3 Biomedical Devices

The excellent flexibility, tunable mechanical properties and sophisticated geometries of 3D flexible electronics enable their conformable integrations with biological systems (e.g. cells, tissues, and organs) in a mechanically imperceptible manner. Figure 9.8e presents a flexible 3D electronic scaffold with arrays of sensing, stimulation and drug-releasing electrodes integrated on a double-floor helical network. By embedding the 3D electronic scaffold into engineered 3D cardiac tissues, this biomedical device is capable of monitoring and controlling cardiac tissue contractions through extracellular potential recordings and stimulation, respectively. On-demand, local release of drugs can be initiated by activating the drug-releasing electrodes [61].

Biomimetic design strategies can be adopted in the design of 3D flexible functional devices to construct electronic systems with programmed curvature distributions, thereby enabling the conformant or replication of the nondevelopable curvy surfaces of biological tissues and organs. For example, based on the bioinspired microlattice design, Cheng et al. developed a spherical-cap-shaped electronic cell scaffold mimicking the rabbit's retina [29]. The 3D electronic cell scaffold is capable of culturing retinal pigment epithelial cells and sensing their real-time, spatial distributions of physiological activities such as growth, migration and apoptosis, which can potentially shed novel insights into the pathogenesis of age-related macular degeneration and other diseases.

Besides, understanding of 3D electrochemical physiological characteristics of organoids has attracted extensive attention. Figure 9.8f shows a 3D flexible electronic framework as compliant, multifunctional neural interfaces to spheroids and to assembloids. The device system has integrated electrical, optical, chemical, and thermal sensors/stimulators. Based on the 3D bio-interface, the spreading of coordinated bursting events across the surface of a cortical spheroid was measured, and the regrowth of bridging tissues across two spheroids was observed [1].

9.4.4 Flexible Robotics

The development of miniaturized 3D flexible electronic robots (e.g. climbing micro-robots and micro-fliers) with integrated functional components has important implications for a diverse range of engineering areas (e.g. minimally invasive surgical devices, targeted drug delivery, and environmental sensing). Figure 9.8g shows an untethered, terrestrial robot with peekytoe crab-like geometries. By adding a layer of one-way shape memory alloy at the crease regions, eight legs of the biomimetic microstructure can deform reversibly. A variety of locomotion modes can be realized through laser-induced local thermal actuation, such as linear/curvilinear crawling, walking, turning, and jumping. By combining soft active materials (e.g. liquid crystal elastomers) with programmed strain distributions and 3D buckling methods, a 3D multigait climbing micro-robot consisting of morphable footpads, smart joints, and a bridge-shaped deformable body was developed [65]. This micro-robot can climb on surfaces with diverse shapes (e.g. flat plane, cylinder, wavy surface, wedge-shaped groove, and sphere), and also transition between two distinct surfaces, thereby facilitating exploration and monitoring in a confined space when carrying an integrated micro-camera.

Inspired by wind-dispersed seeds, miniaturized 3D fliers fabricated through the mechanics-guided assembly were demonstrated (Figure 9.8h), which can carry active electronic and/or colorimetric payloads while maintaining stable rotational falling in the air. The bio-inspired 3D geometry of the electronic flier can create large momentum deficits in wakes induced by free fall, while undergoing rotational movements, thereby promoting high drag forces, low terminal velocities and high directional stability. By integrating the near field communication (NFC) coil, micro-controller and UV sensors, this 3D flier can serve as a monitoring platform for air pollution and ultralow-power technologies for IoT applications [63].

9.5 Concluding Remarks

This chapter summarizes the fundamental aspects of mechanics-guided 3D assembly, covering various design strategies and underlying mechanics of such assembly methods. Because of its intriguing advantages and excellent compatibility, the mechanics-guided 3D assembly method provides a novel avenue for the development of 3D flexible electronics with new functions and/or unprecedented performances. This chapter discusses representative applications in flexible sensors, tunable electromagnetic devices, flexible biomedical devices, and flexible robotics.

Despite the significant progress in the area of mechanics-guided 3D assembly, scientific challenges remain. For example, at the sub-micro/nanoscale, the prominent interface adhesion would hinder the separation of 2D precursors from the elastomer substrate, and therefore, a controlled interface delamination is very difficult to achieve. A successful assembly requires that only selective regions of the film remain bonded to the substrate throughout the assembly, while the other regions are required to be fully separated from the substrate at the end of the process. However, the realization of strong and weak interfaces (with distinct differences in interface energy/strength) at the nanoscale is still very challenging. Another challenge lies in the rapid design and optimization of 2D precursors to yield the desired target shape. In the mechanics-guided 3D assembly, the 2D precursors undergo highly nonlinear postbuckling deformations, which could involve complex interaction behavior with the substrate (e.g. contact, interface delamination, et al.). Full-scale FEA of the entire assembly process could be very computationally costly, and inverse design methods based on the FEA typically require more computational resources. Open opportunities also exist in the development of high-efficiency computational methods of nonlinear postbuckling deformations in the mechanics-guided assembly.

References

1 Park, Y., Franz, C.K., Ryu, H. et al. (2021). Three-dimensional, multifunctional neural interfaces for cortical spheroids and engineered assembloids. *Science Advances* 7 (12): eabf9153.

2 Yan, Z., Han, M., Shi, Y. et al. (2017). Three-dimensional mesostructures as high-temperature growth templates, electronic cellular scaffolds, and self-propelled microrobots. *Proceedings of the National Academy of Sciences* 114 (45): E9455–E9464.

3 Liu, F., Chen, Y., Song, H. et al. (2019). High performance, tunable electrically small antennas through mechanically guided 3D assembly. *Small* 15 (1): 1804055.

4 Xu, S., Yan, Z., Jang, K.-I. et al. (2015). Assembly of micro/nanomaterials into complex, three-dimensional architectures by compressive buckling. *Science* 347 (6218): 154–159.

5 Cho, J.-H., Keung, M.D., Verellen, N. et al. (2011). Nanoscale origami for 3D optics. *Small* 7 (14): 1943–1948.

6 Zhang, F., Liu, F., and Zhang, Y. (2019). Analyses of mechanically-assembled 3D spiral mesostructures with applications as tunable inductors. *Science China Technological Sciences* 62 (2): 243–251.

7 Mei, Y., Huang, G., Solovev, A.A. et al. (2008). Versatile approach for integrative and functionalized tubes by strain engineering of nanomembranes on polymers. *Advanced Materials* 20 (21): 4085–4090.

8 Huang, W., Koric, S., Yu, X. et al. (2014). Precision structural engineering of self-rolled-up 3D nanomembranes guided by transient quasi-static FEM modeling. *Nano Letters* 14 (11): 6293–6297.

9 Choi, C., Choi, M.K., Liu, S. et al. (2017). Human eye-inspired soft optoelectronic device using high-density MoS2-graphene curved image sensor array. *Nature Communications* 8: 1664.

10 Kim, D.-H., Lu, N., Ghaffari, R. et al. (2011). Materials for multifunctional balloon catheters with capabilities in cardiac electrophysiological mapping and ablation therapy. *Nature Materials* 10 (4): 316–323.

11 Song, Y.M., Xie, Y., Malyarchuk, V. et al. (2013). Digital cameras with designs inspired by the arthropod eye. *Nature* 497 (7447): 95–99.

12 Xue, Z., Song, H., Rogers, J.A. et al. (2019). Mechanically-guided structural designs in stretchable inorganic electronics. *Advanced Materials* 32: e1902254.

13 Liu, H.D., Zhang, H.J., Han, W.Q. et al. (2021). 3D Printed flexible strain sensors: from printing to devices and signals. *Advanced Materials* 33 (8): 19.

14 Su, R., Park, S.H., Ouyang, X. et al. (2022). 3D-printed flexible organic light-emitting diode displays. *Science Advances* 8 (1): eabl8798.

15 Kim, B.H., Lee, J., Won, S.M. et al. (2018). Three-dimensional silicon electronic systems fabricated by compressive buckling process. *ACS Nano* 12 (5): 4164–4171.

16 Zhang, Y., Yan, Z., Nan, K. et al. (2015). A mechanically driven form of Kirigami as a route to 3D mesostructures in micro/nanomembranes. *Proceedings of the National Academy of Sciences* 112 (38): 11757–11764.

17 Yan, Z., Zhang, F., Liu, F. et al. (2016). Mechanical assembly of complex, 3D mesostructures from releasable multilayers of advanced materials. *Science Advances* 2 (9): e1601014.

18 Fu, H., Nan, K., Bai, W. et al. (2018). Morphable 3D mesostructures and microelectronic devices by multistable buckling mechanics. *Nature Materials* 17 (3): 268–276.

19 Luo, G., Fu, H., Cheng, X. et al. (2019). Mechanics of bistable cross-shaped structures through loading-path controlled 3D assembly. *Journal of the Mechanics and Physics of Solids* 129: 261–277.

20 Xue, Z., Jin, T., Xu, S. et al. (2022). Assembly of complex 3D structures and electronics on curved surfaces. *Science Advances* 8 (32): eabm6922.

21 Cheng, X. and Zhang, Y. (2019). Micro/nanoscale 3D assembly by rolling, folding, curving, and buckling approaches. *Advanced Materials* 31 (36): e1901895.

22 Guo, X., Xu, Z., Zhang, F. et al. (2018). Reprogrammable 3D mesostructures through compressive buckling of thin films with prestrained shape memory polymer. *Acta Mechanica Solida Sinica* 31 (5): 589–598.

23 Lim, S., Luan, H., Zhao, S. et al. (2020). Assembly of foldable 3D microstructures using graphene hinges. *Advanced Materials* 32 (28): 2001303.

24 Yan, Z., Zhang, F., Wang, J. et al. (2016). Controlled mechanical buckling for origami-inspired construction of 3D microstructures in advanced materials. *Advanced Functional Materials* 26 (16): 2629–2639.

25 Shi, Y., Zhang, F., Nan, K. et al. (2017). Plasticity-induced origami for assembly of three dimensional metallic structures guided by compressive buckling. *Extreme Mechanics Letters* 11: 105–110.

26 Cheng, X., Liu, Z., Jin, T. et al. (2021). Bioinspired design and assembly of a multilayer cage-shaped sensor capable of multistage load bearing and collapse prevention. *Nanotechnology* 32 (15): 155506.

27 Han, M., Wang, H., Yang, Y. et al. (2019). Three-dimensional piezoelectric polymer microsystems for vibrational energy harvesting, robotic interfaces and biomedical implants. *Nature Electronics* 2 (1): 26–35.

28 Fan, Z., Yang, Y., Zhang, F. et al. (2020). Inverse design strategies for 3D surfaces formed by mechanically guided assembly. *Advanced Materials* 32 (14): 1908424.

29 Cheng, X., Fan, Z., Yao, S. et al. (2023). Programming 3D curved mesosurfaces using microlattice designs. *Science* 379 (6638): 1225–1232.

30 Zhao, H., Cheng, X., Wu, C. et al. (2022). Mechanically guided hierarchical assembly of 3D mesostructures. *Advanced Materials* 34 (12): 2109416.

31 Zhang, F., Li, S., Shen, Z. et al. (2021). Rapidly deployable and morphable 3D mesostructures with applications in multimodal biomedical devices. *Proceedings of the National Academy of Sciences of the United States of America* 118 (11): e2026414118.

32 Xu, Z., Fan, Z., Pang, W. et al. (2022). Inverse design strategies for buckling-guided assembly of 3D surfaces based on topology optimization. *Extreme Mechanics Letters* 51: 101582.

33 Luan, H., Cheng, X., Wang, A. et al. (2019). Design and fabrication of heterogeneous, deformable substrates for the mechanically guided 3D assembly. *ACS Applied Materials & Interfaces* 11 (3): 3482–3492.

34 Nan, K., Luan, H., Yan, Z. et al. (2017). Engineered elastomer substrates for guided assembly of complex 3D mesostructures by spatially nonuniform compressive buckling. *Advanced Functional Materials* 27 (1): 1604281.

35 Zhao, H., Li, K., Han, M. et al. (2019). Buckling and twisting of advanced materials into morphable 3D mesostructures. *Proceedings of the National Academy of Sciences of the United States of America* 116 (27): 13239–13248.

36 Guo, X., Wang, X., Ou, D. et al. (2018). Controlled mechanical assembly of complex 3D mesostructures and strain sensors by tensile buckling. *npj Flexible Electronics* 2 (1): 14.

37 Pang, W., Cheng, X., Zhao, H. et al. (2019). Electro-mechanically controlled assembly of reconfigurable 3D mesostructures and electronic devices based on dielectric elastomer platforms. *National Science Review* 7 (2): 342–354.

38 Bai, K., Cheng, X., Xue, Z. et al. (2020). Geometrically reconfigurable 3D mesostructures and electromagnetic devices through a rational bottom-up design strategy. *Science Advances* 6 (30): eabb7417.

39 Pang, W., Liu, L., Xu, S. et al. (2023). Electroadhesion-mediated interface delamination for assembly of reconfigurable 3D mesostructures. *Journal of Applied Mechanics* 90 (6): 061006.

40 Li, R., Li, M., Su, Y. et al. (2013). An analytical mechanics model for the island-bridge structure of stretchable electronics. *Soft Matter* 9 (35): 8476–8482.

41 Su, Y.W., Wu, J., Fan, Z.C. et al. (2012). Postbuckling analysis and its application to stretchable electronics. *Journal of the Mechanics and Physics of Solids* 60 (3): 487–508.

42 Liu, Y., Yan, Z., Lin, Q. et al. (2016). Guided formation of 3D helical mesostructures by mechanical buckling: analytical modeling and experimental validation. *Advanced Functional Materials* 26 (17): 2909–2918.

43 Liu, Y., Xu, Z., Hwang, K.C. et al. (2019). Postbuckling analyses of frame mesostructures consisting of straight ribbons for mechanically guided three-dimensional assembly. *Proceedings of the Royal Society A Mathematical Physical and Engineering Sciences* 475 (2225): 20190012.

44 Zhao, J. (2022). Postbuckling analysis of ultra-low rigidity serpentine structures. *Journal of Applied Mechanics* 89 (4): 041008.

45 Wang, S.D., Xiao, J.L., Song, J.Z. et al. (2010). Mechanics of curvilinear electronics. *Soft Matter* 6 (22): 5757–5763.

46 Song, J., Huang, Y., Xiao, J. et al. (2009). Mechanics of noncoplanar mesh design for stretchable electronic circuits. *Journal of Applied Physics* 105 (12): 123516–112351.

47 Ko, H.C., Shin, G., Wang, S. et al. (2009). Curvilinear electronics formed using silicon membrane circuits and elastomeric transfer elements. *Small* 5 (23): 2703–2709.

48 Fan, Z., Wu, J., Ma, Q. et al. (2017). Post-buckling analysis of curved beams. *Journal of Applied Mechanics* 84 (3): 031007.

49 Fan, Z.C., Hwang, K.C., Rogers, J.A. et al. (2018). A double perturbation method of postbuckling analysis in 2D curved beams for assembly of 3D ribbon-shaped structures. *Journal of the Mechanics and Physics of Solids* 111: 215–238.

50 Lacour, S.P., Jones, J., Wagner, S. et al. (2005). Stretchable interconnects for elastic electronic surfaces. *Proceedings of the IEEE* 93 (8): 1459–1467.

51 Kim, D.-H., Song, J., Choi, W.M. et al. (2008). Materials and noncoplanar mesh designs for integrated circuits with linear elastic responses to extreme mechanical deformations. *Proceedings of the National Academy of Sciences of the United States of America* 105 (48): 18675.

52 Zhao, J., Zhang, F., Guo, X. et al. (2021). Torsional deformation dominated buckling of serpentine structures to form three-dimensional architectures with ultra-low rigidity. *Journal of the Mechanics and Physics of Solids* 155: 104568.

53 Jiang, H., Sun, Y., Rogers, J.A. et al. (2007). Mechanics of precisely controlled thin film buckling on elastomeric substrate. *Applied Physics Letters* 90 (13): 133119.

54 Liu, Y., Wang, X., Xu, Y. et al. (2019). Harnessing the interface mechanics of hard films and soft substrates for 3D assembly by controlled buckling. *Proceedings of the National Academy of Sciences of the United States of America* 116 (31): 15368–15377.

55 Shuai, Y., Zhao, J., Bo, R. et al. (2023). A wrinkling-assisted strategy for controlled interface delamination in mechanically-guided 3D assembly. *Journal of the Mechanics and Physics of Solids* 173: 105203.

56 Li, S.P., Han, M.D., Rogers, J.A. et al. (2019). Mechanics of buckled serpentine structures formed via mechanics-guided, deterministic three-dimensional assembly. *Journal of the Mechanics and Physics of Solids* 125: 736–748.

57 Vella, D., Bico, J., Boudaoud, A. et al. (2009). The macroscopic delamination of thin films from elastic substrates. *Proceedings of the National Academy of Sciences of the United States of America* 106 (27): 10901.

58 Ye, J., Zhang, F., Shen, Z. et al. (2021). Tunable seesaw-like 3D capacitive sensor for force and acceleration sensing. *npj Flexible Electronics* 5 (1): 28.

59 Won, S.M., Wang, H., Kim, B.H. et al. (2019). Multimodal sensing with a three-dimensional piezoresistive structure. *ACS Nano* 13 (10): 10972–10979.

60 Cheng, X., Zhang, F., Bo, R. et al. (2021). An anti-fatigue design strategy for 3D ribbon-shaped flexible electronics. *Advanced Materials* 33 (37): 2102684.

61 Wang, X., Feiner, R., Luan, H. et al. (2020). Three-dimensional electronic scaffolds for monitoring and regulation of multifunctional hybrid tissues. *Extreme Mechanics Letters* 35: 100634.

62 Han, M., Guo, X., Chen, X. et al. (2022). Submillimeter-scale multimaterial terrestrial robots. *Science Robotics* 7 (66): eabn0602.

63 Kim, B.H., Li, K., Kim, J.-T. et al. (2021). Three-dimensional electronic microfliers inspired by wind-dispersed seeds. *Nature* 597 (7877): 503–510.

64 Liu, F., Cheng, X., Zhang, F. et al. (2019). Design and assembly of reconfigurable 3D radio-frequency antennas based on mechanically triggered switches. *Advanced Electronic Materials* 5 (6): 1900256.

65 Pang, W., Xu, S., Wu, J. et al. (2022). A soft microrobot with highly deformable 3D actuators for climbing and transitioning complex surfaces. *Proceedings of the National Academy of Sciences of the United States of America* 119 (49): e2215028119.

10

Harnessing Wrinkling and Buckling Instabilities for Stretchable Devices and Healthcare

Yao Zhao, Fangjie Qi, Haoze Sun, Yanbin Li, Haitao Qing, and Jie Yin

North Carolina State University, Department of Mechanical and Aerospace Engineering, 1840 Entrepreneur Drive, Raleigh, NC 27606, USA

10.1 Introduction

Buckling instabilities are ubiquitous in our daily lives, from the folded hanging curtains [1] to the blisters on painted walls [2], and to the skin wrinkles [3–5]. Folded hanging curtains and blisters are phenomena of global or local buckling, respectively. The constrained free-standing hinging curtains become buckled under compression to form periodic undulating waves or buckles [1]. The painted coating delaminates locally from the rigid wall or substrate to form a blister, driven by the compression-induced buckling of the swollen painting [2]. Skin wrinkles are a phenomenon of periodic wrinkling, where the aged soft tissue shrinks more than the stiff skin, which leads to the compression-induced wrinkling in both the skin and the top layer of the bonded soft tissue [3–6]. Wrinkling can be considered as a special form of buckling, which depends on the bonding strength between the film and substrate, i.e. the interfacial toughness Γ [7, 8]. When Γ is small, the film prefers to buckle out of a plane with a weak bonding and the extreme case is a free-standing thin sheet without a substrate as seen in the example of hanging curtains. When Γ is moderate, the film prefers to undergo buckling-driven localized delamination when rested on a stiff substrate [2, 9], e.g. blisters, or periodic delamination when rested on a soft substrate [7, 8, 10], e.g. periodic buckles. When Γ is large, the film prefers wrinkling by coherently deforming with the soft substrate to form periodic undulating waves due to the strong bonding [11]. The phase diagram on wrinkling or buckling modes for different material systems can be found in the literature [7].

Buckling has historically been viewed as a failure mechanism – the sudden transition of a flat surface to a wavy surface under some form of critical load. The incredible potential of buckling as an efficient method to spontaneously create wavy surface micro-patterning was first articulated by Bowden et al. in 1998 [12] on the wrinkling of thin metal coating on soft elastomers via thermal deposition and cooling. The larger shrinkage in the elastomeric substrates during cooling

Mechanics of Flexible and Stretchable Electronics, First Edition. Edited by Yong Zhu and Nanshu Lu.

leads to the compression-induced periodic wrinkling of the metal coating. Since then, mechanical instability has been pursued as a versatile platform for generating ordered surface patterns [13, 14], as well as harnessing the surface topography for a wide range of controllable and tunable applications in material properties measurement [15], micro-fabrication [16], wetting [16–18], friction [19], smart windows [20, 21], and anti-biofouling [22, 23]. In 2006, the wrinkling strategy was first applied by Khang et al. to achieve stretchable forms of brittle single-crystal silicon on an elastomer substrate via the mechanical pre-stretching method for stretchable electronics [24]. An elastomer substrate is first pre-stretched to a certain large strain, followed by transfer printing of thin silicon ribbons. Releasing the substrate pre-strain leads to the wrinkling of the silicon ribbons to form wavy forms, which can be reversibly stretched and compressed without damaging the silicon. The accumulated strain in the silicon is largely released by the buckling and bending of the silicon ribbons with the peak strain well below its fracture strain. To date, the pioneering work in bridging buckling and stretchability has ignited tremendous work in different forms of stretchable structures [24], including but not limited to origami/kirigami structures [25–35] and slender structures with or without elastomer substrates [20, 36–38], and various stretchable devices for applications in wearable electronics [39], sensors [40], energy storage [41], energy harvesting [42, 43], and healthcare [44].

The strategy of wrinkling and buckling provides an efficient and versatile platform for harnessing wrinkled or buckled morphology for tunable surface-related properties such as wetting, adhesion, friction, optics, stretchability, and combined, especially for wearable and stretchable devices where both functionality and stretchability are important. Despite recent reviews on surface wrinkling or structural buckling, most of them focus on either mechanics [45] or specific applications in surface patterning [16], surface properties [15, 16], or sensors [40]. In this work, we focus on the mini-review of leveraging versatile wrinkling and buckling for multifunctionality in stretchable devices and healthcare. The mini-review is organized into four sections: the Section 10.2 is on the structure designs and related mechanics of the designs of stretchable electronics. The Section 10.3 is on its application in stretchable electronics including stretchable sensors, e.g. strain sensors, pressure sensors, temperature sensors, chemical sensors, stretchable batteries, and other stretchable devices. The Session 10.4 is on its application in healthcare, including biosensors and biological interfaces. The Session 10.5 is on the conclusion and outlook.

10.2 Structural Designs and Mechanics

10.2.1 Structural Designs for Buckling-Enabled Stretchability

Buckling is an effective and efficient way to achieve stretchability in less-stretchable and non-stretchable materials such as metals, e.g. gold and silver, and semiconductors, e.g. silicon. Normally, the non-stretchable materials take various forms of an initial stress-free planar shape, including straight ribbons [24, 46, 47]

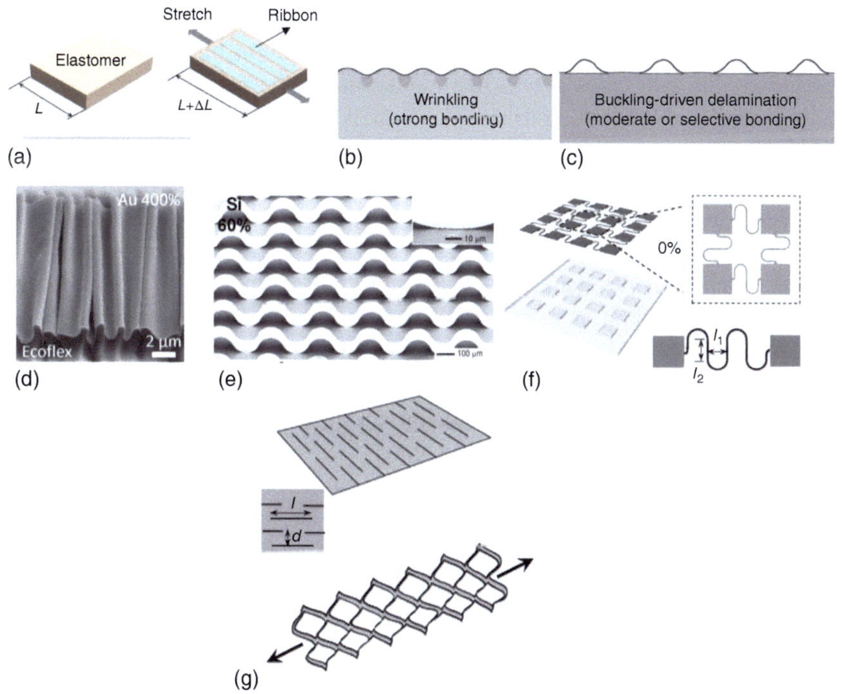

Figure 10.1 Structural designs for buckling-enabled stretchability. (a) Schematics of the pre-strain release induced buckling of non-stretchable stiff ribbons on elastomeric substrates. Stress-free ribbons are first bonded to a pre-stretched elastomer substrate, followed by the substrate strain release, leading to compression-induced buckling. (b) Schematic of the cross-sectional view of periodic wrinkling of stiff ribbons on elastomeric substrates when the bonding is strong. (c) Schematic of a cross-sectional view of periodic buckling-driven delamination of stiff ribbons on elastomeric substrates for moderate bonding or selective bonding at patterned spots. (d) An example of periodically delaminated buckles of gold nano-ribbons on an ecoflex substrate with a pre-strain of 400%. Source: Ref. [46]. Reproduced with permission of © American Chemical Society. (e) An example of controllable delaminated buckles of silicon ribbons on a PDMS substrate with a pre-strain of 60% and selective bonding. Source: Ref. [47]. Reproduced with permission of © Springer Nature. (f) Schematics of stretchable island-bridge designs by leveraging buckling of serpentine interconnection ribbons. (g) Schematics of stretching-induced buckling in a kirigami sheet with patterned parallel slits for high stretchability.

(Figure 10.1a–e), wavy ribbons such as serpentines [48] (Figure 10.1f), and thin kirigami sheets with periodic cut slits (Figure 10.1g). They employ two different out-of-plane buckling strategies to release the strain energy via bending and/or twisting. One is compression-induced buckling, as seen in the examples of straight ribbons (Figure 10.1a–e). The other is tension-induced buckling such as the serpentine interconnects in the island-bridge designs [48] (Figure 10.1f) and the kirigami sheets [49] (Figure 10.1g).

Stretchability achieved by compression-induced buckling, it requires fully or selectively bonding the stress-free planar straight or wavy ribbons with a pre-stretched elastomeric substrate. Releasing the pre-strain in the elastomer leads to compression-induced out-of-plane buckling in the non-stretchable ribbons on the top, favoring bending rather than directly stretching to release the strain. It employs two different types of buckling modes: one is the periodic wrinkling [24] and the other is the buckling-driven delamination [8, 47]. For the case of full bonding, depending on the interfacial bonding strength, the straight ribbon can undergo either periodic wrinkling for strong bonding [24] (Figure 10.1b) or periodic buckle-delamination for moderate bonding [8, 46] (Figure 10.1c,d). For the case of selectively bonding, the ribbons (Figure 10.1e) prefer to delaminate from the nonbonded sites for controllable out-of-plane buckling and pop-up by manipulating the patterning of bonding sites [47].

For stretchability achieved by tension-induced buckling, it does not require pre-stretching the elastomeric substrates or even a substrate. In the island-bridge designs (Figure 10.1f), when stretching the substrate, the stretching is mainly accommodated by the deformation of the bridges, i.e. the serpentine interconnects, which will undergo lateral buckling to release the stretching strain [48]. The ribbons often have a much smaller thickness than their width, i.e. low aspect ratio. Similarly, for the kirigami thin sheets [49] (Figure 10.1g), the stretching is accommodated by the pop-up and bending of the interconnected struts induced by out-of-plane buckling, only leaving the cut tip under stress concentration [50].

10.2.2 Mechanics of Wrinkling and Buckling

With decades of extensive study efforts, the theoretical modeling on the mechanics of wrinkling and buckling for stretchable devices has been explored systematically in the small deformation and finite deformation regimes [45, 51]. Most of the theoretical modeling framework originates from the minimum energy method and aims to obtain the numerical/analytical shape functions capable of predicting the buckled and post-buckled morphologies. Here, the general procedures of theoretical modeling are briefly introduced [45, 52, 53].

Generally, the total elastic energy in different buckled or wrinkled structures can be written as

$$U = U_{\text{stretching}} + U_{\text{bending}} + U_{\text{bonding}} + U_{\text{sub}} \tag{10.1}$$

The first two terms of $U_{\text{stretching}}$ and U_{bending} represent the elastic energy due to the stretching and bending of different structural parts. The last two terms of U_{bonding} and U_{sub} represent the interfacial bonding energy and the strain energy in the substrate if a soft substrate exists. The value of the energy terms in Eq. (10.1) may differ according to their intrinsic deformations. Particularly, more complicated deformations may exist, which induce more elastic energy in Eq. (10.1). The elastic energy of $U_{\text{stretching}}$ and U_{bending} can be derived through

$$U_{\text{stretching}} = 1/2 \int E_{ijkl} u_{i,j} u_{k,l} \mathrm{d}u \tag{10.2}$$

$$U_{\text{bending}} = 1/2 \int M_{\alpha\beta} \kappa_{\alpha\beta} \mathrm{d}\kappa \tag{10.3}$$

E_{ijkl} is the elastic modulus of the related material compositions and u are the displacements of distinct structural parts. $M_{\alpha\beta}$ and $\kappa_{\alpha\beta}$ are the bending moment and bending curvature of the deformed structural profile w (with $\kappa_{\alpha\beta} = w_{,\alpha\beta}$), respectively. The interfacial bonding and substrate elastic energies can be obtained through

$$U_{\text{bonding}} = \int \Gamma \mathrm{d}A \tag{10.4}$$

$$U_{\text{sub}} = \int_V W_{\text{s}} \mathrm{d}V \tag{10.5}$$

Γ an A are the interfacial toughness between different layers and contacting area, respectively. W_{s} is the strain energy density in the substrate. Substituting Eqs. (10.2)–(10.5) into (10.1) can determine the explicit forms of the deformed structural profiles (i.e. u and w) through minimizing U, i.e.

$$\partial_{\text{u}} U = 0, \partial_{\text{w}} U = 0 \tag{10.6}$$

Next, we summarize the theoretical results on the predictions of the critical buckling strain, the profile of the wrinkles or buckles, and the peak strain in the representative structures.

10.2.2.1 Compression-Induced Wrinkling

For the wrinkling of elastic stiff thin films on soft substrates with sinusoidal wrinkle profiles, when only considering the small deformation in the thin films, the critical buckling strain ε_{c}, wrinkling wavelength λ_{o}, and the wrinkle amplitude A_{o} are given by [54]

$$\varepsilon_{\text{c}} = \frac{1}{4}\left(\frac{3\overline{E}_{\text{s}}}{\overline{E}_{\text{f}}}\right)^{2/3} \tag{10.7}$$

$$\lambda_{\text{o}} = 2\pi t\left(\frac{\overline{E}_{\text{f}}}{3\overline{E}_{\text{s}}}\right)^{1/3} \tag{10.8}$$

$$A_{\text{o}} = t\left(\frac{\varepsilon}{\varepsilon_{\text{c}}} - 1\right)^{\frac{1}{2}} \tag{10.9}$$

where the plane strain Young's modulus $\overline{E} = E/(1 - \vartheta^2)$ with the subscript f and s denote the film and substrate, respectively. ϑ is the Poisson's ratio. t is the film thickness. ε is the strain in the film. The above equations hold true only for small deformation in the film under small strain. However, when the film is subjected to moderate strain, finite deformation in the film needs to be considered [52], especially for the case of wrinkling of films on pre-strained elastomer substrates, where a moderate pre-stretching strain ε_{pre} over 40% is often applied. By assuming the sinusoidal wrinkle profile, the new wrinkle wavelength λ and wrinkle amplitude A are modified as [52]

$$\lambda = \frac{\lambda_{\text{o}}}{(1 + \varepsilon_{\text{pre}})(1 + \xi)^{1/3}} \tag{10.10}$$

$$A = \frac{A_{\text{o}}}{(1 + \varepsilon_{\text{pre}})^{1/2}(1 + \xi)^{1/3}} \tag{10.11}$$

where $\xi = 5\varepsilon_{pre}(1+\varepsilon_{pre})/32$. The peak strain ε_{peak} in the buckled film or ribbon can be obtained by

$$\varepsilon_{peak} = \frac{t}{2R_c} \text{ with } R_c = \left|\frac{(1+y'^2)^{3/2}}{y''}\right| \tag{10.12}$$

R_c is the radius of curvature of a buckle or wrinkle at its crest and $y(x)$ is the function of the buckled profile. The peak strain ε_{peak} in the film is given by [52]

$$\varepsilon_{peak} = 2(\varepsilon_{pre}\varepsilon_c)^{1/2}\frac{(1+\xi)^{1/3}}{(1+\varepsilon_{pre})^{1/2}} \tag{10.13}$$

10.2.2.2 Compression-Induced Constrained Buckle-Delamination

For buckling-driven delamination of thin ribbons or films on soft substrate, the periodicity and amplitude of delaminated buckles can be well controlled through selective bonding of the thin ribbons on pre-strained substrate (Figure 10.1e) [47]. For parallel patterned adhesion sites with periodicity w_{free}, i.e. the distance between two neighboring bonding sites, the periodicity λ_b and amplitude A_b of the delaminated buckles are given by [47]

$$\lambda_b = 2\frac{w_{free}}{(1+\varepsilon_{pre})} \tag{10.14}$$

$$A_b = \frac{2\lambda_b}{\pi}\left(\varepsilon_{pre} - \frac{t^2\pi^2}{3\lambda_b{}^2}\right)^{1/2} \tag{10.15}$$

where the profile of the delaminated buckles is assumed to employ a sinusoidal function. The peak strain ε_{peak} in the buckle-delaminated ribbon is given by [47]

$$\varepsilon_{peak} = \frac{2\pi t}{\lambda_b}\left(\varepsilon_{pre} - \frac{t^2\pi^2}{3\lambda_b{}^2}\right)^{1/2} \tag{10.16}$$

10.2.2.3 Compression-Induced Spontaneous Buckle-Delamination

Similar periodic buckling-driven delamination occurs spontaneously for thin ribbons or films on largely pre-stretched elastomer substrates without the need of selective patterned adhesion sites (Figure 10.1c,d) [8, 46]. During the release of the large pre-strain release, the deformation in the film or ribbons transits from periodic delaminated sinusoidal buckles to post-buckled elastica-shaped buckles. For the periodic delaminated buckles with sinusoidal profiles, the buckle wavelength λ_d and amplitude A_d are given by [8]

$$\lambda_d = \pi t\left(\frac{\overline{E}_f t}{\Gamma}\right)^{1/2}\left(\frac{\varepsilon_{pre}}{1+\varepsilon_{pre}}\right)^{1/2} \tag{10.17}$$

$$A_d = t\left(\frac{\overline{E}_f t}{\Gamma}\right)^{1/2}\frac{\varepsilon_{pre}}{(1+\varepsilon_{pre})^{1/2}} \tag{10.18}$$

where Γ is the interfacial toughness. The peak strain in the ribbon is [8]

$$\varepsilon_{peak} = 2\left(\frac{\Gamma(1+\varepsilon_{pre})}{\overline{E}_f t}\right)^{1/2} \tag{10.19}$$

For the post-buckled elastica-shaped buckles, its profile can be described as [8]

$$\lambda_d(s) = -s + \frac{2}{\alpha}E[am(\alpha s, \sin(\overline{\theta}/2), \sin(\overline{\theta}/2))] \tag{10.20}$$

$$A_d(s) = \frac{2\sin(\overline{\theta}/2)}{\alpha}[1 - cn(\alpha s)] \tag{10.21}$$

where $s \in \left[-\frac{l}{2}, \frac{l}{2}\right]$ is the curvilinear coordinate and l is the constant arc length. $\theta(s)$ is the rotation angle of the ribbon axis and $\overline{\theta} = \theta\left(-\frac{l}{4}\right)$. $E(,)$ is the incomplete elliptic integrals of the second kind, am and cn are the Jacobi amplitude and the Jacobi cosine amplitude functions, respectively. α is the dimensionless load parameter. The peak strain in the buckled ribbons with elastic profiles can be obtained numerically by using the formula in Eq. (10.12). Compared to sinusoidal buckles, the elastic-shaped buckles show relatively small peak strain [8]. When $\varepsilon_{\text{pre}} < 43\%$, both sinusoidal and elastic-shaped buckles exhibit similar peak strain with slightly lower peak strain in the elastica profile. However, when $\varepsilon_{\text{pre}} > 43\%$, $\varepsilon_{\text{peak}}$ grows exponentially in the sinusoidal buckles while approximately linearly in the elastica buckles [8]. For example, when $\varepsilon_{\text{pre}} = 400\%$, the peak strain in the sinusoidal buckles is over six times higher than that in elastica buckles. It implies that elastica buckles can release the strain energy more efficiently [8].

10.2.2.4 Tension-Induced Buckling in Serpentine Structures

In the island-bridge structure with serpentine geometry (Figure 10.1f), both straight and arc segments of the ribbons undergo combined shearing and bending deformation, which trigger lateral buckling. The buckling deformation is governed by three normalized parameters: the length-to-space ratio $\alpha = l_2/l_1$ (l_1 and l_2 are the diameter of the semi-circles and the length of a straight line in the unit cell, respectively, Figure 10.1f), the number of unit cells m, and the cross sectional aspect ratio t/w (t is the thickness and w is the width) [48]. The critical buckling strain ε_{cr} can be given by [48]

$$\varepsilon_{\text{cr}} = f(m, \alpha)\left(\frac{G}{E}\right)^{\frac{1}{2}}\left(\frac{t}{w}\right)^2 \tag{10.22}$$

where $f(m, \alpha)$ is a function of m and α [48]. G and E are the shear modulus and Young's modulus of the island material, respectively. ε_{cr} is linearly proportional to $(t/w)^2$. When $\alpha > 3$ with multiple unit cells, i.e. $m > 1$, ε_{cr} increases linearly with α and decreases monotonically with m. The peak strain in the buckled serpentine structure can be given by [48]

$$\varepsilon_{\text{peak}} = g_1(m, \alpha)(\varepsilon_{\text{appl}})^{\frac{1}{2}}\frac{t}{l_1} + g_2(m, \alpha)(\varepsilon_{\text{appl}})^2\frac{w}{l_1} \tag{10.23}$$

where g_1 and g_2 are functions of m and α [48]. The elastic stretchability can be analytically obtained when $\varepsilon_{\text{peak}}$ reaches the yield or fracture strain of the materials ε_y or ε_f. The stretchability increases with the spacing or yield or fracture strain via $\varepsilon_y l_1/w$ while it decreases with the serpentine thickness via t/w.

10.2.2.5 Tension-Induced Buckling in Kirigami Structures

Similar tension-induced buckling can be achieved in the kirigami structure. Despite the different types of cut patterns, we focus on the classical kirigami pattern consisting of parallel cuts shown in Figure 10.1g that achieves high stretchability. The cut pattern is characterized by the parallel spacing d between cuts and the cut length l with $l > d$ (Figure 10.1g). The critical out-of-plane buckling strain ε_{cr} can be obtained as [49]

$$\varepsilon_{cr} = \left(\frac{t}{d}\right)^2 \tag{10.24}$$

where t is the thickness of the thin sheet. The peak strain ε_{peak} in the buckled kirigami structures can be given by [55]

$$\varepsilon_{peak} = f(t/d)(\varepsilon_{appl})^{\frac{1}{2}}\frac{t}{l} \tag{10.25}$$

where f is a function of t/d [55] and ε_{appl} is the applied strain. By letting ε_{peak} equal to the fracture strain of the materials ε_f, the elastic stretchability ε_s can be obtained as [55]

$$\varepsilon_s = \left(\frac{1}{f}\right)^2 \varepsilon_f^2 \left(\frac{l}{t}\right)^2 \tag{10.26}$$

It shows that the elastic stretchability ε_s increases proportionally with the square of the normalized cut length l/t.

10.3 Applications in Stretchable Devices

Based on the mechanics of wrinkling and buckling in different structures, in this section, we will discuss how to combine mechanics and different materials and structure systems to achieve high stretchability in a variety of devices, including sensors, batteries, and other electronics.

10.3.1 Stretchable Sensors

Buckling and wrinkling are often employed to enhance stretchability in stretchable sensors, especially for non-stretchable materials like metals and ceramics [40, 56–58]. For example, Hassani et al. [59] have fabricated a kind of capacitive sensor capable of measuring the strains of bladders. In this work, the wrinkles are introduced with a typical strain mismatch method. Specifically, the electrode layers (Au/PET/Kapton) are transferred onto an acrylic elastomer layer (VHB tape) with large pre-strains. The wrinkles can be formed when the pre-strains are released (from 2200 to 526 μm for each unit). The wrinkles in this sensor not only improve the stretchability of the electrodes but also augment the sensor sensitivity. Remarkably, the sensor is integrated with a kind of shape memory alloy-based actuator that can help to void bladders. In addition to sensors that leverage electrical properties, wrinkling can also find its uniqueness in sensors based on other properties, such as triboelectricity. For instance, Lee et al. [60] have

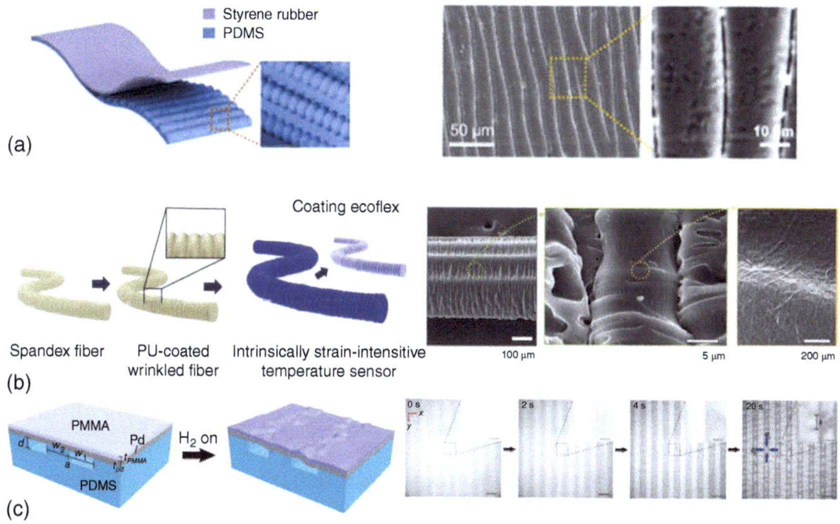

Figure 10.2 Applications in stretchable sensors. (a) The structure of the pressure sensor and the SEM image of the patterned PDMS layer. Source: Reproduced with permission of Ref. [60]. © American Chemical Society. (b) The preparation method and the SEM images for the composite fiber temperature sensor. Source: Reproduced with permission of Ref. [61]. © John Wiley & Sons. (c) The schematic and images for the deformation of the sensor when it is exposed to H_2. Source: Reproduced with permission of Ref. [62]. © American Chemical Society.

developed a kind of wearable triboelectric sensor for pressure measurement. As shown in Figure 10.2a, this stretchable triboelectric pressure sensor is composed of a layer of styrene butadiene rubber and a patterned PDMS layer. The pattern of the PDMS layer combines a higher level of wrinkles and a lower level of domes. The pattern is fabricated with two steps: First, domes can be obtained by curing PDMS on a mold with a dome array pattern. Second, the PDMS is stretched and treated with ultraviolet/ozone. The sensor is integrated with a layer of styrene butadiene rubber by drying polystyrene-block-poly(ethylene-ran-butylene)-block-polystyrene toluene solution on the PDMS pattern and adhering an aluminum foil served as the electrode. Upon pressure, the triboelectricity is generated by contact, and the PDMS and the styrene-butadiene rubber are respectively charged negatively and positively. The wrinkle based pattern can magnify the output voltage since the pattern can enhance the contact area difference of the PDMS and the styrene-butadiene rubber layers at the same pressure.

In addition to sensors for sensing mechanical signals, such as strain and pressure sensors, buckling and wrinkling can also be utilized in other types of sensors, such as temperature sensors and chemical sensors. One example is a type of temperature sensor developed by Lee et al. [61]. As shown in Figure 10.2b, the sensor is in a fiber shape, which contains an elastic core and a conductive shell. The wrinkles are fabricated through the typical strain mismatch method: a pre-stretched spandex fiber is coated with another elastomer polyurethane (PU). The pre-strain is released after the PU is fully cured and the wrinkles can be formed. Then, the conductive layer that contains poly(3,4-ethylenedioxythiophene) polystyrene sulfonate (PEDOT:PSS) and single-walled carbon nanotubes (SWCNTs) is coated on the wrinkled fiber. Subsequently, a thin layer of Ecoflex is coated on the composite fiber as a protective layer. The PEDOT:PSS and SWCNTs mixture are chosen as the conductive materials since the conductivity is largely temperature dependent. It shows that the sensor is not sensitive to strains; thus, it can effectively reduce the influence of the strains. They also believed that these fiber sensors are very promising in smart sportswear for health monitoring.

Another example is the chemical sensor designed specifically for detecting hydrogen gas. Hydrogen gas detection is significant, as it represents a clean energy source. However, it also poses a considerable danger due to its highly explosive nature. Being colorless with a wide flammable concentration range and low ignition energy, hydrogen gas requires careful monitoring and detection. She et al. [62] have designed a kind of hydrogen gas sensor that differs from conventional sensors that utilize wrinkles to enhance stretchability. In this unique design, the sensor's signal is directly generated through the process of wrinkle formation. The structure of the sensor is illustrated in Figure 10.2c. A bilayer structure composed of a thin PMMA layer and a thin Pd layer is suspended on a PDMS substrate with microgroove arrays. When exposed to even a small amount of H_2, the Pd film undergoes volume expansion. Consequently, wrinkles are formed due to the Pd/PMMA bilayer structure and the microgrooves, which results in a significant drop in reflectance. Additionally, the presence of the PMMA layer provides protection to the Pd layer against contaminations and fractures. The images in Figure 10.1c demonstrate the reflectance change upon 4% H_2 flow.

10.3.2 Stretchable Batteries

Stretchable batteries play an important role in powering stretchable and wearable electronics since they provide the flexibility and stretchability to accommodate large deformation. Wrinkles and buckles are commonly utilized to improve the stretchability either in specific components of electrodes or in the overall structures [29, 63–65].

Stretchable electrodes are one of the key components that enable fully deformable and stretchable battery cells. However, there is a tradeoff between the electrochemical performances and mechanical deformation due to the stretching-induced increase in the electrode resistance. To overcome the tradeoff, Ji et al. utilized the micro-crumpled electrodes through the pre-strained elastomer method to create an ultra-stretchable and compact Zn–MnO_2 rechargeable battery [66]. As shown in Figure 10.3a, a thin layer of conductive material (Au) is thermally coated on an elastomer membrane, and the composite is then transferred to the pre-stretched sticky elastomer substrate. Upon releasing the strain, micro-wrinkles are formed. Figure 10.3b shows that the wrinkles completely vanished when subjected to a 525% strain, significantly enhancing the stretchability of the electrodes. It shows that the micro-wrinkled metal films exhibit limited resistance change after over 1000 cycles

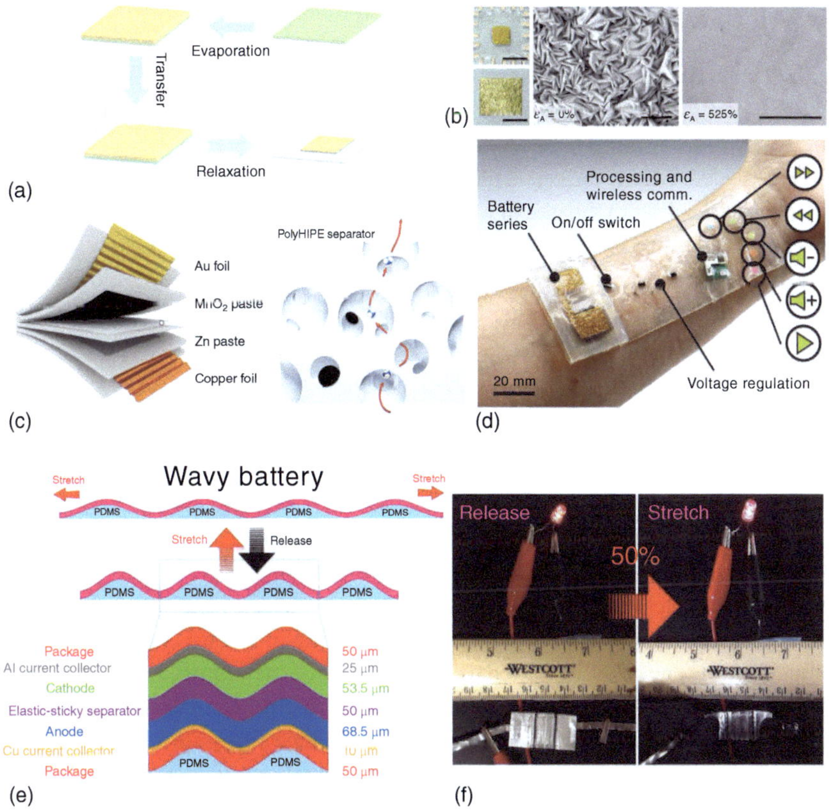

Figure 10.3 Applications in stretchable batteries. (a) The preparation method for the current collector. (b) The optical images and the SEM images of the sample at different strains. Source: Reproduced with permission of Ref. [66].© American Chemical Society. (c) The structure of the stretchable battery. (d) The stretchable battery powers a soft electronic skin fixed on the human arm. Source: Ref. [67]/John Wiley & Sons/CC BY 4.0. (e) The structure of the stretchable battery. (f) The stretchable battery powers an LED at both release and stretch states. Source: Reproduced with permission of Ref. [68]. © John Wiley & Sons.

of stretching and relaxation under 525% areal strain. Danninger et al. [67] employed a similar method to create wrinkled components and subsequently assembled a stretchable battery. As shown in Figure 10.3c, to improve the stretchability of the battery, they utilize the intrinsically compliant anode Zn paste, the cathode MnO_2 paste, and the separator polymerized high internal phase emulsions (polyHIPEs), where the current collectors Au layer and Cu layer are fabricated in the wrinkled states. Figure 10.3d illustrates that the stretchable battery powers a soft electronic skin adhered to a human arm.

Instead of inducing wrinkles solely on specific components of batteries, an alternative approach utilizes global buckling across the entire battery structure. For example, Liu et al. [68] have manufactured a kind of stretchable batteries with wavy structures as illustrated in Figure 10.3e. During the typical manufacturing process, a sticky separator is produced using the electrospinning method, which employs a mixture of PU and polyvinylidene fluoride (PVDF) dissolved in tetrahydrofuran and dimethylformamide solutions. Next, the sticky separator is sandwiched between the $LiCoO_2$ cathode and graphite anode, and the layers can be firmly bonded due to the sticky nature of the separator. Then, the electrolyte is injected into the cell before it is sealed. Lastly, the cell is placed in a wavy mold and PDMS is cured in the valley regions, leading to the global wavy structure. The stretchability of the battery is illustrated in Figure 15.3f. As shown in the images, the battery effectively powers a standard LED in both a releasing mode (0 strain) and a stretching mode (50% strain). The repetitive stretching and releasing of the battery over 60 cycles barely impact its overall performance.

10.3.3 Other Stretchable Electronics

In addition to sensors and batteries, buckling and wrinkling are also utilized in other electronic devices to enhance stretchability, including supercapacitors, solar cells, and electronic skins.

Stretchable supercapacitors have recently attracted growing interest in soft energy storage systems due to their fast charging/discharging processes and robust cycle stability [69, 70]. The two-dimensional (2D) transitional metal carbides and nitrides MXene have become a promising electrode material due to its high capacitance and metallic conductivity. However, the pristine MXene films are very fragile. To enhance its stretchability, Feng et al. [71] fabricated a kind of supercapacitor that consists of two crumpled MXene electrodes and an elastomer as the dielectric layer. The crumpling of the MXene electrodes is generated by bonding them to a biaxially pre-stretched acrylic elastomer. Figure 10.4a demonstrates the supercapacitor at 0% strain and 800% strain, respectively. The crumpled surface morphology of the MXene electrodes is shown in Figure 10.4b. It shows that the supercapacitor not only possesses high stretchability but also shows high electrochemical stability with dynamic deformations.

Solar cells are another green energy source that can directly transfer solar energy into electricity [74–76]. The traditional solar cell that is stationary, rigid, and bulky, whereas the angle of the Sun irradiation varies with time. To enhance the energy harvesting efficiency in the solar cells, Lamoureux et al. [72] proposed leveraging the buckling of kirigami structures to dynamically change the orientation of solar cells by making the solar cells become stretchable, as shown in Figure 10.4c. The stretchable solar cell is composed of two components. The first one is the kirigami substrate that is fabricated with Kapton with parallel cuts. The buckling of the cut struts enables large stretchability of the kirigami structure. The second component is a thin-film solar cell that is attached to the kirigami substrate. Notably, the solar cell has a very small thickness (~50 μm) which provides substantial flexibility. In addition to the large stretchability, the solar cell exhibits effective solar tracking by tuning the tilting orientations of the buckled struts with simple strain, which ensures optimal exposure to sunlight.

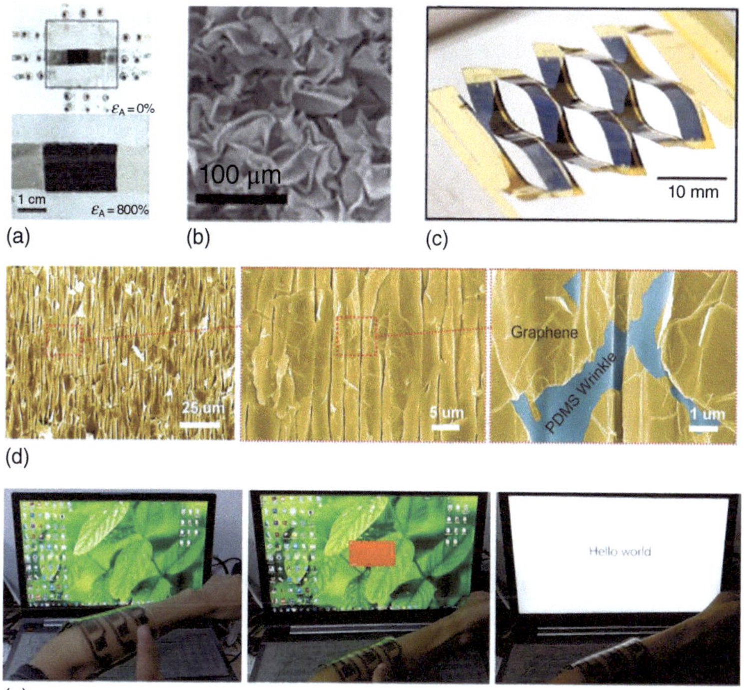

Figure 10.4 Applications in other electronic devices. (a) Optical images of the supercapacitor under biaxial tensile strain with 0% and 800% strains. (b) SEM image of a stretchable MXene electrode with a crumpled surface. Source: Reproduced with permission of Ref. [71]. © American Chemical Society. (c) A thin-film solar cell with a kirigami cut. Source: Ref. [72]/Springer Nature/CC BY 4.0. (d) The SEM images of the graphene on the wrinkled PDMS substrate. (e) The demo shows e-skin communicating and controlling the computer. Source: Reproduced with permission of Ref. [73]. © John Wiley & Sons.

The electronic skin, commonly referred to as e-skin, is a kind of flexible, thin, and lightweight electronic device that is composed of sensors, actuators, and other electronic components to mimic human-skin-like sensory capabilities [77–79]. e-skin has emerged as an attractive research topic, even though it is relatively nascent in its history. This concept was first raised in fictitious movies as early as in the 1970s, and the first e-skin was crafted by General Electric (GE) in 1985 [80]. Recently, the focus on e-skin has evolved to e-skin with extra stretchability and functionality. Similar to other electronics, the strategic introduction of wrinkling and buckling serves as an instrumental technique to enhance stretchability. For example, He et al. [73] recognized the exceptional conductivity and minimal thickness of graphene as an ideal candidate for e-skin. Yet, a notable limitation resided in its restricted stretchability. The wrinkling method is utilized to address this issue. In a typical fabrication process, a PDMS layer is first stretched and treated by Ar plasma with a patterned mask, which can result in a thin and dense layer. After the graphene is transferred to the PDMS, the strain is released, which triggers the wrinkle formation due to the modulus difference between the surface and the interior of the PDMS. Figure 10.4d shows the wrinkled graphene on the PDMS substrate. It shows that the wrinkles cannot only improve the stretchability of the e-skin but also enhance the adhesion between the graphene layer and the PDMS substrate. Figure 10.4e illustrates an e-skin integrated with the wrinkled structure that can recognize the touching movement and then the built-in program instructs the computer to open MS PowerPoint software and print "Hello World" on the computer screen as a human–machine interface.

10.4 Applications in Healthcare

In this section, we will discuss leveraging wrinkling and buckling for healthcare applications, including biosensors that utilize wrinkles and buckles for amplifying sensing signals and enhancing bioadhesion, and stretchable and conformable bio-interfaces.

10.4.1 Biosensors

Devices that measure *in situ* biomedical reactions or body movements call for not only accuracy, reliability, and precision but also continuous measurement of body functions or biomedical signals. For biosensors used in biomedical systems, buckles and wrinkles are often coupled. The buckle allows the sensor to fit the surfaces of organs or skin, but wrinkles usually enhance the sensibility of the sensors. Mickle et al. [81] introduce a compact bio-optoelectronic implant featuring a soft biophysical sensor (Figure 10.5a). This implant incorporates a low-modulus strain gauge (SG) within the neuromodulation system, enabling real-time monitoring of bladder filling and voiding. Microscale inorganic light-emitting diodes (μ-ILEDs) are intricately woven around the bladder to track volume fluctuations and offer optogenetic stimulation to bladder-innervating neurons. The wireless control and power (WCP) module orchestrates the strain gauge's feedback, governs μ-ILEDs' operations, and oversees power management. Power is delivered wirelessly via resonant magnetic coupling facilitated by an antenna transmitter. Figure 10.5b illustrates how the strain gauge is wrapped around the mimic bladder and how stretch is exerted on the strain gauge as the bladder undergoes expansion. A small absorbable suture is carefully threaded through the buckle, securing it to the smooth muscle layer of the bladder, thus effectively fastening the upper band in place. The buckle of the upper band ensures the sensor fits well with the natural dome shape of the bladder.

Another example demonstrated by Han et al. [82] shows how piezoelectric microsystem is of use in biomedical sensing via buckling in between two stable states. In Figure 10.5c, a depiction in an exploded view schematic illustration showcases a prototypical 3D piezoelectric microsystem. This microsystem encompasses a stratum of piezoelectric polymer (specifically, polyvinylidene difluoride or PVDF), with thicknesses measuring 13 and 28 μm in the slender and broader segments correspondingly. Additionally, the structure incorporates metallic layers (composed of Cr/Au, with thicknesses of 10 and 100 nm, respectively) that function as electrodes on both the upper and lower surfaces. The process for assembly of

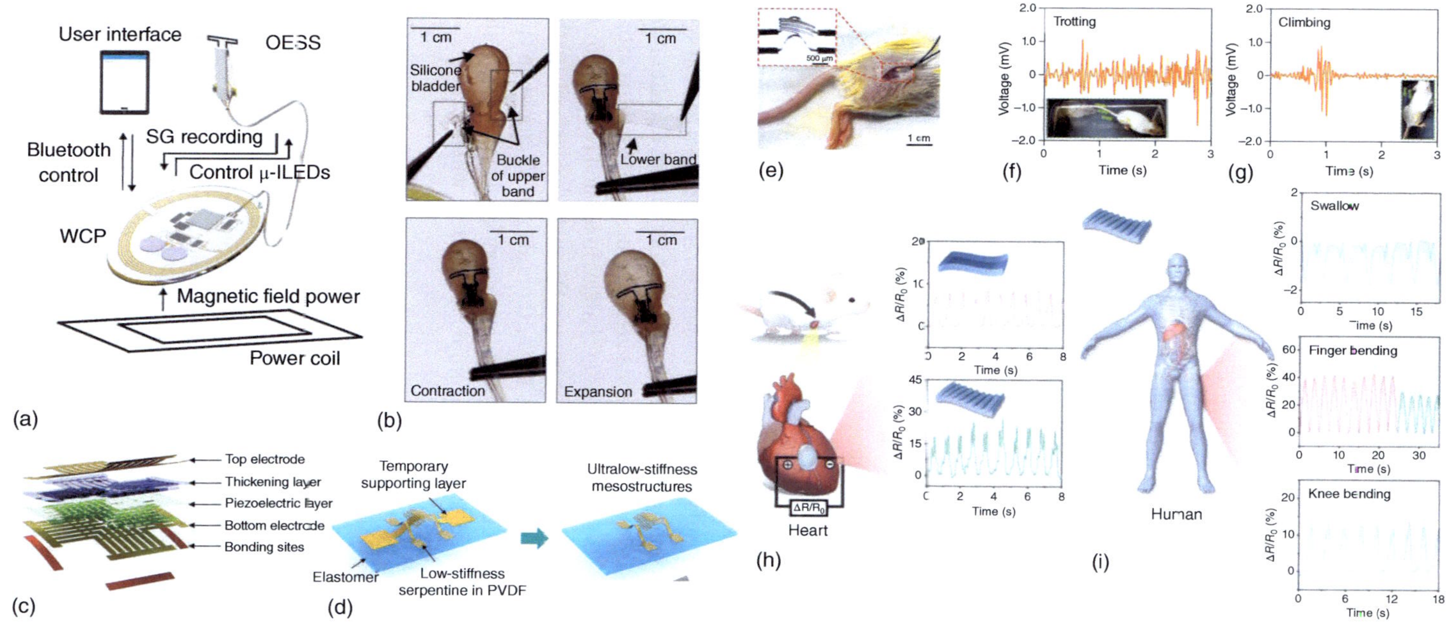

Figure 10.5 Applications in biosensors. (a) Schematic illustrations of a fully implantable, soft optoelectronic system for wireless, closed-loop optogenetic modulation of bladder function. (b) Demonstration of strain gauge illustrating how the two silicone bands wrap around a mimic bladder. Source: Reproduced with permission of Ref. [81]. © Springer Nature. (c) Exploded-view schematic illustrations of a 3D PVDF mesostructure with top and bottom metal electrodes. (d) Process for assembly of mesostructures with ultralow stiffnesses. (e) Optical image of a 3D piezoelectric energy harvester implanted into the hind leg of a mouse. (Insert: image of a magnified device) (f,g) Trotting and climbing movements of the rat and their related signal. Source: Reproduced with permission of Ref. [82]. © Springer Nature. (h) Comparison between the signal detections of normal rat cardiac by flat hydrogel and wrinkled sensor. (i) Wrinkled hydrogel as a strain sensor for the detection of swallowing, finger bending, and knee bending of the human body. Source: Reproduced with permission Ref. [33]. © John Wiley & Sons.

mesostructures initiates with the simultaneous buckling of a bilayer 2D precursor. A strip of polyimide (PI) measuring 12.5 μm in thickness serves as the foundational support layer, accommodating a slender PVDF structure (with a thickness of 9 μm) configured in a low-stiffness, serpentine pattern. Throughout the assembly phase, the PI ribbon is affixed at its endpoints to an elastomer substrate, with a prestrain of 10%. This setup causes the PI ribbon to buckle upward, thereby transmitting forces onto the rear side of the PVDF serpentine structure, inducing out-of-plane movements and eventually leading to the creation of a 3D configuration (Figure 10.5d). With the 3D piezoelectric energy harvester implanted into the hind leg of a mouse (Figure 10.5e), output signals can be monitored as the tested object is trotting (Figure 10.5f) or climbing (Figure 10.5g).

To enhance the ability of adhesion to organ surfaces, wrinkles are introduced in a bioinspired adaptive nanoclay-interlocked soft strain sensor. Based on a highly stretchable and elastic ionic-conductive hydrogel, Zhang et al. [83] demonstrate the wrinkled sensor exhibits the ability to detect tiny variations in the cardiac signal compared to the flat sensor (Figure 10.5h), indicating that the presence of wrinkle enabled the sensor to be more sensitive to minute deformations. This effect could be attributed to even a subtle force inducing substantial deformation in the wrinkled surface. This deformation, in turn, amplifies the contact area between the tissue surface and the hydrogel, resulting in a notable alteration of resistance. As depicted in Figure 10.5i, the strain sensor demonstrates its capability to track substantial motions like joint flexion (knee and finger), as well as minor actions such as the swallowing motion of humans.

Furthermore, Hassani et al. [59] reported a soft sensor for a sensing-actuation system capable of large bladder volume changes. During bladder filling, the alteration in sensor capacitance serves as an indicator of bladder volume. Notably, the sensor's inherent flexibility facilitates its responsiveness to changes in tension resulting from bending. Moreover, the presence of wrinkles enhances the sensor's sensitivity to the perpendicular force exerted by bladder filling. The authors also preformed an in vivo micturition test with a rat. The bladder volume of the rat can be visualized with the sensor, and the user can actuate the system to empty the bladder when it is full.

10.4.2 Biological Interfaces

The introduction of wrinkling and buckling patterns has shown significant enhancement in overall stretchability and conformability. Wrinkled surfaces closely mimic bio-interfaces because of their innate softness, curvature, dynamic nature, and textured quality. Wrinkled biological interfaces integrated electronic devices play an important role in the realm of healthcare, particularly in areas such as medical diagnosis, implantable devices, and electrophysiological interfaces.

The utilization of buckles and wrinkles within bio-interfaces broadens functionality, such as scaffolds within the biomedical field. This extension of capability consequently enhances overall functionality in medical diagnosis. As shown in Figure 10.6a, Zuo et al. [84] utilized the cellulose wrinkled hydrogel with excellent biocompatibility and high specific area as the bio-interface to capture heterogeneous

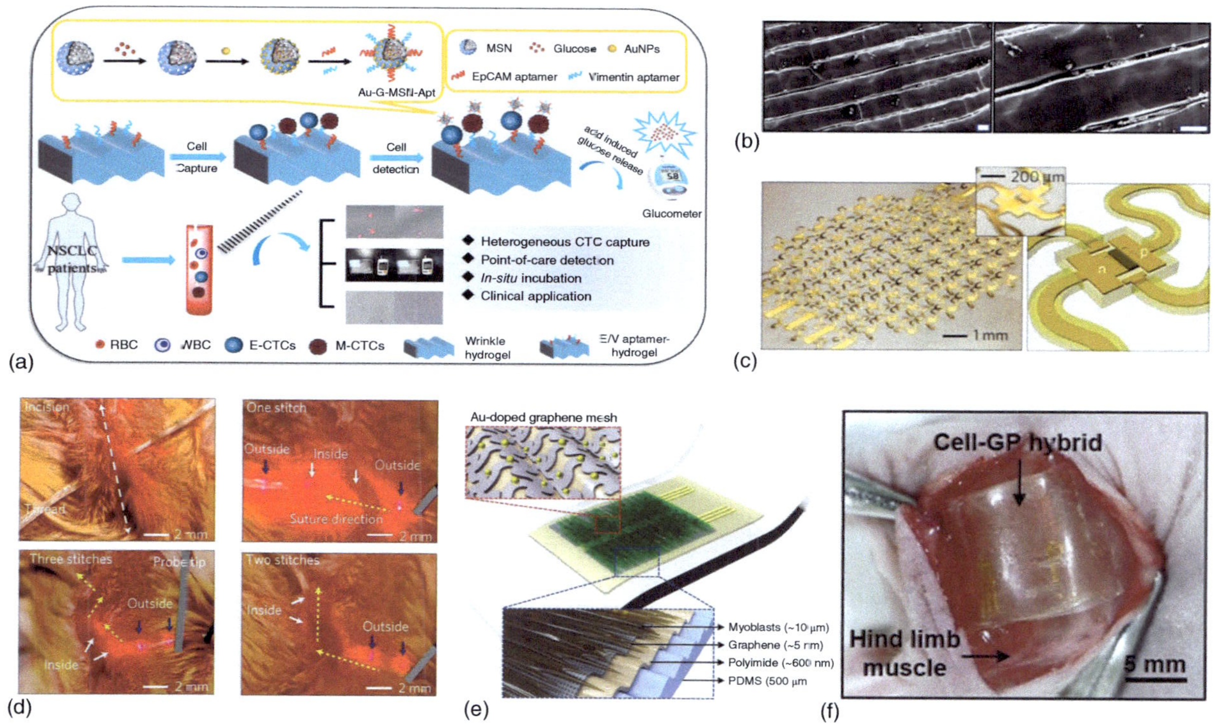

Figure 10.6 Applications in biological interfaces. (a) Fabrication of wrinkled hydrogel and nanoprobes and applications in capture and detection of CTCs in NSCLC patients. (b) SEM images of the wrinkled hydrogel. Scales bars represent 100 μm. Source: Reproduced with permission of Ref. [84]. © American Chemical Society. (c) Optical image of a 6 × 6 array of μ-ILEDs with noncoplanar serpentine bridges on a thin PDMS substrate (left-hand frame). Schematic illustration (right) and corresponding photograph (inset) of a representative device. (d) Light-emitting suture consisting of a 1 × 4 array of μ-ILEDs on a thread, demonstrated in an animal model with a conventional suture needle. Source: Reproduced with permission of Ref. [85]. © Springer Nature. (e) A schematic illustration of the cell-sheet–graphene hybrid. (f) Implantation of the cell-sheet–graphene hybrid onto the target site of a nude mouse *in vivo*. Source: Reproduced with permission of Ref. [86]. © John Wiley & Sons.

circulating tumor cells (CTC). This wrinkled structure had a relatively high specific surface area, which is beneficial for creating multiple binding sites with cell adhesions (Figure 10.6b). Moreover, it was found that the captured cells maintained a higher viability on the hydrogel and could be *in situ* recultured without releasing from the substrate. The hydrogel's efficacy in capturing CTCs was quantifiable through readings obtained from a portable glucose meter. The successful application of this integrated device allowed for the quantification of CTCs in patient blood samples.

In the realm of implantable electronic devices, the implementation of strategic buckles and wrinkles arises as a feasible tactic to realize the overall stretchability. It is reported [85] that microbuckling structures are employed as mechanically optimized arrangements to create arrays of inorganic LEDs within systems capable of accommodating significant mechanical deformations. The integration was achieved on a variety of substrate materials and formats. As shown in Figure 10.6c, the arrays of μ-ILEDs are connected by serpentine-shaped ribbons that serve as either structural bridges or electrical interconnects, transferred to a thin, pre-strained sheet of PDMS. The microbuckling structure could be introduced by mechanical pre-stretching of the elastic substrates. The substrates will shrink after shape transformation, while the electronic materials on top will be compressed, providing additional stretchability and noncoplanar layouts in the serpentines. This design strategy opens new opportunities for the seamless integration of optoelectronics with biomedical and robotic systems. Figure 10.6d illustrates the examples of light-emitting sutures within an animal model. These sutures were manipulated using a conventional suture needle, commencing from the initial incision and progressing through the execution of three stitches. In this scenario, arrays of μ-ILEDs function without encountering any failures. This success can be attributed to two main factors: the presence of a fully encapsulating PDMS layer, which serves as a soft, elastomeric, and biocompatible barrier against the surrounding tissue and biofluids, and the favorable mechanical properties discussed earlier.

Apart from the improved stretchability, buckled and wrinkled structures exhibit the capability to closely replicate physiological environments compared to flat surfaces. This characteristic has the potential to greatly influence cellular proliferation and alignment, resulting in a notable boost to the effectiveness of electrophysiological interfacing. As shown in Figure 10.6e, it is reported [86] that the functions of signal testing and tissue engineering can be integrated for electrophysiological testing and therapy via one cell-adhered buckled electrode. This device uses a cell-sheet–graphene hybrid, which can be implanted to form a high-quality biotic/abiotic interface. The anisotropic buckled structure induces cell alignment and enhances stretchability and softness. In addition, it has enabled the adjustment of various cell functions via electrical stimuli, chemical stimuli, topographic stimuli, and dynamic mechanical stimuli. Figure 10.6f demonstrates that this integrated design can be transplanted onto the target muscle tissue to record electromyographical (EMG) signals and stimulate implanted sites. This integration allows for combined electrophysiological testing and therapeutic applications.

10.5 Conclusion and Outlook

Mechanical instabilities such as buckling and wrinkling in free-standing structures and structures bonded with soft elastomeric substrates enable dramatic shape changes in response to mechanical forces or external stimuli. The buckled or wrinkled surface morphologies or structures can facilitate the significant release of the accumulated strain energy to avoid the yielding or cracking failure of the structures made of less or non-stretchable metals or semiconductors. Thus, stretchability is achieved in various flexible and stretchable electronic devices, such as sensors, batteries, supercapacitors, solar cells, e-skins, and beyond, when transforming from buckled to unbuckled states or vice versa. The reversible shape change also facilitates the adaption to dynamic motions, amplifying the signal generation, the sensing of environmental stimuli and physiological signals, and improving interfacial adhesion and conformability.

Mechanical instability is a powerful platform for integrating with different structure and materials systems for multifunctionality in stretchable devices and healthcare. However, the full potential of mechanical instabilities remains to be explored from both fundamental and application perspectives. Fundamentally, most of the theoretical framework assumes small and moderate deformation regimes in the buckled structures. However, the achieved extreme stretchability often involves large deformation [8, 34, 46, 87, 88]. The high nonlinearity from the large deformation in the buckled structures including wrinkled or buckle-delaminated ribbons, freestanding ribbons, or kirigami sheets remains to be explored to predict the extreme stretchability. From a structural design perspective, buckling-induced 2D-to-3D shape changes enable a tremendous design space to expand the library of structures [89, 90]. Recent advances in compression-induced buckling of 2D-to-3D popped structures at small scales show its promise in applications in healthcare [91, 92]. By selectively bonding 2D designs of purposely designed lattice structures or ribbons to the pre-strained elastomer, these planar structures can buckle and pop-up into complicated 3D architectures [91, 92]. Moreover, these pop-up architectures can potentially transform their shapes or architecture locally or globally, enabling enhanced dynamic adaptivity to changing environments. How to leverage these deployable and reconfigurable 3D structures for healthcare applications remains to be explored.

Acknowledgments

J. Y. acknowledges the funding support from NSF awards # CMMI-2005374 and CMMI-2126072.

References

1 Vandeparre, H., Piñeirua, M., Brau, F. et al. (2011). Wrinkling hierarchy in constrained thin sheets from suspended graphene to curtains. *Physical Review Letters* 106 (22): 224301.

2 Yu, H.-H. and Hutchinson, J.W. (2002). Influence of substrate compliance on buckling delamination of thin films. *International Journal of Fracture* 113: 39–55.

3 Yin, J., Gerling, G.J., and Chen, X. (2010). Mechanical modeling of a wrinkled fingertip immersed in water. *Acta Biomaterialia* 6 (4): 1487–1496.

4 Genzer, J. and Groenewold, J. (2006). Soft matter with hard skin: from skin wrinkles to templating and material characterization. *Soft Matter* 2 (4): 310–323.

5 Yin, J., Cao, Z., Li, C. et al. (2008). Stress-driven buckling patterns in spheroidal core/shell structures. *Proceedings of the National Academy of Sciences* 105 (49): 19132–19135.

6 Yin, J., Chen, X., and Sheinman, I. (2009). Anisotropic buckling patterns in spheroidal film/substrate systems and their implications in some natural and biological systems. *Journal of the Mechanics and Physics of Solids* 57 (9): 1470–1484.

7 Wang, Q. and Zhao, X. (2016). Beyond wrinkles: multimodal surface instabilities for multifunctional patterning. *MRS Bulletin* 41 (2): 115–122.

8 Zhang, Q. and Yin, J. (2018). Spontaneous buckling-driven periodic delamination of thin films on soft substrates under large compression. *Journal of the Mechanics and Physics of Solids* 118: 40–57.

9 Ortiz, M. and Gioia, G. (1994). The morphology and folding patterns of buckling-driven thin-film blisters. *Journal of the Mechanics and Physics of Solids* 42 (3): 531–559.

10 Vella, D., Bico, J., Boudaoud, A. et al. (2009). The macroscopic delamination of thin films from elastic substrates. *Proceedings of the National Academy of Sciences* 106 (27): 10901–10906.

11 Mei, H., Landis, C.M., and Huang, R. (2011). Concomitant wrinkling and buckle-delamination of elastic thin films on compliant substrates. *Mechanics of Materials* 43 (11): 627–642.

12 Bowden, N., Brittain, S., Evans, A.G. et al. (1998). Spontaneous formation of ordered structures in thin films of metals supported on an elastomeric polymer. *Nature* 393 (6681): 146–149.

13 Efimenko, K., Rackaitis, M., Manias, E. et al. (2005). Nested self-similar wrinkling patterns in skins. *Nature Materials* 4 (4): 293–297.

14 Chen, X. and Yin, J. (2010). Buckling patterns of thin films on curved compliant substrates with applications to morphogenesis and three-dimensional micro-fabrication. *Soft Matter* 6 (22): 5667–5680.

15 Chung, J.Y., Nolte, A.J., and Stafford, C.M. (2011). Surface wrinkling: a versatile platform for measuring thin-film properties. *Advanced Materials* 23 (3): 349–368.

16 Yang, S., Khare, K., and Lin, P.C. (2010). Harnessing surface wrinkle patterns in soft matter. *Advanced Functional Materials* 20 (16): 2550–2564.

17 Chung, J.Y., Youngblood, J.P., and Stafford, C.M. (2007). Anisotropic wetting on tunable micro-wrinkled surfaces. *Soft Matter* 3 (9): 1163–1169.

18 Lin, G., Zhang, Q., Lv, C. et al. (2018). Small degree of anisotropic wetting on self-similar hierarchical wrinkled surfaces. *Soft Matter* 14 (9): 1517–1529.

19 Rand, C.J. and Crosby, A.J. (2009). Friction of soft elastomeric wrinkled surfaces. *Journal of Applied Physics* 106 (6).

20 Lin, G., Chandrasekaran, P., Lv, C. et al. (2017). Self-similar hierarchical wrinkles as a potential multifunctional smart window with simultaneously tunable transparency, structural color, and droplet transport. *ACS Applied Materials & Interfaces* 9 (31): 26510–26517.

21 Kim, H.N., Ge, D., Lee, E., and Yang, S. (2018). Multistate and on-demand smart windows. *Advanced Materials* 30 (43): 1803847.

22 Efimenko, K., Finlay, J., Callow, M.E. et al. (2009). Development and testing of hierarchically wrinkled coatings for marine antifouling. *ACS Applied Materials & Interfaces* 1 (5): 1031–1040.

23 Ware, C.S., Smith-Palmer, T., Peppou-Chapman, S. et al. (2018). Marine antifouling behavior of lubricant-infused nanowrinkled polymeric surfaces. *ACS Applied Materials & Interfaces* 10 (4): 4173–4182.

24 Khang, D.-Y., Jiang, H., Huang, Y., and Rogers, J.A. (2006). A stretchable form of single-crystal silicon for high-performance electronics on rubber substrates. *Science* 311 (5758): 208–212.

25 Ning, X., Wang, X., Zhang, Y. et al. (2018). Assembly of advanced materials into 3D functional structures by methods inspired by origami and kirigami: a review. *Advanced Materials Interfaces* 5 (13): 1800284.

26 Song, Z., Wang, X., Lv, C. et al. (2015). Kirigami-based stretchable lithium-ion batteries. *Scientific Reports* 5 (1): 10988.

27 Zhang, Y., Yan, Z., Nan, K. et al. (2015). A mechanically driven form of Kirigami as a route to 3D mesostructures in micro/nanomembranes. *Proceedings of the National Academy of Sciences* 112 (38): 11757–11764.

28 Xue, Z., Song, H., Rogers, J.A. et al. (2020). Mechanically-guided structural designs in stretchable inorganic electronics. *Advanced Materials* 32 (15): 1902254.

29 Zhang, Y., Huang, Y., and Rogers, J.A. (2015). Mechanics of stretchable batteries and supercapacitors. *Current Opinion in Solid State and Materials Science* 19 (3): 190–199.

30 Song, Z., Ma, T., Tang, R. et al. (2014). Origami lithium-ion batteries. *Nature Communications* 5 (1): 3140.

31 Tang, Y., Li, Y., Hong, Y. et al. (2019). Programmable active kirigami metasheets with more freedom of actuation. *Proceedings of the National Academy of Sciences* 116 (52): 26407–26413.

32 Cho, Y., Shin, J.H., Costa, A. et al. (2014). Engineering the shape and structure of materials by fractal cut. *Proceedings of the National Academy of Sciences* 111 (49): 17390–17395.

33 Tang, Y., Lin, G., Yang, S. et al. (2017). Programmable kiri-kirigami metamaterials. *Advanced Materials* 29 (10): 1604262.

34 Guan, Y.S., Zhang, Z., Tang, Y. et al. (2018). Kirigami-inspired nanoconfined polymer conducting nanosheets with 2000% stretchability. *Advanced Materials* 30 (20): 1706390.

35 Hong, Y., Chi, Y., Wu, S. et al. (2022). Boundary curvature guided programmable shape-morphing kirigami sheets. *Nature Communications* 13 (1): 530.

36 Fan, J.A., Yeo, W.H., Su, Y. et al. (2014). Fractal design concepts for stretchable electronics. *Nature Communications* 5 (1): 3266.

37 Zhang, Y.-F., Li, Z., Li, H. et al. (2021). Fractal-based stretchable circuits via electric-field-driven microscale 3D printing for localized heating of shape memory polymers in 4D printing. *ACS Applied Materials & Interfaces* 13 (35): 41414–41423.

38 Ma, Q. and Zhang, Y. (2016). Mechanics of fractal-inspired horseshoe microstructures for applications in stretchable electronics. *Journal of Applied Mechanics* 83 (11): 111008.

39 Khang, D.Y., Rogers, J.A., and Lee, H.H. (2009). Mechanical buckling: mechanics, metrology, and stretchable electronics. *Advanced Functional Materials* 19 (10): 1526–1536.

40 Lee, G., Zarei, M., Wei, Q. et al. (2022). Surface wrinkling for flexible and stretchable sensors. *Small* 18 (42): 2203491.

41 Liu, W., Song, M.S., Kong, B., and Cui, Y. (2017). Flexible and stretchable energy storage: recent advances and future perspectives. *Advanced Materials* 29 (1): 1603436.

42 Wen, Z., Yang, Y., Sun, N. et al. (2018). A wrinkled PEDOT: PSS film based stretchable and transparent triboelectric nanogenerator for wearable energy harvesters and active motion sensors. *Advanced Functional Materials* 28 (37): 1803684.

43 Kang, H., Zhao, C., Huang, J. et al. (2019). Fingerprint-inspired conducting hierarchical wrinkles for energy-harvesting E-skin. *Advanced Functional Materials* 29 (43): 1903580.

44 Liu, Y., Pharr, M., and Salvatore, G.A. (2017). Lab-on-skin: a review of flexible and stretchable electronics for wearable health monitoring. *ACS Nano* 11 (10): 9614–9635.

45 Li, B., Cao, Y.-P., Feng, X.-Q., and Gao, H. (2012). Mechanics of morphological instabilities and surface wrinkling in soft materials: a review. *Soft Matter* 8 (21): 5728–5745.

46 Zhang, Q., Tang, Y., Hajfathalian, M. et al. (2017). Spontaneous periodic delamination of thin films to form crack-free metal and silicon ribbons with high stretchability. *ACS Applied Materials & Interfaces* 9 (51): 44938–44947.

47 Sun, Y., Choi, W.M., Jiang, H. et al. (2006). Controlled buckling of semiconductor nanoribbons for stretchable electronics. *Nature Nanotechnology* 1 (3): 201–207.

48 Zhang, Y., Xu, S., Fu, H. et al. (2013). Buckling in serpentine microstructures and applications in elastomer-supported ultra-stretchable electronics with high areal coverage. *Soft Matter* 9 (33): 8062–8070.

49 Isobe, M. and Okumura, K. (2016). Initial rigid response and softening transition of highly stretchable kirigami sheet materials. *Scientific Reports* 6 (1): 24758.

50 Tang, Y., Lin, G., Han, L. et al. (2015). Design of hierarchically cut hinges for highly stretchable and reconfigurable metamaterials with enhanced strength. *Advanced Materials* 27 (44): 7181–7190.

51 Cerda, E. and Mahadevan, L. (2003). Geometry and physics of wrinkling. *Physical Review Letters* 90 (7): 074302.

52 Jiang, H., Khang, D.Y., Song, J. et al. (2007). Finite deformation mechanics in buckled thin films on compliant supports. *Proceedings of the National Academy of Sciences* 104 (40): 15607–15612.

53 Wang, T., Yang, Y., and Xu, F. (2022). Mechanics of tension-induced film wrinkling and restabilization: a review. *Proceedings of the Royal Society A* 478 (2263): 20220149.

54 Chen, X. and Hutchinson, J.W. (2004). Herringbone buckling patterns of compressed thin films on compliant substrates. *Journal of Applied Mechanics* 71 (5): 597–603.

55 Wang, Y. and Wang, C. (2021). Buckling of ultrastretchable kirigami metastructures for mechanical programmability and energy harvesting. *International Journal of Solids and Structures* 213: 93–102.

56 Yu, C., Wang, Z., Yu, H., and Jiang, H. (2009). A stretchable temperature sensor based on elastically buckled thin film devices on elastomeric substrates. *Applied Physics Letters* 95 (14): 141912.

57 Liu, Z.F., Fang, S., Moura, F.A. et al. (2015). Hierarchically buckled sheath-core fibers for superelastic electronics, sensors, and muscles. *Science* 349 (6246): 400–404.

58 Zhen, Z., Li, Z., Zhao, X. et al. (2018). Formation of uniform water microdroplets on wrinkled graphene for ultrafast humidity sensing. *Small* 14 (15): 1703848.

59 Arab Hassani, F., Jin, H., Yokota, T. et al. Soft sensors for a sensing-actuation system with high bladder voiding efficiency. *Science Advances* 6 (18): eaba0412.

60 Lee, H.J., Chun, K.-Y., Oh, J.H., and Han, C.-S. (2021). Wearable triboelectric strain-insensitive pressure sensors based on hierarchical superposition patterns. *ACS Sensors* 6 (6): 2411–2418.

61 Lee, J., Kim, D.W., Chun, S. et al. (2020). Intrinsically strain-insensitive, hyperelastic temperature-sensing fiber with compressed micro-wrinkles for integrated Textronics. *Advanced Materials Technologies* 5 (5): 2000073.

62 She, X., Yao, Q., Yang, G. et al. (2022). Suspended palladium/polymer bilayer for high-contrast and fast hydrogen sensors. *ACS Sensors* 7 (1): 116–122.

63 Mackanic, D.G., Kao, M., and Bao, Z. (2020). Enabling deformable and stretchable batteries. *Advanced Energy Materials* 10 (29): 2001424.

64 Wang, C., Zheng, W., Yue, Z. et al. (2011). Buckled, stretchable polypyrrole electrodes for battery applications. *Advanced Materials* 23 (31): 3580–3584.

65 Zhai, Q., Xiang, F., Cheng, F. et al. (2020). Recent advances in flexible/stretchable batteries and integrated devices. *Energy Storage Materials* 33: 116–138.

66 Ji, K., Li, S., Lin, Y. et al. (2023). Ultrastretchable and compact Zn-MnO_2 rechargeable battery. *ACS Materials Letters* 5 (4): 955–961.

67 Danninger, D., Hartmann, F., Paschinger, W. et al. (2020). Stretchable polymerized high internal phase emulsion separators for high performance soft batteries. *Advanced Energy Materials* 10 (19): 2000467.

68 Liu, W., Chen, J., Chen, Z. et al. (2017). Stretchable lithium-ion batteries enabled by device-scaled wavy structure and elastic-sticky separator. *Advanced Energy Materials* 7 (21): 1701076.

69 Chen, T., Xue, Y., Roy, A.K., and Dai, L. (2014). Transparent and stretchable high-performance supercapacitors based on wrinkled graphene electrodes. *ACS Nano* 8 (1): 1039–1046.

70 Gu, T. and Wei, B. (2016). High-performance all-solid-state asymmetric stretchable supercapacitors based on wrinkled MnO_2/CNT and Fe_2O_3/CNT macrofilms. *Journal of Materials Chemistry A* 4 (31): 12289–12295.

71 Feng, S., Wang, X., Wang, M. et al. (2021). Crumpled MXene electrodes for ultrastretchable and high-area-capacitance supercapacitors. *Nano Letters* 21 (18): 7561–7568.

72 Lamoureux, A., Lee, K., Shlian, M. et al. (2015). Dynamic kirigami structures for integrated solar tracking. *Nature Communications* 6 (1): 8092.

73 He, J., Zhou, R., Zhang, Y. et al. (2022). Strain-insensitive self-powered tactile sensor arrays based on intrinsically stretchable and patternable ultrathin conformal wrinkled graphene-elastomer composite. *Advanced Functional Materials* 32 (10): 2107281.

74 Zhang, Z., Yang, Z., Deng, J. et al. (2015). Stretchable polymer solar cell fibers. *Small* 11 (6): 675–680.

75 Zhang, L., Zhao, Y., and Dai, Q. (2021). Recent progress in perovskite solar cell: fabrication, efficiency, and stability. In: *Development of Solar Cells: Theory and Experiment* (ed. J.K. Roy, S. Kar, and J. Leszczynski), 1–32. Cham: Springer International Publishing.

76 Fukuda, K., Yu, K., and Someya, T. (2020). The future of flexible organic solar cells. *Advanced Energy Materials* 10 (25): 2000765.

77 You, I., Kim, B., Park, J. et al. (2016). Stretchable E-skin apexcardiogram sensor. *Advanced Materials* 28 (30): 6359–6364.

78 Hua, Q., Sun, J., Liu, H. et al. (2018). Skin-inspired highly stretchable and conformable matrix networks for multifunctional sensing. *Nature Communications* 9 (1): 244.

79 Chou, H.-H., Nguyen, A., Chortos, A. et al. (2015). A chameleon-inspired stretchable electronic skin with interactive colour changing controlled by tactile sensing. *Nature Communications* 6 (1): 8011.

80 Hammock, M.L., Chortos, A., Tee, B.C.K. et al. (2013). 25th anniversary article: the evolution of electronic skin (E-skin): a brief history, design considerations, and recent Progress. *Advanced Materials* 25 (42): 5997–6038.

81 Mickle, A.D., Won, S.M., Noh, K.N. et al. (2019). A wireless closed-loop system for optogenetic peripheral neuromodulation. *Nature* 565 (7739): 361–365.

82 Han, M., Wang, H., Yang, Y. et al. (2019). Three-dimensional piezoelectric polymer microsystems for vibrational energy harvesting, robotic interfaces and biomedical implants. *Nature Electronics* 2 (1): 26–35.

83 Zhang, J., Shen, S., Lin, R. et al. (2023). Highly stretchable and biocompatible wrinkled Nanoclay-composite hydrogel with enhanced sensing capability for precise detection of myocardial infarction. *Advanced Materials* 35 (9): 2209497.

84 Zuo, Y., Lu, W., Xia, Y. et al. (2022). Glucometer readout for portable detection of heterogeneous circulating tumor cells in lung cancer captured on a dual aptamer functionalized wrinkled cellulose hydrogel interface. *ACS Sensors* 8 (1): 187–196.

85 Kim, R.-H., Kim, D.H., Xiao, J. et al. (2010). Waterproof AlInGaP optoelectronics on stretchable substrates with applications in biomedicine and robotics. *Nature Materials* 9 (11): 929–937.

86 Kim, S.J., Cho, K.W., Cho, H.R. et al. (2016). Stretchable and transparent biointerface using cell-sheet–graphene hybrid for electrophysiology and therapy of skeletal muscle. *Advanced Functional Materials* 26 (19): 3207–3217.

87 Zang, J., Ryu, S., Pugno, N. et al. (2013). Multifunctionality and control of the crumpling and unfolding of large-area graphene. *Nature Materials* 12 (4): 321–325.

88 Thomas, A.V., Andow, B.C., Suresh, S. et al. (2015). Controlled crumpling of graphene oxide films for tunable optical transmittance. *Advanced Materials* 27 (21): 3256–3265.

89 Hu, N. and Burgueño, R. (2015). Buckling-induced smart applications: recent advances and trends. *Smart Materials and Structures* 24 (6): 063001.

90 Cheng, X. and Zhang, Y. (2019). Micro/nanoscale 3D assembly by rolling, folding, curving, and buckling approaches. *Advanced Materials* 31 (36): 1901895.

91 Xu, S., Yan, Z., Jang, K.I. et al. (2015). Assembly of micro/nanomaterials into complex, three-dimensional architectures by compressive buckling. *Science* 347 (6218): 154–159.

92 Bo, R., Xu, S., Yang, Y., and Zhang, Y. (2023). Mechanically-guided 3D assembly for architected flexible electronics. *Chemical Reviews* 123 (18): 11137–11189.

Part III

Applications

11

Spherical Indentation Behavior of Soft Electronics

Changxian Wang[1,2], Zequn Cui[1], and Xiaodong Chen[1,3]

[1] Nanyang Technological University, Innovative Center for Flexible Devices (iFLEX) & Max Planck–NTU Joint Lab for Artificial Senses, School of Materials Science and Engineering, 50 Nanyang Avenue, Singapore 639798, Singapore
[2] Jinan University, School of Mechanics and Construction Engineering, Guangzhou 510632, PR China
[3] Nanyang Technological University, Institute for Digital Molecular Analytics and Science (IDMxS), 59 Nanyang Drive, Singapore 636921, Singapore

11.1 Spherical Indentation of the Semi-infinite Solid

In this section, we first introduce the displacement function method for solving three-dimensional problems. For two-dimensional problems, the stress function is commonly used and convenient, while in three dimensions the displacement function is more concise than the stress function. Two typical displacement functions are introduced. We then restrict the three-dimensional problem to a scenario where a semi-infinite elastic solid is indented with a revolutionary indenter. A third displacement function composed of the previous two is introduced. The boundary conditions are also discussed. Finally, the scenario of spherical indentation is presented. The solution first proposed by Segedin is emphasized.

11.1.1 General Solution for Elastic Solid with Displacement Function

In three-dimensional Euclidean space, the equilibrium equations are [1]

$$\left.\begin{aligned}
\frac{\partial \sigma_x}{\partial x} + \frac{\partial \tau_{yx}}{\partial y} + \frac{\partial \tau_{zx}}{\partial z} + f_x = 0\\
\frac{\partial \tau_{xy}}{\partial x} + \frac{\partial \sigma_y}{\partial y} + \frac{\partial \tau_{zy}}{\partial z} + f_y = 0\\
\frac{\partial \tau_{xz}}{\partial x} + \frac{\partial \tau_{yz}}{\partial y} + \frac{\partial \sigma_z}{\partial z} + f_z = 0
\end{aligned}\right\} \tag{11.1}$$

Mechanics of Flexible and Stretchable Electronics, First Edition. Edited by Yong Zhu and Nanshu Lu.

Strain–displacement relations are

$$\left.\begin{aligned}
\varepsilon_x &= \frac{\partial u}{\partial x}, \varepsilon_y = \frac{\partial v}{\partial y}, \varepsilon_z = \frac{\partial w}{\partial z} \\
\gamma_{xy} &= \left(\frac{\partial u}{\partial y} + \frac{\partial v}{\partial x}\right) \\
\gamma_{yz} &= \left(\frac{\partial v}{\partial z} + \frac{\partial w}{\partial y}\right) \\
\gamma_{zx} &= \left(\frac{\partial w}{\partial x} + \frac{\partial u}{\partial z}\right)
\end{aligned}\right\} \tag{11.2}$$

Stress–strain relations are

$$\left.\begin{aligned}
\sigma_x &= \frac{E}{1+\mu}\left(\frac{\mu}{1-2\mu}\theta + \varepsilon_x\right) \\
\sigma_y &= \frac{E}{1+\mu}\left(\frac{\mu}{1-2\mu}\theta + \varepsilon_y\right) \\
\sigma_z &= \frac{E}{1+\mu}\left(\frac{\mu}{1-2\mu}\theta + \varepsilon_z\right) \\
\tau_{xy} &= \frac{E}{2(1+\mu)}\gamma_{xy} \\
\tau_{yz} &= \frac{E}{2(1+\mu)}\gamma_{yz} \\
\tau_{zx} &= \frac{E}{2(1+\mu)}\gamma_{zx}
\end{aligned}\right\} \tag{11.3}$$

Substitute strain–displacement relations (11.2) into stress–strain relations (11.3), we get the stress–displacement relations

$$\left.\begin{aligned}
\sigma_x &= \frac{E}{1+\mu}\left(\frac{\mu}{1-2\mu}\theta + \frac{\partial u}{\partial x}\right) \\
\sigma_y &= \frac{E}{1+\mu}\left(\frac{\mu}{1-2\mu}\theta + \frac{\partial v}{\partial y}\right) \\
\sigma_z &= \frac{E}{1+\mu}\left(\frac{\mu}{1-2\mu}\theta + \frac{\partial w}{\partial z}\right) \\
\tau_{xy} &= \frac{E}{2(1+\mu)}\left(\frac{\partial u}{\partial y} + \frac{\partial v}{\partial x}\right) \\
\tau_{yz} &= \frac{E}{2(1+\mu)}\left(\frac{\partial v}{\partial z} + \frac{\partial w}{\partial y}\right) \\
\tau_{zx} &= \frac{E}{2(1+\mu)}\left(\frac{\partial w}{\partial x} + \frac{\partial u}{\partial z}\right)
\end{aligned}\right\} \tag{11.4}$$

where

$$\theta = \frac{\partial u}{\partial x} + \frac{\partial v}{\partial y} + \frac{\partial w}{\partial z} \tag{11.5}$$

And then substitute stress–displacement relations (11.4) into equilibrium Eq. (11.1), we have

$$\left.\begin{aligned}\frac{E}{2(1+\mu)}\left(\frac{1}{1-2\mu}\frac{\partial\theta}{\partial x}+\nabla^2 u\right)+f_x=0\\ \frac{E}{2(1+\mu)}\left(\frac{1}{1-2\mu}\frac{\partial\theta}{\partial y}+\nabla^2 v\right)+f_y=0\\ \frac{E}{2(1+\mu)}\left(\frac{1}{1-2\mu}\frac{\partial\theta}{\partial z}+\nabla^2 w\right)+f_z=0\end{aligned}\right\} \tag{11.6}$$

If we further assume that the body forces are zero, we have

$$\left.\begin{aligned}\frac{1}{1-2\mu}\frac{\partial\theta}{\partial x}+\nabla^2 u=0\\ \frac{1}{1-2\mu}\frac{\partial\theta}{\partial y}+\nabla^2 v=0\\ \frac{1}{1-2\mu}\frac{\partial\theta}{\partial z}+\nabla^2 w=0\end{aligned}\right\} \tag{11.7}$$

So far, we have obtained equations that involve only the components of the displacements.

The solutions of Eq. (11.7) can be solved with the displacement function [2, 3]. The simplest displacement function representation is that in which the displacement is set equal to the gradient of a scalar function

$$\left.\begin{aligned}u=\frac{1}{2G}\frac{\partial\psi}{\partial x}\\ v=\frac{1}{2G}\frac{\partial\psi}{\partial y}\\ w=\frac{1}{2G}\frac{\partial\psi}{\partial z}\end{aligned}\right\} \tag{11.8}$$

Then, the volumetric strain θ can be expressed with

$$\theta=\frac{\partial u}{\partial x}+\frac{\partial v}{\partial y}+\frac{\partial w}{\partial z}=\frac{1}{2G}\nabla^2\psi \tag{11.9}$$

and further

$$\left.\begin{aligned}\frac{\partial\theta}{\partial x}=\frac{1}{2G}\frac{\partial}{\partial x}\nabla^2\psi=\nabla^2 u\\ \frac{\partial\theta}{\partial y}=\frac{1}{2G}\frac{\partial}{\partial y}\nabla^2\psi=\nabla^2 v\\ \frac{\partial\theta}{\partial z}=\frac{1}{2G}\frac{\partial}{\partial z}\nabla^2\psi=\nabla^2 w\end{aligned}\right\} \tag{11.10}$$

Substitute this equation into Eq. (11.7), and we have

$$\left.\begin{aligned}\frac{\partial}{\partial x}\nabla^2\psi=0\\ \frac{\partial}{\partial y}\nabla^2\psi=0\\ \frac{\partial}{\partial z}\nabla^2\psi=0\end{aligned}\right\} \tag{11.11}$$

which means

$$\nabla^2\psi = C \tag{11.12}$$

The corresponding stress can be represented as

$$\left.\begin{aligned} \sigma_x &= \frac{\partial^2\psi}{\partial x^2}, \sigma_y = \frac{\partial^2\psi}{\partial y^2}, \sigma_z = \frac{\partial^2\psi}{\partial z^2}, \\ \tau_{xy} &= \frac{\partial^2\psi}{\partial x \partial y}, \tau_{yz} = \frac{\partial^2\psi}{\partial y \partial z}, \tau_{zx} = \frac{\partial^2\psi}{\partial z \partial x}. \end{aligned}\right\} \tag{11.13}$$

Other kinds of displacement functions could also be adopted:

$$\left.\begin{aligned} u &= \frac{z}{2G}\frac{\partial\varphi}{\partial x} \\ v &= \frac{z}{2G}\frac{\partial\varphi}{\partial y} \\ w &= \frac{z}{2G}\frac{\partial\varphi}{\partial z} - \frac{3-4\mu}{2G}\varphi \end{aligned}\right\} \tag{11.14}$$

Then, the volumetric strain θ can be expressed with

$$\theta = \frac{\partial u}{\partial x} + \frac{\partial v}{\partial y} + \frac{\partial w}{\partial z} = \frac{z}{2G}\nabla^2\varphi - \frac{1-2\mu}{G}\frac{\partial\varphi}{\partial z} \tag{11.15}$$

and further

$$\left.\begin{aligned} \frac{\partial\theta}{\partial x} &= \frac{z}{2G}\frac{\partial}{\partial x}\nabla^2\varphi - \frac{1-2\upsilon}{G}\frac{\partial^2\varphi}{\partial x\partial z} \\ \frac{\partial\theta}{\partial y} &= \frac{z}{2G}\frac{\partial}{\partial y}\nabla^2\varphi - \frac{1-2\upsilon}{G}\frac{\partial^2\varphi}{\partial y\partial z} \\ \frac{\partial\theta}{\partial z} &= \frac{z}{2G}\frac{\partial}{\partial z}\nabla^2\varphi + \frac{z}{2G}\nabla^2\varphi - \frac{1-2\upsilon}{G}\frac{\partial^2\varphi}{\partial z^2} \\ \nabla^2 u &= \frac{z}{2G}\frac{\partial}{\partial x}\nabla^2\varphi + \frac{1}{G}\frac{\partial^2\varphi}{\partial x\partial z} \\ \nabla^2 v &= \frac{z}{2G}\frac{\partial}{\partial y}\nabla^2\varphi + \frac{1}{G}\frac{\partial^2\varphi}{\partial y\partial z} \\ \nabla^2 w &= \frac{z}{2G}\frac{\partial}{\partial z}\nabla^2\varphi + \frac{1}{G}\frac{\partial^2\varphi}{\partial z^2} - \frac{3-4\mu}{2G}\nabla^2\varphi \end{aligned}\right\} \tag{11.16}$$

Similarly, substitute this equation into Eq. (11.7), and we have

$$\nabla^2\varphi = 0 \tag{11.17}$$

The corresponding stress can be represented as

$$\left.\begin{aligned}
\sigma_x &= z\frac{\partial^2\varphi}{\partial x^2} - 2\mu\frac{\partial\varphi}{\partial z},\\
\sigma_y &= z\frac{\partial^2\varphi}{\partial y^2} - 2\mu\frac{\partial\varphi}{\partial z},\\
\sigma_z &= z\frac{\partial^2\varphi}{\partial z^2} - 2(1-\mu)\frac{\partial\varphi}{\partial z},\\
\tau_{xy} &- \frac{\partial^2\varphi}{\partial x\,\partial y},\\
\tau_{yz} &= z\frac{\partial^2\varphi}{\partial y\,\partial z} - (1-2\mu)\frac{\partial\varphi}{\partial y},\\
\tau_{zx} &= z\frac{\partial^2\varphi}{\partial x\,\partial z} - (1-2\mu)\frac{\partial\varphi}{\partial x}.
\end{aligned}\right\} \tag{11.18}$$

11.1.2 Indentation Behavior of Revolution Indenter

For frictionless indentation problems, the shear stress equals zero at all points in the indentation interface, here we denote the plane with $z = 0$. This problem can be transferred to classical problems solved with displacement function. The key point of this problem is constructing a displacement function that makes the shear stress vanish at plane $z = 0$. This displacement function could be constructed with combined functions ψ and φ. We can construct a single displacement function out of functions ψ and φ [2, 4]:

$$\psi = (1-2\mu)\phi, \quad \varphi = \frac{\partial\phi}{\partial z}, \quad \nabla^2\phi = 0 \tag{11.19}$$

The displacement could be expressed with

$$\left.\begin{aligned}
u &= \frac{z}{2G}\frac{\partial^2\phi}{\partial x\,\partial z} + \frac{1-2\mu}{2G}\frac{\partial\phi}{\partial x}\\
v &= \frac{z}{2G}\frac{\partial^2\phi}{\partial y\,\partial z} + \frac{1-2\mu}{2G}\frac{\partial\phi}{\partial y}\\
w &= \frac{z}{2G}\frac{\partial^2\phi}{\partial z^2} - \frac{1-\mu}{G}\frac{\partial\phi}{\partial z}
\end{aligned}\right\} \tag{11.20}$$

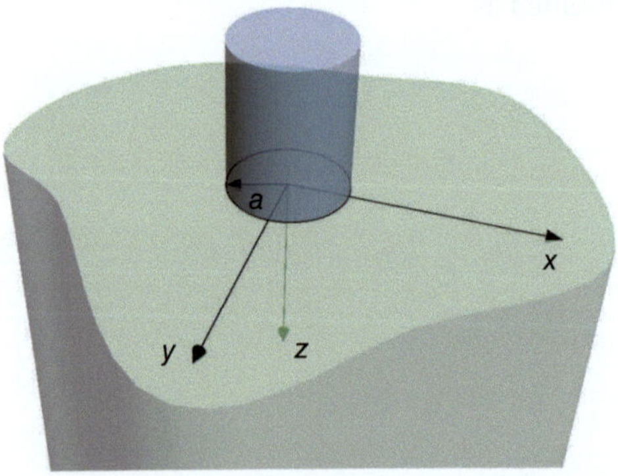

Figure 11.1 A rigid revolutionary indenter indented on an elastic semi-infinite solid.

Substitute this equation into stress–displacement relations (11.4) we get

$$\left.\begin{aligned}
\sigma_x &= z\frac{\partial^3\phi}{\partial x^2\,\partial z} + \frac{\partial^2\phi}{\partial x^2} + 2\mu\frac{\partial^2\phi}{\partial y^2}\\
\sigma_y &= z\frac{\partial^3\phi}{\partial y^2\,\partial z} + \frac{\partial^2\phi}{\partial x^2} + 2\mu\frac{\partial^2\phi}{\partial y^2}\\
\sigma_z &= z\frac{\partial^3\phi}{\partial z^3} - \frac{\partial^2\phi}{\partial z^2}\\
\tau_{xy} &= z\frac{\partial^3\phi}{\partial x\,\partial y\,\partial z} + (1-2\mu)\frac{\partial^2\phi}{\partial x\,\partial y}\\
\tau_{yz} &= z\frac{\partial^3\phi}{\partial y\,\partial z^2}\\
\tau_{zx} &= z\frac{\partial^3\phi}{\partial x\,\partial z^2}
\end{aligned}\right\} \tag{11.21}$$

Obviously, from the stress Eq. (11.21), we note that with the independent variable z going to zero, the shear stress vanishes. The stresses on the plane $z = 0$ reduce to

$$\tau_{yz} = \tau_{zx} = 0, \quad \sigma_z = -\frac{\partial^2\phi}{\partial z^2} \tag{11.22}$$

Now, suppose a perfectly rigid revolutionary solid with a given shape, with a revolutionary axis that coincides with z-axis, is normally indented on the surface $z = 0$ of a semi-infinite elastic medium (see Figure 11.1). During indentation, the revolutionary indenter contacts with the elastic medium, and the deformed surface of the medium will fit the outline. The contact area will be a circle with a circle taken to be a. The z-component of displacement at surface $z = 0$ is prescribed over the region $r \le a$, $z = 0$, due to the prescribed shape of the revolutionary indenter. At the outer area of the contact region, the normal stress is zero. Thus,

$$\left.\begin{aligned}
&w = f(r), && (0 \le r \le a)\\
&\sigma_z = 0, && (r > a)\\
&\tau_{zx} = \tau_{zy} = 0, && (r \ge 0)
\end{aligned}\right\} \tag{11.23}$$

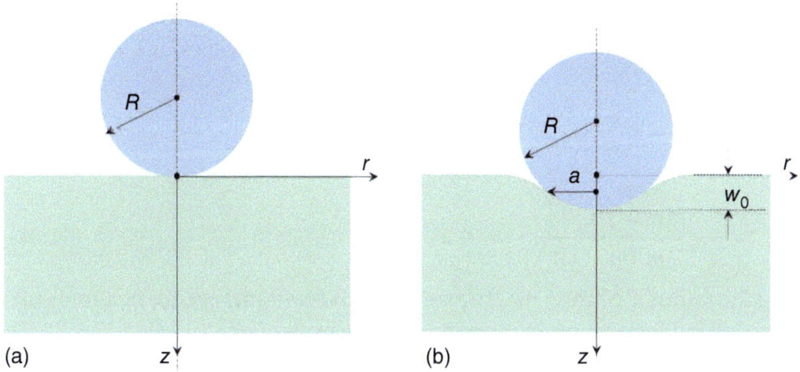

Figure 11.2 Section view of contact between the spherical indenter and the elastic semi-infinite solid: unloaded configuration (a) and loaded configuration (b).

11.1.3 Indentation Behavior of Spherical Indenter

Suppose a rigid sphere, with radius R, indents a depth w_0 below the level of undisturbed boundary $z = 0$, and correspondingly, the radius of the contact area is a. Naturally, a dimensionless quantity was defined $\alpha = a/R$, and obviously, $\alpha \in [0, 1]$. For the indentation placement at the contact dome, we have [4–7]:

$$w(r, 0) = R\left(\sqrt{1 - \left(\frac{r}{R}\right)^2} - 1\right) - w_0, r \in [0, a] \tag{11.24}$$

The z-component displacement could be related to the displacement function

$$w(r, 0) = -\frac{1-\mu}{G}\frac{\partial \phi}{\partial z} \tag{11.25}$$

Since we assume that this is a frictionless contact problem, the shearing stress is zero over the surface of the sample, in Figure 11.2, the surface at $z = 0$; thus, we have the boundary conditions related to force:

$$\left.\begin{aligned} \sigma_z &= 0, \qquad (r > a) \\ \tau_{zr} &= 0, \qquad (r \geq 0) \end{aligned}\right\} \tag{11.26}$$

The total force applied on the spherical indenter could be calculated from the integral on the contacted dome surface:

$$F_p = \int_0^a 2\pi r \sigma_z \mathrm{d}r \tag{11.27}$$

where the compression stress could be calculated from

$$\sigma_z = \frac{E}{2(1-\mu^2)}\frac{\partial w}{\partial z} \tag{11.28}$$

The indentation problem is then reduced to find a function $w(r, z)$, which vanishes at infinity and satisfies the mixed boundary conditions

$$\left.\begin{aligned} & w(r, 0) = f(r), 0 \leq r \leq a, z = 0 \\ & \frac{\partial w}{\partial z} = 0, r > a, z = 0 \end{aligned}\right\} \tag{11.29}$$

Then, the spherical indentation problem is transformed into the finding of an axisymmetric function w. With Taylor expansion of the contact line Eq. (11.24) we get

$$w(r,0) = w_0 - R\sum C_{2m}r^{2m}/R^{2m},\quad 0 \le r \le a \tag{11.30}$$

where

$$C_{2m} = \frac{(2m-2)!}{2^{2m-1}m!(m-1)!},\qquad m \in \mathbb{Z}^+ \tag{11.31}$$

If taking the radius a of the contact area into consideration, Segedin first finds the solution to this indentation problem by finding the form of $w(r, 0)$

$$w(r,0) = R\sum C_{2m}A_{2m}\Phi^{(2m)}(r,0,a)/R^{2m} \tag{11.32}$$

where

$$\left.\begin{aligned} A_{2m} &= \frac{(2^m m!)^2}{(2m-1)!} \\ \Phi^{(2m)}(r,0,a) &= \frac{1}{2m}\left[a^{2m} - r^{2m}\frac{(2m)!}{2^{2m}m!m!}\right] \end{aligned}\right\} \tag{11.33}$$

By solving the Eq. (11.33), the relation between indentation displacement w_0 and radius a of the contact area is obtained

$$w_0 = R\sum (a/R)^{2m}/(2m-1) \tag{11.34}$$

This series is convergent when $(a/R) < 1$, and can be summed into

$$w_0 = a\tanh^{-1}\left(\frac{a}{R}\right) \tag{11.35}$$

where tanh refers to the hyperbolic tangent function, and superscript −1 denotes the inverse function. The total force applied on the spherical indenter is also related to the radius of the contact area:

$$F_{\mathrm{p}} = \frac{ER^2}{1-\mu^2}\left[\left(1+\left(\frac{a}{R}\right)^2\right)\tanh^{-1}\left(\frac{a}{R}\right) - \frac{a}{R}\right] \tag{11.36}$$

11.2 Applications in a Force-Softness Bimodal Sensor Array for Human Body Feature Identification

Identifying human body features through touch is critical in clinical and human–robot interactions. For example, doctors commonly palpate a patient's body by feeling the softness with their hands and fingers to assess the exact location of pain for diagnostic evaluation and to locate specific anatomical structures and abnormalities during surgery. New-generation social robots will need to interact with humans smoothly and naturally, and acquire social-cognitive skills including emotion-based bodily interactions, such as recognizing and interpreting a person's personality and affective information during a handshake, since hand characteristics and grasping forces vary with individuals and psychological states. Digitizing the touch-based identification process on the human body provides quantitative and objective results that would revolutionize subjective and experience-based conventional palpation in the clinic and facilitate the development of human-like social robots. As the indentation behavior of a spherical indenter can explicitly obtain information on the Young's modulus of the matrix material, this phenomenon is well suited for the identification of human body characteristics, especially the physiological characteristics directly related to the softness of tissues and organs.

11.2.1 Design of the Spherical Indenter-Based Force-Softness Sensor

Zequn Cui and Xiaodong Chen et al. first devised a force-softness bimodal sensor (FSS) by utilizing the indentation response phenomenon [8]. The sensor is designed by adding a spherical indenter on the top of a piezoresistive pressure sensor for detecting the information of pressure and softness (Figure 11.3). In general, only the pressure can be detected from a sole pressure sensor (Figure 11.4a, i). The electrical signal is only related to the pressure, which encounters difficulty in distinguishing the samples with varied softness. However, with meticulous adaptation, adding a spherical indenter on the pressure sensor, the resistance experiences two stages (p_1 and p_2) and these two stages represent the combined information of the pressing force (F_{tot}) and Young' s moduli (E) of the samples (Figure 11.4b, ii). In the initial stage, only the indenter is in contact with the sample, for both the softer and harder samples, the same forces are applied on the indenter (F_P, $F_{P\text{-}s} = F_{P\text{-}h}$ = F), and no force is exerted on the substrate (F_S, $F_{S\text{-}s} = F_{S\text{-}h} = 0$), as shown in Figure 11.4c. As the load increases (moment t_0), the softer specimen deforms more and firstly goes into stage II, coming into contact with the substrate (with a touching force $F_{S\text{-}s} = 0$) compared to the harder specimen ($F_{P\text{-}s} = F_{P\text{-}h}$). As the compression forces (moment t_1)

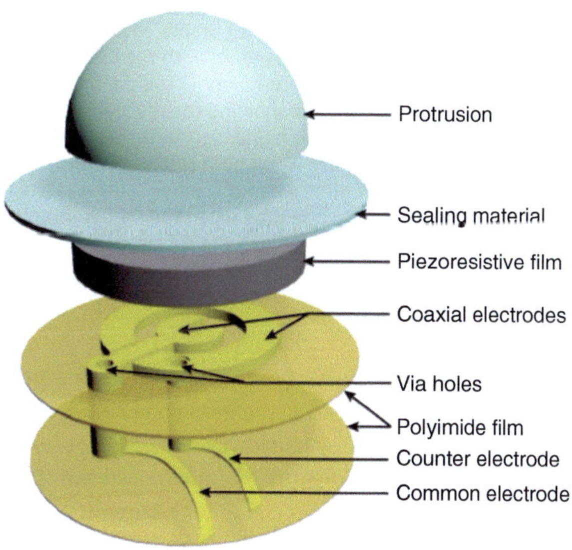

Figure 11.3 Schematic diagram of the force-softness sensor. The diagram shows the architectural design of a single sensing unit. The piezoresistive film adheres to a pair of coaxial circular electrodes, which connect with the common electrode and the counter electrode through holes, respectively. The piezoresistive film is further sealed, and a spherical protrusion adheres on the top.

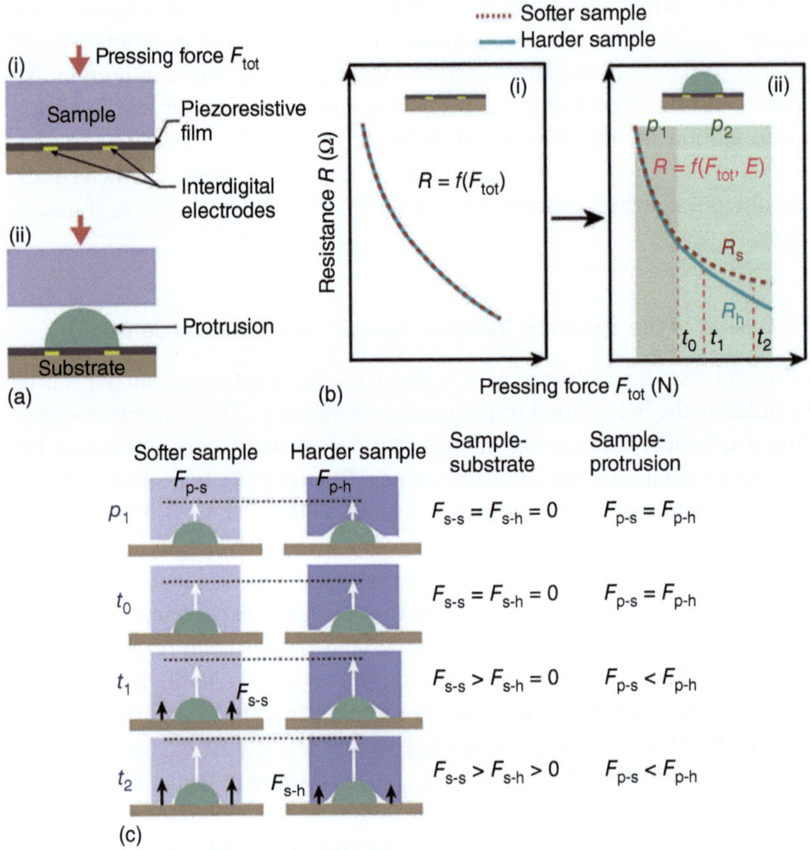

Figure 11.4 The force-softness bimodal sensing mechanism of the sensor. (a) Schematic showing pressure sensors without (i) and with (ii) a hemispherical protrusion on the top, which is pressed by a sample. The pressure sensor consists of a piezoresistive film placed on a pair of interdigital electrodes. (b) Graphs indicate the difference in resistance change between the two sensors when samples with the different softness (Young's modulus) are pressed against them. The FSS can identify the different samples (ii), while the pressure sensor cannot (i). (c) Schematic showing the mechanism of softness identification of the FSS. Softer specimens result in higher sensor resistances than harder ones under the same pressing forces. p_1, the phase when both softer and harder samples only touch protrusion but not substrate. t_0, the time when the softer sample starts to touch the substrate, but the harder sample does not. t_1, the time when the harder sample starts to touch the substrate. t_2, the time after both samples touch the substrate. $F_{S\text{-}s}$, force on substrate pressed by softer sample. $F_{S\text{-}h}$, force on substrate pressed by harder sample. $F_{P\text{-}s}$, force on protrusion pressed by softer sample. $F_{P\text{-}h}$, force on protrusion pressed by harder sample.

continue to increase, more of the compression force on the softer sample is shared by the substrate ($F_{S\text{-}s} > 0$), while the harder specimen comes into contact with the substrate ($F_{S\text{-}h} = 0$), resulting in a lower force on the protrusion for the softer specimen ($F_{P\text{-}s} < F_{P\text{-}h}$) and consequently a higher resistance signal ($R_s > R_h$). The trend continues as the pressing forces (moment t_2) are further increased, resulting in higher resistance for the softer specimen ($F_{P\text{-}s} < F_{P\text{-}h}$, $R_s > R_h$). Therefore, the softness of the

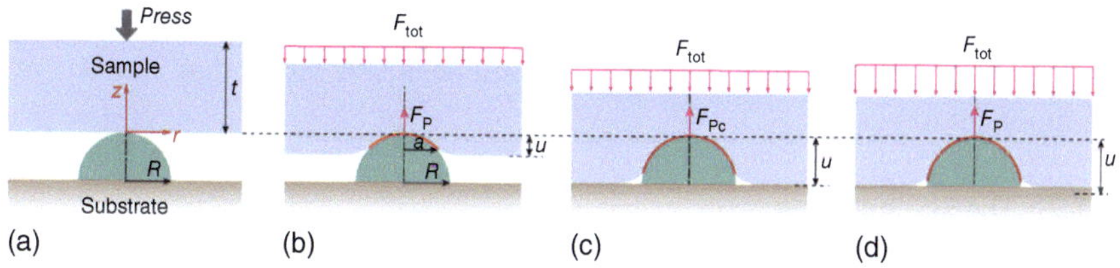

Figure 11.5 Schematic showing the process of a cylindrical elastic sample (radius: l, thickness: t) pressing a hard hemispherical protrusion. (a) The sample just touches the protrusion. (b) The sample presses on the protrusion with a total pressing force of F_{tot} and a displacement of w_0, and the force exerted on the protrusion is F_P. The radius of the projected area of the contact surface is a. (c) The sample just touches the substrate. The force exerted on the protrusion at this moment is marked as F_{Pc}. (d) The sample further presses on the protrusion with more displacement.

sample can be deduced from the change in resistance of the pressure sensor with a protrusion; the softer the sample, the higher the resistance under the same pressing forces. Based on the indentation theory, the explicit relationship between the force applied on the indenter, the total force, and the Young' s modulus could be obtained.

The pressing process begins with the sample just touching the protrusion (Figure 11.5). The whole sample-protrusion pressing can be divided into two processes: (i) The process before (Figure 11.1a,b) and at the moment (Figure 11.5c) that the sample just touches the substrate; (ii) The process after the sample touches the substrate (Figure 11.5d).

11.2.1.1 The First Stage

Before the sample touches the substrate, only the spherical protrusion contacts with and intends on the substrate. Here, we just consider the frictionless contact problem for an axisymmetric rigid spherical indentation on an elastic semi-infinite solid. When the spherical protrusion, with radius R, indented a distance w_0 onto the semi-infinite flat plate with Young's modulus E and Poisson's ratio υ. Geometrically, the contact surface should be a spherical dome. Here, we suppose the radius of the base of the dome is a. Based on the indentation theory, we have:

$$F_p = \frac{ER^2}{1-\upsilon^2}\left[\left(1+\left(\frac{a}{R}\right)^2\right)\tanh^{-1}\left(\frac{a}{R}\right) - \frac{a}{R}\right] \tag{11.37}$$

At this stage, the force applied on the spherical indenter constitutes the total force:

$$F_p = F_{tot} \tag{11.38}$$

As the indentation displacement w_0 reaches the radius of spherical protrusion R, the sample just touches the substrate. At this time, the total force is called critical load. This load divides the loading process into two stages. The critical radius of the contact area a_c could be obtained

$$a_c \tanh^{-1}\left(\frac{a_c}{R}\right) = R \Rightarrow a_c \approx 0.83R \tag{11.39}$$

And thus, the critical load

$$F_{pc} \approx \frac{1.2ER^2}{1-\mu^2} \tag{11.40}$$

After this moment, the total force consists of the force applied on the protrusion F_p and the force applied on the substrate F_s, and obviously larger than any one of the component forces.

11.2.1.2 The Second Stage

After the sample-substrate touch, the sample far away from the protrusion suffers from a vertical compression w_s:

$$w_s = \frac{F_s t}{AE} \tag{11.41}$$

where t is the thickness of the sample, and A refers to the surface area of the substrate. Here, the surface of the substrate is much larger than that of the protrusion, $A \gg \pi R^2$, hence we suppose that $F_s = F_{tot} - F_p \approx F_{tot}$. Thus, the indentation displacement could be calculated with:

$$w_0 = R + w_s \tag{11.42}$$

Combine the above equation with (11.37), we have

$$F_p = \frac{E}{1-\mu^2}\left[\begin{array}{l}\left(R^2 + \left(\phi^{-1}\left(R + w_s\right)\right)^2\right)\tanh^{-1}\left(\dfrac{\phi^{-1}\left(R + w_s\right)}{R}\right)\\ -R\phi^{-1}\left(R + w_s\right)\end{array}\right] \tag{11.43}$$

For simplicity, we make a linear approximation of the above equation with

$$F_p = \beta\frac{ER^2}{1-\mu^2}\left[\frac{F_{tot}t}{AER} + 1\right] \tag{11.44}$$

where β is a solely unknown parameter that could be derived from simulation results.

To establish an equation combined with the two processes, a Heaviside step function is introduced:

$$H(x) = \begin{cases} 0, & x \le 0 \\ 1, & x > 0 \end{cases} \tag{11.45}$$

Finally, combining Eqs. (11.40) and (11.44), the force exerted on the protrusion (F_P) in the whole process can be expressed as the function of the total force (F_{tot}) and the Young's modulus of the sample (E):

$$F_p = F_{tot}H\left(\frac{1.2ER^2}{1-\mu^2} - F_{tot}\right) + \beta\frac{ER^2}{1-\mu^2}\left[\frac{F_{tot}t}{AER} + 1\right]H\left(F_{tot} - \frac{1.2ER^2}{1-\mu^2}\right) \tag{11.46}$$

The parameter β could be obtained with fitting the numerical data.

Numerical simulation is also a powerful method to investigate the indentation behavior. Figure 11.6a–d show the displacement field of an elastic solid indented by a rigid spherical indenter. The reaction forces of the substrate and the protrusion are exacted. In the initial stage, only the indenter contacts with the sample, and the reaction force of the indenter constitutes the total force. With indentation displacement increasing, the substrate contacts with the sample. At this stage, the total force

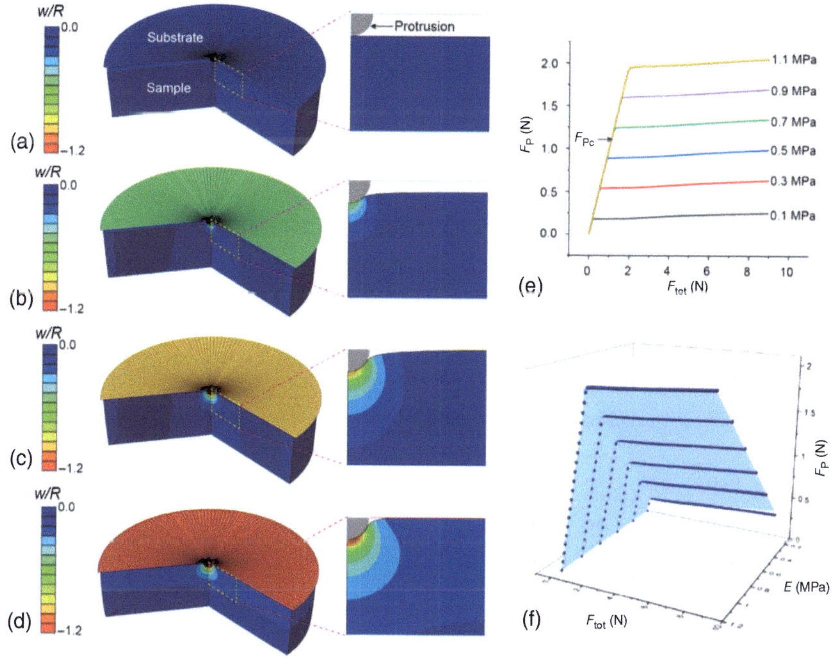

Figure 11.6 Numerical results of the contact between samples with different Young's moduli and the spherical protrusion on a substrate. (a–d) The protrusion with a substrate approaches the sample with a displacement of 0 (a), 0.5R (b), 1.0R (c), and 1.2R (d). R is the radius of the spherical protrusion. (e) The forces exerted on the protrusion (F_P) increased with the total pressing force (F_{tot}) and increased with the Young's moduli of samples. The forces exerted on the protrusion at the moment of samples just touching the substrate (F_{Pc}) exhibited linear response with the Young's moduli of samples. (f) The force exerted on the protrusion (F_P) is strictly correlated with the total pressing force (F_{tot}) and the Young's moduli of samples. The dark blue dots were numerical results, and the light blue curved surface was the corresponding fitted surface expressed by Eq. (11.46) with a coefficient of determination (R^2) of 0.995.

consists of the force applied on the spherical protrusion and the substrate. The theoretical data shows excellent consistency with finite element analysis results with a coefficient of determination (R^2) of 0.995.

Experimental data further proves and realizes the design. As the resistance of the piezoresistive sensor is dependent on the pressure applied on the indenter. The resistance also shows two stages with increasing the total force (Figure 11.7a). For the samples with different Young's moduli, the inflection points are also different. For the pressing of softer samples, the resistance drops slower at the initial stage and always remains higher than that of harder samples, indicating that the resistance signal involves the information of the pressing force and the softness of the samples. The minimum detection forces (~20 mN) are much smaller than those of the pressure sensor (Figure 11.7b). Based on the minimum detection force, a minimum Young's moduli of samples that can be detected is also determined (~12 kPa). The resistances of the FSS at 20 N pressing forces decrease with the samples' Young's moduli (Figure 11.7c), demonstrating the sensing of the softness by the FSS.

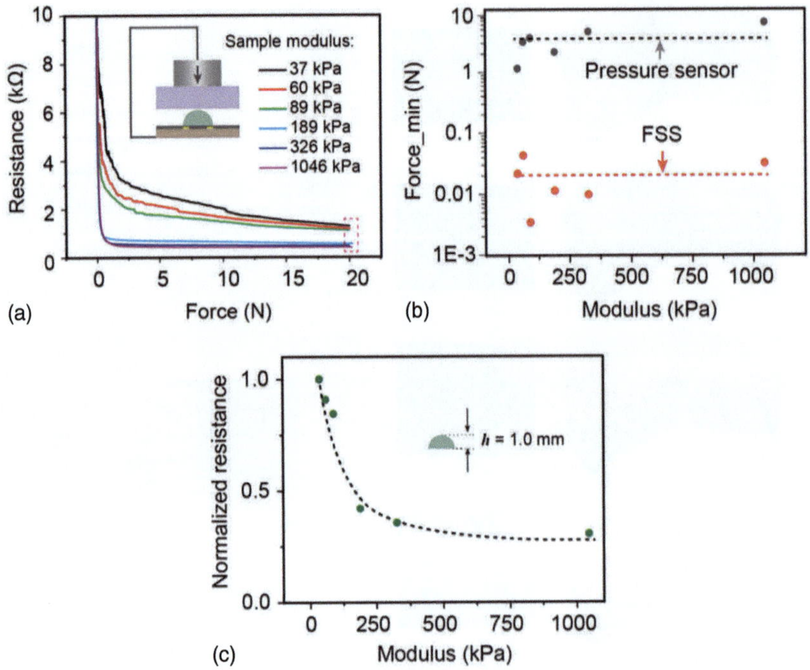

Figure 11.7 Properties of the FSS. (a) Resistance of the FSS decreases differently under the pressing of six different samples controlled by a mechanical testing machine (schematically shown in the inset). Different colored curves indicate samples with different Young's moduli. (b) The minimum response forces of the FSSs are ~2 orders of magnitude lower than those of the pressure sensors. (c) Normalized resistances of the FSS when pressed by different samples at 20 N (red dashed box in (a)) showing that the resistance decreases with the increase of samples' Young's moduli.

11.2.2 Integration of the Force-Softness Sensor Array: Tactile Glove

To detect the force and softness information in a spatial field, the bimodal tactile sensor can be tailored into a hand shape and integrated into a textile glove, yielding a tactile glove (Figure 11.8a). Electrical stable material (polyimide) and hyperelastic material (polyamide, polyurethane) could be used as the substrate to make the glove stretchable and, therefore, conformal. Figure 11.8a displays the scalable and conformal tactile glove with 112 FSSs and a light-weighted massive analog signal channel-based flexible printed circuit board worn on the arm for data collection. For each sensor, the electrodes consist of a pair of exposed coaxial circular electrodes and buried common and counter electrodes (Figure 11.8b). Hot-press was adopted to interlock the structure between the electrode and the piezoresistive sheets (Figure 11.8c inset). Experimental results show that with proper temperature and pressure, hot-press enhances the sensitivity (Figure 11.8c,d). The electrical response of the glove, pressed with "N," "T," and "U" patterns, shows this system is crosstalk-free (Figure 11.8e). Further, the glove can effortlessly uncover the softness mapping when pressed by a heterogeneous material with different Young's modulus (89 and 1000 kPa), as shown in Figure 11.8f. Besides softness mapping,

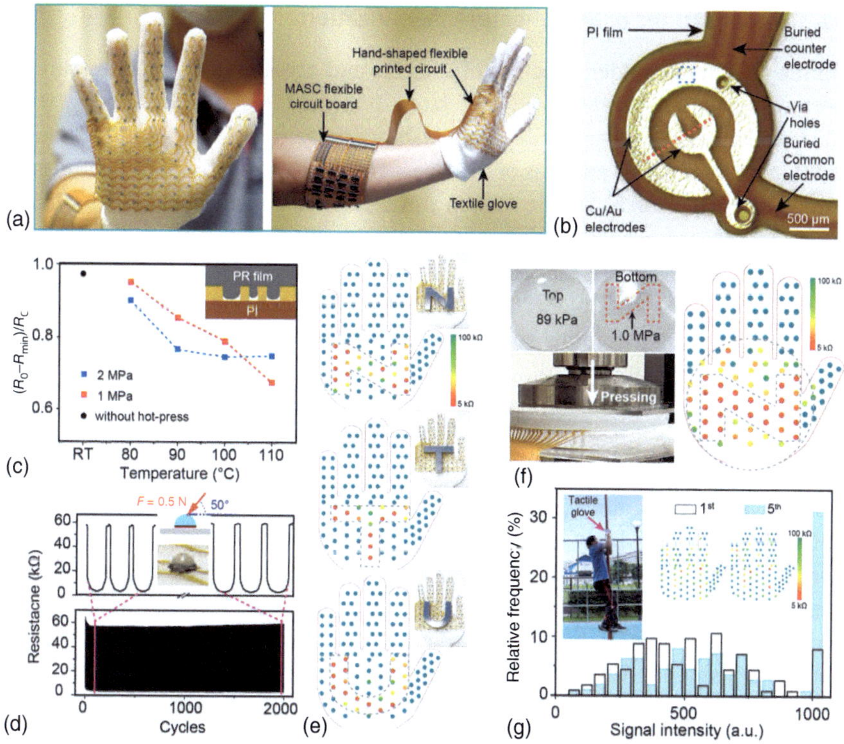

Figure 11.8 Structure and performance of the tactile glove. (a) Photos showing the scalable tactile glove consisting of 112 FSSs and a massive flexible printed circuit board based on analog signal channel worn on the forearm. (b) Photo showing the exposed coaxial circular electrodes, via holes, and buried common electrode and the counter electrode of the FSS for the tactile glove. (c) Resistance variations of the piezoresistive pressure sensor fabricated under different hot-pressing conditions (pressure and temperature). Inset showing the interlocked structure between the piezoresistive film and the electrodes. (d) Dynamic mechanical–electrical response of the FSS under tilted force (0.5 N @ 50°) with 2000 cycles shows strong mechanical robustness. (e) Tactile images show crosstalk-free tactile gloves when pressed by three models. (f) A nonhomogeneous sample (a harder mirrored "N" letter embedded in the bottom of a softer disk sample) was pressed on the tactile glove, showing softness mapping of the tactile glove. (g) The tactile glove can withstand a person (~60 kg) climbing a pole for more than five times gripping action. Tactile images (insets) and signal intensities of all sensors of the tactile glove at 1st (blank bars) and 5th (light green bars) gripping were shown. Signal intensities larger than 1000 indicate non-touched or damaged sensors. The increased relative frequency from 1st (8%) to 5th (31%) grasp suggests that 77% of all sensors are still working well after five times grasps, demonstrating the mechanical robustness of the tactile glove. The color bars in (e,f,g) indicate resistances ranging from 5 kΩ (red) to 100 kΩ (dark green).

the tactile glove is robust enough to withstand frequent high-force pressing to meet the practical requirement. A person (~60 kg) wearing the glove climbing a pole with more than five times gripping action proves that the glove can withstand both large normal and shear forces (Figure 11.8g). The increased relative frequency of signals larger than 1000 indicates that most sensors (77%) remained working after five times gripping.

11.2.3 Applications in Body Feature Identification

With the advantage of identification of softness and pressure, this capacity can be expected to extend into sensing a real person's body features (e.g. chest, belly, arm, shoulder, Figure 11.9a). The tactile glove can be worn by a doctor in palpation assessment (Figure 11.9b). The patterns of the tactile signals were not differentiable either in different human body features (or in different organ models of the simulator, Figure 11.9c). This issue could be addressed by employing the machine learning method (Figure 11.9d–g). A tactile glove with 112 sensors can identify the body features (arm, belly, chest, shoulder) of a real person and healthy/pathological liver and spleen models of an abdominal simulator (liver-P, liver-H, spleen-P, and spleen-H) with high accuracy of 98%. Traditionally, the body features can be identified by doctors in diagnosis with palpation skills to assess the health condition. In this respect, the tactile glove may offer quantitative data of inner body features to support the doctor in making appropriate treatment methods.

The above results demonstrate the effectiveness of the tactile glove in identifying features of the human body and its potential application in the digitalized assessment of palpation. This would be important in clinical practice as conventional palpation is subjective, based on experience, and therefore can be unreliable. For example, when assessing the location of pain in patients with low back pain, the reliability of palpation varies widely in different tissues: kappa values (an indicator of reliability) vary from 0.34 to 0.80 for soft tissues and 0.14 to 0.73 for joint and bony structures. Considering that the identification reliability (Kappa value) of conventional manual palpation on real patients is mostly around 0.4, the Kappa values of our single FFS and tactile glove were calculated for identifying the real body features and the simulator, and the tactile glove achieved a reliability of 0.98 (Figure 11.9f), indicating potential applicability in future healthcare.

11.3 Applications in a Self-Locked Young's Modulus Sensor for Quantifying the Softness of Swollen Tissues in the Clinic

In principle, a single extensometer can only measure the strain or force applied to the sensor. Since the Young's modulus of a sample is usually calculated by measuring the force (stress) and the deformation of the sample (strain), two sensors and a displacement control system are usually required. This usually results in a bulky and complicated system. To minimize the size of the E-modulus measurement system and achieve *in situ* detection, it is essential to simplify the two variables (force and deformation) into one. To achieve this, the self-locking effect can be used with indentation behavior. With the self-locking effect, the information of deformation and applied force could be obtained simultaneously, and the Young's modulus could also be obtained. The self-locking sensor can be used to restore human-like haptic behavior in elasticity sensing and medical palpation.

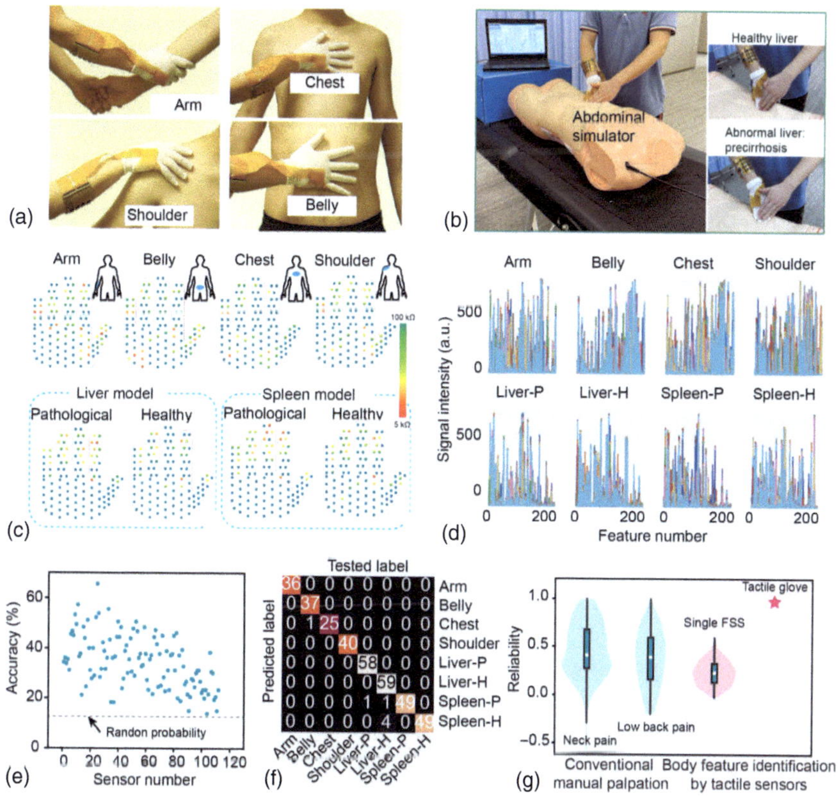

Figure 11.9 Identification of body features. (a,b) Photos showing that the tactile glove was used to identify body parts of a real person (a) and organ models inside an abdominal simulator (b). (c) Tactile signal patterns of four different body parts (arm, belly, chest, shoulder) of the real person (top) and four different organ models (healthy liver, pathological liver, healthy spleen, pathological spleen) of the abdominal simulator obtained by the tactile glove. The color bar indicates resistance ranging from 5 kΩ (red) to 100 kΩ (dark green). (d) Statistic of all 224 feature intensities during one pressing on each object cannot be distinguished from each other. (e) The accuracy of each individual FSS of the tactile glove in identifying the eight objects shows that most of them are much higher than the random probability of 11.5, with a few sensors reaching an accuracy of ~60%, showing the priority of the FSS. (f) Machine learning results show high accuracy (98%) of the tactile glove in identifying the eight items. (g) Comparison of the reliability (Kappa value) of body feature identification among conventional manual palpation, single FSSs, and the tactile glove. Manual palpation typically has lower reliability (~0.4) on real patients for identifying the pain locations of neck pain and lower back pain. The identification reliability of the single FSSs and the whole tactile glove on the four different body parts and the four artificial organ models achieves 0.2–0.3 and 0.98, respectively.

11.3.1 Design of the Fingertip Modulus Sensor

The indentation behavior was applied in the design of a modulus sensor. Zequn Cui and Xiaodong Chen et al. proposed a fingertip modulus sensor (FMS) that can recreate humanoid softness perception relying on the ability of *in situ* Young's modulus measurement (Figure 11.10) [9]. A strain sensor was enclosed into a

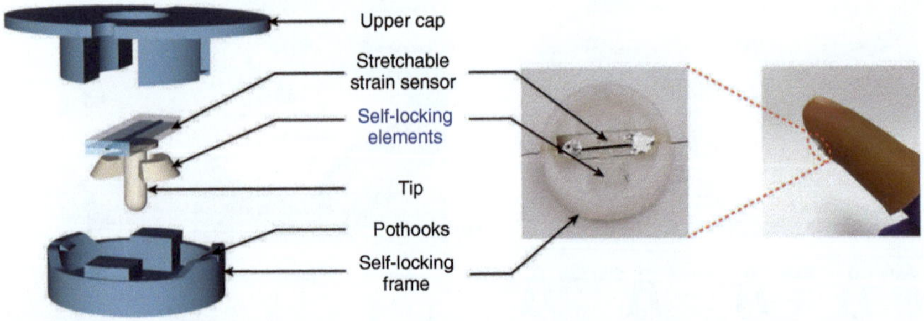

Figure 11.10 Schematic diagram of the fingertip modulus sensor. Schematic showing the architecture of an FMS consisting of an upper cap, a stretchable strain sensor, a tip, a pair of self-locking elements and a self-locking frame. The real photos show the internal structure of the miniaturized FMS and the fingertip wearable FMS inserted into a prosthetic glove and worn on a human finger.

3D-printed architecture to record the indentation data of the spherical indenter. A choreographed self-locking architecture was employed to lock the indenter at a specific moment, at which both the indentation force and the sample deformation can be measured by the same signal of the strain sensor. The self-locking architecture simplifies two variables into one and enables Young's modulus measurement independent of the pressing forces. This further simplifies the components and minimizes the size of the whole system, yielding a fingertip-wearable modulus sensor that can quantify the Young's modulus of a sample in a humanoid softness detection behavior. Although the sensor is very tiny, it can detect the softness of materials in a wide range, from foods to the human body, with just pressing on the target. Especially, the application in the human body can offer quantitative data of softness to support the doctor in palpation to judge whether the tissues or organs are in a healthy state or not.

Theoretically, when a concentrated normal force F is applied on the boundary of an elastic semi-infinite solid, with assuming the displacement function [10]

$$\phi = C[\rho - z\ln(\rho + z)] \tag{11.47}$$

where ρ refers to the distance between the target point and the force bearing point, z refers to the depth beneath the surface of the elastic solid, and C is a constant, the displacement at the surface of the solid could be deduced explicitly

$$w|_{z=0} = \frac{(1-\mu^2)F}{\pi E\rho} \tag{11.48}$$

When the applied force was replaced with a distributed force P, the displacement would be

$$w|_{z=0} = \frac{(1-\mu^2)}{\pi E\rho}\int q\mathrm{d}A \tag{11.49}$$

For a spherical indenter normally indented on the semi-infinite solid (Figure 11.11a), the displacement would be

$$w = w_\mathrm{i} + w_\mathrm{s} = \left(\frac{1-{\mu_\mathrm{i}}^2}{\pi E_\mathrm{i}\rho} + \frac{1-{\mu_\mathrm{s}}^2}{\pi E_\mathrm{s}\rho}\right)\int q\mathrm{d}A \tag{11.50}$$

Hertz model can interpret this contact condition between a spherical indenter and a sample. The radius of contact area a is related to the contact force $F_{\text{i-s}}$ and the mechanical properties of contacting materials, given by [11]:

$$a = \left(\frac{3}{4}\frac{F_{\text{i-s}}R}{E^*}\right) \tag{11.51}$$

where R is the radius of the indenter, E^* is the effective Young's modulus consisting of that of the indenter E_i and the sample E_s:

$$\frac{1}{E^*} = \frac{1-\mu_i^2}{E_i} + \frac{1-\mu_s^2}{E_s} \tag{11.52}$$

where υ_i and υ_s are the Poisson's ratio of the indenter and the sample, respectively. For a rigid indenter and a soft material contact, $E_i \gg E_s$, Eq. (11.52) is simplified and rearranged as follows

$$\frac{1}{E^*} = \frac{1-\mu_s^2}{E_s} \tag{11.53}$$

The contact radius a is related to the indentation depth of the sample h_s, given by

$$a = \sqrt{\rho h_s} \tag{11.54}$$

Substituting Eqs. (11.53) and (11.54) into Eq. (11.51), we obtain:

$$E_s = \frac{3}{4}\left(1-\mu_s^2\right) F_{\text{i-s}}\rho^{-1/2}h_s^{-3/2} \tag{11.55}$$

In the above equation, only the loading force $F_{\text{i-s}}$ and indentation displacement h_s are unknown variables (Figure 11.11a). The loading force can be measured by a resistive stretchable strain sensor, and the indentation displacement can be measured by the tip length outside the frame when the frame just touches the sample surface. The self-locking architecture can lock the tip and maintain the tip length outside the frame unchanged once the frame touches the sample (Figure 11.11b). Before the frame touches the sample (period P_1 in Figure 11.11c), the tip moves upward to stretch the strain sensor, and both tip displacement and sensor strain vary; once the frame touches the sample (moment M_1 in Figure 11.11c), the suddenly increased contact area leads to rapidly increased touch force, which is large enough to push the self-locking elements to move horizontally and start to generate friction force on the tip to lock tip movement; a larger touch force generates a larger friction force on the tip, keeping tip location and maintaining the strain at ε_l and the resistance at R_1 (period P_2 in Figure 11.11c). Relying on the stable value ε_l, both the applied force and indentation displacement at moment M_1 are obtained.

The measurement of standard materials further confirms the correctness and accuracy of this sensor. A series of samples of PDMS and Eco-flex with different proportions and different curing conditions were fabricated. The Young's modulus values measured using the mechanical tester (MTS Model 42) were adopted as the standard index. For FMS, the directly measured data are resistance (Figure 11.12a), and then the resistance was transformed into Young's modulus. Great correlation between the values measured by FMS and MTS suggests that the FMS is highly reliable in detecting the Young's modulus (Figure 11.12b). In daily life, a person feels the

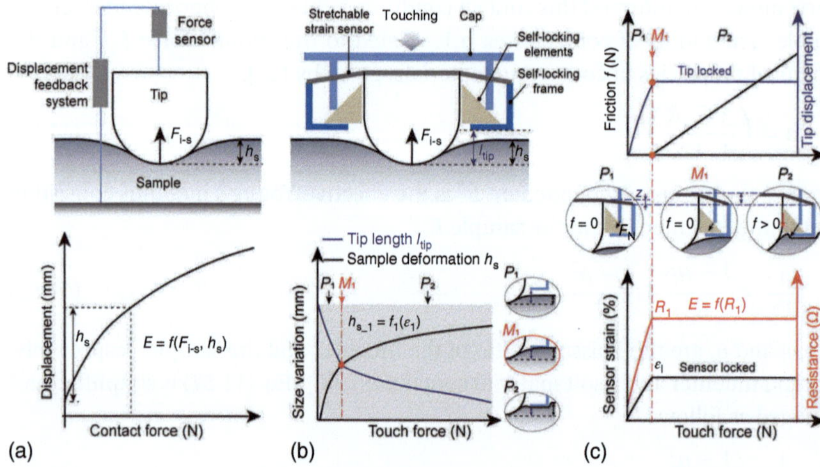

Figure 11.11 Development of locking methodology for simple and rapid Young's modulus measurements. (a) Schematic showing conventional methods based on the Hertz model using a force sensor and a displacement feedback system to simultaneously record the contact force and displacement of the tip (top). The Young's modulus of a stabilized sample is calculated from the two variables (bottom). (b) Schematic showing the architecture design of the locking architecture using a self-locking effect to eliminate the need for displacement feedback control. The architecture consists of a stretchable strain sensor to measure the contact force $F_{i\text{-}s}$ and self-locking structures to obtain sample deformation h_s (top). During touching, the tip length l_{tip} decreases and sample deformation h_s increases; at the moment that the self-locking frame just touches the sample surface (moment M_1), the sample deformation h_{s_1} is equal to the tip length l_{tip_1}, and can be obtained from the strain ε_l of the strain sensor (bottom). (c) Schematic showing the self-locking effect, which keeps the tip location (top) and the sensor strain (bottom) by friction since moment M_1. Insets: Schematic showing the tip pushed up in P_1 (left); schematic showing the tip location and sensor stretching status are maintained from M_1 (middle) to P_2 (right), leading to stable resistance R_1.

surrounding object by tough feeling. Especially when a person picks foods, information of softness from haptic feeling is kind of important information that supports them in judging the foods. Conventionally, in food science, a texture analyzer (TA) is a commonly adopted instrument to measure softness. A TA defines food elasticity by the maximum contact force (peak force) obtained when the probe indents the food at a given depth. To demonstrate the application potential of FMS, a series of foods, including ham sausage, softs, cake bread, jelly drop, marshmallow, glutinous rice cake, glue pudding, and jelly, are measured using the proposed FMS (Figure 11.12c). For comparison, the peak forces measured by the TA were adopted as a benchmark. The experimental data measured by TA and FMS shows great consistency in trend, further proving the reliability of the proposed FMS (Figure 11.12d).

11.3.2 Applications in Softness Measurement of Swollen Tissues in the Clinic

With the advantage of identification of Young's modulus, this capacity can be expected to extend into sensing the pathological characteristics of living tissues or

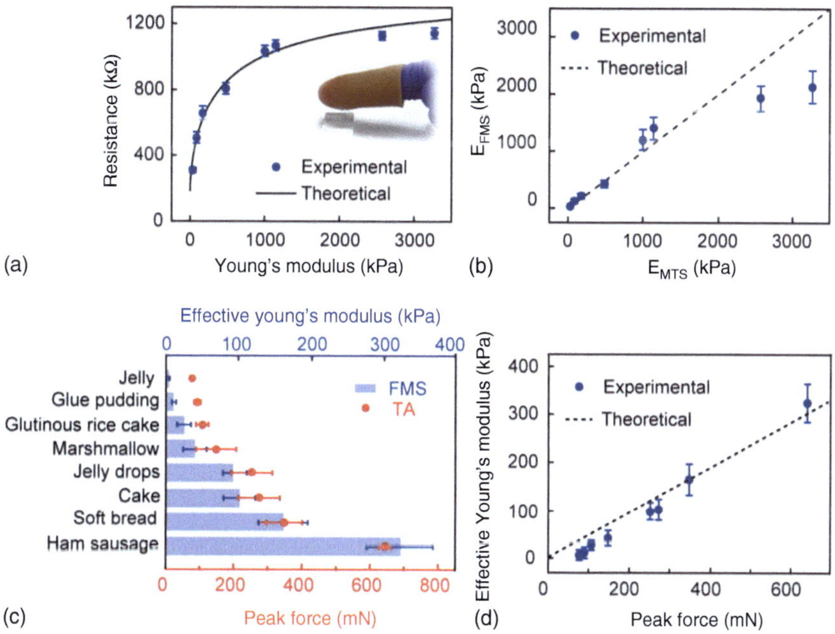

Figure 11.12 Theoretical analysis and experimental verification of the reliability of the self-locked sensor. (a) Experimental results (blue dots) and theoretical analysis (black curve) of the relationship between resistance and the Young's modulus agree well. Inset: Photo showing an FMS is integrated on a prosthetic glove to measure a sample by touching. (b) Good correlation between Young's moduli measured by FMS and those obtained using commercial MTS shows our FMS is accurate. The dotted line is a theoretical prediction of the correlation between the two methods. The error bars in (a) and (b) are the calculated standard deviations from nine measurements. (c) Effective Young's moduli (blue bars) measured by the FMS-armed prosthesis and peak forces (red dots) measured by a commercial texture analyzer (TA) of various foods show the same trend in elasticity. (d) Good correlation between effective Young's moduli obtained by our FMS-armed prosthesis and peak forces measured by a TA demonstrate the accuracy of our FMS-armed prosthesis. Blue dots are experimental data from (c). The dashed line is the theoretical results based on the Hertz model. Error bars in (c) and (d) are standard deviations calculated from nine measurements.

organs. A typical and potential application case is the assessment of the softness of swollen tissues. Assessing the severity of swollen tissues is vital as skin suture problems would occur when performing an operation on the swollen tissues. The swollen injury would cause dramatic variation in the softness of the tissue. Hence this change of softness could be adopted as a kind of indicator that can reflect the severity degree of swelling. In clinics, a doctor could assess the severity by palpation. Therefore, the results largely rely on the experience of the attending doctor. In this manner, the FMS may offer quantitative values of Young's moduli of the swollen tissues that may support the doctor in the clinic.

To make the sensor more compact and portable, a Bluetooth module, a microcontroller unit, and a mini power bank module were adopted to construct the wireless system (Figure 11.13a). Such a simple and compact wireless system is owing

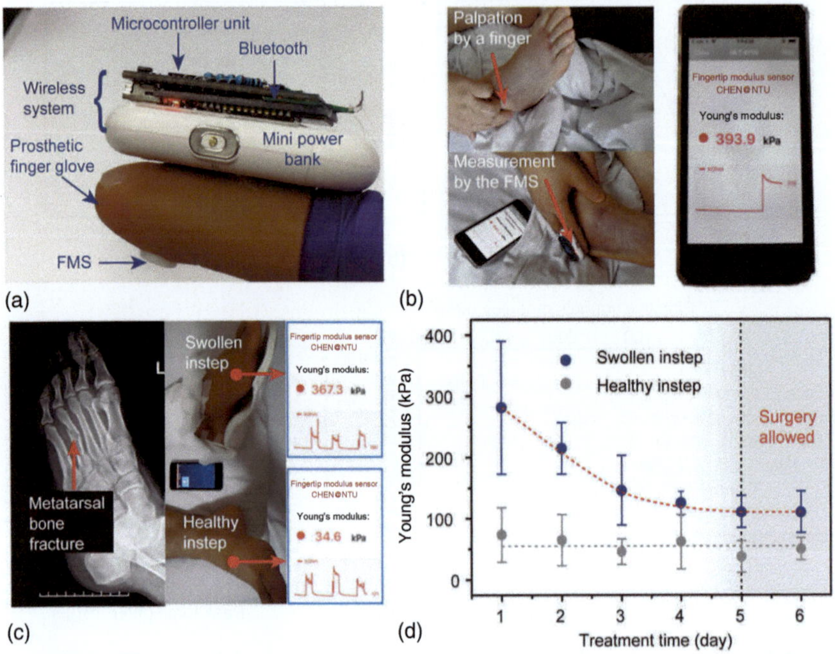

Figure 11.13 A fully mobile FMS-armed prosthesis enables human-like palpation skills. (a) Photograph showing the fully mobile FMS-armed prosthesis containing an FMS, a wireless module and a mini power bank mounted on a prosthetic finger glove. (b) Photographs of a doctor palpating a patient's swollen right ankle (top left) and using the FMS-armed prosthesis to measure Young's modulus of the swollen ankle (bottom left). The smartphone app interface immediately displays the Young's modulus values. (c) FMS-armed prosthesis measurements on patient 2. Radiograph showing the fracture of the fourth metatarsal of the left foot (left). The swollen left instep and the healthy right instep were measured on the first day of the fracture (center) and the Young's modulus values were shown in real-time (right). (d) Young's modulus measurements of the swollen and healthy instep of patient 2 obtained with our FMS-armed prosthesis show swelling subsided within four days and surgery is recommended since day 5, in line with the doctor's assessment. Error bars are standard deviations calculated from a minimum of five measurements.

to the sole requirement of a resistance signal, which is easy to be measured and transmitted. In comparison to palpation, the measured Young's modulus is more precise and quantitative. Moreover, this data can be displayed on the smartphone app in real-time and be stored in the memorizer permanently. Last but not the least, a new person without palpation experience can also manipulate this sensor after a brief training and further share the data with the doctor. A demonstration of a patient (75-year-old female) who suffers from calcaneus comminuted fracture on the right ankle showed that in comparison with the healthy tissue, the swollen injury can significantly increase the Young's modulus (Figure 11.13b). To further show the capability of FMS in clinical diagnosis, a long-term monitoring case (33-year-old male, metatarsal bone fracture on the left foot) that lasts for six days is also demonstrated (Figure 11.13c). At the first day, the Young's modulus of the left swollen instep is more than 10 times larger than that of the right healthy instep. And at the initial stage the Young's modulus of the swollen instep is decreased heavily, and in the following stage the modulus value decreases in a gentle way. Based on Young's modulus readings and the criterion, surgery was recommended after day 5, which agreed well with the doctor's assessment (Figure 11.13d).

11.4 Conclusions

Soft electronics have attracted overwhelming attention due to their potential applications in a wide range of industries such as soft robotics, electronic skins, biomedical engineering, and health monitoring. Many mechanical phenomena are widely used in soft electronics, such as the use of bulking of strips in the application of flexible conductors, such as the use of cracks of conductive film in the application of strain sensors. In this chapter, we first discussed the basic mechanical theories of indentation behavior, from the general three-dimensional problem to the revolutionary indentation problem of the indenter, and finally to the specific scenario of a semi-infinite isotropic elastic solid being indented by a spherical indenter. Finally, two sensors, the bimodal force-softness sensor array and the self-locking Young's modulus sensor, were outlined. The sensor can be used in various medical devices for digitized health monitoring and haptic interfaces for advanced humanoid robots.

References

1 Sadd, M.H. (2009). *Elasticity: Theory, Applications, and Numerics*. Academic Press.

2 Green, A.E. and Zerna, W. (1992). *Theoretical Elasticity*. Courier Corporation.

3 Timoshenko, S. and Goodier, J.N. (1951). *Theory of Elasticity* (ed. S. Timoshenko and J.N. Goodier). McGraw-Hill.

4 Segedin, C. (1957). The relation between load and penetration for a spherical punch. *Mathematika* 4 (2): 156–161.

5 Bunzai, A.N. (1959). Elasticity problem in the case a rigid sphere is pressed on the plane surface of semi-infinite elastic solid: the effect of friction on the contact surface. *Bulletin of JSME* 2 (6): 244–251.

6 Barber, J. (1973). Indentation of the semi-infinite elastic solid by a hot sphere. *International Journal of Mechanical Sciences* 15 (10): 813–819.

7 Harding, J. and Sneddon, I. (ed.) (1945). The elastic stresses produced by the indentation of the plane surface of a semi-infinite elastic solid by a rigid punch. In: *Mathematical Proceedings of the Cambridge Philosophical Society*, 16–26. Cambridge University Press.

8 Cui, Z., Wang, W., Xia, H. et al. (2022). Freestanding and scalable force-softness bimodal sensor arrays for haptic body-feature identification. *Advanced Materials* 34 (47): e2207016.

9 Cui, Z., Wang, W., Guo, L. et al. (2022). Haptically quantifying Young's modulus of soft materials using a self-locked stretchable strain sensor. *Advanced Materials* 34 (25): e2104078.

10 Argatov, I. and Mishuris, G. (2018). *Indentation Testing of Biological Materials.* Springer.

11 Johnson, K.L. and Johnson, K.L. (1987). *Contact Mechanics.* Cambridge University Press.

12

Mechanics of Wet Adhesion

Jiawei Yang[1] *and Ruobing Bai*[2]

[1] *Worcester Polytechnic Institute, 100 Institute Road, Worcester, MA 01609-2280, USA*
[2] *Northeastern University, Department of Mechanical & Industrial Engineering, Boston, MA 02115, USA*

12.1 Introduction

Wet materials such as synthetic hydrogels and biological tissues are three-dimensional hydrophilic polymer networks in a water-rich environment. They are soft and wet, with elastic modulus ranging from a few pascals to megapascals and water content above 60 wt%. Synthetic hydrogels can mimic biological tissues chemically, mechanically, electrically, and physiologically and have thus enabled fast-expanding applications in biology and biomedicine such as tissue scaffolds [1, 2], wound dressings [3–7], and drug delivery [8–12]. They are also being developed for new transformative applications such as ionotronics [13–15], bioelectronics [16], soft robots [13, 17, 18], and soft power sources [19–22]. In particular, flexible or stretchable electronic devices that intimately integrate with synthetic hydrogels, living organs, or tissues have demonstrated diagnostic and therapeutic capabilities for treating a broad range of diseases and disorders [16, 23–28]. In such a device, the wet material (a hydrogel or living organ/tissue) is required to integrate with various materials, either abiotic or biotic, to form wet adhesion at their interfaces with specified features such as being strong, tough, or sometimes compliant. This emerging demand for fully controllable wet adhesion for various materials promotes intensive research in areas including mechanics, polymer science, electrical engineering, and bioengineering.

Two materials adhered together are called *adherends*. Their adhesion is often, but not always, via a third material, called an adhesive. In general, we call the adhesion *wet adhesion* when at least one of the adherends is a hydrogel (in this chapter, we use a hydrogel to represent either a synthetic hydrogel or a biological tissue). A hydrogel consists of a stretchable polymer network with a mesh size on the order of 10 nm, much larger than the water molecules (size of ~1 nm) it absorbs. The abundant water molecules in the hydrogel change neighbors readily and transmit force negligibly,

Part of materials in this chapter was published by J.Y., R.B., B.C., and Z.S. in a review article "Hydrogel Adhesion: A Supramolecular Synergy of Chemistry, Topology, and Mechanics" in Advanced Functional Materials, 2019.

Mechanics of Flexible and Stretchable Electronics, First Edition. Edited by Yong Zhu and Nanshu Lu.

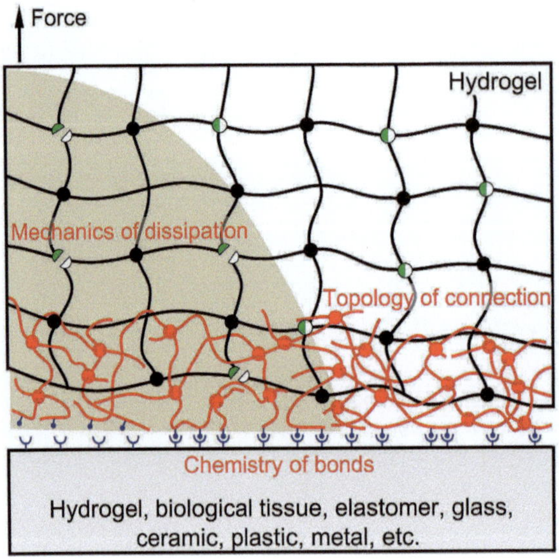

Figure 12.1 Illustration of a synergy of chemistry of bonds, topology of connection, and mechanics of dissipation for strong wet adhesion. Source: Ref. [29]/John Wiley & Sons.

hence posing a fundamental challenge to adhesion: a strong dry adhesion may easily form using an adhesive (e.g. epoxy) between two dry materials (e.g. elastomers, plastics, metals, ceramics, etc.) without special chemical bonds as long as the interfacial physical bonds are sufficiently dense; by contrast, a strong wet adhesion *cannot* rely on the water molecules at a wet interface due to their negligible force transmission but needs special ways to connect the polymer network of the wet material to the other adherend. Indeed, hydrogels typically do not adhere strongly to any material, including themselves.

Last few years have seen transformative advances in achieving strong wet adhesion based on a generic principle to elicit the synergy of chemistry of bonds, topology of connection, and mechanics of dissipation (Figure 12.1) [29]. The polymer network of a wet material is connected to another material through covalent bonds or noncovalent complexes together with a special topology of connection. Separating the adherends dissipates energy through cascading events across multiple length scales including bond cleavage, chain retraction, and bulk hysteresis, leading to strong adhesion. Weak adhesion inevitably originates from missing one or more parts of this synergy.

12.2 Characterization of Adhesion

Adhesion is usually characterized by adhesion strength and adhesion energy (toughness). The former (in unit of Pa) measures the maximum stress to separate two adherends and the latter (in unit of J/m^2) measures the total energy needed

Table 12.1 Representative adhesion strength and adhesion energy for different wet adhesion systems.

Wet adhesion	Lap-shear strength (kPa)	Adhesion energy (J/m^2)
Hydrogels with any materials	0–1	0.1–10
Tissue adhesives	10–50	10–100
Connective tissues with bones	100–1000	~1000

to advance an interface separation (crack). Commonly used mechanical tests are briefly discussed below and illustrated in Figure 12.2.

The adhesion strength can be measured by probe-pull (tack) [30–32] or lap-shear [33–37] tests. In such a test, the two adherends are pulled or sheared apart. The adhesion strength is measured as the maximum recorded force divided by the area of contact.

The adhesion energy can be measured by peel or bilayer stretch tests. Among various types of peel tests, the most widely used are the 90° peel (applicable for separating one flexible adherend from one stiff adherend) and the 180° peel (applicable for separating two flexible adherends) [38–43]. In a peel test, two adherends are peeled apart at a constant speed. The peel force as a function of displacement is measured. As the peeling process enters a steady state, the force reaches a plateau. The steady state adhesion energy Γ is calculated as the plateau force divided by the sample width for 90° peel, $\Gamma = F/w$, and two times the plateau force divided by the sample for 180° peel, $\Gamma = 2F/w$.

Bilayer stretch test is used less often than peel [44, 45]. In such a test, the critical force and critical stretch of debonding are measured. The adhesion energy is calculated as $\Gamma = P(\lambda - 1) - U(\lambda)$, where P is the critical force per unit sample width in the undeformed state, λ the critical stretch, and $U(\lambda)$ the integration of the P–λ curve.

Burst pressure tests are commonly used for testing the sealing ability of tissue adhesives that prevent air leakage or bleeding [46–48]. In such a test, a fluid (water or air) is pumped into the adhesive-sealed tissue to exert force. The burst pressure is defined as the peak pressure before the tissue burst and is used for measuring the adhesion property.

The American Society for Testing and Materials (ASTM) provides standard guidelines for evaluating the mechanical performance of adhesives. For example, ASTM F2258, ASTM F2255, ASTM F2256, ASTM D2861, and ASTM F2392 describe the standard tensile test, lap-shear test, peel test, and burst pressure test. Some representative values of adhesion strength and adhesion energy in wet adhesion are listed in Table 12.1.

12.3 General Principles for Strong Wet Adhesion

Strong wet adhesion is achieved through the synergy of chemistry of bonds, topology of connection, and mechanics of dissipation. At the molecular level,

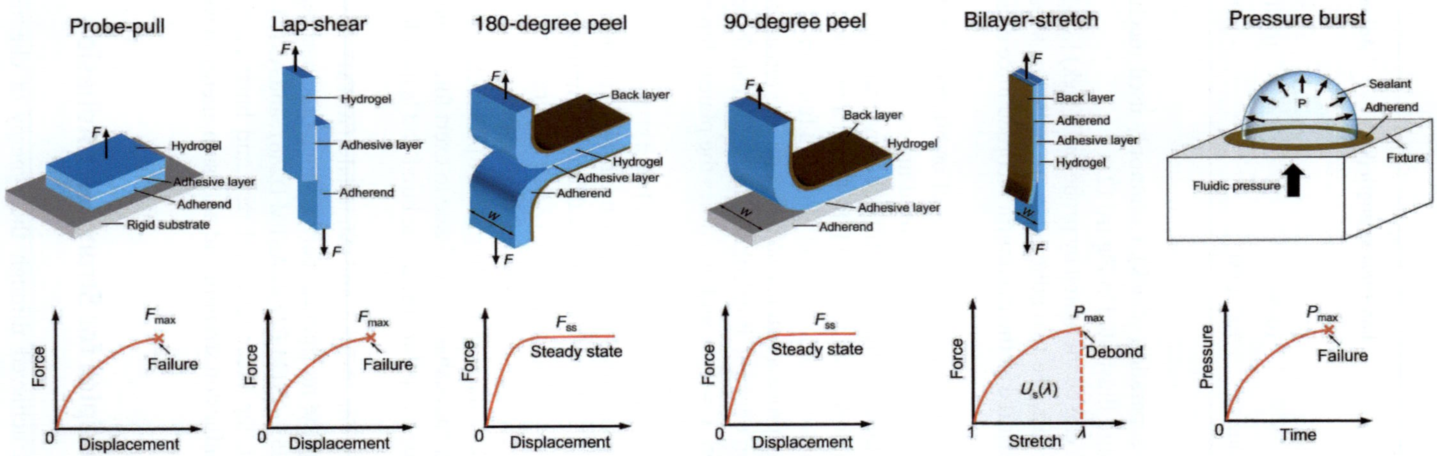

Figure 12.2 Commonly used mechanical tests for measuring the adhesion strength and adhesion energy.

the polymer network of a hydrogel adherend must bond to the other adherend through strong covalent bonds or noncovalent complexes. This bonding further enables the connection between two adherends through a certain topology at the mesoscale (e.g. polymer chains and networks), such as direct interfacial bonding between two nonporous materials [49–51], polymer chains linking to nanoparticles [44, 52–54], bridging polymer chains [7], and additional polymer networks providing topological entanglements [55–61]. Finally, a mechanism for great amount of energy dissipation during debonding is needed to significantly enhance the adhesion.

12.3.1 Chemistry of Bonds

Covalent bonds are strong while noncovalent bonds are usually weak. An aggregate of noncovalent bonds, however, can be strong. Polymer chains can interact strongly with each other through aggregates of noncovalent bonds, forming a supramolecular complex [62, 63]. Hydrogel-tissue adhesives have long been developed based on covalent bonds, such as CoSeal, DuraSeal, and BioGlue, which are widely practiced in surgery [3, 4, 64]. Supramolecular complexes are also widely used in hydrogel adhesion such as mucoadhesion [65], underwater adhesion [66], polyampholyte adhesion [67], reversible adhesion [68, 69], and topological adhesion [55, 57]. A brief overview of various bonding chemistries as well as their molecular structures is provided in Figure 12.3.

12.3.1.1 Static Covalent Bonds

Common static covalent bonds include carbon–carbon, peptide, thioester, siloxane, and urea linkage. These bonds are relatively strong and irreversible once broken. The two adherends should naturally, or chemically, be modified to have complementary functional groups to form bonds.

12.3.1.2 Dynamic Covalent Bonds

These bonds can reversibly dissociate and reform under specific conditions such as pH, heat, light, or redox conditions. The bond energy is usually comparable to or lower than those of static covalent bonds. The bond healing kinetics vary from minutes to hours. Common dynamic covalent bonds include disulfide, imine, acylhydrazone, boronate ester bond, and Diels–Alder reaction.

12.3.1.3 Ionic Bonds

An ionic bond is formed by electrostatic interaction between anions and cations, one or both of which are from polyelectrolyte chains. For example, alginate chains bear negatively charged carboxylate groups which can form ionic bonds with divalent cations such as Ca^{2+}, Br^{2+}, and Mg^{3+} [70]. Ionic bonds may or may not be fully irreversible. Their strength and reversibility depend on the specific bond type.

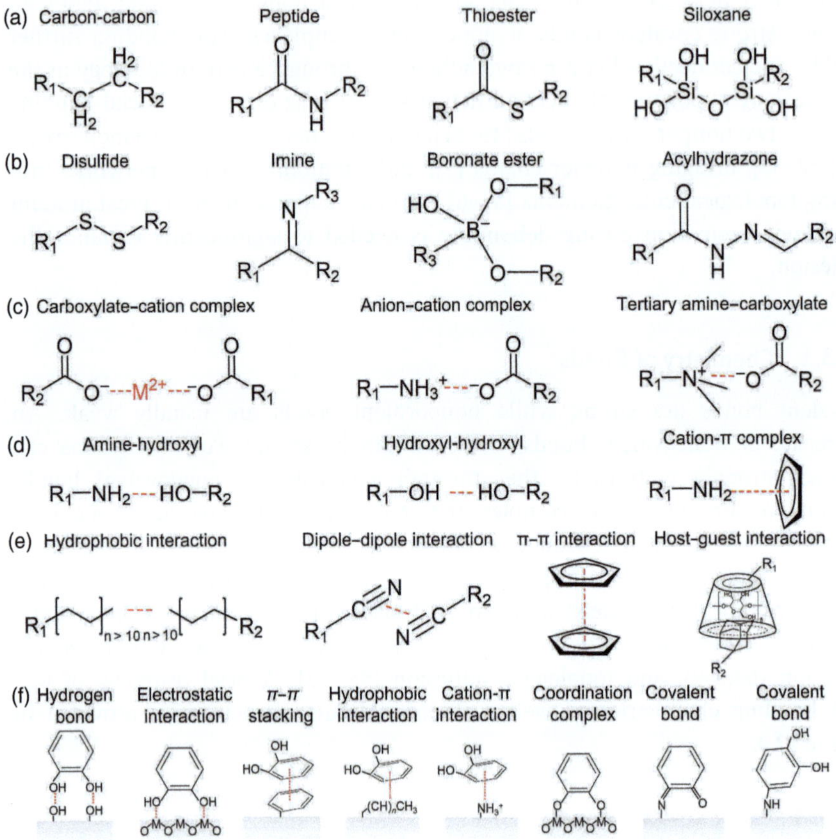

Figure 12.3 Various bonding chemistries as well as their molecular structures for wet adhesion. Source: Ref. [29]/John Wiley & Sons.

12.3.1.4 Hydrogen Bonds

A hydrogen bond is formed between a hydrogen atom bound to a more electronegative atom and another electronegative atom with a lone pair of electrons. A single hydrogen bond is weak, but a dense array of hydrogen bonds can be as strong as covalent bonds, enabling hydrogen-bonded complexes for adhesion [71, 72]. Various chemical groups, including hydroxyl, carboxylic acid, and amines have been exploited to make hydrogen-bonded complexes for wet adhesion. Hydrogen bonds between nucleobases, like those in DNA and RNA, have also been used to enable wet adhesion [73].

12.3.1.5 Hydrophobic Interaction

Hydrophobic molecules tend to aggregate in water. When hydrophobic moieties are copolymerized into hydrophilic polymer chains, the polymer chains aggregate at the hydrophobic sites, which serve as crosslinks to form a polymer network. Hydrophobic interaction has been used to make tough and self-healing hydrogels [74–76] but has not been used to create strong wet adhesion.

12.3.1.6 π–π Stack

π–π stack refers to noncovalent interactions between aromatic rings that functionalize as crosslink or interlink sites. π–π stack has been used in forming hydrogels, conductive polymers, and wet adhesion [55, 57, 77–79].

12.3.1.7 Dipole–Dipole Interaction

Dipole–dipole interaction is an electrostatic interaction between the permanent or transient dipoles on the polymer chain. A well-known example is the instant glue made of cyanoacrylate [80, 81]: it forms a glassy polymer network through interchain nitrile dipole–dipole interaction.

12.3.1.8 Host–Guest Interaction

Host–guest interaction refers to two or more molecules or ions that are held together by noncovalent bonds. Binding between host and guest is highly specific based on molecular recognition. Host–guest interactions have been used to make hydrogels [82–84] and supramolecular Velcro for reversible underwater adhesion [85].

12.3.1.9 Catechol Chemistry

Catechol chemistry is inspired by mussel adhesion [86–88]. Catechol has a benzene ring with two adjacent hydroxyl groups. The hydroxyl groups form hydrogen bonds with hydrophilic surfaces and electrostatic interaction with metal oxide surfaces. The benzene ring forms π–π stack or hydrophobic interaction with hydrophobic surface and cation–π interaction with positively charged surface. In addition, when catechols are oxidized to quinones, they form a strong coordination complex with metal oxide surfaces and covalent bonding with nucleophiles (e.g. amine and thiol) through possible Michael addition. Catechols even strongly adhere to inert metals such as gold. Catechols have been introduced to hydrogels to enable wet adhesion with various materials [50, 89, 90].

12.3.2 Topology of Connection

Topology of connection means a way to connect two adherends. The connection can be made through bonds, polymer chains, polymer networks, nanoparticles, mechanical interlocks, or their combinations, with length scales ranging from molecular to macroscopic.

Macroscopic mechanical interlock. Mechanical interlock connects two adherends geometrically at the macroscopic scale. Mechanical interlock can be either a *key-lock* or a *thread-hole* topology as shown in Figure 12.4a. A key-lock topology typically does not provide strong adhesion due to the softness of the synthetic hydrogel or biological tissue [44, 52]. A thread-hole topology may provide stronger adhesion [91], but such a method depends on the casting of a synthetic hydrogel precursor on a surface of connected pores, which greatly limits its application scenario.

Molecular topology. *Bond* and *stitch* are two basic molecular topologies (Figure 12.4b). When two adherends have complementary functional groups,

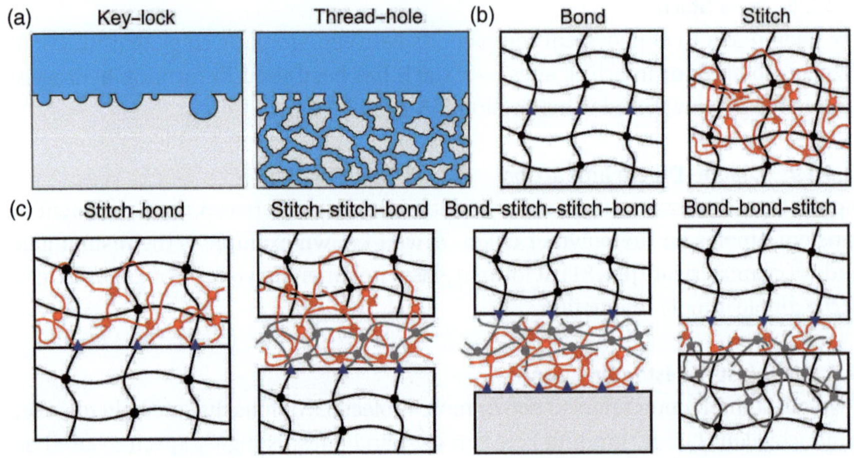

Figure 12.4 Various topologies of connection for wet adhesion. (a) Macroscopic mechanical interlock. (b) Molecular topologies of bond and stitch. (c) Combination of topologies.

they can connect through bonding [92–94]. When two adherends do not have complementary functional groups, a third material having the required functional groups for both adherends is needed for connection [7]. For example, a third group of polymer chains can bond both adherends and bridge them across the interface. This bond–bond topology can also be formed using other materials such as nanoparticles [44, 52–54].

When two adherends of polymer networks do not have complementary functional groups for bonding, a stitch topology can be used for connection. For example, a polymer precursor can diffuse into both adherends of preformed polymer networks and crosslink into another polymer network *in situ*, in topological entanglement with both adherends. Stitch does not require two adherends to form any bond since the connection is entirely through topological entanglement. The additional polymer network acts like a molecular suture, stitching the two adherend polymer networks together [55].

Combination of topologies. The basic topologies can be further combined to create more topologies to bond diverse materials of different forms (Figure 12.4c). When one adherend has functional groups for bonding but the other adherend has a polymer network with no functional group for bonding, a stitch–bond topology can be used for strong adhesion. More complex topologies include a stitch–stitch–bond topology, a bond–stitch–stitch–bond topology, and a bond–bond–stitch topology.

12.3.3 Mechanics of Dissipation

The separation of two adherends causes either the fracture of the interface (*adhesive failure*) or the fracture of adherends (*cohesive failure*). The fracture involves events at multiple length scales: bond cleavage at the atomic scale, chain

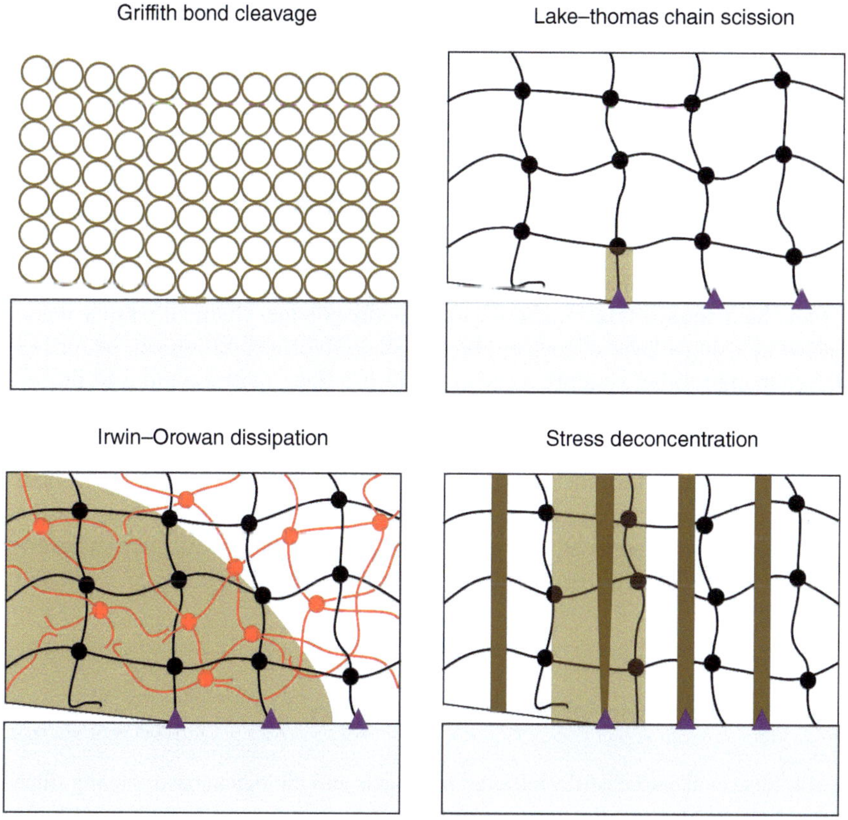

Figure 12.5 Separating adherends causes cascading events of energy dissipation. Source: Ref. [29]/John Wiley & Sons.

scission at the molecular scale, and bulk dissipation at the macroscopic scale. The energy dissipation mechanisms from these fracture events are discussed below (Figure 12.5).

Griffith bond cleavage. Two brittle solids can adhere through atomic bonds at the interface. From the classic Griffith picture of fracture [95], during separation, a crack cleaves one layer of atomic bonds, while all other atomic bonds in the adherends remain intact and do not dissipate energy. Taking the bond energy as 1 eV (10^{-19} J) and the area of a single bond as 10^{-19} m^2, the adhesion energy due to the bond cleavage mechanism is estimated to be 1 J/m^2.

Lake–Thomas chain scission. When the separation of adherends involves at least one polymer network, following the classic Lake–Thomas picture [96], a crack not only cleaves one layer of interfacial bonds but also stretches polymer chains ahead of the crack tip and releases the elastic energy in the entire polymer chain during separation. The adhesion energy is amplified by the number of monomer units in the polymer chain (degree of polymerization) beyond the Griffith bond cleavage mechanism.

Irwin–Orowan dissipation. An adherend can contain sacrificial bonds or other molecular processes that dissipate additional energy during a crack propagation following the Irwin–Orowan mechanism [97, 98]. During separation, the strong and stretchable polymer chains at the crack front transmit the stress to the bulk material and break sacrificial bonds in a large region. The size of bulk dissipation can be macroscopic, and the adhesion energy can be as high as 1000–10 000 J/m^2.

Stress deconcentration. Fracture is accommodated by stress concentration at the crack tip. As a result, deconcentrating the stress can enhance the crack resistance. The Lake–Thomas mechanism deconcentrates the stress at the level of a single polymer chain: the stress is transmitted to the entire polymer chain rather than being localized at a single layer of covalent bonds. Stress deconcentration can be further achieved in many other systems. As an example, in a fiber–matrix composite, during crack propagation, the matrix shears greatly and the fiber remains intact, thus transmitting the crack tip stress over a large segment of fiber. If the fiber–matrix interface is weak, the fiber pins the crack and the crack deflects into the matrix along the fiber direction rather than propagating straight; if the fiber–matrix interface is strong, the fiber ruptures and releases all the stored energy [99, 100]. Recently, stress deconcentration has been used to enhance the adhesion between a hydrogel and a solid [101, 102].

12.4 Methods for Strong Wet Adhesion

In 2014, Rose et al. successfully adhered hydrogels and biological tissues using silica nanoparticles with the adhesion energy enhanced from <1 to 10–20 J/m^2 [52]. In 2016, Yuk et al. achieved adhesion energy of about 1000 J/m^2 between hydrogels and nonporous materials for the first time [49]. Thereafter, new methods for strong wet adhesion between a hydrogel and another material have been reported in rapid succession. In the following, we describe representative methods for creating strong wet adhesion (Figure 12.6).

Direct attachment. In occasional cases, one adherend happens to have complementary functional groups that can directly bond to the other adherend [49–51]. These bonds are usually noncovalent such as hydrogen bonds, electrostatic interactions, and catechol bonds. Nevertheless, a high density of these noncovalent bonds can form adhesion as strong as covalent bonds. As one example, the rich carboxylic acid groups in a poly(acrylic acid) (PAA) hydrogel form hydrogen bonds with many functional groups such as amine, amide, and hydroxyl groups, while the carboxylic acid themselves do not form strong bonds. Upon simply attaching a PAA hydrogel to other hydrogels or tissues bearing those functional groups, strong adhesion is established rapidly [72].

Adhesive islands. Strong adhesion can be created through discrete adhesive islands that are distributed over the entire interface. Cyanoacrylate, a commonly used glue that forms a stiff adhesive layer when triggered by water, can be diluted in an organic solvent and spread at the interface between two stretchable adherends to create glassy adhesive islands [103, 104]. These glassy islands act as sparse

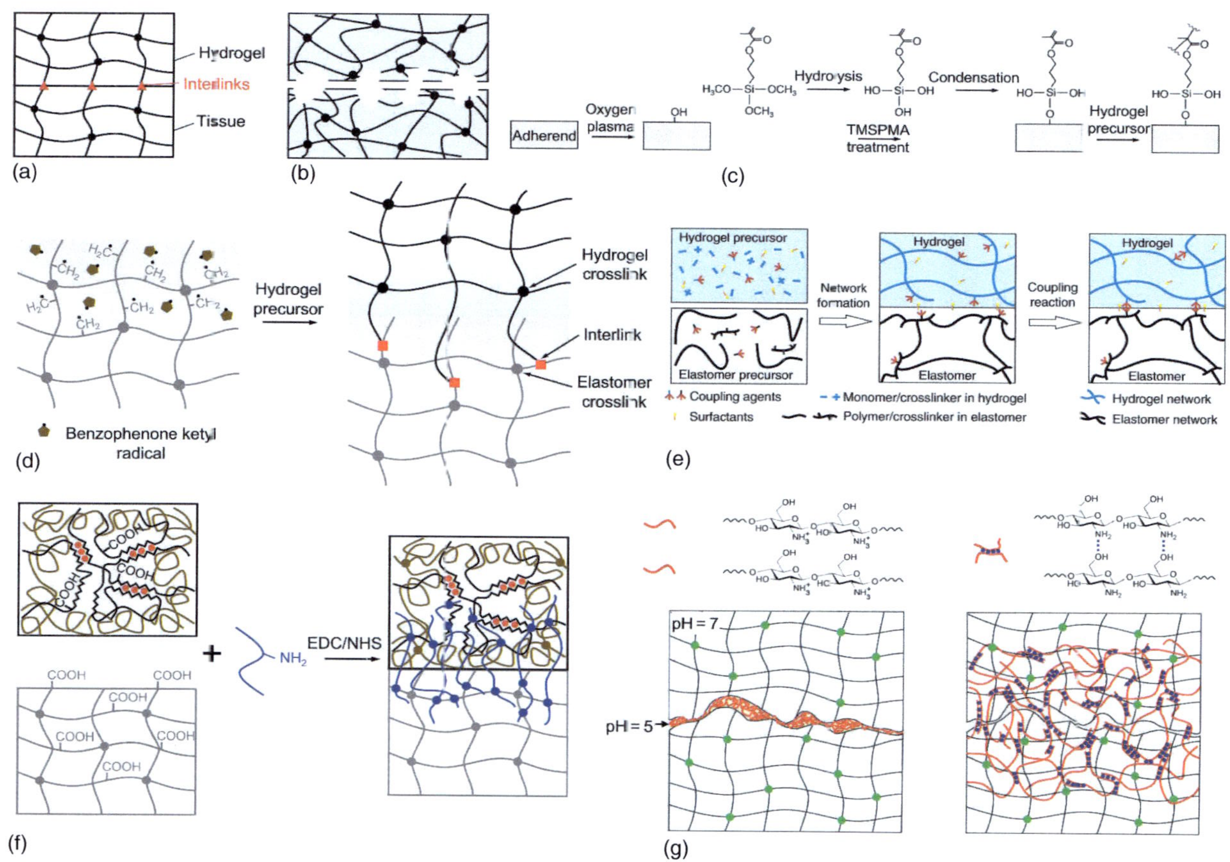

Figure 12.6 Methods for strong wet adhesion. (a) Direct attachment. (b) Adhesive islands. (c) Surface modification. (d) Surface initiation. Source: (a-d, f) Ref. [29]/John Wiley & Sons. (e) Bulk modification. Source: (e) Ref. [93]/Springer Nature/CC BY 4.0. (f) Bridging polymers. (g) Topological adhesion. Source: (g) Ref. [55]/John Wiley & Sons.

and strong interlinks to ensure both strong adhesion and high stretchability of adherends.

Surface modification. This method requires the surface chemical modification of a solid (e.g. glass, ceramic, aluminum) so that it possesses complementary functional groups to form covalent interlinks with a hydrogel network. Such method is often used during the *in situ* polymerization of a hydrogel precursor on a prefabricated solid. Among many methods, silanes are commonly used to modify surfaces with a variety of reactive groups (e.g. methylacrylate, amine, and epoxy) [105]. A solid is typically first treated with oxygen plasma to generate hydroxyl groups on its surface and then treated with silanes to convert hydroxyl groups to reactive groups based on silane coupling chemistry. Afterward, a hydrogel precursor may be cast on the silane-modified surface to *in situ* polymerize and be chemically grafted on the surface [49].

Surface initiation. Surface modification may not be applicable to hydrogel adhesion with elastomers, as elastomers are permeable to oxygen, inhibiting the free radical polymerization of hydrogels [106, 107]. One way to circumvent this issue is to introduce additional initiators bounded on elastomer surfaces. These initiators are typically hydrophobic photo- or thermo-initiators (e.g. benzophenone) prepared in organic solvents. Such initiators not only serve as oxygen scavengers to alleviate oxygen inhibition but also allow the hydrogel network to be linked to the elastomer network. The initiator solution diffuses into the elastomer network. Upon exposure to UV light or elevated temperature, the initiators react with the network to generate anchored radicals on the network as well as hydrophobic free radicals. A hydrogel precursor is subsequently cast on the elastomer, diffuses into the radical-activated surface layer, and polymerizes into a hydrogel network [92]. The elastomer network and the hydrogel network form covalent interlinks, and the two networks are likely in topological entanglement over some thickness.

Bulk modification. For preformed hydrogel and tissue adherends, to allow adhesion, bulk modification of adherends during their synthesis can be employed. For example, functional silanes can be added to the precursor and copolymerized into the polymer network. When two silane-modified polymer networks are in contact, the water molecules in the hydrogel rapidly hydrolyze silane molecules to form silanol groups which further condensate to form siloxane bonds, chemically interlinking the two networks [93]. Similarly, amine-coupling molecules can be incorporated in a hydrogel network to directly bond with the abundant primary amines on tissue surfaces [108].

Bridging polymers. Biological tissues have many functional groups such as hydroxyl groups, primary amines, thiols, and carboxyl groups, which can be exploited to form bonds. To create strong tissue adhesion, additional bridging polymers can be introduced. As an example, a carboxylic acid and a primary amine can react to form peptide bonds in the presence of catalysts [109, 110]. Therefore, polymers bearing primary amines can diffuse into both a hydrogel with carboxylic acids and a tissue. The polymers react with carboxylic acids to form peptide bonds with both networks, thus chemically bridging them across the interface.

Strong adhesion is achieved with various organs with adhesion energy beyond 1000 J/m^2 [7].

Topological adhesion. Strong adhesion can also be achieved by an additional polymer network forming topological entanglement with pre-existing adherend networks [55]. A topological adhesion does not require any functional group from the adherends. In a typical adhesion process, a polymer solution is spread at the interface of two polymer networks and the spread polymer chains diffuse into both networks, crosslink into a third polymer network triggered by an environmental stimulus such as pH, and form topological entanglements with both existing polymer networks across the interface. The third polymer network acts as a suture, stitching the two pre-existing networks together at the molecular scale. The measured adhesion energies of topological adhesion approach the fracture energies of the adherends and can be as high as 1000 J/m^2. By controlling the crosslinking kinetics or amount of polymer solution, the time for adhesion varies from a few minutes to a few hours [111].

Combined methods of adhesion. The above methods can be combined to accommodate adhesion in more complex systems and working scenarios. For example, the surface of a dry and nonporous material can be chemically modified to exhibit functional groups such that they form direct bonds with stitching polymers, which diffuse into another hydrogel network and form topological entanglements [57]. With combined methods, one can design and tailor a series of adhesion methods to realize specific requirements of multi-material integration. For example, a topological-primer and a dual-topological-primer have been developed to integrate hydrogels and elastomers in various manufacturing processes such as coating, painting, and printing through combining bonding and other topologies [93, 112].

12.5 Mechanics of Wet Interfaces

Wet adhesion connects two adherends across length scales. Under mechanical loads, the interface exhibits complex mechanical behaviors in coupling with the adherends in a spatiotemporal manner. Rheologies of both the interface and adherends, material sizes, loading conditions, and chemical reactions will lead to complex failure modes, rate-dependent debonding, degradation, and fatigue.

12.5.1 Adhesive Failure and Cohesive Failure

When two adherends are separated, a crack will propagate either along the interface or inside one of the adherends. These two failure modes are called *adhesive failure* and *cohesive failure*, respectively. The evolution of different failure modes, their transitions, and coupling, have been extensively studied in stiff materials such as stiff thin films [113–116], but have only started to be systematically investigated in soft materials in recent years [117–119]. Factors that affect this process include fracture energies and strengths of the adherends, the adhesion energy and strength, the sample geometry and size, elastic moduli, and others. For example, when a soft film of lower fracture energy compared to the interfacial adhesion energy is covalently bonded to a stiff substrate, the separation causes the crack to run inside the film, often leaving a thin residual film on the substrate after peel. If the interface is created by a material of low fracture energy but high strength such as polycyanoacrylate, a crack still kinks into the film rather than breaking the strong adhesive layer [103].

12.5.2 Fractocohesive Length- and Size-Dependency

Fracture mechanics defines a material length scale called *fractocohesive length*. Take Γ as the fracture energy in unit of J/m^2 measured using a sample with a long precut crack, and W_f as the work of fracture in unit of J/m^3 measured as the integration of stress-strain curve of a "perfect" sample containing no crack up to bulk fracture. The fractocohesive length is defined as Γ/W_f [120, 121]. This length is considered a material constant in an elastic or elastic–plastic material and may depend on the loading rate in a viscoelastic or viscoplastic material.

The fractocohesive length can be further defined in adhesion, but its use depends on the specific fracture mode. In the case of adhesion, Γ should be defined as the relevant fracture energy of the failure mode, either cohesive or adhesive, and W_f may be defined using the bulk material of the adherend that is involved when assuming the adhesive layer has minimum thickness.

The fractocohesive length defines the size of crack processing or inelastic zone around a crack tip. When the feature size such as crack length or thickness (h) of the adherend or interfacial adhesive is larger than the corresponding fractocohesive length, small-scale inelasticity takes place, and the measured fracture energy or adhesion energy (Γ) is a material constant independent of the feature size (Figure 12.7). When the feature size is smaller than the fractocohesive length, large-scale inelasticity takes place, the inelastic zone is constrained by the feature size, and the measured fracture energy is size-dependent (Figure 12.7). Therefore, the fractocohesive length also describes the flaw sensitivity of a material, i.e. the critical flaw size below which the existence of the flaw does not knock down the measured strength of the material or adhesion [120].

As one application of the fractocohesive length, when two soft adherends are adhered by an array of stiff adhesive islands, the size of the islands should be smaller

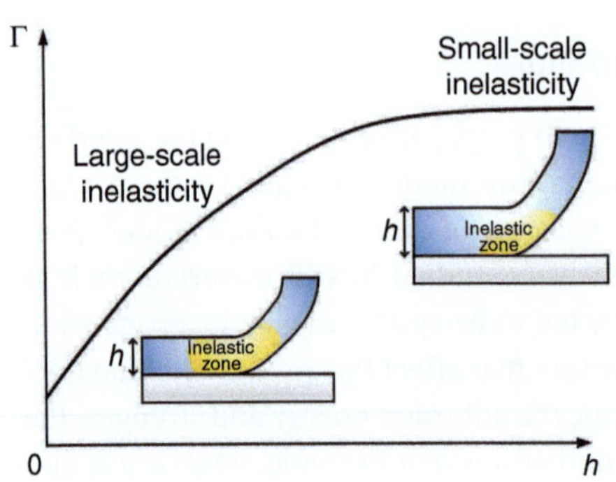

Figure 12.7 The measured fracture energy or adhesion energy is size-dependent under the condition of large-scale inelasticity but is size-independent under the condition of small-scale inelasticity.

than the corresponding fractocohesive length to avoid flaw sensitivity and therefore maintain both strong adhesion and good stretchability of adherends [103].

In certain adhesion methods, the thickness of the adhesive also indirectly affects the adhesion strength and adhesion energy. As an example, specific chemical species such as ions may need to diffuse across the adhesive layer to the adherends, react to gelation, and establish adhesion. The thickness h thus controls the time scale of formation of adhesion following $t \sim h^2/D$, governed by diffusion, where D is the diffusivity of the chemical species. This is the case of a pH-triggered topological adhesion, where the protons diffuse across the thickness of the adhesive layer such as chitosan to induce the pH change, which leads to the crosslinking of the stitching network. Reducing the thickness of the adhesive layer leads to a decrease in the time for adhesion from hours to minutes [59, 111].

12.5.3 Rate Dependency

The diverse bonding chemistries and polymer chain dynamics at both interfaces and bulk adherends lead to complex rheology, giving rise to rate-dependent behaviors of debonding. Following Gent, de Gennes, et al., the adhesion energy can be expressed as $\Gamma(v) = \Gamma_0[1 + f_d(v)]$ [122–125], where Γ_0 is the *intrinsic adhesion energy* determined entirely by the local process in the crack processing zone, and $f_d(v)$ is an amplifying factor due to any energy dissipation in the bulk material, such as the viscoelastic dissipation due to a finite crack speed v. For tough adhesion, $f_d(v)$ can be larger than unity by orders of magnitude. Several experimental methods can be used to measure the $\Gamma(v)$ curve such as peel, which provides valuable insights on the molecular process during debonding. Successful examples include identifying the bond dissociation origin [55] as well as the molecular process such as polymer chain pull-out during a crack propagation [126–128].

Figure 12.8 illustrates a representative curve of $\Gamma(v)$ with the crack speed v plotted on the vertical axis. In general, $\Gamma(v)$ as well as the rate-dependent dissipation decreases as the crack speed decreases (but not always, as can be seen in the case of peeling a viscoelastic pressure-sensitive adhesive [32]). As the crack speed approaches zero, the rate-dependent dissipation vanishes, and the intrinsic adhesion energy Γ_0 is obtained, below which no crack propagation occurs. For fracture and debonding of elastomers, Γ_0 has been historically understood as a material constant that can be described by the Lake–Thomas model (to be discussed in Section 12.5.5) [96]. However, recent studies by Bai et al. have shown that Γ_0 can

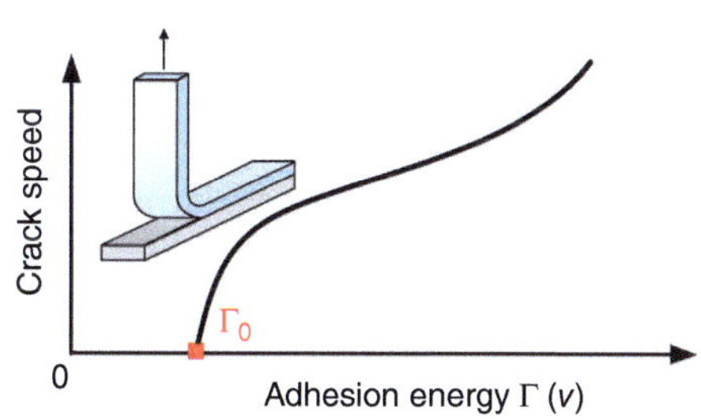

Figure 12.8 Measuring the rate-dependent adhesion energy $\Gamma(v)$ using the 90-degree peel. The intrinsic adhesion energy Γ_0 can be obtained from the curve by extrapolation of data.

be further size-dependent due to the large-scale inelasticity from *rate-independent* energy dissipation that can be constrained by sample thickness [129].

Polymer chain disentanglement and pull-out is another case of rate-dependent debonding. Polymer chains are physically entangled with polymer networks of adherends across the interface. During the separation, polymer chains undergo disentanglement, sliding, and pull-out processes. Fast loading cannot disentangle polymer chains immediately, and the entangled polymer chains behave like a crosslinked solid, leading to network fracture. Slow loading can disentangle the polymer chains where the unentangled polymer chains are subsequently pulled out from the polymer network. The polymer chains slide against each other and the surrounding solvent molecules during the pull-out following a viscous process and dissipate energy. For example, quantitative modeling and experiments of a gelatin gel show that the adhesion energy varies from 2.5 to about $100\,J/m^2$ as the crack speed increases from μm/s to cm/s [126].

12.5.4 Fatigue

When subjected to prolonged cyclic loads, a material or an interface often suffers accumulated internal damage or cyclic crack growth, defined as *fatigue*. Both the fracture and fatigue of elastomers have been studied for decades since Rivlin, Thomas, Lake, et al. [96, 130, 131]. Also see recent reviews in [122, 132–138].

Among various phenomena in fatigue, we denote *fatigue damage* as the gradually accumulative internal damage such as bond breaking in a material without a crack growth, and denote *fatigue fracture* as the accumulative crack growth over prolonged cyclic loads [121]. In many elastomers such as natural rubbers, fatigue damage is minimal due to the negligible hysteresis or energy dissipation in the mostly elastic polymer network. However, fatigue fracture still occurs if the polymer network is subjected to a mechanical load above the *fatigue threshold*, defined as the minimum energy needed to propagate a crack under cyclic loads in units of J/m^2. Lake and Thomas first noted that the experimentally determined fatigue threshold of several elastomers under cyclic loads is about $50\,J/m^2$, orders of magnitude lower than their fracture energy (about $10^4\,J/m^2$) [96, 131]. They hypothesized that at this threshold with vanishingly small crack extension per cycle, the energy dissipation due to inelastic deformation from the bulk material is negligible, and the fatigue threshold is entirely due to the scission of a single layer of polymer chains along the crack plane (Figure 12.9). They further hypothesized that the rupture of each covalent bond dissipates the energy of the entire stretched chain. As a result, the Lake–Thomas model estimates the threshold as $\Gamma_0 = \alpha l\sqrt{n}J/V$, where J, V, and l are the covalent bond energy, volume, and length of one monomer or Kuhn segment, n is the number of total Kuhn segments in the chain, $l\sqrt{n}$ is the average end-to-end distance of a polymer chain, and α is a pre-factor of order unit. The Lake–Thomas model gives a satisfying estimate of the fatigue thresholds of many elastomers at the qualitative and even quantitative levels [121, 122].

The studies on fracture and fatigue of soft materials embraced new opportunities and rapid progress in the past decade due to the emerging development of hydrogels

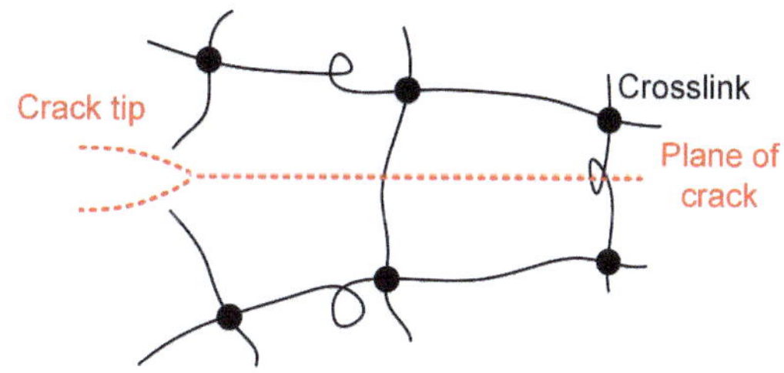

Figure 12.9 The Lake–Thomas model for fatigue threshold.

and their transformative applications. Pioneered by Gong et al. [139], hydrogels as tough as natural rubbers with fracture energy higher than 10^4 J/m^2 have been widely available. The high toughness is attributed to large energy dissipation as discussed in Figure 12.5. However, just like what Lake and Thomas found in elastomers, this toughening mechanism based on large energy dissipation from sacrificial bonds in the bulk material fails to enhance the fatigue threshold. Recent studies have shown that hydrogels with identical covalent networks have approximately the same fatigue threshold following the Lake–Thomas model despite different bulk energy dissipations [140–142]. This has motivated several recent breakthroughs in developing fatigue-resistant soft materials [99, 100, 143, 144] with significantly enhanced thresholds over 10^3 J/m^2.

Following these new advancements in fatigue of elastomers and gels, the research on fatigue of wet adhesion is rapidly growing. As example, interfacial fatigue fracture has been studied for hydrogel-tissue adhesion [45]. Principles for fatigue-resistant adhesion have also been developed, inspired by the same principle for fatigue-resistant hydrogels [101, 102, 145]. It is expected that more progress in this area will soon emerge such as combining interfacial fatigue damage and fracture with complex rheology in soft synthetic materials or tissues.

12.5.5 Stretchable Adhesion

Dry adhesion typically sustains small deformation and their interfaces do not require stretchability. By contrast, wet adhesion between soft materials usually requires the adherends to sustain large deformation, and so does the interface. A representative example is a soft device adhered to a beating heart where the interface is constantly deformed. A nonstretchable interface limits the deformation and may lead to interfacial failure.

Stretchable adhesion needs to ensure the interface remains well bonded while being deformed or after deformation of any kind (e.g. monotonic, cyclic, and multi-axial). Two general strategies have been employed to design a stretchable interface [60]: (i) the interlinking bonds should be neither too dense to harden the interface nor too sparse to cause weak adhesion, and the bond density should ideally match the crosslink density of adherends; and (ii) if a glassy adhesive is used, it can be designed as adhesive islands with size smaller than the flaw sensitivity length of the system to ensure stretchability of adherends [103].

12.6 Summary and Outlook

Dry adhesion has been practiced for decades for rubbers, plastics, ceramics, and metals and has found broad and indispensable applications in industry and households [146–149]. By contrast, wet adhesion is a relatively new subject motivated by new materials and applications. Strong wet adhesion is a supramolecular process through a synergy of chemistry of bonds, topology of connection, and mechanics of dissipation. Both covalent bonds and noncovalent complexes are involved in creating interlinks. Various topologies of connecting adherends are possible and can be designed to meet specific needs. Separating two adherends dissipates energy through cascading events across many length scales including bond cleavage, chain scission, and bulk hysteresis. Mechanical behaviors of wet interfaces are rich and complex, coupling rheologies of both the interface and adherends and demonstrating rate- and size-dependency. In addition to providing strong adhesion, wet interfaces may further bear certain functions to augment human–machine interactions and meet requirements of practical applications. Despite the transformative advances of wet adhesion in recent years, challenges remain in materials development, large-scale manufacturing, tissue compatibility, and physiological compatibility in both fundamental science and practical application.

Maintaining strong adhesion under extreme conditions, e.g. combined mechanical and environmental, places a particular challenge to the design of interface. One familiar case is the body environment where the temperature, ions, chemicals, and enzymes can collectively attack an interface, causing chemical reactions and subsequent degradation of mechanical properties. Meanwhile, an external stress or deformation applied to the interface may accelerate the chemical reaction. This mechanochemical effect results in premature interfacial failures. For example, adhesion built on certain bonds such as hydrogen bonds, siloxane bonds, and peptide bonds are particularly susceptible to temperature degradation, hydrolysis, and biodegradation by proteases. As a result, interdisciplinary efforts are needed for developing new materials and adhesion methods with extreme mechanical properties such as strength and toughness which also tolerate extreme environments. It is also desirable that these mechanical properties can degrade on demand when the material or interface is exposed to certain stimuli to allow full control of the material behavior in the long term.

Biohybrid or synthetic hydrogel-hybrid devices fabricated using wet adhesion have proven successful and functional in small-scale lab settings, but still face grand challenges to meet large-scale industrial needs. The lack of rapid prototyping becomes a major bottleneck hampering mass production. Current fabrications of these devices mainly rely on manual assembly which prohibits miniaturization. Advanced manufacturing technologies offer an opportunity to fabricate highly integrated structures at the resolution of sub-millimeters with potentially low cost. However, multi-material printing especially involving wet materials such as hydrogels and tissues with strong interfacial adhesion remains difficult.

Existing methods of wet adhesion for tissues mainly focus on strong adhesion and biocompatibility but overlook the mechanical mismatch between a tissue and the other adherend. For example, a brain tissue is soft and brittle, while a cartilage is strong and tough. It is critical to design adhesion methods that match tissue-specific properties such as elastic modulus, viscoelasticity, and toughness, to provide them with similar mechanical environments as their natural extracellular matrixes. Experimental evidence has suggested that the mechanical and rheological properties of an extracellular matrix material strongly affect the cell spreading, differentiation, and migration [150–152]. As another example, viscoelastic surface electrode arrays are developed to form conformal contact with soft biological tissues for stimulation and recording [153].

In wet adhesion, an interface itself can assist device functions or create new functions. For example, a functional interface can change its microstructure, permeability, and phase to regulate mass exchanges between the environment and tissue. A conductive interface can form 3D interpenetrating networks with tissues to facilitate deep tissue bioelectric therapies. Such an interface can also encapsulate wireless electronic devices to continuously monitor and record biosignals and perform therapies upon remote control. Finally, it is possible to integrate these functions into a single interface to offer multifaceted interactions and communications with the human body.

The mechanics of wet adhesion provide a unique lens to probe the multiscale molecular processes during debonding. For example, the rate-dependent peel characterizes the viscoelastic dissipation and the molecular level fracture behavior such as bond breaking and chain pull-out, the cyclic fatigue fracture characterizes the *intrinsic* crack resistance of a polymer network, and the monotonic fracture test characterizes the synergistic energy dissipation from the local crack tip to the bulk sacrificial bond breaking. It is foreseeable that the research on wet adhesion will rejuvenate the ancient art of adhesion, with new mechanics, chemistry, physics, materials, and applications.

References

1 Haque, M.A., Kurokawa, T., and Gong, J.P. (2012). Super tough double network hydrogels and their application as biomaterials. *Polymer* 53 (9): 1805–1822.

2 Lee, K.Y. and Mooney, D.J. (2001). Hydrogels for tissue engineering. *Chemical Reviews* 101 (7): 1869–1880.

3 Bouten, P.J., Zonjee, M., Bender, J. et al. (2014). The chemistry of tissue adhesive materials. *Progress in Polymer Science* 39 (7): 1375–1405.

4 Mehdizadeh, M. and Yang, J. (2013). Design strategies and applications of tissue bioadhesives. *Macromolecular Bioscience* 13 (3): 271–288.

5 Annabi, N., Yue, K., Tamayol, A., and Khademhosseini, A. (2015). Elastic sealants for surgical applications. *European Journal of Pharmaceutics and Biopharmaceutics* 95: 27–39.

6 Grinstaff, M.W. (2007). Designing hydrogel adhesives for corneal wound repair. *Biomaterials* 28 (35): 5205–5214.

7 Li, J., Celiz, A., Yang, J. et al. (2017). Tough adhesives for diverse wet surfaces. *Science* 357 (6349): 378–381.

8 Prausnitz, M.R. and Langer, R. (2008). Transdermal drug delivery. *Nature Biotechnology* 26 (11): 1261.

9 Amjadi, M., Sheykhansari, S., Nelson, B.J., and Sitti, M. (2018). Recent advances in wearable transdermal delivery systems. *Advanced Materials* 30 (7): 1704530.

10 Li, J. and Mooney, D.J. (2016). Designing hydrogels for controlled drug delivery. *Nature Reviews Materials* 1 (12): 16071.

11 Hoare, T.R. and Kohane, D.S. (2008). Hydrogels in drug delivery: progress and challenges. *Polymer* 49 (8): 1993–2007.

12 Qiu, Y. and Park, K. (2001). Environment-sensitive hydrogels for drug delivery. *Advanced Drug Delivery Reviews* 53 (3): 321–339.

13 Keplinger, C., Sun, J.Y., Foo, C.C. et al. (2013). Stretchable, transparent, ionic conductors. *Science* 341 (6149): 984–987.

14 Yang, C. and Suo, Z. (2018). Hydrogel ionotronics. *Nature Reviews Materials* 3, 125.

15 Lee, H.-R., Kim, C.-C., and Sun, J.-Y. (2018). Stretchable ionics – a promising candidate for upcoming wearable devices. *Advanced Materials* 30 (42): 1704403.

16 Yuk, H., Lu, B., and Zhao, X. (2019). Hydrogel bioelectronics. *Chemical Society Reviews* 48: 1642–1667.

17 Xu, C., Li, B., Xu, C., and Zheng, J. (2015). A novel dielectric elastomer actuator based on compliant polyvinyl alcohol hydrogel electrodes. *Journal of Materials Science - Materials in Electronics* 26 (11): 9213–9218.

18 Acome, E., Mitchell, S., Morrissey, T. et al. (2018). Hydraulically amplified self-healing electrostatic actuators with muscle-like performance. *Science* 359 (6371): 61–65.

19 Pu, X., Liu, M., Chen, X. et al. (2017). Ultrastretchable, transparent triboelectric nanogenerator as electronic skin for biomechanical energy harvesting and tactile sensing. *Science Advances* 3 (5): e1700015.

20 Parida, K., Kumar, V., Jiangxin, W. et al. (2017). Highly transparent, stretchable, and self-healing ionic-skin triboelectric nanogenerators for energy harvesting and touch applications. *Advanced Materials* 29 (37): 1702181.

21 Xu, W., Huang, L.B., Wong, M.C. et al. (2017). Environmentally friendly hydrogel-based triboelectric nanogenerators for versatile energy harvesting and self-powered sensors. *Advanced Energy Materials* 7 (1): 1601529.

22 Schroeder, T.B., Guha, A., Lamoureux, A. et al. (2017). An electric-eel-inspired soft power source from stacked hydrogels. *Nature* 552 (7684): 214.

23 Rogers, J.A., Someya, T., and Huang, Y. (2010). Materials and mechanics for stretchable electronics. *Science* 327 (5973): 1603–1607.

24 Kim, D.-H., Lu, N., Ma, R. et al. (2011). Epidermal electronics. *Science* 333 (6044): 838–843.

25 Kim, D.-H., Ghaffari, R., Lu, N., and Rogers, J.A. (2012). Flexible and stretchable electronics for biointegrated devices. *Annual Review of Biomedical Engineering* 14 (1): 113–128.

26 Yao, S. and Zhu, Y. (2015). Nanomaterial-enabled stretchable conductors: strategies, materials and devices. *Advanced Materials* 27 (9): 1480–1511.

27 Akinwande, D., Brennan, C.J., Bunch, J.S. et al. (2017). A review on mechanics and mechanical properties of 2D materials—graphene and beyond. *Extreme Mechanics Letters* 13: 42–77.

28 Yao, S., Swetha, P., and Zhu, Y. (2018). Nanomaterial-enabled wearable sensors for healthcare. *Advanced Healthcare Materials* 7 (1): 1700889.

29 Yang, J., Bai, R., Chen, B., and Suo, Z. (2019). Hydrogel adhesion: a supramolecular synergy of chemistry, topology, and mechanics. *Advanced Functional Materials* 30 (2): 1901693.

30 Creton, C. (2003). Pressure-sensitive adhesives: an introductory course. *MRS Bulletin* 28 (6): 434–439.

31 Sun, S., Li, M., and Liu, A. (2013). A review on mechanical properties of pressure sensitive adhesives. *International Journal of Adhesion and Adhesives* 41: 98–106.

32 Chopin, J., Villey, R., Yarusso, D. et al. (2018). Nonlinear viscoelastic modeling of adhesive failure for polyacrylate pressure-sensitive adhesives. *Macromolecules* 51 (21): 8605–8610.

33 Kafkalidis, M.S. and Thouless, M.D. (2002). The effects of geometry and material properties on the fracture of single lap-shear joints. *International Journal of Solids and Structures* 39 (17): 4367–4383.

34 Hui, C.-Y., Liu, Z., Minsky, H. et al. (2018). Mechanics of an adhesive tape in a zero degree peel test: effect of large deformation and material nonlinearity. *Soft Matter* 14 (47): 9681–9692.

35 da Silva, L.F.M., Rodrigues, T.N.S.S., Figueiredo, M.A.V. et al. (2006). Effect of adhesive type and thickness on the lap shear strength. *The Journal of Adhesion* 82 (11): 1091–1115.

36 Wang, Y., Yang, X., Nian, G., and Suo, Z. (2020). Strength and toughness of adhesion of soft materials measured in lap shear. *Journal of the Mechanics and Physics of Solids* 143: 103988.

37 Wang, Y., Nian, G., Yang, X., and Suo, Z. (2021). Lap shear of a soft and elastic adhesive. *Mechanics of Materials* 158: 103845.

38 Gent, A. and Lai, S.M. (1994). Interfacial bonding, energy dissipation, and adhesion. *Journal of Polymer Science Part B: Polymer Physics* 32 (8): 1543–1555.

39 Baumard, T.L.M., Thomas, A.G., and Busfield, J.J.C. (2012). Fatigue peeling at rubber interfaces. *Plastics, Rubber and Composites* 41 (7): 296–300.

40 Gent, A. and Hamed, G. (1977). Peel mechanics of adhesive joints. *Polymer Engineering & Science* 17 (7): 462–466.

41 Kim, K.S. and Aravas, N. (1988). Elastoplastic analysis of the peel test. *International Journal of Solids and Structures* 24 (4): 417–435.

42 Wei, Y. and Hutchinson, J.W. (1998). Interface strength, work of adhesion and plasticity in the peel test. In: *Recent Advances in Fracture Mechanics: Honoring Mel and Max Williams* (ed. W.G. Knauss and R.A. Schapery), 315–333. Dordrecht: Springer Netherlands.

43 Bartlett, M.D., Case, S.W., Kinloch, A.J., and Dillard, D.A. (2023). Peel tests for quantifying adhesion and toughness: a review. *Progress in Materials Science* 137: 101086.
44 Tang, J., Li, J., Vlassak, J.J., and Suo, Z. (2016). Adhesion between highly stretchable materials. *Soft Matter* 12 (4): 1093–1099.
45 Ni, X., Chen, C., and Li, J. (2020). Interfacial fatigue fracture of tissue adhesive hydrogels. *Extreme Mechanics Letters* 34: 100601.
46 Laterreur, V., Ruel, J., Auger, F.A. et al. (2014). Comparison of the direct burst pressure and the ring tensile test methods for mechanical characterization of tissue-engineered vascular substitutes. *Journal of the Mechanical Behavior of Biomedical Materials* 34: 253–263.
47 Naito, Y., Lee, Y.-U., Yi, T. et al. (2013). Beyond burst pressure: initial evaluation of the natural history of the biaxial mechanical properties of tissue-engineered vascular grafts in the venous circulation using a murine model. *Tissue Engineering Parts A* 20 (1–2): 346–355.
48 Latimer, C.A., Nelson, M., Moore, C.M., and Martin, K.E. (2014). Effect of collagen and elastin content on the burst pressure of human blood vessel seals formed with a bipolar tissue sealing system. *The Journal of Surgical Research* 186 (1): 73–80.
49 Yuk, H., Zhang, T., Lin, S. et al. (2016). Tough bonding of hydrogels to diverse non-porous surfaces. *Nature Materials* 15 (2): 190–196.
50 Cheng, H., Yue, K., Kazemzadeh-Narbat, M. et al. (2017). Mussel-inspired multifunctional hydrogel coating for prevention of infections and enhanced osteogenesis. *ACS Applied Materials & Interfaces* 9 (13): 11428–11439.
51 Takahashi, R., Shimano, K., Okazaki, H. et al. (2018). Tough particle-based double network hydrogels for functional solid surface coatings. *Advanced Materials Interfaces* 5: 1801018.
52 Rose, S., Prevoteau, A., Elziere, P. et al. (2014). Nanoparticle solutions as adhesives for gels and biological tissues. *Nature* 505 (7483): 382.
53 Meddahi-Pellé, A., Legrand, A., Marcellan, A. et al. (2014). Organ repair, hemostasis, and in vivo bonding of medical devices by aqueous solutions of nanoparticles. *Angewandte Chemie, International Edition* 53 (25): 6369–6373.
54 Shin, K., Choi, J.W., Ko, G. et al. (2017). Multifunctional nanoparticles as a tissue adhesive and an injectable marker for image-guided procedures. *Nature Communications* 8: 15807.
55 Yang, J., Bai, R., and Suo, Z. (2018). Topological adhesion of wet materials. *Advanced Materials* 30 (25): 1800671.
56 Gao, Y., Wu, K., and Suo, Z. (2018). Photodetachable adhesion. *Advanced Materials* 31: 1806948.
57 Yang, J., Bai, R., Li, J. et al. (2019). Design molecular topology for wet–dry adhesion. *ACS Applied Materials & Interfaces* 11 (27): 24802–24811.
58 Jia, D., Pang, Y., and Liang, X. (1994). Mechanism of adhesion of polyurethane/polymethacrylate simultaneous interpenetrating networks adhesives to polymer substrates. *Journal of Polymer Science Part B: Polymer Physics* 32 (5): 817–823.

59 Steck, J., Kim, J., Yang, J. et al. (2020). Topological adhesion. I. Rapid and strong topohesives. *Extreme Mechanics Letters* 39: 100803.

60 Yang, J., Steck, J., Bai, R., and Suo, Z. (2020). Topological adhesion II. Stretchable adhesion. *Extreme Mechanics Letters* 40: 100891.

61 Steck, J., Yang, J., and Suo, Z. (2019). Covalent topological adhesion. *ACS Macro Letters* 8 (6): 754–758.

62 Bekturov, E.A. and Bimendina, L.A. (1981). *Interpolymer Complexes. Speciality Polymers*, 99–147. Springer.

63 Bekturov, E.A. and Bimendina, L.A. (1997). Complexes of water-soluble polymers. *Journal of Macromolecular Science, Part C: Polymer Reviews* 37 (3): 501–518.

64 Slaughter, B.V., Khurshid, S.S., Fisher, O.Z. et al. (2009). Hydrogels in regenerative medicine. *Advanced Materials* 21 (32–33): 3307–3329.

65 Smart, J.D. (2005). The basics and underlying mechanisms of mucoadhesion. *Advanced Drug Delivery Reviews* 57 (11): 1556–1568.

66 Hofman, A.H., van Hees, I.A., Yang, J., and Kamperman, M. (2018). Bioinspired underwater adhesives by using the supramolecular toolbox. *Advanced Materials* 30 (19): 1704640.

67 Roy, C.K., Guo, H.L., Sun, T.L. et al. (2015). Self-adjustable adhesion of polyampholyte hydrogels. *Advanced Materials* 27 (45): 7344–7348.

68 Heinzmann, C., Weder, C., and de Espinosa, L.M. (2016). Supramolecular polymer adhesives: advanced materials inspired by nature. *Chemical Society Reviews* 45 (2): 342–358.

69 Liu, X., Zhang, Q., Gao, Z. et al. (2017). Bioinspired adhesive hydrogel driven by adenine and thymine. *ACS Applied Materials & Interfaces* 9 (20): 17645–17652.

70 Yang, C.H., Wang, M.X., Haider, H. et al. (2013). Strengthening alginate/polyacrylamide hydrogels using various multivalent cations. *ACS Applied Materials & Interfaces* 5 (21): 10418–10422.

71 Feldstein, M.M., Dormidontova, E.E., and Khokhlov, A.R. (2015). Pressure sensitive adhesives based on interpolymer complexes. *Progress in Polymer Science* 42: 79–153.

72 Wang, Y., Jia, K., Xiang, C. et al. (2019). Instant, tough, noncovalent adhesion. *ACS Applied Materials & Interfaces* 11 (43): 40749–40757.

73 Liu, X., Zhang, Q., and Gao, G. (2017). Bioinspired adhesive hydrogels tackified by nucleobases. *Advanced Functional Materials* 27 (44): 1703132.

74 Tuncaboylu, D.C., Argun, A., Sahin, M. et al. (2012). Structure optimization of self-healing hydrogels formed via hydrophobic interactions. *Polymer* 53 (24): 5513–5522.

75 Tuncaboylu, D.C., Argun, A., Algi, M.P., and Okay, O. (2013). Autonomic self-healing in covalently crosslinked hydrogels containing hydrophobic domains. *Polymer* 54 (23): 6381–6388.

76 Jeon, I., Cui, J., Illeperuma, W.R. et al. (2016). Extremely stretchable and fast self-healing hydrogels. *Advanced Materials* 28 (23): 4678–4683.

77 Li, F., Zhu, Y., You, B. et al. (2010). Smart hydrogels co-switched by hydrogen bonds and π–π stacking for continuously regulated controlled-release system. *Advanced Functional Materials* 20 (4): 669–676.

78 Jayawarna, V., Ali, M., Jowitt, T.A. et al. (2006). Nanostructured hydrogels for three-dimensional cell culture through self-assembly of fluorenylmethoxycarbonyl–dipeptides. *Advanced Materials* 18 (5): 611–614.

79 Smith, A.M., Williams, R.J., Tang, C. et al. (2008). Fmoc-diphenylalanine self assembles to a hydrogel via a novel architecture based on π–π interlocked β-sheets. *Advanced Materials* 20 (1): 37–41.

80 Melody, D.P. (1989). Advances in room temperature curing adhesives and sealants—a review. *Polymer International* 21 (2): 175–179.

81 Singer, A.J., Quinn, J.V., and Hollander, J.E. (2008). The cyanoacrylate topical skin adhesives. *The American Journal of Emergency Medicine* 26 (4): 490–496.

82 Li, C., Rowland, M.J., Shao, Y. et al. (2015). Responsive double network hydrogels of interpenetrating DNA and CB [8] host–guest supramolecular systems. *Advanced Materials* 27 (21): 3298–3304.

83 Kakuta, T., Takashima, Y., and Harada, A. (2013). Highly elastic supramolecular hydrogels using host–guest inclusion complexes with cyclodextrins. *Macromolecules* 46 (11): 4575–4579.

84 Nakahata, M., Takashima, Y., Yamaguchi, H., and Harada, A. (2011). Redox-responsive self-healing materials formed from host–guest polymers. *Nature Communications* 2: 511.

85 Ahn, Y., Jang, Y., Selvapalam, N. et al. (2013). Supramolecular velcro for reversible underwater adhesion. *Angewandte Chemie* 125 (11): 3222–3226.

86 Waite, J.H. (2017). Mussel adhesion–essential footwork. *The Journal of Experimental Biology* 220 (4): 517–530.

87 Ahn, B.K. (2017). Perspectives on mussel-inspired wet adhesion. *Journal of the American Chemical Society* 139 (30): 10166–10171.

88 Lee, H., Dellatore, S.M., Miller, W.M., and Messersmith, P.B. (2007). Mussel-inspired surface chemistry for multifunctional coatings. *Science* 318 (5849): 426–430.

89 Barrett, D.G., Bushnell, G.G., and Messersmith, P.B. (2013). Mechanically robust, negative-swelling, mussel-inspired tissue adhesives. *Advanced Healthcare Materials* 2 (5): 745–755.

90 Liao, M., Wan, P., Wen, J. et al. (2017). Wearable, healable, and adhesive epidermal sensors assembled from mussel-inspired conductive hybrid hydrogel framework. *Advanced Functional Materials* 27 (48): 1703852.

91 Kurokawa, T., Furukawa, H., Wang, W. et al. (2010). Formation of a strong hydrogel–porous solid interface via the double-network principle. *Acta Biomaterialia* 6 (4): 1353–1359.

92 Yuk, H., Zhang, T., Parada, G.A. et al. (2016). Skin-inspired hydrogel–elastomer hybrids with robust interfaces and functional microstructures. *Nature Communications* 7: 12028.

93 Liu, Q., Nian, G., Yang, C. et al. (2018). Bonding dissimilar polymer networks in various manufacturing processes. *Nature Communications* 9 (1): 846.

94 Tian, K., Bae, J., Bakarich, S.E. et al. (2017). 3D Printing of transparent and conductive heterogeneous hydrogel–elastomer systems. *Advanced Materials* 29 (10): 1604827.

95 Griffith, A. (1920). The phenomena of flow and rupture in solids. *Philosophical Transactions of the Royal Society of London. Series A* 221: 163–198.

96 Lake, G. and Thomas, A. (1967). The strength of highly elastic materials. *Proceedings of the Royal Society of London, Series A* 300 (1460): 108–119.

97 Orowan, E. (1949). Fracture and strength of solids. *Reports on Progress in Physics* 12 (1): 185.

98 Irwin, G.R. (1948). *Fracture Dynamics, Fracturing of Metals*, 152. American Society for Metals.

99 Wang, Z., Xiang, C., Yao, X. et al. (2019). Stretchable materials of high toughness and low hysteresis. *Proceedings of the National Academy of Sciences of the United States of America* 201821420.

100 Xiang, C., Wang, Z., Yang, C. et al. (2020). Stretchable and fatigue-resistant materials. *Materials Today* 34: 7–16.

101 Zhang, W., Gao, Y., Yang, H. et al. (2020). Fatigue-resistant adhesion I. Long-chain polymers as elastic dissipaters. *Extreme Mechanics Letters* 39: 100813.

102 Liu, J., Lin, S., Liu, X. et al. (2020). Fatigue-resistant adhesion of hydrogels. *Nature Communications* 11 (1): 1071.

103 Chen, B., Yang, J., Bai, R., and Suo, Z. (2019). Molecular staples for tough and stretchable adhesion in integrated soft materials. *Advanced Healthcare Materials* 8 (19): 1900810.

104 Wirthl, D., Pichler, R., Drack, M. et al. (2017). Instant tough bonding of hydrogels for soft machines and electronics. *Science Advances* 3 (6): e1700053.

105 Plueddemann, E.P. (1991). Chemistry of silane coupling agents. In: *Silane Coupling Agents*, 31–54. Boston, MA: Springer. https://doi.org/10.1007/978-1-4899-2070-6_2.

106 Van Amerongen, G.J. (1946). The permeability of different rubbers to gases and its relation to diffusivity and solubility. *Journal of Applied Physics* 17 (11): 972–985.

107 Le Floch, P., Meixuanzi, S., Tang, J. et al. (2018). Stretchable seal. *ACS Applied Materials & Interfaces* 10 (32): 27333–27343.

108 Yuk, H., Varela, C.E., Nabzdyk, C.S. et al. (2019). Dry double-sided tape for adhesion of wet tissues and devices. *Nature* 575 (7781): 169–174.

109 Nakajima, N. and Ikada, Y. (1995). Mechanism of amide formation by carbodiimide for bioconjugation in aqueous media. *Bioconjugate Chemistry* 6 (1): 123–130.

110 Gilles, M.A., Hudson, A.Q., and Borders, C. Jr., (1990). Stability of water-soluble carbodiimides in aqueous solution. *Analytical Biochemistry* 184 (2): 244–248.

111 Yang, J., Steck, J., and Suo, Z. (2020). Gelation kinetics of alginate chains through covalent bonds. *Extreme Mechanics Letters* 40: 100898.

112 Yao, X., Liu, J., Yang, C. et al. (2019). Hydrogel paint. *Advanced Materials* 31 (39): 1903062.

113 Hutchinson, J.W. and Suo, Z. (1991). Mixed mode cracking in layered materials. *Advances in Applied Mechanics* 29: 63–191.

114 Dauskardt, R., Lane, M., Ma, Q., and Krishna, N. (1998). Adhesion and debonding of multi-layer thin film structures. *Engineering Fracture Mechanics* 61 (1): 141–162.

115 Suo, Z. (2001). Fracture in thin films. In: *Encyclopedia of Materials: Science and Technology* (eds. K.H. Jürgen Buschow, Robert W. Cahn, Merton C. Flemings et al.), 3290–3296. ISBN 9780080431529, https://doi.org/10.1016/B0-08-043152-6/00587-8 (https://www.sciencedirect.com/science/article/pii/B0080431526005878).

116 Cordero, N., Yoon, J., and Suo, Z. (2007). Channel cracks in a hermetic coating consisting of organic and inorganic layers. *Applied Physics Letters* 90 (11): 111910.

117 Men, L., Yu, Y., Hou, Z. et al. (2023). Cracking modes in layered hyperelastic structures. *Journal of the Mechanics and Physics of Solids* 174: 105254.

118 Shao, X., Cai, Y., Yin, S. et al. (2022). Mechanics of interfacial delamination in deep-sea soft robots under hydrostatic pressure. *Journal of Applied Mechanics* 90 (2): 021009.

119 Cai, Y., Ma, J., Shen, Z. et al. (2023). Enhancing the fracture resistance of hydrogels by regulating the energy release rate via bilayer designs: theory and experiments. *Journal of the Mechanics and Physics of Solids* 170: 105125.

120 Chen, C., Wang, Z., and Suo, Z. (2017). Flaw sensitivity of highly stretchable materials. *Extreme Mechanics Letters* 10: 50–57.

121 Bai, R., Yang, J., and Suo, Z. (2019). Fatigue of hydrogels. *European Journal of Mechanics - A/Solids* 74: 337–370.

122 Creton, C. and Ciccotti, M. (2016). Fracture and adhesion of soft materials: a review. *Reports on Progress in Physics* 79 (4): 046601.

123 de Gennes, P.G. (1996). Soft adhesives. *Langmuir* 12 (19): 4497–4500.

124 Gent, A. (1996). Adhesion and strength of viscoelastic solids. Is there a relationship between adhesion and bulk properties? *Langmuir* 12 (19): 4492–4496.

125 Gent, A.N. and Schultz, J. (1972). Effect of wetting liquids on the strength of adhesion of viscoelastic material. *The Journal of Adhesion* 3 (4): 281–294.

126 Baumberger, T., Caroli, C., and Martina, D. (2006). Solvent control of crack dynamics in a reversible hydrogel. *Nature Materials* 5 (7): 552–555.

127 Baumberger, T. and Ronsin, O. (2009). From thermally activated to viscosity controlled fracture of biopolymer hydrogels. *The Journal of Chemical Physics* 130 (6): 061102. https://doi.org/10.1063/1.3078267.

128 Naassaoui, I., Ronsin, O., and Baumberger, T. (2018). A poroelastic signature of the dry/wet state of a crack tip propagating steadily in a physical hydrogel. *Extreme Mechanics Letters* 22: 8–12.

129 Bai, R., Chen, B., Yang, J., and Suo, Z. (2019). Tearing a hydrogel of complex rheology. *Journal of the Mechanics and Physics of Solids* 125: 749–761.

130 Rivlin, R. and Thomas, A.G. (1953). Rupture of rubber. I. Characteristic energy for tearing. *Journal of Polymer Science, Part A: Polymer Chemistry* 10 (3): 291–318.

131 Thomas, A. (1958). Rupture of rubber. V. Cut growth in natural rubber vulcanizates. *Journal of Polymer Science* 31 (123): 467–480.

132 Mars, W. and Fatemi, A. (2004). Factors that affect the fatigue life of rubber: a literature survey. *Rubber Chemistry and Technology* 77 (3): 391–412.

133 Mars, W. and Fatemi, A. (2002). A literature survey on fatigue analysis approaches for rubber. *International Journal of Fatigue* 24 (9): 949–961.

134 Lake, G.J. (2003). Fracture mechanics and its application to failure in rubber articles. *Rubber Chemistry and Technology* 76 (3): 567–591.

135 Long, R. and Hui, C.-Y. (2015). Crack tip fields in soft elastic solids subjected to large quasi-static deformation—a review. *Extreme Mechanics Letters* 4: 131–155.

136 Creton, C. (2017). 50th Anniversary perspective: networks and gels: soft but dynamic and tough. *Macromolecules* 50 (21): 8297–8316.

137 Long, R., Hui, C.-Y., Gong, J.P., and Bouchbinder, E. (2021). The fracture of highly deformable soft materials: a tale of two length scales. *Annual Review of Condensed Matter Physics* 12: 71–94.

138 Long, R. and Hui, C.-Y. (2016). Fracture toughness of hydrogels: measurement and interpretation. *Soft Matter* 12 (39): 8069–8086.

139 Gong, J.P., Katsuyama, Y., Kurokawa, T., and Osada, Y. (2003). Double-network hydrogels with extremely high mechanical strength. *Advanced Materials* 15 (14): 1155–1158.

140 Tang, J., Li, J., Vlassak, J.J., and Suo, Z. (2017). Fatigue fracture of hydrogels. *Extreme Mechanics Letters* 10: 24–31.

141 Bai, R., Yang, Q., Tang, J. et al. (2017). Fatigue fracture of tough hydrogels. *Extreme Mechanics Letters* 15: 91–96.

142 Bai, R., Yang, J., Morelle, X.P. et al. (2018). Fatigue fracture of self-recovery hydrogels. *ACS Macro Letters* 312–317.

143 Hua, M., Wu, S., Ma, Y. et al. (2021). Strong tough hydrogels via the synergy of freeze-casting and salting out. *Nature* 590 (7847): 594–599.

144 Lin, S., Liu, X., Liu, J. et al. (2019). Anti-fatigue-fracture hydrogels. *Science Advances* 5 (1): eaau8528.

145 Zhang, W., Hu, J., Yang, H. et al. (2021). Fatigue-resistant adhesion II: swell tolerance. *Extreme Mechanics Letters* 43: 101182.

146 Skeist, I. (2012). *Handbook of Adhesives*. Springer Science & Business Media.

147 Kinloch, A.J. (2012). *Adhesion and Adhesives: Science and Technology*. Springer Science & Business Media.

148 Kendall, K. (2007). *Molecular Adhesion and Its Applications: The Sticky Universe*. Springer Science & Business Media.

149 Ebnesajjad, S. and Landrock, A.H. (2014). *Adhesives Technology Handbook*, 3e. USA: William Andrew.

150 Chaudhuri, O., Gu, L., Klumpers, D. et al. (2016). Hydrogels with tunable stress relaxation regulate stem cell fate and activity. *Nature Materials* 15 (3): 326–334.

151 Chaudhuri, O., Koshy, S.T., Branco da Cunha, C. et al. (2014). Extracellular matrix stiffness and composition jointly regulate the induction of malignant phenotypes in mammary epithelium. *Nature Materials* 13 (10): 970–978.
152 Chaudhuri, O., Cooper-White, J., Janmey, P.A. et al. (2020). Effects of extracellular matrix viscoelasticity on cellular behaviour. *Nature* 584 (7822): 535–546.
153 Tringides, C.M., Vachicouras, N., de Lázaro, I. et al. (2021). Viscoelastic surface electrode arrays to interface with viscoelastic tissues. *Nature Nanotechnology* 16 (9): 1019–1029.

13

Electromechanics of Soft Resistive and Capacitive Tactile Sensors

Zhengjie Li[1], Sangjun Kim[2], Zheliang Wang[1], Zhengtao Zhu[3], and Nanshu Lu[1,2,4,5,6]

[1] *The University of Texas at Austin, Department of Aerospace Engineering and Engineering Mechanics, Austin, TX 78712, USA*
[2] *The University of Texas at Austin, Department of Mechanical Engineering, Austin, TX 78712, USA*
[3] *South Dakota School of Mines & Technology, Department of Chemistry, Biology, and Health Sciences, Rapid City, SD 57701, USA*
[4] *The University of Texas at Austin, Department of Electrical and Computer Engineering, Austin, TX 78712, USA*
[5] *The University of Texas at Austin, Department of Biomedical Engineering, Austin, TX 78712, USA*
[6] *The University of Texas at Austin, Texas Materials Institute, Austin, TX 78712, USA*

13.1 Introduction

13.1.1 Overview of Tactile Sensors

In the fascinating realm of human physiology, the skin stands out as the largest and one of the most versatile organs [1], endowed with remarkable properties that have inspired technological innovation, particularly in the development of electronic skins (e-skins). This extraordinary organ boasts a tactile-sensing capability with a force detection threshold as delicate as 1 mN [2], coupled with a high sensitivity of nearly 2 kPa^{-1}, extending up to forces of 10 N [3] or pressures of 70 kPa [4]. Beyond its touch sensitivity, the human skin exhibits an impressive stretchability, capable of fully recovering after elongating up to 40% of its original length [5, 6]. This elasticity is complemented by the skin's ability to discern pressure even amidst the complex interplay of simultaneous tension, shear, and twist forces. These intricate sensory and mechanical characteristics of human skin not only highlight its biological sophistication but also serve as a blueprint for the design and engineering of advanced tactile-sensing e-skins, aiming to replicate the skin's multifaceted functionalities in artificial systems.

E-skins have a broad array of applications centered around both humans and robotics [2, 7–10]. Selected examples are displayed in Figure 13.1 [11–19]. For humans, the applications span pulse sensing [12, 20, 21], prosthetics [13, 22], human–machine interaction (HMI) [14, 23], and biometric authentication such as fingerprint and typing style recognition [15, 24]. For robotic applications, e-skins can aid in identifying the shape or stiffness of objects [16, 17, 25], facilitating precise

Mechanics of Flexible and Stretchable Electronics, First Edition. Edited by Yong Zhu and Nanshu Lu.

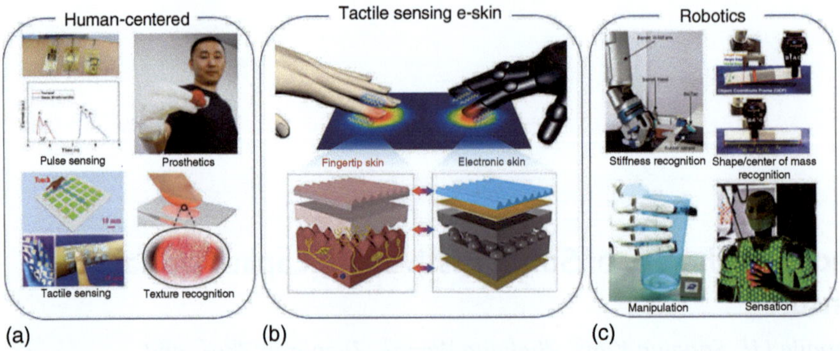

Figure 13.1 Tactile-sensing e-skin and its applications. (b): Tactile-sensing human skin versus e-skin. Source: Reproduced with permission from Ref. [11]. Copyright 2015 AAAS. (a): human-centered applications including pulse sensing. Source: Reproduced with permission from Ref. [12]. Copyright 2022 John Wiley & Sons, prosthesis. Reproduced with permission from Ref. [13]. Copyright 2022, Springer; tactile sensing. Reproduced with permission from Ref. [14], Copyright 2020 Elsevier; and texture recognition. Reproduced with permission from Ref. [15]/Springer Nature/CC BY 4.0. (c): robotics applications including stiffness recognition. Source: Reproduced with permission from Ref. [16]/Frontiers Media S.A./CC BY 3.0. Reproduced with permission from Ref. [17]. Copyright 2019 Springer; manipulation. Reproduced with permission from Ref. [18]. Copyright 2022 Wiley; sensation. Reproduced with permission from Ref. [19]. Copyright 2020 MDPI.

manipulation [18, 26], and endowing robots with somatosensory capabilities [19, 27].

As a central component of e-skins, mechanically soft tactile sensors have been the subject of extensive research for decades [28–32]. Recent advances in tactile sensors commonly adopt pressure-sensing mechanisms, including resistive [33–37], capacitive [38–44], iontronic [45], piezoelectric [46–48], triboelectric [49], optical [50, 51], and magnetoelastic [52] readouts, as summarized in Table 13.1. Under compression, different pressure-sensing materials generate a change in electrical resistance ΔR, capacitance ΔC, capacitance of electrical double layer ΔC, charge Δq, transmittance of light ΔI, or magnetic flux density ΔB, respectively. Each mechanism possesses its own set of advantages and disadvantages, as outlined in Table 13.1. This chapter only focuses on resistive tactile sensors (RTSs) and capacitive tactile sensors (CTSs) because they rely on basic materials and readout circuits for pressure sensing with a long history. In addition, they can maintain a stable performance over extended time [28, 53]. Compared with resistive and capacitive tactile sensors, iontronic tactile sensors only emerged a decade ago. They leverage the supercapacitive nature of electrical double layers achieved through the movement of ions within an ionically conductive material. Iontronic sensors can be ultrathin, optically transparent, mechanically soft, and highly sensitive over a large pressure range. But they are sensitive to environmental factors such as temperature, humidity, and exposure to chemicals and may suffer from ion migration over time [54]. Moreover, many electrolytic liquids used in intronic sensors are not biocompatible [55]. Piezoelectric and triboelectric sensors do not require input power, but they are only suitable for

Table 13.1 Summary of tactile sensors: mechanism, advantages, and disadvantages.

Sensor type		Advantages	Limitations
Piezoresistive (ΔR)		• Simple structure, facile fabrication • High pressure sensitivity • Easy signal readout and collection	• Large hysteresis • Sensitive to temperature • DC power consumption
Capacitive (ΔC)	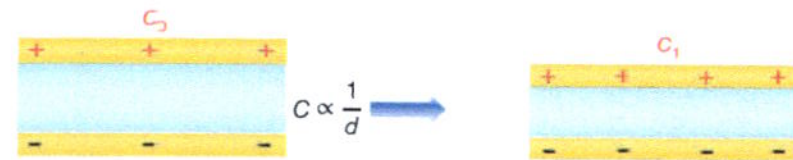	• Simple structure, facile fabrication • High pressure sensitivity • Temperature insensitivity • Low power consumption	• Decayed sensitivity with increasing pressure • Sensitive to parasitic capacitance • Sensitive to body loading
Iontronic (ΔC)	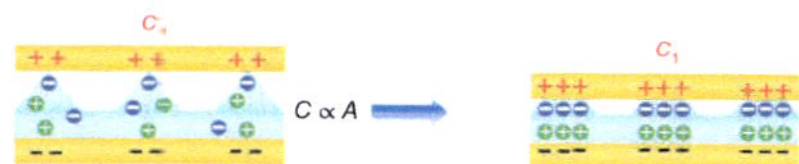	• High sensitivity over larger pressure range • Optical transparency	• Concerns in electrochemical stability • Sensitive to temperature and humidity • Frequency relaxation due to slow motion of ions under AC field
Piezoelectric/ Triboelectric (Δq)		• Self-powered • Ideal for vibration sensing • Fast response	• Not suitable for static pressure detection • Complex readout circuits • Triboelectric devices are sensitive to humidity
Optical (ΔI)		• Simple wiring • Low drift	• High power consumption • Contamination from ambient light • Signal attenuation due to misalignment
Magnetoelastic (ΔB)	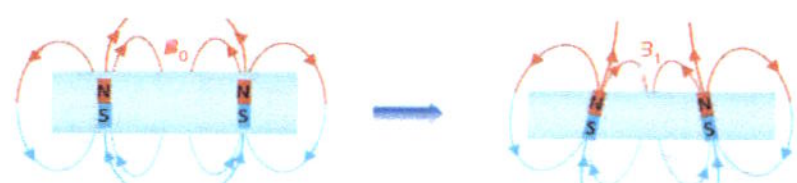	• No encapsulation required for chemical environment • Durability	• Complex magnetization effect in the sensor • Less sensitive to pressure due to weak magnetoelastic effects

measuring dynamic pressures such as pulses or vibrations, and triboelectric sensors are sensitive to humidity and temperature [56]. Optical sensors are simple to wire up and almost immune from signal drift, but they require a high power input, and optical signals may be polluted by ambient light [57]. Magnetoelastic sensors can be used in chemical environments with high durability, but they have limited sensitivity due to weak magnetoelastic effects [52].

Traditional resistive and capacitive tactile sensors suffer from a lack of flexibility and stretchability—unlike natural human skin. This results in some limitations when used on surfaces that are curved or highly deformable. The lack of stretchability means these traditional sensors can either separate from the surface they are attached to or even break when they are stretched excessively [58, 59]. To overcome these limitations, it is critical to incorporate thinness and stretchability into the sensor design to transform conventional tactile sensors into e-skins. Such a design ensures the sensors maintain conformal, or matching, contact with the surface. This not only boosts the sensor's accuracy and reliability in data collection and feedback but also imparts unique adaptability. This adaptability enables the stretchable surface to adjust to various shapes and movements, enhancing its responsiveness to different tasks [60–62]. Therefore, stretchable e-skins surpass the challenges posed by traditional sensors. It aligns e-skin technology closer to the mechanical properties of human skin while enabling its use on surfaces with various shapes and contours to perform a wide range of functions.

13.1.2 Bottlenecks of Soft Resistive and Capacitive Tactile Sensors

Years of research have significantly improved the fundamental capabilities of resistive and capacitive pressure sensors. This includes enhancements in sensitivity, sensing range, linearity, and response time through the materials and structural innovations, which have already been well summarized in a few recent review articles [57, 63–66]. Despite these advancements, Figure 13.2 illustrates two outstanding obstacles with soft resistive and capacitive pressure sensors: a decrease in sensitivity as pressure increases [67–69], and interference caused by lateral deformation [43, 64, 70]. Figure 13.2a exhibits the simplest resistive and capacitive sensors set up with parallel electrodes. The area of these electrodes is represented by "A", and "d" represents the distance between them. These sensors are subjected to a normal pressure labeled as "P". The lateral tensile strain is labeled as "ε" in Figure 13.2b.

For a RTS, the resistance of the piezoresistive material sandwiched between the electrodes can be expressed as

$$R = \rho \frac{d_R}{A_R} \tag{13.1}$$

where R is the resistance of the piezoresistive material and ρ is its resistivity. Since resistance is measured under a constant DC voltage "U" and given the well-known Ohm's Law:

$$I = \frac{U}{R} = \frac{UA_R}{\rho d_R} \tag{13.2}$$

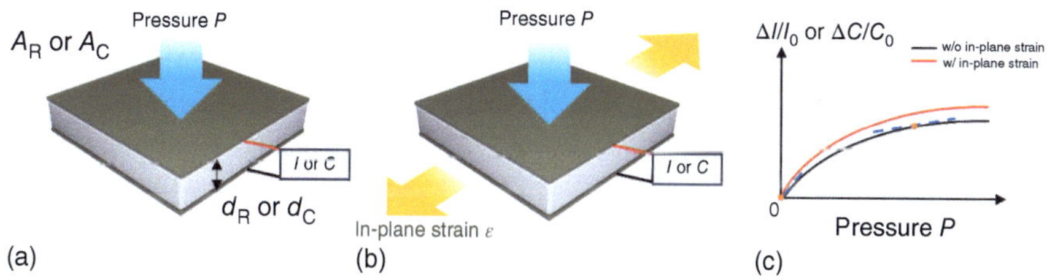

Figure 13.2 Two long-lasting bottlenecks in the stretchable resistive and capacitive tactile sensors. (a) Schematics of the resistive or capacitive tactile sensor subjected to pressure. (b) Schematics of the resistive (capacitive) tactile sensor upon pressure and in-plane stretch. (c) Relative change of current or capacitance as a function of pressure without and with in-plane stretch.

where I is the measured current response, the output of piezoresistive sensors is often plotted as I versus pressure instead of R versus pressure, as shown in Figure 13.2c.

The sensitivity of RTS is therefore the slope of the curve, which is given by:

$$S = \frac{\mathrm{d}\left(\frac{\Delta I}{I_0}\right)}{\mathrm{d}P} \tag{13.3}$$

where I_0 represents the initial current measured before pressure P is applied.

For a CTS with parallel electrodes sandwiching a dielectric material, the capacitance can be expressed as:

$$C = k\varepsilon_0 \frac{A_C}{d_C} \tag{13.4}$$

where k is the dielectric constant and ε_0 is the permittivity of vacuum. The sensitivity of CTS is therefore the slope of the curve, which is given by:

$$S = \frac{\mathrm{d}\left(\frac{\Delta C}{C_0}\right)}{\mathrm{d}P} \tag{13.5}$$

where C_0 is the initial capacitance of the CTS before P is applied.

Both RTS and CTS exhibit nonlinear pressure responses, as illustrated in Figure 13.2c. While the curve may start with a large slope at small pressure, it can quickly bend with diminishing sensitivity. This phenomenon is called the pressure-sensitivity trade-off, which is attributable to a marked increase in the effective stiffness of the pressure-sensing material under large pressure, accompanied by boundary confinement from electrodes [64, 68].

The second bottleneck is called stretch interference, meaning that when pressure and lateral stretch are simultaneously applied to the sensor, as illustrated in Figure 13.2b, the lateral stretch may skew the pressure response, as indicated by the red curve in Figure 13.2c. As a result, the lateral stretch would cause inaccurate pressure readings. Both RTS and CTS suffer from this bottleneck because both the R given in Eq. (13.1) and the C given in Eq. (13.4) depend on the sensor's geometric dimensions, A and d, both of which are affected by the lateral stretch (d is affected due to the Poisson's effect).

To mitigate issues related to pressure-sensitivity trade-off and stretch interference, diverse structural [42, 44, 71, 72] and material strategies [71, 73, 74] have been researched for soft RTS and CTS. Notable structural designs are micro-patterned surfaces [44], nanowire networks [71], foams [72], and textiles [42]. Common materials used include conductive composites [73], polymers [74], and nanomaterials [71]. Comprehensive documentation of the materials, structures, sensor performance, and applications of these sensors exists in several review papers [75, 76]. Therefore, this book chapter does not focus on those aspects. Despite a wealth of publications, there still lacks a systematic electromechanics understanding of why these strategies are effective and where the future research on RTS and CTS should head. So, the objective of this chapter is to outline the existing electromechanical models in soft resistive and capacitive tactile sensors, including simplified circuit models, resistivity and dielectric constant of soft composites, and how these properties change under applied pressure. Additionally, the chapter will discuss the impacts of different structural designs. The hope is that this basic discussion could facilitate cross-disciplinary description and understanding of soft RTS and CTS among materials scientists, mechanical engineers, and electrical professionals. Also, this chapter seeks to guide researchers in assessing and predicting sensor performance and identify directions for future sensor research and development to ultimately overcome the challenges of building human-like e-skins.

13.2 Resistive Tactile Sensors

13.2.1 Introduction of Resistive Tactile Sensors

Flexible RTS can be categorized into three types based on the materials/structures of the sensor – microstructured contact, bulk conductive nanocomposites, and porous conductive materials, as shown in Figures 13.3a–c. The microstructured contact-based RTS is highly sensitive at low pressure but has a limited pressure-sensing range [77], while the RTS based on the bulk conductive nanocomposites exhibits relatively low sensitivity but can operate over a wide pressure range [78]. The device based on the porous conductive materials can have varied piezoresistive response performance [79]. Popular materials for these sensors are composites of elastomers and conductive fillers [80], although fabrics [81] and network of nanowires [82] or 2D materials [83] have also been explored. The complexity of the material parameters increases from microstructured contact to bulk conductive nanocomposites and porous conductive material.

The resistance of the RTS can be modeled by the equivalent circuit of contact resistance and bulk resistance connected in series (Figure 13.3d), i.e.

$$R = 2R_c + R_b \tag{13.6}$$

where R is the total resistance of the sensor, R_c is the contact resistance, and R_b is the bulk resistance of the sensing layer. The piezoresistive response of the sensor can be analyzed and modeled by changes in contact resistance R_c and sensing layer resistance R_b under compressive pressure.

Figure 13.3 Schematics of three representative types of RTS. (a) Microstructured contact. (b) Bulk conductive nanocomposites. (c) Porous conductive materials. (d) A simple equivalent circuit of RTS.

13.2.2 Resistive Tactile Sensors Based on Microstructured Contact

For RTS based on the microstructured contact, the bulk resistance (R_b) of the conductive elastomer can be negligible, and the total resistance (R) of the device equals the contact resistance (R_c). In this scenario, the filler concentration of the conductive elastomer needs to be sufficiently high to achieve negligibly low bulk resistance, so that the device performance is not affected by the filler concentration. The piezoresistive response with applied pressure largely depends on the area of the contact and Young's modulus of the conductive elastomer.

The relative resistance changes of the RTS based on the microstructured contact can be expressed as follows:

$$\frac{\Delta R}{R} = \frac{(2R_{c0} + R_{b0}) - (2R_c + R_b)}{2R_{c0} + R_{b0}} = \frac{R_{c0} - R_c}{R_{c0}} = 1 - \frac{R_c}{R_{c0}} \tag{13.7}$$

where $\Delta R = R_0 - R$, R_0, R_{c0}, and R_{b0} are the initial total resistance, initial contact resistance, and initial bulk resistance, respectively, R, R_c, and R_b are the total resistance, contact resistance, and bulk resistance under pressure, respectively.

The piezoresistive response can be modeled using contact theory [84–89]. Figure 13.4a shows two conductive elastic spheres composed of conductive filler (e.g. silver particles) and elastomer (e.g. polydimethylsiloxane [PDMS]) in contact with each other under a normal force (N). Both spheres undergo a local contact deformation, creating a flat circular contact surface of radius a. Considering the high filler concentration in the composite, the resistivity of the spheres, ρ_b, is significantly small, and the resistance of the system is governed by the contact resistance. At the interfacial contact area, as shown in Figure 13.4b, the contact resistance arises from two possible mechanisms – constriction resistance R_{cr} and tunneling resistance R_t.

$$R_c = R_{cr} + R_t \tag{13.8}$$

where R_c is the contact resistance, R_{cr} is the constriction resistance, and R_t is the tunneling resistance.

Constriction resistance originates from the roughness of the interfacial contact area at the micro- or nanoscale, which results in an effective interfacial contact area less than the cross-section of the interface [90]. The constriction resistance for a circular contact can be derived as [88],

$$R_{cr} = \frac{\rho_b}{2a} = \frac{\rho_b}{2\sqrt{\frac{A}{\pi}}} = \frac{\rho_b}{2}\left(\frac{\pi}{A}\right)^{1/2} \tag{13.9}$$

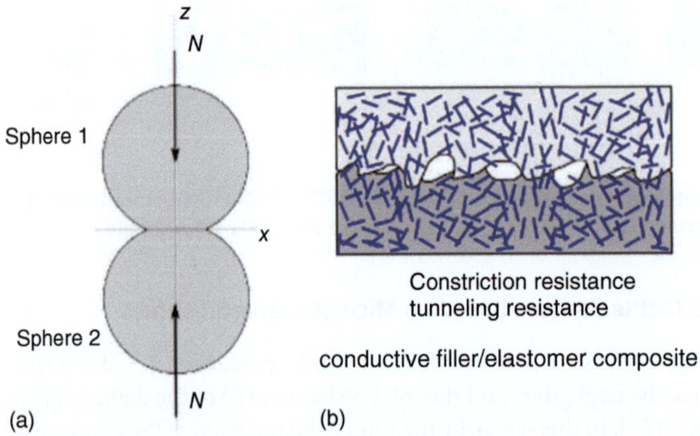

Figure 13.4 (a) Schematic diagram of two spheres contacting under normal force *N*. (b) Schematic diagram showing the rough contact interfaces between two pieces of conductive composites.

where ρ_b is the bulk resistivity, A is the (effective) contact area, and a is the radius of the circular contact. Tunneling resistance is associated with the tunneling current between the conductive fillers at the interfacial contact areas. The tunneling resistance is given as [87]:

$$R_t = \frac{\rho_t}{A} \tag{13.10}$$

where ρ_t is the tunneling resistivity, depending on the barrier thickness and barrier height by the following equation [89],

$$\rho_t = \frac{V}{J} \text{ and } J = \frac{e}{2\pi h \Delta s^2}\left[\begin{array}{c}\bar{\phi}\exp\left(-\frac{4\pi\Delta s\sqrt{2m}}{h}\sqrt{\bar{\phi}}\right)\\ -(\bar{\phi}+eV)\exp\left(-\frac{4\pi\Delta s\sqrt{2m}}{h}\sqrt{\bar{\phi}+eV}\right)\end{array}\right] \tag{13.11}$$

where V is the applied voltage, J is the current density, e is the elementary charge, h is Planck's constant, Δs is the effective barrier thickness, $\bar{\phi}$ is the barrier height, and m is the electron mass.

Combining Eqs. (13.9), (13.10), and (13.11), the contact resistance R_c is

$$R_c = R_{cr} + R_t = \frac{\rho_b}{2}\left(\frac{\pi}{A}\right)^{1/2} + \frac{\rho_t}{A} \tag{13.12}$$

The contact resistance in Eq. (13.12) depends on the interfacial areas of the contact A, the resistivity of the bulk material ρ_b, and the tunneling resistivity ρ_t at the interfacial contact area. The contact areas of the system shown in Figure 13.4a can be analyzed using the Hertz normal contact theory [84], which gives the contact surface radius a and areas A under normal force F as follows,

$$a = \left(\frac{3Fr}{4K}\right)^{\frac{1}{3}} \text{ and } A = \pi a^2 = \pi\left(\frac{3Fr}{4K}\right)^{\frac{2}{3}} \tag{13.13}$$

where F is the normal force, K is the effective stiffness, and r is the effective radius of curvature of two spheres in contact.

The effective stiffness is given by

$$\frac{1}{K} = \frac{1-\nu_1^2}{E_1} + \frac{1-\nu_2^2}{E_2} \tag{13.14}$$

where E_1 and E_2 are Young's moduli and ν_1, ν_2 are the Poisson's ratios of the two spheres.

The effective radius is given by

$$\frac{1}{r} = \frac{1}{r_1} + \frac{1}{r_2} \tag{13.15}$$

where r_1 and r_2 are the radii of the two spheres.

Substituting Eq. (13.12) for A in Eq. (13.13) gives,

$$R_c = \frac{\rho_b}{2}\left(\frac{\pi}{\pi\left(\frac{3Fr}{4K}\right)^{\frac{2}{3}}}\right)^{1/2} + \frac{\rho_t}{\pi\left(\frac{3Fr}{4K}\right)^{\frac{2}{3}}} = \frac{\rho_b}{2}\left(\frac{4K}{3Fr}\right)^{\frac{1}{3}} + \frac{\rho_t}{\pi}\left(\frac{4K}{3Fr}\right)^{\frac{2}{3}} \tag{13.16}$$

If a bias voltage of V is applied, the current I is,

$$I = \frac{V}{R_c} = \frac{V}{\left(\frac{\rho_b}{2}\left(\frac{4K}{3Fr}\right)^{\frac{1}{3}} + \frac{\rho_t}{\pi}\left(\frac{4K}{3Fr}\right)^{\frac{2}{3}}\right)} \tag{13.17}$$

Equation (13.17) can be reduced to the following form if constriction resistance is dominant or tunneling resistance is dominant:

$$I \sim \begin{cases} \frac{2V}{\rho_b}\left(\frac{3Fr}{4K}\right)^{\frac{1}{3}}, \text{ if constriction resistance is dominant;} \\ \frac{\pi V}{\rho_t}\left(\frac{3Fr}{4K}\right)^{\frac{2}{3}}, \text{ if tunneling resistance is dominant.} \end{cases} \tag{13.18}$$

The sensitivity of the piezoresistive response of the contact resistance between two spheres can be estimated as:

$$S = \frac{d\left(\frac{\Delta R}{R_0}\right)}{dP} \approx \frac{d\left(\frac{R}{R_0}\right)}{dF} \sim \frac{1}{R_0}\left(\frac{\rho_b}{6}\left(\frac{4K}{3r}\right)^{\frac{1}{3}}\left(\frac{1}{F}\right)^{\frac{4}{3}} + \frac{2\rho_t}{3\pi}\left(\frac{4K}{3r}\right)^{\frac{2}{3}}\left(\frac{1}{F}\right)^{\frac{5}{3}}\right) \tag{13.19}$$

$$S = \frac{\mathrm{d}\left(\frac{\Delta I}{I_0}\right)}{\mathrm{d}P} \approx \frac{\mathrm{d}\left(\frac{I}{I_0}\right)}{\mathrm{d}F} \sim \begin{cases} \frac{2R_0}{3\rho_b}\left(\frac{3r}{4K}\right)^{\frac{1}{3}}\left(\frac{1}{F}\right)^{\frac{2}{3}}, \text{ if constriction resistance is dominant;} \\ \frac{2\pi R_0}{3\rho_t}\left(\frac{3r}{4K}\right)^{\frac{2}{3}}\left(\frac{1}{F}\right)^{\frac{1}{3}}, \text{ if tunneling resistance is dominant.} \end{cases} \tag{13.20}$$

The model aforementioned describes the piezoresistive response of two spheres in contact. For the device with arrays of spheres in contact, the equivalent circuit can be considered as the parallel contact resistor, which gives the device resistance as $R_{device} = R_c/N$, in which R_{device} is the device resistance, and N is the number of spheres in contact. The piezoresistive response of the device still follows the similar relationship described in Eqs. (13.16)–(13.20).

Assuming that ρ_b, and ρ_t are intrinsic material properties that do not change when the system is subjected to a normal pressure, the change of the contact resistance would be dominant by the change of the contact areas under pressure. This assumption is valid only for conductive elastomeric composites with conductive filler concentration well above the percolation threshold. In this case, the piezoresistive response is governed by Eq. (13.12) and is related to the pressure, the mechanical properties of the device, and the geometry of the device by Eqs. (13.13), (13.16), and (13.17). The sensitivity of the device, as shown in Eqs. (13.19) and (13.20), decreases with an increase in the applied pressure, following a negative power law relationship with applied pressure. The variation of the sensitivity (i.e. the linearity of the piezoresistive response) depends on the structure and stiffness of the device.

Both constriction resistance and tunneling resistance contribute to the contact resistance (Eq. (13.12)). As Eqs. (13.16) and (13.18) indicate, the constriction resistance and tunneling resistance follow the power law relationship of different exponent values with respect to pressure. The plot of $\log(R_c)$ versus $\log(P)$ may help understand the contribution of constriction resistance and tunneling resistance. In the composite composed of the conductive fillers embedded in a matrix, the size of the filler is typically much smaller than the contact area. Therefore, the constriction resistance is negligible, and the contact resistance is dominated by the tunneling resistance [87]. For the tunneling resistance, the material parameters essential to the tunneling current are the effective barrier thickness Δs and the barrier height $\overline{\phi}$, as indicated in Eq. (13.11). These parameters are related to the electrical properties of the filler, the distribution and concentration of fillers in the composite, and the interactions between the filler and matrix, which may be challenging to control precisely in composites.

The change of contact areas in the RTS based on the structural contact depends on the dimensions and shapes of the microstructure. For the two elastic spheres, the contact areas increase with the applied pressure and the radius of the sphere but decrease with the stiffness of the material, all following the 2/3 power law relationship. Different microstructures, such as hemispheres [89], micro-pyramids [74, 91], micro-pillars [92], hierarchical structures [93], and epidermis-inspired microstructures [94], have been explored for the RTS. While analytical models for the change of the contact areas in these microstructures under applied pressure may be challenging, finite-element simulations are routinely used to simulate and understand the mechanics of these microstructures under pressure.

Park et al. reported an RTS based on interlocked micro-domes (or hemispheres) arrays of carbon nanotube (CNT)/PDMS composite, as shown in Figure 13.5a(i). Finite-element analysis (FEA) was used to simulate the local stress and contact areas under applied pressure. By varying the elastic modulus of composites with different CNT concentrations, the simulation results indicate that composites with low elastic moduli will undergo large deformation. The geometric contact area A_{dome} obtained by the simulation was refined to accurately represent the electrical contact area of CNTs, formulated as $A_{CNT} = \left(\frac{V_f}{2\pi}\right)^2$. Here, V_f symbolizes the volume fraction of the CNT in the composite and $\frac{V_f}{2\pi}$ represents the areal fraction in the cross-sectional area of the composite. Figure 13.5a(ii) plots the simulated A_{CNT} as a function of the applied pressure for composite with different CNT doping ratios. The result can be fitted with the power law function $y = ax^b$ with exponent $b = 0.7$ for all samples. The value is close to 2/3 in Eq. (13.13), suggesting that the contact between the interlocked micro-domes is elastic within the simulated pressure range. The electric contact area increases with the increase of the fraction of CNT in the composite. For the device shown in Figure 13.5a(i), the tunneling resistance was calculated by $R_t = R_{device} - R_b$, omitting the constriction resistance. The tunneling resistance versus pressure can be fitted using Eq. (13.11), as shown in Figure 13.5b(viii). The fitting results reveal that the initial barrier thickness decreases with the increase of the fraction of CNT in the composite.

Pang et al. reported the RTS based on the epidermis microstructures of graphene/PDMS composite. The mechanism of the sensor is shown in Figure 13.5b(i–iii) [94]. The simulation of pressure distribution under 5 kPa pressure load for pyramid, hemisphere, nanowire, and randomly distributed spinosum (RDS) is visualized in Figure 13.5b(iv, v, vi, vii). Pyramid and hemisphere structures experience pressure concentration on the local contact area, leading to a pronounced increase in the contact area and heightened sensitivity at lower pressures, but this sensitivity sharply declines with increased pressure. The micro-pillar nanowire array has a uniform pressure distribution, while the RDS structure can transmit the concentrated stress at the peak to the root segment, thereby achieving a more homogenous pressure distribution. To further elucidate these observations, a MATLAB model was established incorporating the simulated contact areas and an equivalent circuit model of the parallel resistors in contacted structures. The resulting piezoresistive responses under pressure loading for different microstructures are shown in Figure 13.5b(viii). The results underscore an insight: randomly distributed microstructures can offer synergistic effects of high sensitivity and wider linear ranges for the RTS based on microstructured contact [94].

Ji et al. presented theoretical analysis and experimental design to optimize the sensitivity and linearity of RTS based on microstructured contact [86]. Figure 13.5c (i, ii, iii) shows the working principles of the piezoresistive response of the single conductive layer (SCL) devices composed of PDMS micro-domes spray-coated with carbon black PDMS (CPDMS) or dip-coated with silver nanowire (AgNW), and double conductive layer (DCL) device composed of PDMS micro-domes covered by both CPDMS and AgNWs double conductive layer (CPDMS/AgNWs DCL). Detailed theoretical analysis of these device structures reveals that the pass-through resistance (R_p in Figure 13.5c, denoted as orange resistors.) of the DCL structure balances the nonlinear variation in contact resistance under pressure, which contributes to

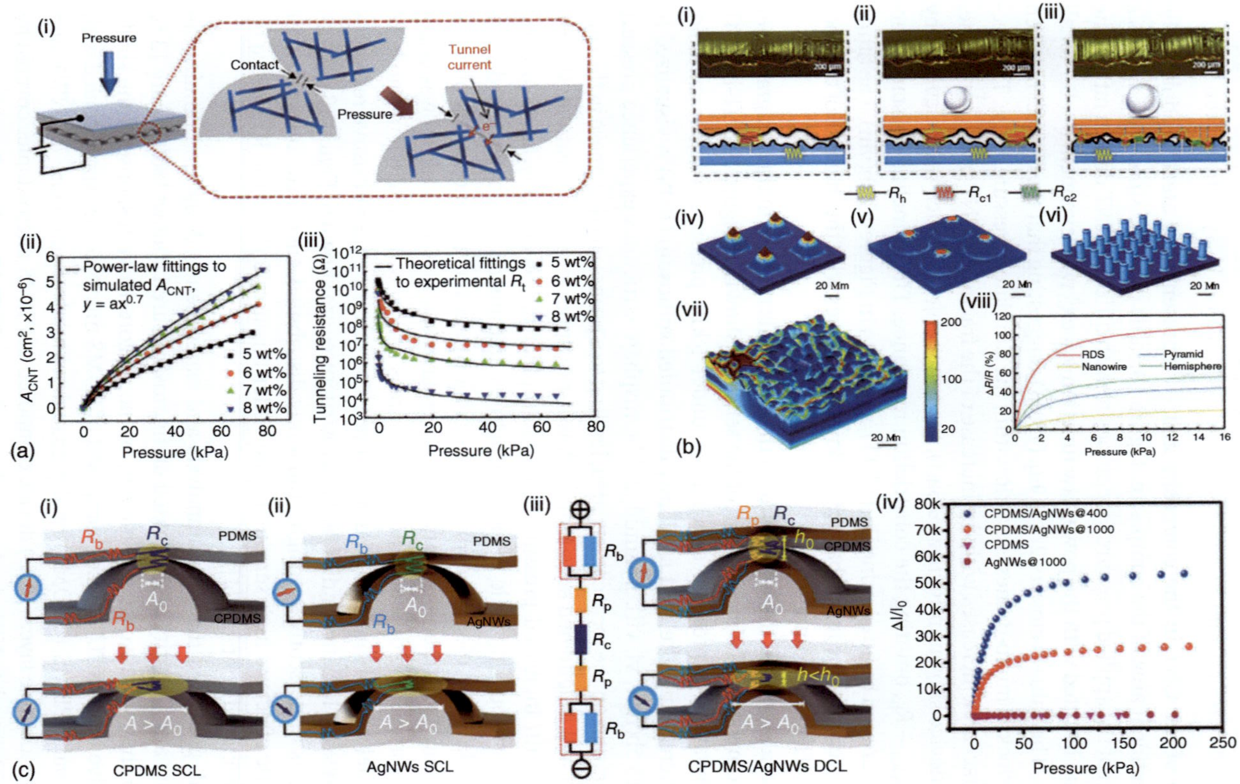

(i)
Pressure
Contact
Pressure
Tunnel current
(ii)
Power-law fittings to simulated A_{CNT}, $y = ax^{0.7}$
5 wt%
6 wt%
7 wt%
8 wt%
A_{CNT} (cm², ×10⁻⁶)
Pressure (kPa)
(iii)
Theoretical fittings to experimental R_t
Tunneling resistance (Ω)
Pressure (kPa)
(a)
(i)
(ii)
(iii)
R_h
R_{c1}
R_{c2}
(iv)
(v)
(vi)
(vii)
(viii)
RDS
Nanowire
Pyramid
Hemisphere
ΔR/R (%)
Pressure (kPa)
(b)
(i)
R_b
R_c
PDMS
CPDMS
A_0
$A > A_0$
CPDMS SCL
(ii)
AgNWs
AgNWs SCL
(iii)
R_p
h_0
$h<h_0$
CPDMS/AgNWs DCL
(iv)
CPDMS/AgNWs@400
CPDMS/AgNWs@1000
CPDMS
AgNWs@1000
ΔI/I₀
Pressure (kPa)
(c)

Figure 13.5 RTS based on microstructured contact. (a) (i) Schematic showing the working principle of the micro-dome-based RTS. (ii) The electrical contact area (A_{CNT}) as a function of pressure for different CNT wt% (5–8 wt%). A_{CNT} is estimated by considering the CNT area fraction over the geometrical contact area, which is obtained by using FEM simulation. The solid lines represent power law fits to A_{CNT} with an exponent of 0.7. (iii) Experimental tunneling resistances (dots) of the interlocked micro-dome arrays for different CNT concentrations (5–8 wt%) fitted to the tunneling resistance model (curves). Source: Reproduced with permission from Ref. [89]. Copyright 2014 American Chemical Society. (b) Working mechanism of the randomly distributed spinosum (RDS) graphene pressure sensor. Photographs and schematic illustrations of circuit models corresponding to (i) the initial state of unloading, (ii) light loading, and (iii) heavy loading. The pressure distribution of the simulation results for different geometries: (iv) pyramid, (v) hemisphere, (vi) nanowire, and (vii) RDS microstructure at an external loading pressure of 5 kPa. (viii) The simulation results of resistance variation versus applied pressure for different surface microstructures. Source: Reproduced with permission from Ref. [94]. Copyright 2018 American Chemical Society. (c) Schematic illustration of the principle of (i) CPDMS single conductive layer (SCL); (ii) AgNWs SCL, and (iii) CPDMS/AgNW double conductive layer (DCL); (iv) Relative current variation of the sensor fabricated with different conductive layers based on the solid dome array structure under external pressures. Source: Reproduced with permission from Ref. [86]. Copyright 2020 American Chemical Society.

highly sensitive and broad linear response (0 to 6 kPa) (Figure 13.5c(iv)) [86]. The device structure was further optimized by the synergistic effect of DCL and porous PDMS dome array.

13.2.3 Resistive Tactile Sensors Based on Bulk Conductive Nanocomposites

The piezoresistive properties of bulk nanocomposites, which are composed of elastomeric matrices and conductive fillers such as metal nanoparticles, graphenes, and CNTs, have been extensively investigated for their potential applications in tactile sensing [95]. Mechanisms of the piezoresistive properties of these materials are treated within the framework of percolation theory and an effective conductive path model in the literature. Figure 13.6 shows the schematic of several key aspects of the piezoresistivity of nanocomposites. The conductivity of the nanocomposite composed of conductive fillers depends critically on the concentration of the filler in the composite. Below the percolation threshold (insulation zone), the fillers cannot form a conductive path, and the nanocomposite has very low conductivity. The conductivity of the composite increases drastically when the filler concentration is around the percolation threshold (percolation zone), at which the conductive paths start to form; above the threshold (conduction zone), the conductivity continuously increases but with a small increment and eventually levels off. For the piezoresistive response, applied pressure may have a similar effect as the concentration of the conductive filler. Conductivity can increase drastically when the applied pressure leads to the formation of continuous conductive paths. Beyond the percolation concentration, the number of conductive paths can be described by the effective conductive path model. Additionally, the relaxation of the elastomeric matrices of the piezoresistive composite would affect the hysteresis, the response time,

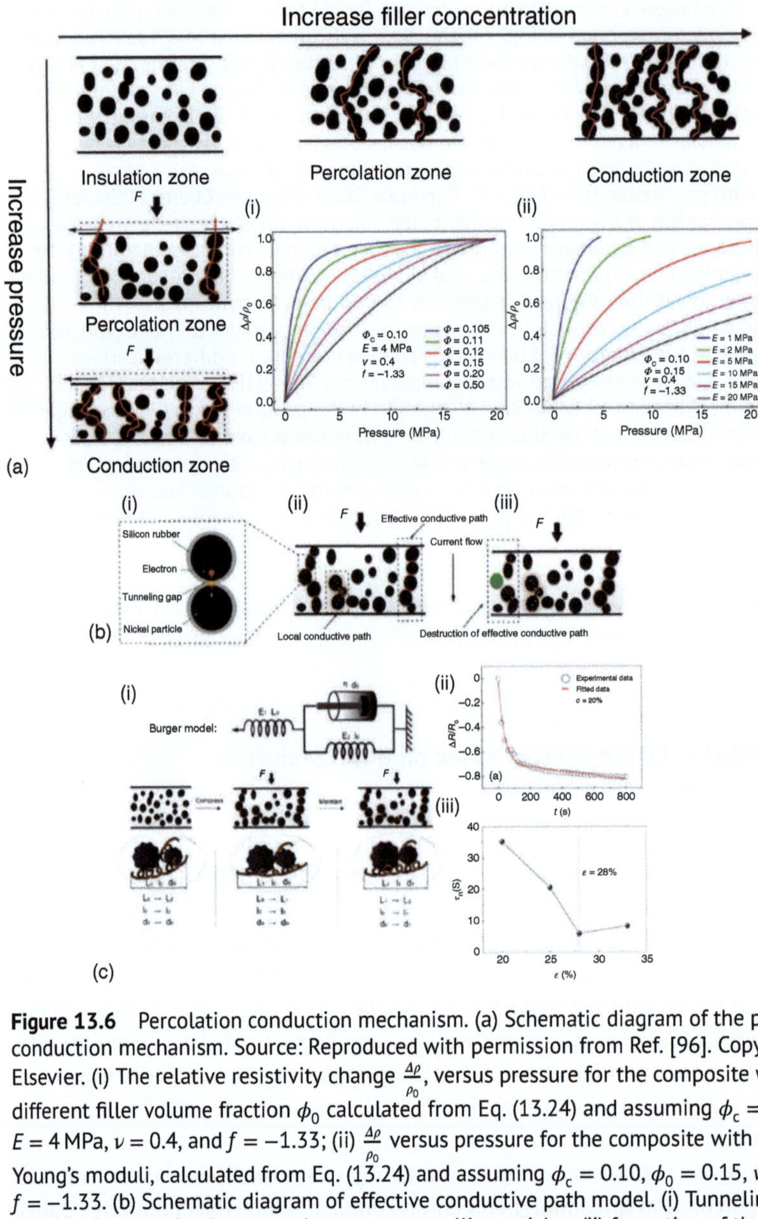

Figure 13.6 Percolation conduction mechanism. (a) Schematic diagram of the percolation conduction mechanism. Source: Reproduced with permission from Ref. [96]. Copyright 2017 Elsevier. (i) The relative resistivity change $\frac{\Delta\rho}{\rho_0}$, versus pressure for the composite with different filler volume fraction ϕ_0 calculated from Eq. (13.24) and assuming $\phi_c = 0.1$, $E = 4\,\text{MPa}$, $\nu = 0.4$, and $f = -1.33$; (ii) $\frac{\Delta\rho}{\rho_0}$ versus pressure for the composite with different Young's moduli, calculated from Eq. (13.24) and assuming $\phi_c = 0.10$, $\phi_0 = 0.15$, $\nu = 0.4$, and $f = -1.33$. (b) Schematic diagram of effective conductive path model. (i) Tunneling through the thin film existing between the nearest metallic particles, (ii) formation of the effective conductive path, and (iii) destruction of the effective conductive path. Source: Reproduced with permission from Ref. [96]. Copyright 2017 Elsevier. (c) (i) Schematic diagrams of Burger's model and the change process of the microstructure of the percolation network for composites under static compression; (ii) Fractional electrical resistance change $\Delta R/R_0$ as a function of time t during stress relaxation at 20% strain for T4/123-4 FAQTC, "FAQTC" denotes the field-assisted quantum tunneling composites, "T4/123-4" represents the composite prepared by silicon T4 and type 123 spiky spherical nickel powders with a weight ratio of nickel powder to silicon rubber of 4 : 1 and cured; (iii) Fitted mean relaxation time as a function of strain. Source: Reproduced with permission from Ref. [96]. Copyright 2017 Elsevier.

and other aspects of the piezoresistive response. The piezoresistive responses of these materials can decrease or increase with applied pressure, depending on the material's composition, processing conditions, and other factors.

The resistivity of the nanocomposite can be described by the percolation scaling law [96],

$$\rho = X(\phi - \phi_c)^f \quad (\phi > \phi_c) \tag{13.21}$$

where ρ is the resistivity, X is the fitting constant, ϕ is the volume fraction of the conductive filler, ϕ_c is the percolation threshold volume fraction, and f is the percolation exponent. The value of f depends on the dimensionality of the filler. Note that the resistivity decreases with the increase of the conductive filler, so the value of f is negative. For 3D, $f = -2$ and for 2D, $f = -1.33$. Below ϕ_c, the composite can be considered as an insulator. Besides the volume fraction of the conductivity filler, the resistivity around the percolation threshold ϕ_c can change significantly by many other factors, including pressure, temperature (heating), and film thickness [97].

Assuming the nanocomposite is elastic under small uniaxial strain, the volume of the composite can be expressed by:

$$V_s = V_{s,0}\left(1 - \frac{P(1-2\nu)}{E}\right) \tag{13.22}$$

where V_s is the volume under pressure, $V_{s,0}$ is the initial volume, P is the applied pressure, ν is the Poisson's ratio, and E is the Young's modulus of the composite.

The filler volume fraction ϕ under pressure becomes

$$\phi = \frac{V_f}{V_s} = \frac{V_f}{V_{s,0}\left(1 - \frac{P(1-2\nu)}{E}\right)} = \frac{\phi_0}{1 - \frac{P(1-2\nu)}{E}} \tag{13.23}$$

where V_f is the volume of filler, ϕ_0 is the initial volume fraction of the conductive filler. Here, the filler is assumed to be incompressible so V_f is a constant under pressure.

Combining Eq. (13.22) and Eq. (13.23),

$$\frac{\Delta\rho}{\rho_0} = \frac{\rho_0 - \rho}{\rho_0} = 1 - \left(\frac{\frac{\phi_0}{1 - \frac{P(1-2\nu)}{E}} - \phi_c}{\phi_0 - \phi_c}\right)^f \tag{13.24}$$

where ρ is the resistivity under pressure, ρ_0 is the initial resistivity, and $\Delta\rho = \rho_0 - \rho$ is the change of the resistivity. The percolation threshold volume fraction ϕ_c is assumed to be unaffected by the pressure.

Equation (13.23) is only valid when the volume fraction is above ϕ_c. If the initial volume fraction is below ϕ_c, the resistivity of the nanocomposite does not change with the pressure until the percolation threshold pressure P_c is reached. P_c can be calculated by setting ϕ in Eq. (13.24) equal to ϕ_c.

$$P_c = \left(1 - \frac{\phi_0}{\phi_c}\right)\frac{E}{1-2\nu} \tag{13.25}$$

The piezoresistive response described in Eqs. (13.23) and (13.25) assumes the linear elasticity and small strain conditions. While the model may be oversimplified, it provides some insights into the piezoresistive response of the nanocomposite with conductive fillers.

- The composition of the nanocomposite is significant in determining the response of the RTS. In the scenario where the volume fraction of the conductive filler is lower than the percolation volume fraction, the resistivity of the sensor will remain unchanged until the pressure reaches the percolation threshold pressure P_c; above P_c, the resistivity of the device can decrease significantly, resulting in a highly sensitive piezoresistive response. This behavior has been observed in spiky nickel nanoparticle-filled rubber composites [96]. For example, the electrical resistivity of a material composed of silicon T4 and nickel powders with a weight ratio of nickel powder to silicon rubber of 4 : 1 was above 10^9 Ω cm; when the pressure was above 0.6 MPa, the resistivity dropped quickly by orders of magnitude. To utilize these materials in applications for tactile sensing, Eq. (13.24) suggests that the composition close to the percolation threshold concentration and with a low Young's modulus would lower the pressure detection limit of the device.
- The composition of the piezoresistive composite plays a crucial role in the tradeoff between the sensitivity and the detection range. Figure 13.6a(i) plots the relative resistivity change $\frac{\Delta\rho}{\rho_0}$ as a function of pressure for the composites with different filler volume fraction ϕ_0, calculated from Eq. (13.22) and assuming $\phi_c = 0.1$, $E = 4\,\text{MPa}$, $\nu = 0.4$, and $f = -1.33$. When ϕ_0 approximated to ϕ_c, $\frac{\Delta\rho}{\rho_0}$ exhibits a rapid change at small pressure and then levels off, indicating high sensitivity and narrow operation range of the sensor. With a rise of the conductive filler concentration (i.e. increase of ϕ_0), the sensitivity decreases while the quasi-linear range becomes wider.
- Young's modulus of the piezoresistive composite is also linked to the tradeoff between the sensitivity and the upper detection range. Figure 13.6a(ii) plots the relative resistivity change $\frac{\Delta\rho}{\rho_0}$ versus pressure for the composites with different Young's modules, derived from Eq. (13.22) and assuming $\phi_c = 0.10$, $\phi_0 = 0.15$, $\nu = 0.4$, and $f = -1.33$. The plot clearly indicates that a lower Young's modulus in the composite yields high sensitivity but restricts the working range. Conversely, a composite with a higher Young's modulus can function across a broader pressure range, albeit at a reduced sensitivity. Note that the pressure range of the plot is the result of $V_s = 0$ in Eq. (13.23) of the oversimplified elastic model and small strain assumptions, which are certainly not applicable for $V_s = 0$. Nevertheless, the result suggests controlling the Young's modulus of the composite may be important for optimization of the RTS based on the bulk conductive nanocomposites.

The resistivities of bulk conductive nanocomposites can be attributed to constriction resistance at the filler contact, tunneling resistance at the filler contact, and the intrinsic resistance of the filler [87]. Based on the model by Ruschau et al.

$$\rho_c^{\text{plastic}} = 0.89\rho_i\left(\frac{\lambda Hwt}{F}\right)^{\frac{1}{2}} + \frac{\rho_t \lambda Hwt}{Fd} + \frac{\rho_i}{0.524} \tag{13.26}$$

$$\rho_{\mathrm{c}}^{\mathrm{ceramic}} = 0.57\rho_{\mathrm{i}}\left(\frac{Ewt}{F}\right)^{\frac{1}{3}} + \left(\frac{0.26\rho_{\mathrm{t}}}{d}\right)\left(\frac{Ewt}{F}\right)^{\frac{2}{3}} + \frac{\rho_{\mathrm{i}}}{0.524} \tag{13.27}$$

where $\rho_{\mathrm{c}}^{\mathrm{plastic}}$ and $\rho_{\mathrm{c}}^{\mathrm{ceramic}}$ represent resistivities of the composites with plastic fillers (polymer or metal) and ceramic fillers, respectively, ρ_{i} is the intrinsic resistivity of the filler, ρ_{t} is the tunneling resistivity, H denotes the contact hardness, and E represents Young's modulus of the composites, respectively, F is the applied force, w is the width of the composite, t is the thickness of the composite, d is the diameter of the filler, and λ is the empirical elasticity factor here.

The model anticipates that factors such as hardness, elastic modulus, filler particle size, and the overall dimensions of the composite sample significantly influence its piezoresistive properties. Specifically, fillers characterized by a higher modulus (or hardness) and smaller size are found to enhance the piezoresistivity of the composite. It is worth noting that Eq. (13.26) for the piezoresistive bulk nanocomposites and Eq. (13.16) for the contact-based RTS are essentially the same. In both models,

$$\rho_{\mathrm{device}} \sim \left(\frac{1}{F}\right)^{\frac{1}{2}} \text{ for the constriction resistance} \tag{13.28}$$

$$\rho_{\mathrm{device}} \sim \left(\frac{1}{F}\right)^{\frac{2}{3}} \text{ for the tunneling resistance} \tag{13.29}$$

Therefore, experimentally, the slope of the $\log(\rho_{\mathrm{device}})$ versus $\log(F)$ plot may help understand the contribution of the constriction resistance and the tunneling resistance to the overall piezoresistive response.

In general, according to Eqs. (13.26) and (13.27), the tunneling resistance dominates the magnitude of the overall resistance of the composite. The effective conductive path model, which provides a microscopic conduction mechanism of the piezoresistive response of the bulk conductive nanocomposites [96, 98, 99], considers only the tunneling conduction in the composite. Figure 13.6b shows the schematic diagram of the effective conductive path. The model posits that applied pressure changes the distance between conductive fillers within the nanocomposite. A sufficiently reduced inter-filler distance enables electron tunneling between adjacent fillers, thereby establishing localized conductive connections. When these connections collectively form a continuous conductive pathway between electrodes through the matrix, the nanocomposite's resistivity decreases. Conversely, compressive forces may cause the fillers to slip, disrupting these effective conductive pathways and consequently increasing the resistivity. The nanocomposite's piezoresistive behavior emerges from the interplay of these two microscopic phenomena, quantifiable through the variation in the number of effective conductive paths during compression. The tunneling current between the two adjacent conductive particles is given by [98],

$$I^{i,j} = \frac{3e^2V^{i,j}\sqrt{2m\phi^{i,j}}}{2h^2D^{i,j}}\exp\left(-\frac{4\pi D^{i,j}\sqrt{2m\phi^{i,j}}}{h}\right) \tag{13.30}$$

where $V^{i,j}$, $\phi^{i,j}$, $D^{i,j}$ are the external voltage, the height of the potential barrier, and the thickness of the insulating layer (i.e. tunneling gap in Figure 13.6b) between the conductive filler i and j, m is the mass of electron, e is the elementary charge, and h is Planck's constant.

Then, the resistance of the ith effective conductive path can be given by considering the M_i number of conductive particles in series,

$$R_i = \sum_{j=1}^{M_i-1} R^{i,j} = \sum_{j=1}^{M_i-1} \frac{V^{i,j}}{I^{i,j}} = \sum_{j=1}^{M_i-1} \frac{2h^2 D^{i,j}}{3e^2 s^{i,j} \sqrt{2m\phi^{i,j}}} \exp\left(\frac{4\pi D^{i,j}\sqrt{2m\phi^{i,j}}}{h}\right) \tag{13.31}$$

where $s^{i,j}$ is the area of the cross-section of the insulating film between the conductive filler i and j.

To simplify the model, assuming the average insulating layer thickness D, the average number of conductive particles in one effective path M, the average cross-sectional area s, and the average barrier height ϕ, then Eq. (13.31) becomes:

$$R_i = (M-1)\frac{2h^2 D}{3e^2 s\sqrt{2m\phi}} \exp\left(\frac{4\pi D\sqrt{2m\phi}}{h}\right) \tag{13.32}$$

The total resistance of the composite R can be calculated by considering the N effective conductive paths in parallel,

$$R = \frac{1}{N}\left[(M-1)\frac{2h^2 D}{3e^2 S\sqrt{2m\phi}} \exp\left(\frac{4\pi D\sqrt{2m\phi}}{h}\right)\right] \tag{13.33}$$

In Eq. (13.33), the number of effective conductive paths N and the insulating film thickness between the fillers D depend on the applied pressure P. The change in the number of effective conductive paths, $\frac{N(P)}{N(P=0)}$, can be modeled using different empirical formulas [96, 98]. Ding et al. assumed that the number of the conductive paths increases exponentially with the applied pressure [96],

$$\begin{aligned}\frac{N(P)}{N(P=0)} &= \exp(A\epsilon + B\epsilon^2 + C\epsilon^3 + D\epsilon^4) \\ &= \exp\left(A\left(\frac{P}{E_c}\right) + B\left(\frac{P}{E_c}\right)^2 + C\left(\frac{P}{E_c}\right)^3 + D\left(\frac{P}{E_c}\right)^4\right)\end{aligned} \tag{13.34}$$

where $N(P)$ is the number of conductive paths under applied stress P, $N(P = 0)$ is the number of conductive paths with no stress, ϵ is the compressive strain, E_c is the compressive modulus of the composite, and A, B, C, and D are empirical parameters.

Assuming the conductive fillers are incompressible, the change in the insulating layer thickness between the adjacent conductive fillers under pressure can be expressed as follows:

$$\frac{D(P)}{D(P=0)} = (1-\varepsilon) = 1 - \frac{P}{E_c} \tag{13.35}$$

Combining Eqs. (13.34), (13.35), and (13.36), the ratio of resistance under pressure and no pressure can be expressed as

$$\frac{R(P)}{R(P=0)} = \left(1-\frac{P}{E_c}\right)\exp\left(\begin{aligned}&\left(A + \frac{4\pi\sqrt{2m\phi}}{h}D(\sigma=0)\right)\left(\frac{P}{E_c}\right)\\ &+B\left(\frac{P}{E_c}\right)^2 + C\left(\frac{P}{E_c}\right)^3 + D\left(\frac{P}{E_c}\right)^4\end{aligned}\right) \tag{13.36}$$

Ding et al. fitted the piezoresistive response of nickel particle-filled elastomer composite using Eq. (13.36), giving a reasonably good agreement between the experimental data and the model. This model uses several fitting parameters, including A, B, C, and D in Eq. (13.36) and k_1, k_2, and k_3 for stress-dependent compressive modulus.

The model aforementioned posits an exponential increase in the number of conductive paths, as detailed in Eq. (13.34), leading to a monotonic decrease in resistivity with escalating compressive stress. Conversely, the applied compressive stress might also reduce the number of conductive paths owing to slippage and disconnection among adjacent filler particles, potentially elevating resistance [98, 100]. To account for these contrasting impacts of pressure on piezoresistive response, Wang et al. have delineated the relative alteration in the number of conductive paths with the subsequent equation [98],

$$\frac{N(P)}{N(P=0)} = \frac{1}{W\mathrm{e}^{-X \cdot P} + Y\mathrm{e}^{Z \cdot P}} \tag{13.37}$$

where $W + Y = 1$, $W, X, Y, Z > 0$. In this case, the two terms in the denominator represent the two opposite effects of the pressure P on the number of conductive paths.

Combining Eqs. (13.33), (13.35), and (13.37), the ratio of resistance under pressure and no pressure can be expressed as

$$\frac{R(P)}{R(P=0)} = (W\mathrm{e}^{-X \cdot P} + Y\mathrm{e}^{Z \cdot P})\left(1 - \frac{P}{E_c}\right)\mathrm{e}^{\left(\frac{4\pi\sqrt{2m\phi}}{h} D(P=0)\right)} \tag{13.38}$$

Since the distance between the adjacent filler particles, $D(P = 0)$, can be correlated to the volume fraction and sizes of the filler particles in the composite, Eq. (13.38) serves as a model to elucidate the influence of various factors on the piezoresistive response of conductive nanocomposites. These factors include the volume fraction and size of the conductive filler, as well as the composite's compressive modulus. Wang et al. [98] successfully used Eq. (13.38) to interpret the diverse piezoresistive behaviors exhibited by silicon rubber infused with carbon black at varying concentrations. Their findings indicated that at low carbon black concentrations, resistance consistently decreases as pressure increases. In contrast, at moderate concentrations, resistance initially decreases and then rises with escalating pressure. Conversely, at high carbon black concentrations, the resistance consistently increases in response to rising pressure.

For RTS based on the bulk conductive nanocomposites, the matrix materials are typically elastomers (e.g. PDMS, silicon rubber, or polyurethane). These materials exhibit notable viscoelastic characteristics, including stress relaxation and creep, which can significantly influence the piezoresistive response of sensors incorporating conductive nanocomposites. A key challenge is that the piezoresistive response may change over time under constant strain or stress, leading to an inaccurate measurement in the practical application of these sensors. Despite its importance, such viscoelastic behavior has not been extensively explored in existing research [96, 101]. Ding et al. investigated the relaxation of the resistance of the nickel-particle-filled PDMS composite under stress [96]. Their experiments revealed that the composite's resistance diminishes over time until reaching a steady state during stress relaxation. To model this phenomenon under pressure, the conventional Burger model was applied, as depicted in Figure 13.6c,

$$\frac{\Delta R}{R_0} = \frac{R - R_0}{R_0} = a + b \exp\left(-\frac{t}{\tau_R}\right) + ct \tag{13.39}$$

where R_0 denotes the initial resistance at time $t = 0$, R signifies the resistance at a subsequent time t, τ_R is the relaxation time, and a, b, and c are constants. Eq. (13.39) fits the resistance relaxation data at different strains well, as shown in Figure 13.6c(ii). The relaxation time τ_R of the device varies with the strain (Figure 13.6c(iii)), which is attributed to the rigid structure of the nickel particles.

In summary, the piezoresistive response of the RTS based on bulk conductive composites is modeled by percolation theory and an effective conductive path model. These models offer valuable insights into the underlying mechanisms governing the piezoresistive response. They highlight that filler concentration and composite modulus are critical parameters in the design of composite materials for RTS. However, the impact of stress relaxation on the piezoresistive response of these composite materials warrants further investigation, both theoretically and experimentally. Notably, current models, including the effective conductive path model, do not yet incorporate factors such as the influence of stress relaxation on the number of conductive paths and tunneling resistance, underscoring an area for future research and development.

13.2.4 Resistive Tactile Sensors Based on Porous Conductive Materials

The RTS based on porous conductive materials typically demonstrates a negative piezoresistive effect, characterized by a decrease in sensor resistance under compression. This behavior can be attributed to the formation of contact areas or points within porous material's structure under compression. Such compression facilitates the creation of additional conductive paths, consequently lowering the electrical resistance. Although this mechanism is widely accepted in the literature to elucidate the piezoresistive response of porous conductive materials, there is a scarcity of quantitative analyses that could offer practical guidance for the optimization of materials and structural designs specific to tactile sensors. This gap indicates a need for more detailed studies to inform and refine sensor design.

Highly porous materials exhibit hyper-elastic behavior. Their stress–strain response is characterized by the elastic, plateau, and densification regimes with an increase of the strain. The crucial parameters of a porous material are porosity (η), the relative density of the porous material (DP), and the density of the solid forming the porous material (d). Phenomenological models have been developed to understand the stress–strain response of the porous materials (such as polyurethane foams) and the effects of porosity, foam density, and loading rate [102]. These phenomenological models may be useful when the density and porosity of the porous conductive materials are considered for RTS applications.

Zhang et al. reported an analytical model of the piezoresistive behavior of highly compressible sensors made of microporous nanocomposites (Figure 13.7a)

[103]. In these materials, the porosity η decreases under applied stress/strain. The relationship between porosity and strain in highly porous materials may be complex. The complexity stems from predicting how porous materials will deform under compression. Such predictions are significantly influenced by both the porosity and the specific geometric characteristics of the porous materials. To offer a simplified model, the decrease in the porosity is assumed to be positively correlated with the deformation by $(1 - \eta) = (1 - \eta_0)/\lambda$, resulting in [103]:

$$\eta = 1 - \frac{(1 - \eta_0)}{\lambda} \tag{13.40}$$

where η is the porosity under compressive strain ε, η_0 is the original porosity, and λ is the stretch. Note that $\lambda = 1$ when the material is not under compressive strain and decreases with increase in the compressive strain/stress.

Following the highly compressible hyper-elastic model [106], the stress can be approximately correlated with the porosity η by the following equation:

$$\sigma = \frac{1}{\lambda}\sum_{i=1}^{n}\frac{2\mu_i}{\alpha_i}\left(\frac{1-\eta}{1-\eta_0}\right)^{\alpha_i}(\lambda^{\alpha_i} - 1) \tag{13.41}$$

where σ is the stress, n is the order of the model, and μ_i and α_i are the material parameters. The stress–strain curve of the carbon black/polyurethane (PU) porous nanocomposite (PNC) with different compositions and mechanical properties was fit to Eq. (13.41), and good agreement was obtained.

Following Archie's porosity–resistivity model [107], the conductivity of the porous conductive materials can be expressed as

$$k = k_0(1 - \eta)^f \tag{13.42}$$

where k is the conductivity of the porous material, k_0 is the conductivity of the solid that forms the porous material, and f is the exponent related to the microstructures of the pores.

Combining Eq. (13.40) and Eq. (13.42) leads to

$$k = \frac{k_0(1 - \eta_0)^f}{\lambda^f} \tag{13.43}$$

In the porous material based on the conductive filler/elastomer composite, quantum tunneling effects need to be considered when the material is compressed. A modified model can be written as:

$$k = \frac{k_0(1 - \eta_0)^f}{\lambda^a} \tag{13.44}$$

where a is a coefficient related to the microstructures, conductive fillers, and substrate material.

Equation (13.44) was used to fit the conductivity–porosity curve and conductivity–elongation curves of the porous composites with different contents of carbon black, as shown in Figure 13.7a(ii). The good fitting suggests that Eq. (13.44) can accurately represent the piezoresistive behaviors of highly porous medium. It is worth noting that the fitting parameters f and a are empirical, and the microstructure of

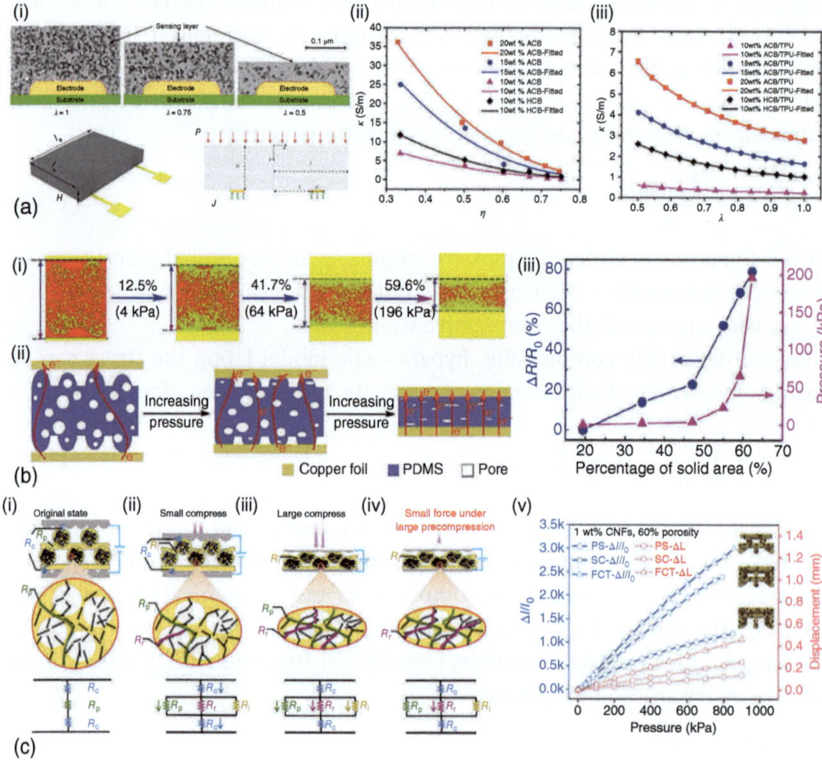

Figure 13.7 RTS based on porous conductive materials. (a) (i) The change of effective electric contact resistance at the microporous interface by pressing, schematic exploded view of a pressure sensor, and simplified 2D analytical model of the sensor; (ii) Conductivity–porosity curves of microporous nanocomposites with different contents of acetylene black (ACB) and highly conductive carbon black (HCB); (iii) Conductivity–elongation curves of microporous nanocomposites with different contents of ACB and HCB. Source: Reproduced with permission from Ref. [103]. Copyright 2021 John Wiley & Sons. (b) (i) Optical microscopic images acquired using an ultra-depth 3D microscope showing the cross-section of the sensor based on the B-micro/pore-3 structure under deformation. In the image, red, green, and yellow colors represent the pores, the solid PDMS region, and the metal holder, respectively; (ii) the proposed mechanism of the RTS based on the B-micro/pore-3 structure; (iii) pressure and $\Delta R/R_0$ (%) of the sensor under different percentages of solid area calculated from the cross-sectional images in Figure 13.7b(i). Source: Reproduced with permission from Ref. [104]. Copyright 2021 American Chemical Society. (c) Schematic illustration and circuit diagram of sensing mechanism of hierarchical porous piezoresistive sensor (i) in its original state, (ii) under small compression, (iii) under large compression. (iv) A tiny pressure change under a large preload is used to illustrate the high-pressure resolution performance; (v) Current response and sensitivity of 1 wt% carbon nanofibers, 60% porosity pressure sensor with different lattice structures (Parallel Stacked [PS], Simple cubic [SC], and Face-Centered Tetragonal [FCT]). Source: Reproduced with permission from Ref. [105]/Springer Nature/CC BY 4.0.

the porous material is missing. k_0 is related to the conductivity of the nonporous composite material.

Zhang et al. further treated the contact resistance at the microporous interface with electrode in terms of the equivalent contact area per unit area A_{ged}, porosity η_0, and elongation λ. The contact resistance can be written as [103],

$$R_{cr} = \frac{\sqrt{\pi}\lambda^{a+b}}{4k_0(1-\eta_0)^{f-0.5}\sqrt{A_{ged}dW}} \tag{13.45}$$

where A_{ged} is the effective contact area per unit area, and d and W are the thickness and width of electrodes, respectively.

The contact resistance in Eq. (13.45) is similar to the constriction resistance in Eq. (13.9) but includes the parameters accounting for the porosity of the materials. Eq. (13.45) is useful for estimating the contribution of the contact resistance to the overall piezoresistive response of the porous conductive materials. While in most of the experimental studies of the piezoresistive porous materials, the contact resistance is not explicitly investigated, the change in the contact resistance can be significant due to the rough interface between the porous material and the electrode and may need more quantitative evaluation.

Analytical models of the microporous pressure sensor with a pair of electrodes or interdigitated electrodes were presented in Ref. [103]. Analytical solutions of the sensitivities to deformation and pressure were obtained and verified by the 3D FEA. The electromechanical behavior was simulated using the static thermomechanical modules because of the similarity between thermal governing equations and electrical governing equations. A scaling law for the device was also derived. These solutions were mathematically complex but provided a systematic analysis of the effect of the electrode configuration, the sensing layer dimension, and contact on the sensor performance. It is worth noting that the theoretical treatment here is for the co-planar electrodes and may not apply to the typical sensor configuration in which the two electrodes are on opposite sides of the sensing layer. The treatment also does not include many of the key parameters (e.g. pore size, density, concentration of filler, etc.) of the porous materials that govern the hyper-elastic mechanical properties.

One of the strategies to improve the sensitivity and detection range is to design a piezoresistive material combining the contact microstructures and the porous materials. Such porous material is based on patterning surface microstructures on the PDMS/MWCNT (multiple wall carbon nanotubes) porous structure by nonwoven fabric [104]. As a result of the synergistic interaction between the porous structure and surface microstructure, the flexible sensor exhibited a highly sensitive piezoresistive behavior across a broad pressure range from 1 Pa to 100 kPa. The high sensitivity at lower pressure was attributed to changes in the contact areas between the electrodes and surface microstructures. At higher pressure, up to 100 kPa, the sensor maintained its sensitivity, primarily due to the inherent porous nature of the piezoresistive PDMS/MWCNT material. As shown in Figure 13.7b, the mechanism was in general described as an increase of conductive paths under applied pressure. To better understand the sensing mechanism and the effects of porosity/contact, the cross-sectional images of the sensor under pressure were collected by an ultra-depth three-dimensional microscope, and then the percentage of the solid component in the image was calculated as the percentage of solid area in the image. Figure 13.7b(iii) shows the relative resistance change or pressure as a function of the percentage of the solid area. The result provides the direct experimental data of the porosity change with the piezoresistive response or pressure, which may be useful for future correlation of the experimental material design and theoretical modeling of the RTS.

Xu et al. recently reported an innovative approach to fabricate piezoresistive pressure sensors with high sensitivity and broad linearity range [105]. A hierarchical in situ filling porous structure as the sensing layer was fabricated by direct writing printing and curing of carbon nanofibers (CNFs)/PDMS emulsion. First, the emulsified water droplets filled with CNFs were dispersed in the PDMS emulsion. The CNFs/water/PDMS emulsion could be extruded uniformly and continuously and stacked layer-by-layer to form a 3D lattice structure. After curing the PDMS, water evaporated, and left pores filled with CNFs. Compared with the conventional porous conductive materials, in which the conductive fillers are only in the solid and the edge of the pore, the unique fabrication process ensures that the conductive fillers not only adhere to the edges of the pores but also fill the pores *in situ*. Figure 13.7c(i) shows the relative change of the current as a function of pressure for four types of sensing layers of 1 wt% CNFs and 60% porosity. The synergetic effect of the internal CNF-filled pores and the lattice structure results in the device with the highest sensitivity of 2.4 kPa^{-1} and a high linearity ($R^2 = 0.993$) up to 1 MPa. Different porosities, concentrations of CNFs, and lattice structures were also investigated. The mechanism of the improved sensing performance is schematically described in Figure 13.7c(ii). Under compression, the lattice structure can readily deform, resulting in large contact area changes, effective stress distribution, and generating more conductive paths. When subjected to slight pressure changes, the CNF networks embedded within the internal porous structure can easily establish contact with each other, creating freestanding conductive paths within the internal pores. Moreover, the CNF networks act as a "bridge," connecting the isolated conductive fillers dispersed at the pore edges and forming additional conductive paths. This results in the generation of conductive paths within the elastomer, pores, and elastomer-pore interface. Consequently, a more substantial resistance change can be induced across the entire sensing range.

New strategies for achieving high sensitivity and wide-linear range RTS have rapidly evolved in recent years [108–110]. Chen et al. reported using an elastic spacer surrounding the sensing layer to improve the sensitivity of an RTS over a wide linear range (up to 1.4 MPa) [108]. By introducing the stiffness control of the elastic spacer, effect of micro-pyramid geometry on the sensitivity can be canceled by the stiffening constant, resulting in a wide linear range of the sensor. In a different approach, a broad linear range and highly sensitive pressure sensor were achieved by the design of multigradient materials (including modulus, conductivity, and microstructure gradient) [110]. Fine control of the mechanical and structural properties of piezoresistive sensing materials and devices will need to have better analytical and numerical models for devices with high sensitivity and linearity over a wide pressure range for future applications in e-skins and other practical applications.

13.2.5 Stretchable Resistive Tactile Sensors

Most studies on flexible RTS based on soft conductive materials do not consider how to decouple the response of the devices induced by pressure or lateral stretch. Research on stretchable RTS is still at an early stage [111–113]. To incorporate stretchability into the RTS, strategies for stretchable conductive pressure-sensing material, stretchable electrodes, and stretchable structural engineering have been proposed. Several examples of stretch-insensitive RTS are discussed in this section. This section begins by introducing how to mathematically quantify "strain insensitivity" [114], followed by strategies to mitigate stretch interference. The term "strain insensitivity" was first introduced in Ref. [114], herein, we further refine the mathematical definition of "strain insensitivity".

As illustrated in Figure 13.2b, lateral stretch geometrically affects both the area and the thickness of an RTS, resulting in a change in the resistance according to Eq. (13.1). Figure 13.8a offers an example [113]: the resistance at 10 kPa pressure and a concurrent 50% uniaxial strain, marked by the yellow triangle, is much higher than the resistance at the same pressure without any lateral strain, indicated by the yellow square. This lateral-stretch-induced offset in resistance would result in unreliable pressure measurement when both pressure and stretch are present. To quantitatively evaluate stretch interference, Figure 13.8b presents a plot prepared by us. In this plot, resistances without lateral strain and with 50% uniaxial strain from Figure 13.8a are plotted as the x-axis and y-axis, respectively. If the RTS is completely stretch insensitive, then the experimental results should all fall on the diagonal dash-dot line represented by $y = x$. However, the stretch interference causes the actual data points

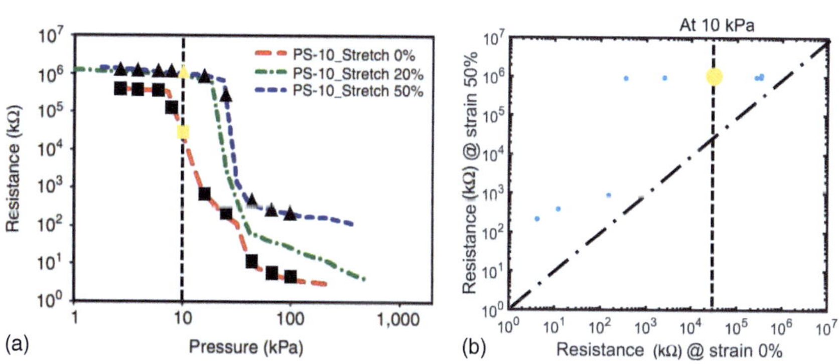

Figure 13.8 Definition of strain insensitivity. (a) The resistance change of an RTS under pressure without strain (red), under pressure, and concurrent 20% (green) or 50% (blue) uniaxial strain. Source: Adapted from Ref. [113]. (b) Resistive response under both pressure and 50% uniaxial strain versus resistive response under only pressure. Markers closer to the dash-dotted line indicate higher strain insensitivity.

to deviate from this line, as illustrated by the blue markers in the graph. To quantify this deviation, we apply two statistical metrics: the root mean square error (RMSE)

$$\text{RMSE} = \sqrt{\frac{\sum_{i=1}^{n}\left(R_{\text{P+S}}^{i} - R_{\text{P}}^{i}\right)^2}{n}} \tag{13.46}$$

Here $R_{\text{P+S}}^{i}$ is the resistance under concurrent pressure and stretch for the ith observation in the dataset, and R_{P}^{i} is the resistance under only pressure for the ith observation in the dataset. n is the sample size of the experiment data set, which equals 10 in the experimental data given in Figure 13.8a. The normalized root mean square error (NRMSE):

$$\text{NRMSE} = \text{RMSE}/\overline{R}_{\text{P+S}} \tag{13.47}$$

The average resistance under both pressure and stretch conditions, denoted as $\overline{R}_{\text{P+S}}$, is calculated as $\overline{R}_{\text{P+S}}=\frac{\sum_{i=1}^{n} R_{\text{P+S}}^{i}}{n}$. The NRMSE is a statistical metric that ranges from 0 to infinity, providing insight into the stretch insensitivity. Specifically, an NRMSE value greater than 1 signifies a substantial discrepancy between $R_{\text{P+S}}^{i}$ and R_{P}^{i}, indicating significant stretch interference. An NRMSE between 0 and 1 suggests some degree of stretch interference, but not overwhelmingly so. When NRMSE is 0, it indicates that stretch does not introduce any interference into the measurement. This scenario represents an ideal situation for stretch-insensitive tactile sensors.

In this framework, strain insensitivity is defined as 1 – NRMSE. A strain insensitivity value below 0 implies that the tactile sensors are severely compromised in their ability to accurately measure pressure when subjected to stretch. On the other hand, a value approaching 1 indicates a high level of strain insensitivity, meaning the sensor's performance is minimally affected by stretch. Referring back to our example, the strain insensitivity index based on Figure 13.8b [113] is −2.13, indicating a significant stretch interference of resistance reading. Consequently, the sensors in Ref. [113] are deemed incapable of providing accurate readings under simultaneous pressure and stretch conditions.

Before the recent proposal of the metric of strain insensitivity for quantifying stretch interference, several strategies were developed to mitigate the issue of stretch interference. One approach involved spacers, as illustrated in Figure 13.9a(i), in which the RTS consists of four components [115]: the Ecoflex matrix, the PDMS spacer, and the top and bottom electrodes, along with the porous sensing materials. The relatively stiff PDMS spacer insulates against in-plane strain, while the soft Ecoflex matrix exhibited significant deformation. This design effectively isolated strain in both the x and y directions, resulting in resistive responses under concurrent pressure and stretch that closely resembled the response of sensors subjected only to pressure (Figure 13.9a(ii)). Figure 13.9b(i) introduces another strategy known as fabric RTS [112]. These sensors are composed of conductive fibers (depicted as the red helixes in Figure 13.9b(ii)) wrapped around nonconductive fibers (yellow strands), creating a fabric-like structure. When the sensor is stretched, the contacts between conductive fibers remain more or less unchanged, whereas new contact points form under compression. As a result, its stretch insensitivity was found to be 0.99.

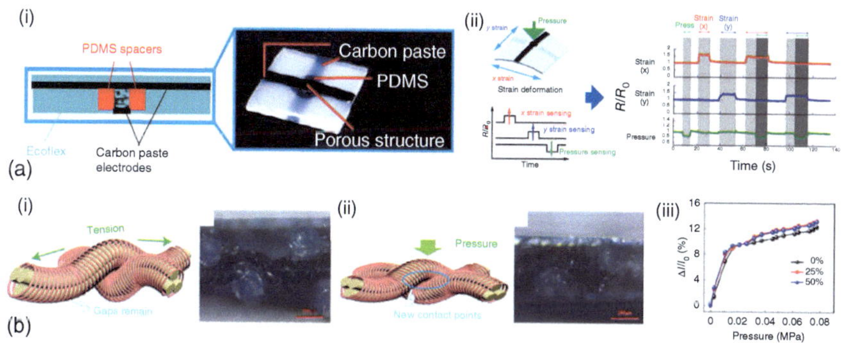

Figure 13.9 Strategies for reducing the stretch interference in resistive pressure sensors. (a) (i) Illustration of the spacer design of RTS. (ii) The normalized resistance versus pressure under spontaneous stretches along different directions. Source: Reproduced with permission from Ref. [115]/Springer Nature/CC BY 4.0. (b) Fabric design of RTS under stretch (left) (i) and pressure (ii); (iii) The normalized current change versus pressure under no stretch (black) and spontaneous 25% (red) and 50% (blue) uniaxial stretch. Source: Reproduced with permission from Ref. [112]. Copyright 2020 American Chemical Society

To provide a comprehensive comparison, Table 13.2 summarizes several representative stretchable RTS in terms of sensing mechanism, sensor structure, material, sensitivity, pressure range, and calculated strain insensitivity. This analysis reveals a notable gap: a sensor that simultaneously exhibits high sensitivity and high strain insensitivity is yet to be developed. This highlights the ongoing challenge in sensor technology to create devices that can accurately measure pressure without being influenced by external factors like stretching.

13.3 Capacitive Tactile Sensors

13.3.1 Introduction of Capacitive Tactile Sensors

Conventional flexible CTS are formed by sandwiching a mechanically compliant dielectric layer between two thin electrodes, as depicted in Figure 13.10a. The corresponding circuit is just a simple capacitor (Figure 13.10b), where the capacitance C can be expressed as:

$$C = k\varepsilon_0 \frac{A}{d} \tag{13.48}$$

where k is the dielectric constant of the dielectric, which is a measure of a material's response to an electric field, ε_0 is the permittivity of vacuum, A is the area of the electrodes, and d is the gap in between.

Upon compression, the electrode gap shrinks, but the electrode area does not change. Therefore, the relative change in capacitance is

$$\frac{\Delta C}{C_0} = \frac{k}{k_0}\frac{d_0}{d} - 1 \tag{13.49}$$

Table 13.2 Summary of Representative Flexible and Stretchable RTS.

References	Structure	Material	Mechanism	Sensitivity (kPa^{-1}) (Sensing range [kPa])	Stretch insensitivity (stretchability)
[112]	Fabric	Resistive rubber	Contact resistance	~0.008 (0–10) ~0.001 (10–80)	0.952 (50%)
[60]	Film	PVDF/ Ti_3AlC_2	Contact resistance	~ 817.4 (0.072–0.74) ~2213.3 (0.74–3)	NM
[69]	Micro-domes	CNT/PDMS	Tunneling resistance/ contact resistance	~15.1 (0–0.5) ~1000 (0.5–2.5)* ~0.1 (2.5–30)*	NM
[116]	Micro-domes	AgNW/ PDMS	Contact resistance	~3 (0–2.5)* ~0.33 (2.5–22.5)*	0.394 (100%)
[74]	Micro-pyramids	PEDOT: PSS/PUD	Contact resistance	~4.88 (0–3) ~1.54 (3–8)	0.537 (40%)
[113]	Micro-pyramids	SWCNT	Contact resistance	~0.75 (0–11)* ~134 (11–100)*	−0.105 (50%)
[111]	Micro-pyramids	AgNF/ graphene	Contact resistance	~1.6 (0–200)*	−2.13 (50%)
[117]	Sponge	CNF/ PAN/PI NF	Porous structure contact	~84.9 (0.05–5) ~140.9 (6–22)	NM
[118]	Sponge	Chitosan/ MXene/PU	Contact resistance/ interior contact	~147 (0–5) ~442 (5–17)	NM
[98]	Sponge	MXene	Porous structure contact	~0.1 (0–15)	NM
[115]	Porous structure	CNT/PDMS	Porous structure contact or contact resistance (not clearly stated)	~1.6 (0–5) ~0.2 (5–25)	0.99 (50%)
[83]	Tetragonal structure	CNF/PDMS	Porous structure contact	~4.7 (0–1000)	NM

"PVDF" denotes polyvinylidene fluoride. "CNT" denotes carbon nanotubes. "PEDOT:PSS" denotes poly(3,4-ethylenedioxythiophene) polystyrene sulfonate. "PUD" denotes aqueous polyurethane dispersion. "AgNF" denotes silver nanofiber. "CNF" denotes carbon nanofiber. "PAN" denotes polyacrylonitrile. "PI NF" denotes polyimide nanofiber. "PU" denotes polyurethane. The * denotes the sensitivity determined using a two-point slope within the corresponding sensing range. "NM" indicates "not mentioned."
Source: Refs. [73, 74, 89, 105, 111–113, 115–119].

(a) (b) (c) (d)

Figure 13.10 Schematics of representative types of CTS. (a) Flat electrodes sandwiching a solid dielectric layer. (b) The circuit model of CTS. (c) CTS with engineered dielectrics. (d) CTS with engineered electrodes.

The initial sensitivity S is therefore

$$S = \frac{\mathrm{d}\left(\frac{\Delta C}{C_0}\right)}{\mathrm{d}P} = \frac{1}{E_0} + \frac{1}{k_0}\frac{\partial k}{\partial P} \tag{13.50}$$

This equation indicates that the initial sensitivity of CTS can be enhanced through two primary strategies:

- **Employing a soft dielectric layer**. This method suggests choosing a dielectric material with a lower effective Young's modulus (E), as it could facilitate a greater decrease in the distance "d" with the same amount of pressure applied.
- **Leveraging the dielectrostriction effect.** Dielectrostriction is a phenomenon that describes how the effective dielectric constant changes under pressure, denoted as a nonzero value of "$\partial k/\partial P$" [120]. A good example to illustrate this is a porous material filled with a combination of high-k material and air. As the material gets compressed, the air is replaced with more of the high-k material, resulting in an increase in the overall effective dielectric constant.

Building on these observations, extensive research has focused on improving the structure, mechanical characteristics, and electrical attributes of dielectrics. These efforts have ushered in what we term "engineered dielectrics," as depicted in Figure 13.10c. In addition, "engineered electrodes," shown in Figure 13.10d, have also been investigated. This section provides a summary of the established electromechanical understandings related to "engineered dielectrics" and "engineered electrodes."

13.3.2 Capacitive Tactile Sensors with Engineered Dielectrics

CTS with micro-patterned dielectrics is sensitive within subtle and low-pressure regimes with a fast response time. Mannsfeld et al. first introduced micro-patterned CTS in 2010 [44]. Micro-pyramid dielectric patterns are fabricated by casting PDMS onto a pyramid silicon mold. The dielectric micro-pyramids enabled CTS to achieve a sensitivity of 0.55 kPa^{-1} within the 0–7 kPa pressure range, with swift recovery times measured to be just a few dozen milliseconds – 30 times faster than CTS with solid dielectrics.

To improve the sensitivity, micro-patterned structures, including micro-pyramids [123], micro-domes [124], and micro-pillars [125], have been designed to consider various geometric parameters, such as sizes, shapes, and spacing between microstructures. Finite-element modeling (FEM) [123, 126] has been employed to optimize sensor sensitivity. FEM allows for the computational simulation and analysis of the physical behavior of these sensors, thereby reducing the reliance on physical experiments for optimization. It provides a means to virtually experiment with different design parameters, such as base length, geometry height, and other structural aspects of micro-patterns like micro-pyramids. While FEM is available for understanding and predicting the performance of specific structures like micro-pyramids, it typically allows only limited explorations of sensor performance by adjusting structural parameters of a specific structure like micro-pyramids. One of the limitations of this approach is that the crucial design parameters for a broad range of applications are not always identified. In addition, even if some crucial design parameters are identified in FEM, it is not clear for scientists to know how crucial design parameters influence sensor performance. These gaps in the methodology restrict the widespread applicability of FEM in the design of CTS.

To overcome the limitations of FEM in the design and optimization of CTS, theoretical models of CTS have been explored. One such approach is seen in the work of Sara et al., who introduced a comprehensive model for microstructured CTS [39]. This model systematically investigated sensor performances by varying structural and material parameters, offering a more holistic understanding of how different design choices impact sensor functionality. The capacitance of these micro-patterned CTS can be calculated using the equivalent circuit illustrated in Figure 13.11b(i):

$$C = \frac{C_{\text{air}} C_{\text{pyramid}}}{C_{\text{air}} + C_{\text{pyramid}}} \tag{13.51}$$

where capacitances of each component are defined as:

$$C_{\text{air}} = \frac{f_{\text{air}} A \varepsilon_{\text{air}}}{t_{\text{pyramids}}}, C_{\text{pyramids}} = \frac{f_{\text{p}} A \varepsilon_{\text{pyramids}}}{t_{\text{pyramids}}} \tag{13.52}$$

Here, the f_{air} represents the volumetric fraction of air, while f_{p} represents volumetric fraction of pyramids,

Upon compression, the dielectric materials are highly simplified as a spring with their stiffness k:

$$k = \frac{EA}{t} \tag{13.53}$$

where E is the dielectric material's modulus, A is the contact area between the electrode and pyramid, and t is the total height of the pyramid as illustrated in Figure 13.12b(ii). It is assumed to have a small pressure range so that the volumetric fraction and the shape change of microstructures are negligible. According to Hook's law, $F = kx$, where F is the applied force and x is the deflection. Rearranging Eq. (13.53), the deflection is expressed as

$$x = \frac{Ft}{EA} \tag{13.54}$$

Using known geometric relationships of trapezoids, the effective $1/A$ for micro-pyramids is expressed:

$$\frac{1}{A} = \int_0^t \frac{t}{(ad - a\delta + b\delta)^2} \mathrm{d}\delta \tag{13.55}$$

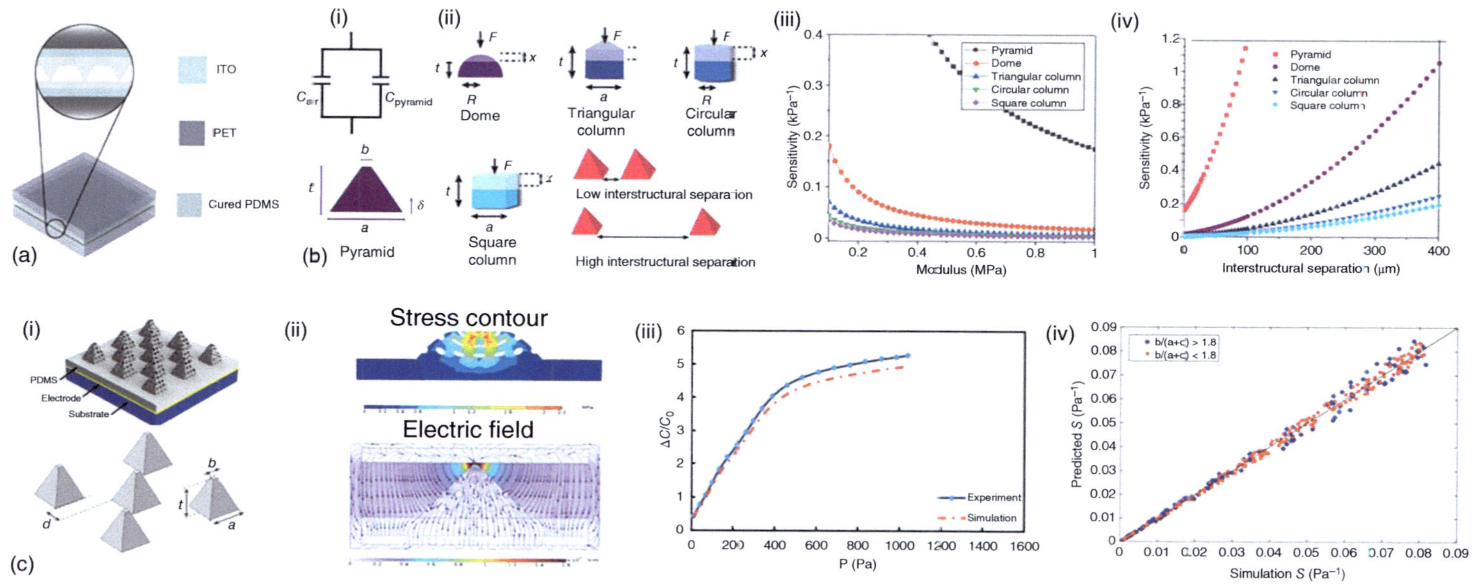

Figure 13.11 CTS with micro-patterned dielectrics. (a) Schematics of CTS with micro-pyramids patterned on the surface of the dielectric. Source: Reproduced with permission from Ref. [39]. Copyright 2019 Wiley. (b) (i) Equivalent circuit of micro-pyramid CTS (top) and design parameters of the pyramids (bottom). Source: Reproduced with permission from Ref. [39]. Copyright 2019 John Wiley & Sons. (ii) Schematics of other surface micro-patterns. The dependence of sensitivity on the (iii) dielectric modulus and (iv) interstructural separation. Source: Reproduced with permission from Ref. [121]. Copyright 2020 American Chemical Society. (c) (i) CTS with porous micro-pyramids. (ii) The stress distribution (top) and the electrical field distribution in the porous micro-pyramid under compression. (iii) Comparison of the relative change in capacitance between simulation and experimental results. (iv) The predicted sensitivity versus the simulated sensitivity. Source: Reproduced with permission from Ref. [122]/Springer Nature/CC BY 4.0.

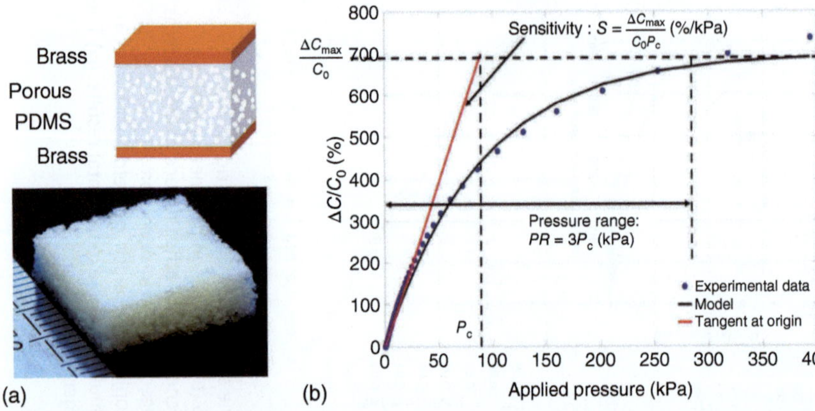

Figure 13.12 CTS with foam dielectrics. (a) Schematics of CTS with foam dielectrics (top) and a photograph of the foam (bottom). (b) An empirical model for the relative capacitance changes versus pressure. Source: Reproduced with permission from Ref. [127]/MDPI/CY BY 4.0.

where δ is the distance from each layer to the bottom, a is base length of pyramids, and t is the height of pyramids. Combining Eqs. (13.54) and (13.55) gives us:

$$x = \frac{F}{E}\int_0^t \frac{t^2}{(ad - a\delta + b\delta)^2}\mathrm{d}\delta \tag{13.56}$$

This analytical expression matches well with the experiment results for micro-pyramid dielectric CTS [39]. Subsequently, Sara et al. expanded this expression to other micro-patterned sensors by substituting different effective $1/A$ for domes, triangular columns, circular columns, and square columns [121] (as depicted in Figure 13.11b(ii)). In the study, a pressure of 1 Pa and the corresponding applied force on sensor F were utilized to update the deflection x according to Eq. (13.54) followed by calculation of the capacitance change based on Eqs. (13.51) and (13.52) and sensitivity.

To identify the crucial design parameter, Sara et al. systematically varied parameters, such as dielectric materials' modulus E, dielectric constant k, base length a, height t, and interstructural separation for five different structures through theoretical modeling (Figure 13.11b(ii)). An important aspect of their study is the consideration of different effective $1/A$ values and varying volumetric fractions of air for each structure. According to their investigation, modulus E and interstructural separation are two crucial factors for sensitivity enhancement (Figure 13.11b(iii,iv)).

For all five structures, amplifying the modulus of dielectrics by an order of magnitude consistently led to a roughly equivalent decrease in initial sensitivity, as depicted in Figure 13.11b(iii). An enhancement of interstructural separation boosts sensitivity (Figure 13.11b(iv)), which effectively diminishes the volumetric ratio of dielectrics, consequently reducing initial capacitance while augmenting the volumetric ratio of air. Remarkably, compared with other structures, the pyramid and dome configurations stand out. This superior performance is largely ascribed to their comparatively lower initial modulus (Figure 13.11b(iii,iv)) [121].

The impact of the dielectric constant on sensitivity is markedly overshadowed by Young's modulus and interstructural separation. An increase in the dielectric constant k from 1 to 60 results in only a modest enhancement of sensitivity from 4 to 11 kPa^{-1} (Ref. [121]). It was also found that enlarging the base length or diameter of microstructure (dome, triangular, circular, and square column) led to a decline in the sensitivity, while this trend has only been demonstrated experimentally with the dome structures [128] and using FEM for the circular columns [129]. Base length and height also contributed to the sensitivity. An increase in base length leads to reduced sensitivity across all structures due to the resultant higher volumetric ratio of dielectrics. In contrast, for triangular, circular, and square column structures, an increase in height correlates with enhanced sensitivity, since increase in height indicates a larger increase in distance between two electrodes [129]. Additionally, it was discerned that comparable changes in the base length and height yield similar effects on sensitivity change. Yet, in contrast to Young's modulus, these dimensions exert a relatively minor influence on sensitivity. This revelation is pivotal for sensor design, underscoring that mere alterations in structural dimensions fail to significantly boost sensitivity and are limited by the constraints of molding techniques. Therefore, the research focus should pivot toward a more profound exploration and refinement of the materials used in CTS.

In addition to the theoretical model for solid micro-patterns offered by Sara et al., Reza et al. proposed porous micro-pyramids for CTS using FEM to simulate electromechanical behaviors (Figure 13.11c(i)) [122]. This design includes several key dimensions: d is the interstructural separation, t is the height of the pyramid, a is the length of the pyramid base, and b is the length of the top of the pyramid. Subsequently, the Neo-Hookean model was employed to describe the mechanical behavior of the dielectrics:

$$W = C_{10}(I_1 - 3) \tag{13.57}$$

where C_{10} is the material constant and I_1 is the first invariant of the left Cauchy–Green deformation tensor. By applying pressure on the top electrodes, the deformation and stress contour can be illustrated in Figure 13.11c(ii). The governing equations for the electric field in the CTS can be described by Gauss's law,

$$\nabla \cdot (k\varepsilon_0 \nabla V) = \rho_v \tag{13.58}$$

where k represents the relative dielectric constant of the domain, V represents the electric potential field, ρ_v represents the volumetric free charge density, and ε_0 is the permittivity of vacuum. The electrical field $\vec{E}$ can be expressed as:

$$\vec{E} = -\nabla V \tag{13.59}$$

Therefore, the electric field can be observed in Figure 13.12c(ii). The electrical simulation results showed strong alignment with the experimental data, as depicted in Figure 13.12c(iii), demonstrating the accuracy of their FEM. Having confirmed the validity of FEM, the author generates a sensitivity FEM dataset by varying sensor parameters such as base height, interstructural separation, and porosity [122]. Subsequently, all of the dataset points are fitted to a theoretical relationship. The original expression, which is too complex to be presented here, is replaced with a simplified coefficient multiplied by a function f for explanatory purposes:

$$S \sim \frac{\left(1 + {}^{d}\!/_{a}\right)^2}{\nu\left(\frac{b}{a}\right)E} f\left(\frac{d}{a}, k, \frac{b}{a}\right) \tag{13.60}$$

where δ symbolizes the porosity of the dielectric layer. As illustrated in Figure 13.12c(iv), the simulated values for S closely aligned with the predicted values presented in the paper. Furthermore, $S \sim \frac{\left(1+{}^{d}\!/_{a}\right)^2}{\left(\frac{b}{a}\right)E}$ indicates that the sensitivity S is inversely related to modulus E and tip width-to-base ratio $\frac{b}{a}$. One order change of the Young's modulus E boosts one order of the sensitivity. The interstructural separation distance d or distance ratio $\frac{d}{a}$ is proportional to S. A change in the interstructural separation distance by one order can lead to two orders increase in sensitivity. These observations align with earlier conclusions from Sara et al., emphasizing the significance of modulus E and interstructural separation distance d. A detailed investigation into the function $f\left(\frac{d}{a}, k, \frac{b}{a}\right)$ would also elucidate the analysis of the effect of structures and materials on the sensitivity.

Based on the preceding discussion, we can summarize the following points for micro-patterned dielectric CTS:

- The modulus E of dielectrics and the interstructural separation distance d have been identified as two controlling factors that can significantly affect sensitivity in the past research. Notably, a change in the modulus E by an order of magnitude can induce a corresponding order of magnitude change in the initial sensitivity. An increase in the interstructural separation contributes to a higher air volume ratio, resulting in a reduction of the initial capacitance and enhancement of compressibility of dielectrics.
- The geometry of the micropattern dielectrics can fine-tune the sensitivity. In the case of micro-domes, both FEM and experiments indicate that decreasing the dome diameter might be a good way to enhance sensitivity [128]. For micro-pyramids, a smaller tip-width-to-base ratio b/a has been shown to increase sensitivity, primarily by augmenting the compressibility of the dielectrics [39]. For structures such as triangular, circular, and square columns, a heightened sensitivity is observed with an increase in their height. This correlation is attributed to the fact that a greater height results in a more pronounced distance change between two electrodes when compressed, thereby influencing sensitivity [129]. These recommendations highlight the importance of optimizing micro-patterned CTS to improve sensitivity and pressure response, providing valuable guidance for future research and development in this field.

  However, it is important to note that there are several limitations in the existing models for micro-patterned dielectric CTS:
- Current models primarily focus on scenarios involving small pressure ranges (<10 kPa), and the sensitivity enhancement strategies from model are only valid for initial sensitivity.

- The trade-off between pressure range and sensitivity was not addressed. Modeling the pressure response of CTS over a wider range of pressures remains unclear.

In addition to micro-patterned dielectric CTS, CTS with foam dielectrics [127] have also gained popularity as common choices for CTS, as depicted in Figure 13.12a. The foam dielectrics are porous polymers with stochastic structures, which are usually fabricated by foaming templates. For example, porous PDMS or Ecoflex can be fabricated by mixing sugar or salt into PDMS or Ecoflex precursor, curing the mixture, and then dissolving sugar or salt. To model the CTS with foam dielectrics using Eq. (13.48) and obtain the sensitivity enhancement strategies based on Eq. (13.50), analytical expressions for the effective dielectric constant and effective Young's modulus of the foam are needed.

The foam can be considered as a two-phase composite of air and polymer, whose effective dielectric constant can be calculated using effective medium theories [130]. For isotropic foam, its effective dielectric constant is found to be within the upper and lower Hashin–Shtrikman bounds [131]:

$$k_{\mathrm{p}} + \frac{3(1-\eta)(k_{\mathrm{p}}-1)}{3+\eta(k_{\mathrm{p}}-1)} < k_{\mathrm{eff}} < k_{\mathrm{p}} - \frac{3k_{\mathrm{p}}\eta(k_{\mathrm{p}}-1)}{3k_{\mathrm{p}}-(1-\eta)(k_{\mathrm{p}}-1)} \tag{13.61}$$

where η is the porosity, k_{p} is the dielectric constant of the polymer, and k_{eff} is the effective dielectric constant of the porous polymer.

Assuming incompressibility of the polymer, the porosity η under pressure can be estimated as [103]

$$\eta = 1 - \frac{1-\eta_0}{1-\epsilon} \tag{13.62}$$

where η_0 is the initial porosity and ϵ is the compressive strain. Although Hashin–Shtrikman bounds also exist for the modulus of two-phase composites, the modulus–porosity relationships of the polymer foams are more commonly estimated via phenomenological models [132] or analytical models based on certain unit cells [133, 134]. The simplest model for open-cell foams with high porosity is [135, 136]:

$$\frac{E}{E_{\mathrm{s}}} = (1-\eta)^2 \tag{13.63}$$

where E_{s} is the modulus of the solid polymer.

Note that even thought there are mature theoretical studies toward effective Youngs' modulus (Eq. (13.63)) and effective dielectric constant (Eq. (13.62)) toward foam dielectrics, no one combined these equations in Eqs. (13.48) and (13.50) to get simple analytical equations for CTS with foam dielectrics. Instead, an empirical formula, Eq. (13.64), as illustrated in Figure 13.12b, has been proposed to fit experiment data of sensors' performance:

$$S = \frac{\Delta C_{\mathrm{max}}}{C_0}\left(1 - \mathrm{e}^{-P/P_{\mathrm{c}}}\right) \tag{13.64}$$

Here, ΔC_{max} represents the saturated capacitance at large pressure, C_0 is the initial capacitance. P is the applied pressure. P_{c} is the characteristic pressure at which the capacitance change reaches 63% of its maximum value. In addition, the pressure measurement range (PR) corresponded to 95% of the maximum capacitance change, according to Figure 13.12b, PR = $3P_c$.

- There is a fair agreement between empirical expression and the experiment with a coefficient of determination R^2 higher than 0.99. However, it is important to highlight that the empirical expression remains somewhat phenomenological, lacking specific guidance for scientists seeking to optimize sensor design.

Nanowire dielectrics are another type of CTS with engineered dielectrics. There are almost no models of nanowire-dielectric CTS. The challenge in modeling nanowire-dielectric CTS stems from its inherent multiscale nature. Addressing this complexity necessitates a homogenization process that transitions nanowire dielectrics from the nano- or microscale to the macroscale. In addition, the intricacy of nanowire dielectrics at the nanoscale is amplified by factors such as geometry, orientation, aspect ratio, distribution, and aggregation of the nanowires, each adding layers of complexity to the homogenization process [75].

13.3.3 Capacitive Tactile Sensors with Engineered Electrodes

CTS with engineered electrodes often incorporate functional fibers or fabric to improve flexibility and conformability for large-area deployment, making them particularly suitable for applications like wearable technology and soft robotics. Qu et al. provided a deeper understanding of textile-based CTS, particularly focusing on their sensitivity [137]. A polymer-based functional fiber was developed using a circular electrode encapsulated by a dielectric layer (Figure 13.13a).

The contact area of the two fibers is assumed to be a double-layer parallel capacitor (Eq. (13.48)), as shown in Figure 13.13b. Due to the relative area of circular electrode change during compression, Eq. (13.50) is revised:

$$S = \frac{\mathrm{d}\left(\frac{\Delta C}{C_0}\right)}{\mathrm{d}P} = \frac{d_0}{k_0 A_0}\left[\frac{A}{d}\left(\frac{\partial k}{\partial P}\right) + \frac{k}{d}\left(\frac{\partial A}{\partial P}\right) - \frac{k}{A}\left(\frac{\partial d}{\partial P}\right)\right] \tag{13.65}$$

The contact area A_0 that is assumed to be a square with side length a_0 ($A_0 = a_0{}^2$) becomes A with side length a ($A = a^2$) under compression (Figure 13.13c). Under linear elastic regime, the change in diameter L in compression direction and the thickness of the assumed double-layer parallel capacitor d are described as follows:

$$d = d_0(1 - \varepsilon_{\text{compress}}), L = L_0(1 - \varepsilon_{\text{compress}}), a = a_0(1 + \varepsilon_{\text{lateral}}) \tag{13.66}$$

where $\varepsilon_{\text{lateral}}$ and $\varepsilon_{\text{compress}}$ is lateral and compressive strain, respectively. Assuming electrodes and dielectric layer have the same Poisson's ratio $\delta = -\varepsilon_{\text{lateral}}/\varepsilon_{\text{compress}}$,

$$\frac{\partial d}{\partial P} = \frac{d_0\partial\left(\frac{d}{d_0}\right)}{\partial P} = -\frac{d_0\partial\varepsilon_{\text{compress}}}{\partial P} = -\frac{d_0}{E_{\text{dielectic}}} \tag{13.67}$$

$$\frac{\partial A}{\partial P} = \frac{2a\partial a}{\partial P} = \frac{2aa_0\partial\left(\frac{a}{a_0}\right)}{\partial P} = \frac{2aa_0\partial(\varepsilon_{\text{lateral}})}{\partial P} = \frac{2aa_0\nu}{E_{\text{electrode}}} \tag{13.68}$$

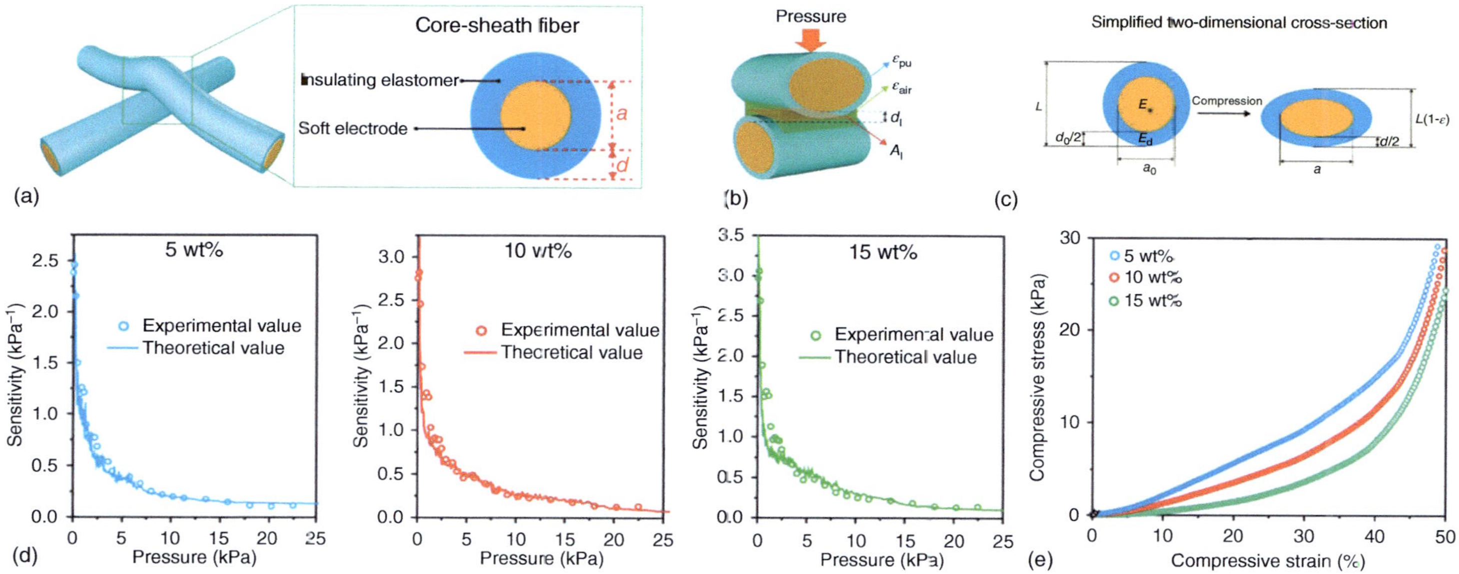

Figure 13.13 CTS with textile electrodes. (a) Schematic of textile CTS. (b) Illustration of the deformation of textile CTS upon compression. (c) Cross-sectional view of the deformation. (d) Comparison of sensitivity versus pressure between experimental and theoretical results for different doping concentrations (5, 10, 15 wt%). (e) Stress–strain curves of the electrodes with different doping concentrations (5, 10, 15 wt%). Source: Reproduced with permission from Ref. [137]. Copyright 2023 American Chemical Society.

Here, $E_{\text{electrode}}$ and $E_{\text{dielectic}}$ stand for Young's modulus of the electrodes and dielectric material, respectively. Hence, the sensitivity can be expressed:

$$S = \frac{\mathrm{d}\left(\frac{\Delta C}{C_0}\right)}{\mathrm{d}P} = \frac{d_0}{k_0 A_0}\left[\frac{A}{d}\left(\frac{\partial k}{\partial P}\right) + \frac{k}{d}\frac{2aa_0\nu}{E_{\text{electrode}}} + \frac{k}{A}\frac{d_0}{E_{\text{dielectic}}}\right] \tag{13.69}$$

Electrodes doped with CNT in a PU matrix were synthesized through a process involving the mixing of CNT, PU, dimethylformamide (DMF), and water. By altering the doping concentrations to 5, 10, and 15 wt%, electrodes of varying stiffness were obtained, as indicated by different values of $E_{\text{electrode}}$. The theoretical model was found to be consistent with the experimental results, as shown in Figure 13.13d. The experiments yielded following findings: electrodes with a 15 wt% doping concentration displayed the greatest sensitivity, whereas those with a 5 wt% concentration exhibited the least. According to Eq. (13.69), it is deduced that electrodes doped with 15 wt% concentration are expected to be the softest, while those with a 5 wt% doping concentration are anticipated to be the stiffest. This theoretical derivation is in alignment with the experimental results presented in Figure 13.13e.

Based on the sensitivity expression Eq. (13.69), it can be concluded that one strategy to enhance sensitivity is to employ softer electrodes, which is similar to aforementioned strategies.

The exploration of CTS with engineered micro-domes on the electrodes represents another innovative direction in sensor design. As depicted in Figure 13.14a, the electrodes are structured as micro-domes with a rectangular base, while the dielectric materials are flat [138]. This design introduces several key geometric parameters that influence the sensor's performance: h_0 represents the thickness of the dielectric layers, R is the radius of the micro-dome, h_1 is the height of the micro-dome, and h_2 is the height of the rectangle base. The electrical fields passing through the void and dielectric micro-dome are represented by $\vec{E_1}$ and $\vec{E_2}$, respectively. The capacitance of the sensor C is described as $C = C_{01} + C_{02}$. The capacitance is calculated through the expression:

$$C = \frac{-q}{U} \tag{13.70}$$

where q is the total charge quantity and U is the electrical potential. The total electrical quantity q can be determined through the following expression:

$$q = \int \sigma_0 \mathrm{d}A \tag{13.71}$$

where σ_0 represents charge density. The charge corresponding to C_{01} and C_{02} is denoted as q_{01} and q_{02} respectively. The charge can be calculated as:

$$q_{01} = \int \sigma_0 \mathrm{d}A_1, q_{02} = \int \sigma_0 \mathrm{d}A_2 \tag{13.72}$$

where A_1 is the overlapping area between two electrodes and void, A_2 is the surface area of dome respectively. It is assumed that the electrical fields passing through the void $\vec{E_1}$ and the dielectric materials $\vec{E_2}$ are uniform. In addition, they posited that the electrical field lines are perpendicular to the dielectric and enter vertically into the dome structures, forming an arc shape, as illustrated in

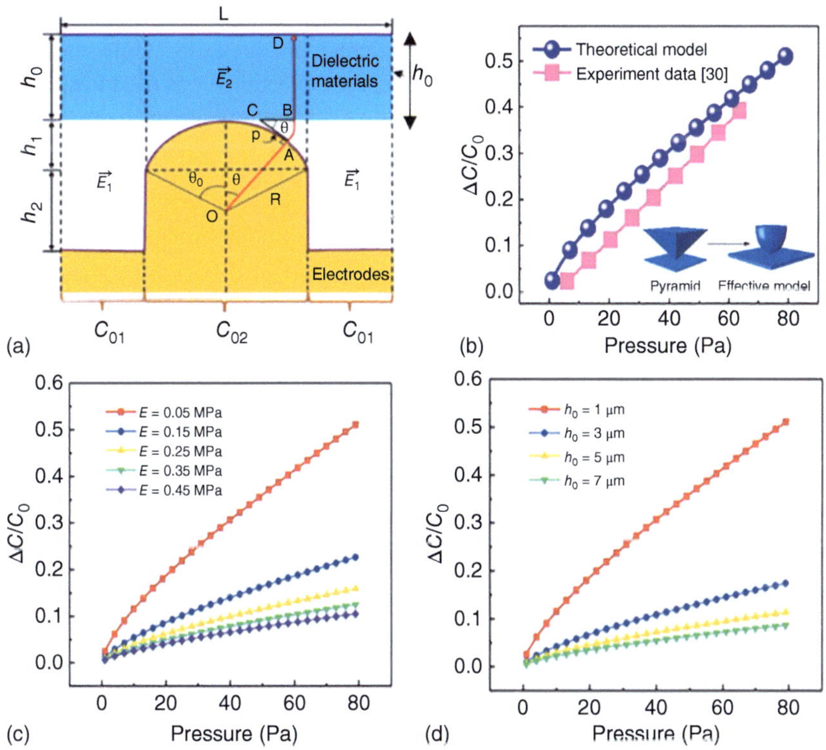

Figure 13.14 CTS with micro-dome electrodes. (a) Schematic of micro-dome electrodes. (b) Theoretical and experimental curves for the relative capacitance change versus pressure. (c) Effects of the Young's modulus of the dielectric material on the relative capacitance change. (d) Effects of the thickness of the dielectric material on the relative capacitance change. Source: Reproduced with permission from Ref. [138]. Copyright 2022 IOP Publishing.

Figure 13.14a. The corresponding electrical potential for C_{01} and C_{02} is U_{01} and U_{02}, respectively.

$$U_{01} = \int_{h_1+h_2} \vec{E_1}\, d\vec{l} + \int_{h_0} \vec{E_2}\, d\vec{l}, U_{02} = U_{01} \tag{13.73}$$

Hence, the two components of capacitance are described as:

$$C_{01} = \frac{q_{01}}{U_{01}}, C_{02} = \frac{q_{02}}{U_{02}} \tag{13.74}$$

Pressure and sensitivity can be calculated by considering how compression affects the sensor's geometry and electrical properties. When pressure is applied, there is a reduction in h_1 (the height of the micro-dome) and corresponding U_{01}(the electrical potential related to C_{01}). This occurs because the micro-dome electrodes come into contact with the dielectric, altering the electrical field intensity and thus U_{02} (the electrical potential related to C_{02}). Consequently, C_{01}, C_{02} are affected. The relationship between the applied force F and the deflection δ is described by contact mechanics:

$$F = \frac{4}{3} K R^{\frac{1}{2}} \delta^{\frac{3}{2}} \tag{13.75}$$

where R is the radius of micro-domes, K is the contact modulus illustrated in Eq. (13.14). The total contact force is used to calculate the pressure. While sensitivity expression is not explicitly given in the work, the deflection δ was used to update the force and the pressure on the sensors and their corresponding sensitivity.

Due to the absence of direct experimental data for such sensors, the existing data from pyramid electrode-based CTS was adapted to an effective dome shape, thus allowing them to test their theoretical predictions against this modified dataset (Figure 13.14b). Subsequently, they adjusted parameters in the theory to identify the critical parameters affecting sensitivity. In Figure 13.14c, they varied Young's modulus of dielectric materials from 0.05 to 0.45 MPa and plotted relative capacitance changes versus pressure. They found that softer dielectric materials resulted in higher sensitivity. This observation is consistent with previous studies and aligns with the understanding that softer materials allow for greater deformation under pressure, leading to larger changes in capacitance. Additionally, they varied the thickness of the dielectric material from 1 to 7 μm and plotted relative capacitance changes versus pressure. They observed that thinner dielectric materials were associated with higher sensitivity (Figure 13.14d). The authors attributed this phenomenon to the observation for parallel plate capacitors that capacitance changes more rapidly when the distance between the two electrodes is reduced [138].

A common conclusion about softer dielectric materials resulting in higher sensitivity was obtained. The authors assumed the electrical field to be an arc line, which was a good trial to simplify the problem, while a more rigorous study needs to be conducted to validate the assumption. Due to a lack of experimental data for CTS with micro-dome electrodes, they simplified micro-pyramid into an effective micro-dome to validate the theory, which is not that rigorous.

CTS with porous electrodes represents another innovative approach to the design of CTS. These sensors typically consist of a flat dielectric layer sandwiched between a porous elastic electrode and a flat electrode. The unique structure of these sensors makes them particularly sensitive to small pressures.

Zhang et al. fabricated a CTS with an elastic sponge, as shown in Figure 13.15a [139]. The sensor has a sensitivity of 2.25 kPa^{-1} up to 5 kPa (Figure 13.15b). They simplified the porous electrode into a honeycomb structure for the simulation. As illustrated in Figure 13.15c, the electrical field in this configuration does not follow a simple vertical path from the top to the bottom electrode. Instead, it concentrates in areas where the porous electrode makes contact with the dielectric layer. This concentrated electrical field implies that the capacitance change is primarily driven by variations in the contact area between the elastic porous electrode and the dielectric layer. Initially, this contact area is small but increases rapidly under slight pressure and eventually saturates at higher pressures. This behavior explains the ultrahigh sensitivity of these sensors to low pressure.

The COMSOL simulation results (Figure 13.15c) suggest that the contact area between the porous electrode and the dielectric layer is a critical factor in sensor performance. Engineering the contact interface could, therefore, be a key strategy for further enhancing the sensitivity of these sensors. However, the analytical modeling of CTS with porous electrodes presents significant challenges, primarily due to

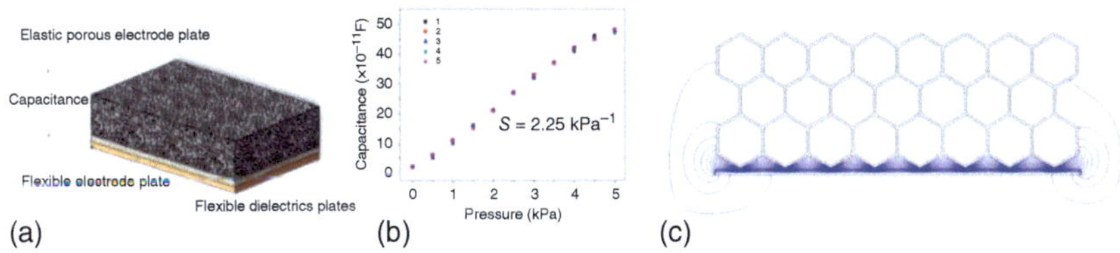

Figure 13.15 CTS with porous electrodes. (a) Schematic of CTS with porous electrodes. (b) Capacitance versus pressure plots of five repetitive tests. (c) COMSOL simulation of the electrical field of CTS with porous electrode. Source: Reproduced with permission from Ref. [139]. Copyright 2018 IOP Publishing.

the non-vertical orientation of the electrical field. Traditional equations that assume a simple, vertical electrical field may not be applicable in this context. This complexity necessitates the development of new theoretical models that can accurately capture the intricate behavior of the electrical field in these porous structures.

13.3.4 Stretchable Capacitive Tactile Sensors

The concept of strain insensitivity initially applied to stretchable RTS can be also extended to stretchable CTS such that we can have a quantitative comparison of the influence of lateral stretch on stretchable CTS.

The mathematical formulation for strain insensitivity in stretchable CTS involves calculating the RMSE and the NRMSE as follows:

$$\begin{aligned} \text{RMSE} &= \sqrt{\frac{\sum_{i=1}^{n}\left(C^{i}_{\text{P+S}} - C^{i}_{\text{P}}\right)^2}{n}}, \text{NRMSE} \\ &= \text{RMSE}/\overline{C}_{\text{P+S}}, \ \text{Strain insensitivity} = 1 - \text{NRMSE} \end{aligned} \tag{13.76}$$

Here $C^i_{\text{P+S}}$ is the capacitance under both pressure and stretch for the ith observation in the dataset and C^i_{P} is the capacitance under pressure for the ith observation in the dataset. n is the sample size of the experiment data set. For example, $n = 7$ in Figure 13.16b.

A popular strategy to mitigate the stretch interference is to employ stiff islands to isolate the lateral strain applied to the substrate from the CTS, as exemplified in Figure 13.16a. The CTS comprised three components: the Ecoflex substrate, the

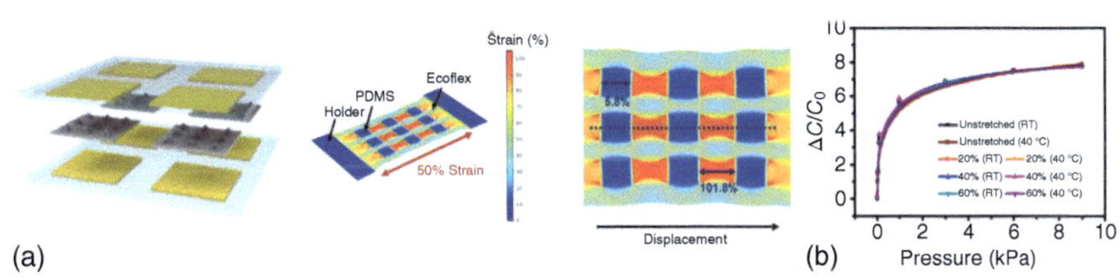

Figure 13.16 Strain isolation through island design. (a) Illustration and strain distribution in the CTS with the stiff island design. (b) The normalized capacitance changes versus pressure under spontaneous stretch and temperature interference. Source: Reproduced with permission from Ref. [140]. Copyright 2019 American Chemical Society.

Table 13.3 Summary of Representative Flexible and Stretchable CTS.

References	Structure	Materials	Mechanism	Sensitivity (kPa^{-1}) (sensing range [kPa])	Stretch insensitivity (stretchability)
[141]	Film dielectrics	Hydrogel (PAM-LiCl)	High-*k* dielectrics	~0.0046 (0–27.5)	NM (480%)
[142]	Film dielectrics	SWCNT/ PDMS	High-*k* dielectrics	~0.0059 (0–200)	NM (95%)
[143]	Film dielectrics	VHB Tape	High-*k* dielectrics	~0.00051 (0–100)	NM (60%)
[29]	Film dielectrics	Ecoflex	High-*k* dielectrics	~0.0077 (0–20)	NM (10%)
[58]	Nanowire stack	Silicone nanowires	Air gap	~ 1.173 (0–10) ~0.5 (10–50)	NM
[144]	Film dielectrics (Micro-pyramid electrodes)	Graphene/ HPU(AgNWs/ HPU)	High-*k* dielectrics/ air gap	~2 (0–3) ~0.6 (3–18)	NM
[145]	Micro-pyramid dielectrics	SiNP/PDMS	Air gap	~1 (0–2) ~0.375 (2–10)	NM
[59]	Porous elastomer dielectrics	CNT elastomer	Air gap	~0.601 (0–10) ~0.4 (10–30) ~0.077 (30–140)	NM
[42]	Film dielectrics (Textile electrode)	Fluorosilicone	Air gap	~0.28 (0–180)	NM
[146]	Wrinkle dielectrics	Ecoflex	Air gap	~0.107 (0–10)	NM
[147]	Film dielectrics	PDMS	Dielectrics	~0.002 (0–500)	NM (5%)
[148]	Film dielectrics	PGS and POMaC	Dielectrics	~0.13 (0–100)	NM (15%)
[149]	Porous dielectrics	Silicone foam	Air gap	~0.0036 (0–250)	0.787 (15%)
[43]	Porous dielectrics w/surface textile	PDMS foam	Air gap	~0.11 (0–25)	NM (30%)
[44]	Porous micro-pyramid	PDMS foam	Air gap	~1.54 (0–1) ~0.601 (1–5) ~0.2 (5–10)	0.98 (60%)
[55]	Film dielectrics	PDMS	Dielectrics	~0.00023 (0–800)	NM (50%)
[150]	Micro-domes	Liquid metal /PDMS	Air gap	~0.17 (0–25)	0.877 (94%) (45%)*

Source: Refs. [38, 42, 43, 70–72, 140–150].

PDMS stiff islands, and the holder. The micro-pyramid-based CTS were located on the PDMS islands, which experienced only a strain of 5.8% when the substrate was stretched by 50%. Figure 13.16b verifies that the curves of normalized capacitance versus pressure nearly coincide, regardless of whether the lateral strain is at 20%, 40%, or 60%.

However, it is important to note that while the island design effectively reduces stretch interference, it can also introduce complexities in design and manufacturing, which might lead to higher production costs and could potentially affect other sensor characteristics, such as sensitivity. As indicated in Table 13.3, CTS employing stretch-mitigation strategies may exhibit lower sensitivity, highlighting a trade-off between sensitivity and stretch insensitivity. Therefore, while strategies like the stiff island design can significantly reduce stretch interference in CTS, they may also bring additional challenges in terms of complexity and sensitivity, which must be carefully considered in the sensor design process.

"PAM-LiCl" denotes poly(acrylamide) – LiCl hydrogel. "SWCNT" denotes single-wall carbon nanotubes. "VHB" is a type of tape in 3M company. "HPU" denotes healable polyurethane. "AgNWs" denotes silver nanowires. "SiNP" denotes silicone nanoparticle. "PGS" denotes poly(glycerol sebacate). "POMaC" denotes poly(octamethylene maleate (anhydride) citrate). "NM" denotes not mentioned. In Ref. [150], the device's stretchability is stated to be 94%. However, because the graph showing the capacitance versus pressure curve is only provided at a 45% strain, the calculation for strain insensitivity is based on this curve.

13.4 Resistive–Capacitive Hybrid Response Tactile Sensors

13.4.1 Flexible Hybrid Response Tactile Sensors

After decades of development, the performance of RTS and CTS is approaching its limits. However, bottlenecks such as sensitivity-pressure tradeoff and stretch interference persist. The emergence of resistive–capacitive hybrid response tactile sensors (HRTS) has shed some light on possible remedies [151]. The first HRTS was formed by sandwiching a PNC between two parallel electrodes with an ultrathin dielectric layer inserted between the PNC and the electrodes (as illustrated in Figure 13.17a) [151]. The PNC was fabricated from CNT-doped Ecoflex, using commercial nickel foam as a sacrificial template, resulting in an open-cell structure with tubular ligaments and an overall porosity of 86%. The simplified equivalent circuit of HRTS is shown in Figure 13.17b. Due to the resistive nature of the ligaments in the PNC and the parasitic capacitance generated by the air pores, the PNC exhibits a combined response involving both piezoresistivity and piezocapacitance. The ultrathin dielectric layer made of polymethyl methacrylate (PMMA) ensures that the overall sensor remains capacitive. HRTS demonstrates exceptional sensitivity, even under high pressure – a remarkable improvement of up to 423% over conventional CTS within the 10–30 kPa range [151].

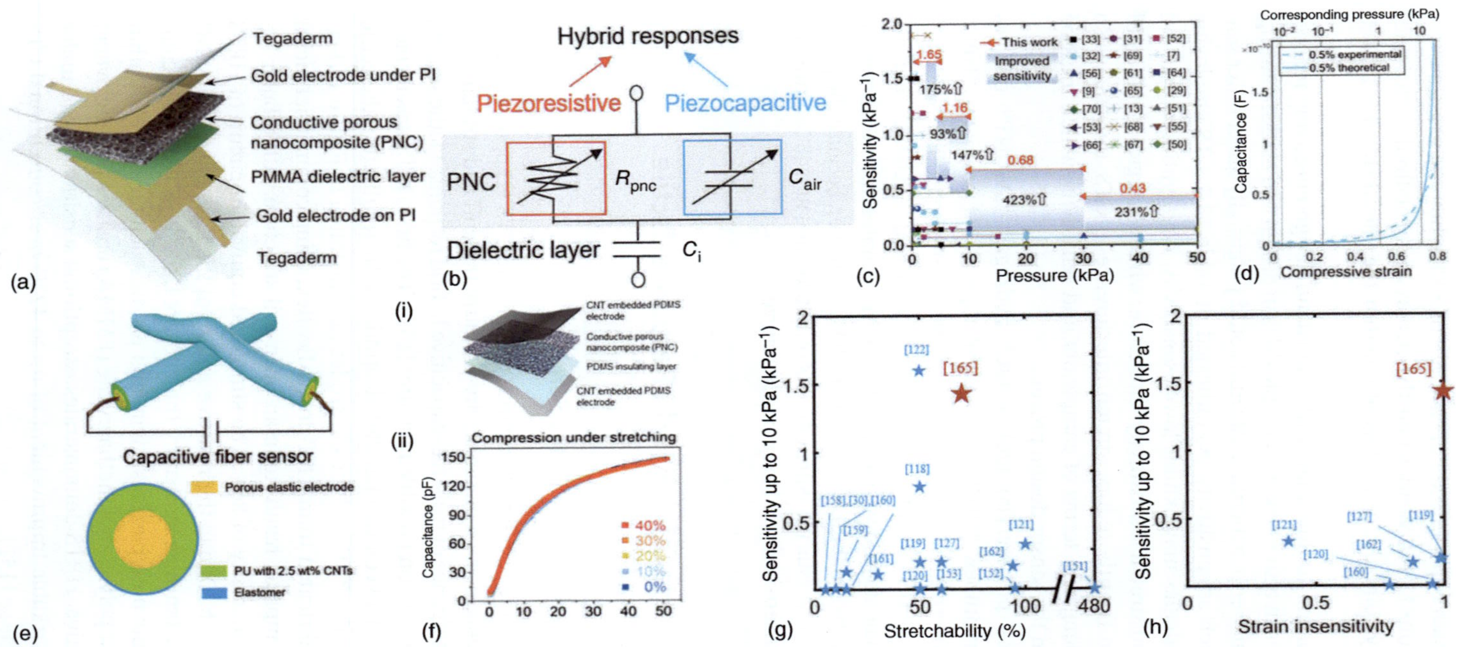

Figure 13.17 HRTS and SHRTS. (a) Schematic of the hybrid response tactile sensors. (b) The equivalent circuit of hybrid response tactile sensor. (c) Comparison of the sensitivity of HRTS with capacitive tactile sensors reported in the literature in the pressure ranges of 1–5, 5–10, 10–30, and 30–50 kPa. (d) The plot of capacitance versus compressive strain for the HRPS with the optimum CNT doping ratio of 0.50 wt%. Source: Reproduced with permission from Ref. [151]. Copyright 2021 John Wiley & Sons. (e) Schematic of the hybrid response fabric CTS. Source: Reproduced with permission from Ref. [152]. Copyright 2023 American Chemical Society. (f) (i) The schematics of stretchable hybrid response tactile sensor. (ii) The capacitance changes versus pressure under spontaneous 0%, 10%, 20%, 30%, and 40% uniaxial stretch [153]. (g) Ashby plot of sensitivity up to $10\,kPa^{-1}$ of stretchable CTS and stretchable RTS versus stretchability. (h) Ashby plot of sensitivity up to $10\,kPa^{-1}$ of stretchable CTS and stretchable RTS versus strain insensitivity.

Modeling the electromechanics of HRTS started with modeling the piezoresistive and piezocapacitive behaviors of the PNC. Due to the difficulties of piezoresistive modeling of highly porous and deformable PNC doped at the percolation threshold, the piezoresistive model (R_{PNC} in Figure 13.17b) was just obtained by fitting the experimental measurements of the PNC. The piezocapacitive model (C_{air} in Figure 13.17b) for the PNC assumes the electrical field is uniform and could be simply modeled as a double parallel plate capacitor.

$$C_{air} = k\varepsilon_0 \frac{A}{d} \tag{13.77}$$

where k is the dielectric constant, ε_0 is the permittivity of vacuum, A is the relative area of the double parallel plate capacitor, and d is the distance between two electrodes. Plugging the piezoresistive and piezocapacitive models into the simplified circuit given in Figure 13.17b yielded a reasonable prediction of HRTS response up to a pressure of 10 kPa, as shown in Figure 13.17d. This model fails to track the HRTS experiment at large pressure because piezocapacitive model cannot capture the capacitance change when PNC gets more conductive and electrical field is distorted.

Furthering this innovation, Qu et al. reported a fabric HRTS [152], which is composed of an electrode layer (yellow in Figure 13.17e), a porous conductive composite (green in Figure 13.17e), and an insulating layer (blue in Figure 13.17e) [152]. This fabric HRTS has improved the sensitivity by approximately 1081% within the 10–25 kPa range compared to existing fabric CTS [152]. The significant enhancement of sensitivity as demonstrated in the two cases establishes the HRTS as a viable solution to overcoming the sensitivity-pressure trade-off.

13.4.2 Stretchable Hybrid Response Tactile Sensors

The gold electrodes and PMMA dielectric layer in the first HRTS inherently lack stretchability. To create a stretchable hybrid response tactile sensor (SHRTS), the gold electrodes were replaced by CNT spray-coated on PDMS, and the PMMA was replaced by PDMS, as illustrated in Figure 13.17f(i) [153]. These modifications have enabled the SHRTS to achieve a stretchability of up to 70%. When the SHRTS was subjected to combined out-of-plane pressure and in-plane stretching, as depicted in Figure 13.17f(ii) [153], its pressure response curves across varying tensile strains (0–40%) almost overlapped, indicating that the sensor is almost immune from stretch interference. To further highlight the exceptional performance of SHRTS, Figure 13.17g presents an Ashby plot that compares the sensitivity and stretchability of various CTS and RTS, as cataloged in Tables 13.2 and 13.3. In this context, 10 kPa is identified as the pressure relevant to gentle human touch [154]. Marked by a red star on the plot, the SHRTS stands out for its high stretchability, coupled with significant sensitivity, making it highly suitable for nondevelopable and/or inflatable surfaces. Additionally, Figure 13.17h features an Ashby plot of sensitivity versus strain insensitivity for the same range of sensors. Here again, the SHRTS, indicated by the red star, demonstrates its minimal sensitivity to stretching interference while maintaining high sensitivity. These two Ashby plots highlight the promising future of SHRTS as a stretchable but relatively stretch-insensitive tactile sensor. The electromechanics of SHRTS remain elusive given the large porosity and deformability of the PNC, difficulties in modeling the contact resistance and interface capacitance, as well as the AC effects. We identify these challenges to be a major frontier of the electromechanics of resistive and capacitive pressure sensors with rich opportunities.

13.5 Conclusion and Outlook

13.5.1 Summary of the Existing Achievements

Resistive and capacitive sensors are among the most basic and commonly used modes for tactile sensing in various applications, from robotics to consumer electronics. Over the past two decades, remarkable advancements have been made in engineering these sensors with flexibility and stretchability using soft conductive materials and dielectrics. Many innovative strategies have been employed to bring sensitivity close to that of human skin (around $2\,kPa^{-1}$) at small pressures (i.e. 10 kPa) [155, 156]. These strategies include engineering the contact and tunneling resistance in RTS [29, 157], utilizing high-k materials, and/or incorporating air gaps in CTS [157, 158]. In limited cases, exceptional sensitivity ($4.7\,kPa^{-1}$) has been achieved in high-pressure regimes (10–100 kPa) [111]. This chapter outlines existing electromechanical models to facilitate the understanding and future development of RTS and CTS, which are briefly summarized as follows:

- **Resistive Tactile Sensors (RTS)**: We have categorized the pressure-sensing components in RTS into three major mechanisms – microstructured contacts, bulk conductive nanocomposites, and porous conductive materials. In the first case, if contact resistance is dominating, decreasing Young's modulus of sensing materials increases the sensitivity. Compared with regular structures like micro-pyramids and micro-domes, RDS may have more contact area change. For bulk conductive nanocomposite-based RTS, the concentration of the filler and the modulus of the composite are two essential parameters for the sensor design. Porous conductive materials are basically a combination of the first two mechanisms, which could lead to an enhancement of the sensitivity and sensing range beyond the individual ones.
- **Capacitive Tactile Sensors (CTS)**: We have divided CTS into engineered dielectrics and engineered electrodes. For CTS with engineered dielectrics, it is well recognized that the Young's modulus and the volume fraction of air of the dielectric layer are critical design parameters. A decrease in modulus or an increase in air volume fraction can enhance sensitivity but may compromise the sensing range. In comparison, structural characteristics might be less influential on these sensors. For CTS with engineered electrodes, the Young's moduli of both the dielectric and the electrodes matter. Since the electric field in those CTS

differs from those of parallel plate capacitors, identifying crucial parameters can be challenging.

- **Hybrid Response Tactile Sensors (HRTS)**: By merging resistive and capacitive responses in one sensor, HRTS, as described in Section 13.4, has demonstrated the potential to overcome some major bottlenecks of RTS and CTS, such as sensitivity-pressure tradeoff and stretch interference [68, 151]. However, HRTS are still in their infancy, and the electromechanical understandings are still very limited. A major difference between HRTS and conventional resistive and CTS lies in their AC frequency-dependent behaviors [159], which opens a new avenue of device tuning and readout circuit engineering.

13.5.2 Remaining Challenges and Future Prospects

While extensive research efforts have been invested in constructing and enhancing the materials and structures used in RTS and CTS, important device details are often insufficiently documented. These include parameters such as electrode and sensing material sizes, initial resistance or capacitance before deformation, electrode contact (separate or integrated with the sensing material), and specifics of the test apparatus. Likewise, inconsistencies in reporting and comparing sensor performance data, including the use of $\Delta I/I_0$ or $\Delta R/R_0$ to present RTS results, and sensitivity evaluated at different pressure magnitudes or ranges, need to be addressed. By harmonizing experimental setups and data reporting, it could pave the way to greater use of machine learning and artificial intelligence for sensor design, sparking innovation in resistive and capacitive tactile sensors.

Additionally, while methods of modeling and simulation to explore the electromechanics of these sensors are commonly used, the details of these methods often lack clarity. There have been only a limited number of comprehensive attempts to elucidate the relationship between factors like Young's modulus, sensor structures, and sensor performance. Some models, like those based on tunneling effects, can be too intricate for immediate use by sensor developers, while others, such as those involving porous materials with stochastic structures, are difficult to simplify and quantify. For example, even in instances of simple structures like micro-domes and micro-pyramids, existing models typically apply to specific cases rather than providing a broader set of guidelines for future research.

To advance the field, it is urgently required to develop a foundational and universally applicable theoretical and numerical modeling framework. This should be grounded in physics, yet feasibly be adopted by engineers. One of the main challenges lies in creating simplified electromechanical models that encapsulate the primary mechanisms and control parameters of resistive and capacitive tactile sensors. This is inherently complex due to the intricate interplay between material properties, sensor structures, and multi-directional mechanical stress. However, simplifications informed by experiments and multiscale modeling in mechanics can be expanded to the research on the electromechanics of RTS and CTS.

Acknowledgments

NL acknowledges the U.S. National Science Foundation (NSF) Grant ASCENT-2133106, the U.S. Army Research Office (ARO) under Cooperative Agreement W911NF-19-2-0333, and the U.S. Office of Naval Research (ONR) Grant N00014-20-1-2112, the U.S. Defense Health Agency (DRA) STTR Grant HT942523C0039. ZZ acknowledges the National Aeronautics and Space Administration (NASA) of the United States under Cooperative Agreement 80NSSC18M0022.

References

1 Johnson, K.O. (2001). The roles and functions of cutaneous mechanoreceptors. *Current Opinion in Neurobiology* 11 (4): 455–461.

2 Chortos, A., Liu, J., and Bao, Z.A. (2016). Pursuing prosthetic electronic skin. *Nature Materials* 15 (9): 937–950.

3 Dargahi, J. and Najarian, S. (2004). Human tactile perception as a standard for artificial tactile sensing—a review. *The International Journal of Medical Robotics and Computer Assisted Surgery* 1 (1): 23–35.

4 McLellan, K., Petrofsky, J.S., Zimmerman, G. et al. (2009). The influence of environmental temperature on the response of the skin to local pressure: the impact of aging and diabetes. *Diabetes Technology & Therapeutics* 11 (12): 791–798.

5 Tepole, A.B., Gosain, A.K., and Kuhl, E. (2012). Stretching skin: the physiological limit and beyond. *International Journal of Non-Linear Mechanics* 47 (8): 938–949.

6 Yu, B., Kang, S.Y., Akthakul, A. et al. (2016). An elastic second skin. *Nature Materials* 15 (8): 911–918.

7 Wang, X., Dong, L., Zhang, H. et al. (2015). Recent progress in electronic skin. *Advanced Science* 2 (10): 1500169.

8 Lee, W.W., Tan, Y.J., Yao, H. et al. (2019). A neuro-inspired artificial peripheral nervous system for scalable electronic skins. *Science Robotics* 4 (32): eaax2198.

9 Wang, M., Luo, Y., Wang, T. et al. (2021). Artificial skin perception. *Advanced Materials* 33 (19): 2003014.

10 Kim, Y., Chortos, A., Xu, W. et al. (2018). A bioinspired flexible organic artificial afferent nerve. *Science* 360 (6392): 998–1003.

11 Park, J., Kim, M., Lee, Y. et al. (2015). Fingertip skin–inspired microstructured ferroelectric skins discriminate static/dynamic pressure and temperature stimuli. *Science Advances* 1 (9): e1500661.

12 Chen, J., Zhang, J., Hu, J. et al. (2022). Ultrafast-response/recovery flexible piezoresistive sensors with DNA-like double helix yarns for epidermal pulse monitoring. *Advanced Materials* 34 (2): e2104313.

13 Gu, G., Zhang, N., Xu, H. et al. (2023). A soft neuroprosthetic hand providing simultaneous myoelectric control and tactile feedback. *Nature Biomedical Engineering* 7 (4): 589–598.

14 Zhou, K., Zhao, Y., Sun, X. et al. (2020). Ultra-stretchable triboelectric nanogenerator as high-sensitive and self-powered electronic skins for energy harvesting and tactile sensing. *Nano Energy* 70: 104546.

15 Lee, B., Oh, J.Y., Cho, H. et al. (2020). Ultraflexible and transparent electroluminescent skin for real-time and super-resolution imaging of pressure distribution. *Nature Communications* 11 (1): 663.

16 Su, Z., Fishel, J.A., Yamamoto, T., and Loeb, G.E. (2012). Use of tactile feedback to control exploratory movements to characterize object compliance. *Frontiers in Neurorobotics* 6: 7.

17 Kaboli, M., Yao, K., Feng, D., and Cheng, G. (2018). Tactile-based active object discrimination and target object search in an unknown workspace. *Autonomous Robots* 43 (1): 123–152.

18 Huang, X., Guo, W., Liu, S. et al. (2022). Flexible mechanical metamaterials enabled electronic skin for real-time detection of unstable grasping in robotic manipulation. *Advanced Functional Materials* 32 (23): 2109109.

19 Bergner, F., Dean-Leon, E., and Cheng, G. (2020). Design and realization of an efficient large-area event-driven e-skin. *Sensors* 20 (7): 1965.

20 Park, D.Y., Joe, D.J., Kim, D.H. et al. (2017). Self-powered real-time arterial pulse monitoring using ultrathin epidermal piezoelectric sensors. *Advanced Materials* 29 (37): 1702308.

21 Lou, Z., Chen, S., Wang, L. et al. (2017). Ultrasensitive and ultraflexible e-skins with dual functionalities for wearable electronics. *Nano Energy* 38: 28–35.

22 Dahiya, R., Yogeswaran, N., Liu, F. et al. (2019). Large-area soft e-skin: the challenges beyond sensor designs. *Proceedings of the IEEE* 107 (10): 2016–2033.

23 Yin, R., Wang, D., Zhao, S. et al. (2020). Wearable sensors-enabled human–machine interaction systems: from design to application. *Advanced Functional Materials* 31 (11): 2008936.

24 Kang, H., Zhao, C., Huang, J. et al. (2019). Fingerprint-inspired conducting hierarchical wrinkles for energy-harvesting e-skin. *Advanced Functional Materials* 29 (43): 1903580.

25 Byun, J., Lee, Y., Yoon, J. et al. (2018). Electronic skins for soft, compact, reversible assembly of wirelessly activated fully soft robots. *Science Robotics* 3 (18): eaas9020.

26 Bao, L., Han, C., Li, G. et al. (2023). Flexible electronic skin for monitoring of grasping state during robotic manipulation. *Soft Robotics* 10 (2): 336–344.

27 Zhang, C., Liu, S., Huang, X. et al. (2019). A stretchable dual-mode sensor array for multifunctional robotic electronic skin. *Nano Energy* 62: 164–170.

28 Pyo, S., Lee, J., Bae, K. et al. (2021). Recent progress in flexible tactile sensors for human-interactive systems: from sensors to advanced applications. *Advanced Materials* 33 (47): 2005902.

29 Pierre Claver, U. and Zhao, G. (2021). Recent progress in flexible pressure sensors based electronic skin. *Advanced Engineering Materials* 23 (5): 2001187.

30 Liu, Y., Bao, R., Tao, J. et al. (2020). Recent progress in tactile sensors and their applications in intelligent systems. *Science Bulletin* 65 (1): 70–88.

31 Dahiya, R.S., Mittendorfer, P., Valle, M. et al. (2013). Directions toward effective utilization of tactile skin: a review. *IEEE Sensors Journal* 13 (11): 4121–4138.

32 Wan, Y., Wang, Y., and Guo, C.F. (2017). Recent progresses on flexible tactile sensors. *Materials Today Physics* 1: 61–73.

33 He, K., Hou, Y., Yi, C. et al. (2020). High-performance zero-standby-power-consumption-under-bending pressure sensors for artificial reflex arc. *Nano Energy* 73, 104743.

34 Li, Z., Qiang, F., Dai, S.W. et al. (2019). Construction of sandwich-like porous structure of graphene-coated foam composites for ultrasensitive and flexible pressure sensor. *Nanoscale* 11: 10229–10238.

35 Lee, Y., Park, J., Cho, S. et al. (2018). Flexible ferroelectric sensors with ultra-high pressure sensitivity and linear response over exceptionally broad pressure range. *ACS Nano* 12 (4): 4045–4054.

36 Liu, M., Pu, X., Jiang, C. et al. (2017). Large-area all-textile pressure sensors for monitoring human motion and physiological signals. *Advanced Materials* 29 (41): 1703700.

37 Lee, S., Reuveny, A., Reeder, J. et al. (2016). A transparent bending-insensitive pressure sensor. *Nature Nanotechnology* 11 (5): 472–478.

38 Shi, H., Al-Rubaiai, M., Holbrook, C.M. et al. (2019). Screen-printed soft capacitive sensors for spatial mapping of both positive and negative pressures. *Advanced Functional Materials* 29 (23): 1809116.

39 Ruth, S.R.A., Beker, L., Tran, H. et al. (2019). Rational design of capacitive pressure sensors based on pyramidal microstructures for specialized monitoring of biosignals. *Advanced Functional Materials* 30: 1903100.

40 Wan, Y., Qiu, Z., Huang, J. et al. (2018). Natural plant materials as dielectric layer for highly sensitive flexible electronic skin. *Small* 14 (35): e1801657.

41 Núñez, C.G., Navaraj, W.T., Polat, E.O. et al. (2017). Energy-autonomous, flexible, and transparent tactile skin. *Advanced Functional Materials* 27 (18): 1606287.

42 Viry, L., Levi, A., Totaro, M. et al. (2014). Flexible three-axial force sensor for soft and highly sensitive artificial touch. *Advanced Materials* 26 (17): 2659–2664.

43 Park, S., Kim, H., Vosgueritchian, M. et al. (2014). Stretchable energy-harvesting tactile electronic skin capable of differentiating multiple mechanical stimuli modes. *Advanced Materials* 26 (43): 7324–7332.

44 Mannsfeld, S.C., Tee, B.C., Stoltenberg, R.M. et al. (2010). Highly sensitive flexible pressure sensors with microstructured rubber dielectric layers. *Nature Materials* 9 (10): 859–864.

45 Li, R.Y., Si, Y., Zhu, Z. et al. (2017). Supercapacitive iontronic nanofabric sensing. *Advanced Materials* 29 (36): 1700253.

46 Chorsi, M.T., Curry, E.J., Chorsi, H.T. et al. (2019). Piezoelectric biomaterials for sensors and actuators. *Advanced Materials* 31 (1): 1802084.

47 Chun, J., Lee, K.Y., Kang, C.Y. et al. (2014). Embossed hollow hemisphere-based piezoelectric nanogenerator and highly responsive pressure sensor. *Advanced Functional Materials* 24 (14): 2038–2043.

48 Choi, W., Lee, J., Kyoung Yoo, Y. et al. (2014). Enhanced sensitivity of piezoelectric pressure sensor with microstructured polydimethylsiloxane layer. *Applied Physics Letters* 104 (12): 123701.

49 Garcia, C., Trendafilova, I., de Villoria, R.G., and del Rio, J.S. (2018). Self-powered pressure sensor based on the triboelectric effect and its analysis using dynamic mechanical analysis. *Nano Energy* 50: 401–409.

50 Ramuz, M., Tee, B.C., Tok, J.B., and Bao, Z. (2012). Transparent, optical, pressure-sensitive artificial skin for large-area stretchable electronics. *Advanced Materials* 24 (24): 3223–3227.

51 Yamada, Y., Morizono, T., Umetani, Y., and Takahashi, H. (2005). Highly soft viscoelastic robot skin with a contact object-location-sensing capability. *IEEE Transactions on Industrial Electronics* 52 (4): 960–968.

52 Zhou, Y., Zhao, X., Xu, J. et al. (2021). Giant magnetoelastic effect in soft systems for bioelectronics. *Nature Materials* 20 (12): 1670–1676.

53 Yang, T.T., Xie, D., Li, Z., and Zhu, H. (2017). Recent advances in wearable tactile sensors: materials, sensing mechanisms, and device performance. *Materials Science & Engineering R: Reports* 115: 1–37.

54 Chang, Y., Wang, L., Li, R. et al. (2021). First decade of interfacial iontronic sensing: from droplet sensors to artificial skins. *Advanced Materials* 33 (7): e2003464.

55 Xiong, Y., Han, J., Wang, Y. et al. (2022). Emerging iontronic sensing: materials, mechanisms, and applications. *Research* 2022: 9867378.

56 Sezer, N. and Koç, M. (2021). A comprehensive review on the state-of-the-art of piezoelectric energy harvesting. *Nano Energy* 80: 105567.

57 Meng, K., Xiao, X., Wei, W. et al. (2022). Wearable pressure sensors for pulse wave monitoring. *Advanced Materials* 34 (21): e2109357.

58 Wang, B. and Facchetti, A. (2019). Mechanically flexible conductors for stretchable and wearable e-skin and e-textile devices. *Advanced Materials* 31 (28): e1901408.

59 Rus, D. and Tolley, M.T. (2015). Design, fabrication and control of soft robots. *Nature* 521 (7553): 467–475.

60 Shepherd, R.F., Ilievski, F., Choi, W. et al. (2011). Multigait soft robot. *Proceedings of the National Academy of Sciences of the United States of America* 108 (51): 20400–20403.

61 Rich, S.I., Wood, R.J., and Majidi, C. (2018). Untethered soft robotics. *Nature Electronics* 1 (2): 102–112.

62 Sachyani Keneth, E., Kamyshny, A., Totaro, M. et al. (2021). 3D printing materials for soft robotics. *Advanced Materials* 33 (19): e2003387.

63 Duan, Y., He, S., Wu, J. et al. (2022). Recent progress in flexible pressure sensor arrays. *Nanomaterials* 12 (14): 2495.

64 He, J., Zhang, Y., Zhou, R. et al. (2020). Recent advances of wearable and flexible piezoresistivity pressure sensor devices and its future prospects. *Journal of Materiomics* 6 (1): 86–101.

65 Nie, Z., Kwak, J.W., Han, M., and Rogers, J.A. (2022). Mechanically active materials and devices for bio-interfaced pressure sensors—a review. *Advanced Materials* e2205609.

66 Li, J., Bao, R., Tao, J. et al. (2018). Recent progress in flexible pressure sensor arrays: from design to applications. *Journal of Materials Chemistry C* 6 (44): 11878–11892.

67 Li, R., Zhou, Q., Bi, Y. et al. (2021). Research progress of flexible capacitive pressure sensor for sensitivity enhancement approaches. *Sensors and Actuators A: Physical* 321: 112425.

68 Ha, K.-H., Huh, H., Li, Z., and Lu, N. (2022). Soft capacitive pressure sensors: trends, challenges, and perspectives. *ACS Nano* 16 (3): 3442–3448.

69 Li, S., Zhang, Y., Wang, Y. et al. (2019). Physical sensors for skin-inspired electronics. *InfoMat* 2 (1): 184–211.

70 Lipomi, D.J., Vosgueritchian, M., Tee, B.C. et al. (2011). Skin-like pressure and strain sensors based on transparent elastic films of carbon nanotubes. *Nature Nanotechnology* 6 (12): 788–792.

71 Cheng, W., Yu, L., Kong, D. et al. (2018). Fast-response and low-hysteresis flexible pressure sensor based on silicon nanowires. *IEEE Electron Device Letters* 39 (7): 1069–1072.

72 Kwon, D., Lee, T.I., Kim, M.S., et al. (2015). Porous dielectric elastomer based ultra-sensitive capacitive pressure sensor and its application to wearable sensing device. In *2015 Transducers-2015 18th International Conference on Solid-State Sensors, Actuators and Microsystems (TRANSDUCERS)*. IEEE. pp. 299–302.

73 Li, L., Fu, X., Chen, S. et al. (2020). Hydrophobic and stable MXene-polymer pressure sensors for wearable electronics. *ACS Applied Materials & Interfaces* 12 (13): 15362–15369.

74 Choong, C.L., Shim, M.B., Lee, B.S. et al. (2014). Highly stretchable resistive pressure sensors using a conductive elastomeric composite on a micropyramid array. *Advanced Materials* 26 (21): 3451–3458.

75 Li, J., Fang, L., Sun, B. et al. (2020). Review—recent progress in flexible and stretchable piezoresistive sensors and their applications. *Journal of the Electrochemical Society* 167 (3): 037561.

76 Mishra, R.B., El-Atab, N., Hussain, A.M., and Hussain, M.M. (2021). Recent progress on flexible capacitive pressure sensors: from design and materials to applications. *Advanced Materials Technologies* 6 (4): 2001023.

77 Ma, C., Xu, D., Huang, Y.C. et al. (2020). Robust flexible pressure sensors made from conductive micropyramids for manipulation tasks. *ACS Nano* 14 (10): 12866–12876.

78 Doshi, S.M. and Thostenson, E.T. (2018). Thin and flexible carbon nanotube-based pressure sensors with ultrawide sensing range. *ACS Sensors* 3 (7): 1276–1282.

79 Jason, N.N., Ho, M.D., and Cheng, W. (2017). Resistive electronic skin. *Journal of Materials Chemistry C* 5 (24): 5845–5866.

80 Dai, H. and Thostenson, E.T. (2019). Large-area carbon nanotube-based flexible composites for ultra-wide range pressure sensing and spatial pressure mapping. *ACS Applied Materials & Interfaces* 11 (51): 48370–48380.

81 Zhang, L., Li, H., Lai, X. et al. (2018). Thiolated graphene@polyester fabric-based multilayer piezoresistive pressure sensors for detecting human motion. *ACS Applied Materials & Interfaces* 10 (48): 41784–41792.

82 Mao, Y., Ji, B., Chen, G. et al. (2019). Robust and wearable pressure sensor assembled from AgNW-coated PDMS micropillar sheets with high sensitivity and wide detection range. *ACS Applied Nano Materials* 2 (5): 3196–3205.

83 Liu, S., Wu, X., Zhang, D. et al. (2017). Ultrafast dynamic pressure sensors based on graphene hybrid structure. *ACS Applied Materials & Interfaces* 9 (28): 24148–24154.

84 Dintwa, E., Tijskens, E., and Ramon, H. (2008). On the accuracy of the Hertz model to describe the normal contact of soft elastic spheres. *Granular Matter* 10: 209–221.

85 Persson, B. (2022). On the electric contact resistance. *Tribology Letters* 70 (3): 88.

86 Ji, B., Zhou, Q., Wu, J. et al. (2020). Synergistic optimization toward the sensitivity and linearity of flexible pressure sensor via double conductive layer and porous microdome array. *ACS Applied Materials & Interfaces* 12 (27): 31021 31035.

87 Ruschau, G., Yoshikawa, S., and Newnham, R. (1992). Resistivities of conductive composites. *Journal of Applied Physics* 72 (3): 953–959.

88 Khalili, N., Naguib, H., and Kwon, R. (2016). A constriction resistance model of conjugated polymer based piezoresistive sensors for electronic skin applications. *Soft Matter* 12 (18): 4180–4189.

89 Park, J., Lee, Y., Hong, J. et al. (2014). Giant tunneling piezoresistance of composite elastomers with interlocked microdome arrays for ultrasensitive and multimodal electronic skins. *ACS Nano* 8 (5): 4689–4697.

90 Greenwood, J.A. (1966). Constriction resistance and the real area of contact. *British Journal of Applied Physics* 17 (12): 1621.

91 Zhu, B., Niu, Z., Wang, H. et al. (2014). Microstructured graphene arrays for highly sensitive flexible tactile sensors. *Small* 10 (18): 3625–3631.

92 Park, H., Jeong, Y.R., Yun, J. et al. (2015). Stretchable array of highly sensitive pressure sensors consisting of polyaniline nanofibers and Au-coated polydimethylsiloxane micropillars. *ACS Nano* 9 (10): 9974–9985.

93 Bae, G.Y., Pak, S.W., Kim, D. et al. (2016). Linearly and highly pressure-sensitive electronic skin based on a bioinspired hierarchical structural array. *Advanced Materials* 28 (26): 5300–5306.

94 Pang, Y., Zhang, K., Yang, Z. et al. (2018). Epidermis microstructure inspired graphene pressure sensor with random distributed spinosum for high sensitivity and large linearity. *ACS Nano* 12 (3): 2346–2354.

95 Kanoun, O., Bouhamed, A., Ramalingame, R. et al. (2021). Review on conductive polymer/CNTs nanocomposites based flexible and stretchable strain and pressure sensors. *Sensors* 21 (2): 341.

96 Ding, S., Han, B., Dong, X. et al. (2017). Pressure-sensitive behaviors, mechanisms and model of field assisted quantum tunneling composites. *Polymer* 113: 105–118.

97 Roldughin, V. and Vysotskii, V. (2000). Percolation properties of metal-filled polymer films, structure and mechanisms of conductivity. *Progress in Organic Coatings* 39 (2–4): 81–100.

98 Luheng, W., Tianhuai, D., and Peng, W. (2009). Influence of carbon black concentration on piezoresistivity for carbon-black-filled silicone rubber composite. *Carbon* 47 (14): 3151–3157.

99 Luheng, W., Tianhuai, D., and Peng, W. (2007). Effects of conductive phase content on critical pressure of carbon black filled silicone rubber composite. *Sensors and Actuators A: Physical* 135 (2): 587–592.

100 Hou, Y., Wang, D., Zhang, X.M. et al. (2013). Positive piezoresistive behavior of electrically conductive alkyl-functionalized graphene/polydimethylsilicone nanocomposites. *Journal of Materials Chemistry C* 1 (3): 515–521.

101 Zhang, X.W., Pan, Y., Zheng, Q., and Yi, X.S. (2000). Time dependence of piezoresistance for the conductor-filled polymer composites. *Journal of Polymer Science Part B: Polymer Physics* 38 (21): 2739–2749.

102 Kirpluks, M., Cabulis, U., Andersons, J. et al. (2018). Modeling the effect of foam density and strain rate on the compressive response of polyurethane foams. *SAE International Journal of Materials and Manufacturing* 11 (2): 131–138.

103 Zhang, J., Wang, Z., and Peng, Z. (2021). Analytical model of the piezoresistive behavior of highly compressible sensors made of microporous nanocomposites. *Advanced Theory and Simulations* 4 (12): 2100247.

104 Li, W., Jin, X., Han, X. et al. (2021). Synergy of porous structure and microstructure in piezoresistive material for high-performance and flexible pressure sensors. *ACS Applied Materials & Interfaces* 13 (16): 19211–19220.

105 Xu, J., Li, H., Yin, Y. et al. (2022). High sensitivity and broad linearity range pressure sensor based on hierarchical in-situ filling porous structure. *npj Flexible Electronics* 6 (1): 62.

106 Storåkers, B. (1986). On material representation and constitutive branching in finite compressible elasticity. *Journal of the Mechanics and Physics of Solids* 34 (2): 125–145.

107 Balberg, I. (1986). Excluded-volume explanation of Archie's law. *Physical Review B* 33 (5): 3618.

108 Chen, R., Luo, T., Wang, J. et al. (2023). Nonlinearity synergy: an elegant strategy for realizing high-sensitivity and wide-linear-range pressure sensing. *Nature Communications* 14 (1): 6641.

109 Bang, J., Chun, B., Lim, J. et al. (2023). Ultra-broad linear range and sensitive flexible piezoresistive sensor using reversed lattice structure for wearable electronics. *ACS Applied Materials & Interfaces* 15 (28): 34120–34131.

110 Chen, J., Chen, K., Jin, J. et al. (2023). Outstanding synergy of sensitivity and linear range enabled by multigradient architectures. *Nano Letters* 23: 11958–11967.

111 Zhang, Y., Lu, Q., He, J. et al. (2023). Localizing strain via micro cage structure for stretchable pressure sensor arrays with ultralow spatial crosstalk. *Nature Communications* 14 (1): 1252.

112 Ding, X., Zhong, W., Jiang, H. et al. (2020). Highly accurate wearable piezoresistive sensors without tension disturbance based on weaved conductive yarn. *ACS Applied Materials & Interfaces* 12 (31): 35638–35646.

113 Chou, H.-H., Nguyen, A., Chortos, A. et al. (2015). A chameleon-inspired stretchable electronic skin with interactive colour changing controlled by tactile sensing. *Nature Communications* 6 (1): 8011.

114 Su, Q., Zou, Q., Li, Y. et al. (2021). A stretchable and strain-unperturbed pressure sensor for motion interference—free tactile monitoring on skins. *Science Advances* 7 (48): eabi4563.

115 Matsuda, R., Mizuguchi, S., Nakamura, F. et al. (2020). Highly stretchable sensing array for independent detection of pressure and strain exploiting structural and resistive control. *Scientific Reports* 10 (1): 12666.

116 Lee, G., Son, J., Kim, D. et al. (2022). Crocodile-skin-inspired omnidirectionally stretchable pressure sensor. *Small* 18 (52): e2205643.

117 Xu, T., Ding, Y., Wang, Z. et al. (2017). Three-dimensional and ultralight sponges with tunable conductivity assembled from electrospun nanofibers for a highly sensitive tactile pressure sensor. *Journal of Materials Chemistry C* 5 (39): 10288–10294.

118 Li, X., Li, X., Liu, T. et al. (2021). Wearable, washable, and highly sensitive piezoresistive pressure sensor based on a 3D sponge network for real-time monitoring human body activities. *ACS Applied Materials & Interfaces* 13 (39): 46848–46857.

119 Yue, Y., Liu, N., Liu, W. et al. (2018). 3D hybrid porous Mxene-sponge network and its application in piezoresistive sensor. *Nano Energy* 50: 79–87.

120 Yang, J., Tang, D., Ao, J. et al. (2020). Ultrasoft liquid metal elastomer foams with positive and negative piezopermittivity for tactile sensing. *Advanced Functional Materials* 30 (36): 2002611.

121 Ruth, S.R.A. and Bao, Z. (2020). Designing tunable capacitive pressure sensors based on material properties and microstructure geometry. *ACS Applied Materials & Interfaces* 12 (52): 58301–58316.

122 Javidi, R., Moghimi Zand, M., and Alizadeh Majd, S. (2023). Designing wearable capacitive pressure sensors with arrangement of porous pyramidal microstructures. *Micro and Nano Systems Letters* 11 (1): 13.

123 Tee, B.C.K., Chortos, A., Dunn, R.R. et al. (2014). Tunable flexible pressure sensors using microstructured elastomer geometries for intuitive electronics. *Advanced Functional Materials* 24 (34): 5427–5434.

124 Boutry, C.M., Negre, M., Jorda, M. et al. (2018). A hierarchically patterned, bioinspired e-skin able to detect the direction of applied pressure for robotics. *Science Robotics* 3 (24): eaau6914.

125 Luo, Y.S., Shao, J., Chen, S. et al. (2019). Flexible capacitive pressure sensor enhanced by tilted micropillar arrays. *ACS Applied Materials & Interfaces* 11 (19): 17796–17803.

126 Peng, S.H., Blanloeuil, P., Wu, S., and Wang, C.H. (2018). Rational design of ultrasensitive pressure sensors by tailoring microscopic features. *Advanced Materials Interfaces* 5 (18): 1800403.

127 Bilent, S., Dinh, T.H.N., Martincic, E., and Joubert, P.Y. (2019). Influence of the porosity of polymer foams on the performances of capacitive flexible pressure sensors. *Sensors* 19 (9): 1968.

128 Zhang, Y., Hu, Y., Zhu, P. et al. (2017). Flexible and highly sensitive pressure sensor based on microdome-patterned PDMS forming with assistance of colloid self-assembly and replica technique for wearable electronics. *ACS Applied Materials & Interfaces* 9 (41): 35968–35976.

129 Zhang, W., Sun, W., Xiao, W. et al. (2019). Numerical simulation analysis of microstructure of dielectric layers in capacitive pressure sensors. *IEEE Sensors Journal* 19 (9): 3260–3266.

130 Choy, T.C. (2015). *Effective Medium Theory: Principles and Applications*, vol. 165. Oxford University Press.

131 Hashin, Z. and Shtrikman, S. (1962). A variational approach to the theory of the effective magnetic permeability of multiphase materials. *Journal of Applied Physics* 33 (10): 3125–3131.

132 Progelhof, R. and Throne, J.L. (1979). Young's modulus of uniform density thermoplastic foam. *Polymer Engineering & Science* 19 (7): 493–499.

133 Gong, L., Kyriakides, S., and Jang, W.Y. (2005). Compressive response of open-cell foams. Part I: morphology and elastic properties. *International Journal of Solids and Structures* 42 (5–6): 1355–1379.

134 Choi, J. and Lakes, R.S. (1995, 1995). Analysis of elastic modulus of conventional foams and of re-entrant foam materials with a negative Poisson's ratio. *International Journal of Mechanical Sciences* 37 (1): 51–59.

135 Moore, D., Couzens, K., and Iremonger, M.J. (1974). The deformational behaviour of foamed thermoplastics. *Journal of Cellular Plastics* 10 (3): 135–139.

136 Gibson, I.J. and Ashby, M.F. (1982). The mechanics of three-dimensional cellular materials. *Proceedings of the Royal Society of London. A. Mathematical and Physical Sciences* 382 (1782): 43–59.

137 Qu, X., Xie, R., Zhou, Z. et al. (2023). Highly sensitive capacitive fiber pressure sensors enabled by electrode and dielectric layer regulation. *ACS Applied Materials & Interfaces* 15 (47): 54966–54976.

138 Yang, W., Ding, W., Liu, M. et al. (2021). A theoretical model of a flexible capacitive pressure sensor with microstructured electrodes for highly sensitive electronic skin. *Journal of Physics D: Applied Physics* 55 (9): 094001.

139 Zhang, Y., Lin, Z., Huang, X. et al. (2019). Highly sensitive capacitive pressure sensor with elastic metallized sponge. *Smart Materials and Structures* 28 (10): 105023.

140 Yang, J.C., Kim, J.O., Oh, J. et al. (2019). Microstructured porous pyramid-based ultrahigh sensitive pressure sensor insensitive to strain and temperature. *ACS Applied Materials & Interfaces* 11 (21): 19472–19480.

141 Larson, C., Peele, B., Li, S. et al. (2016). Highly stretchable electroluminescent skin for optical signaling and tactile sensing. *Science* 351 (6277): 1071–1074.

142 Wang, X., Li, T., Adams, J., and Yang, J. (2013). Transparent, stretchable, carbon-nanotube-inlaid conductors enabled by standard replication technology for capacitive pressure, strain and touch sensors. *Journal of Materials Chemistry A* 1 (11): 3580–3586.

143 Hu, W., Niu, X., Zhao, R. et al. (2013). Elastomeric transparent capacitive sensors based on an interpenetrating composite of silver nanowires and polyurethane. *Applied Physics Letters* 102 (8): 083303.

144 Liu, F., Han, F., Ling, L. et al. (2018). An omni-healable and highly sensitive capacitive pressure sensor with microarray structure. *Chemistry* 24 (63): 16823–16832.

145 Kim, H., Kim, G., Kim, T. et al. (2018). Transparent, flexible, conformal capacitive pressure sensors with nanoparticles. *Small* 14 (8): 1703432.

146 Baek, S., Jang, H., Kim, S.Y. et al. (2017). Flexible piezocapacitive sensors based on wrinkled microstructures: toward low-cost fabrication of pressure sensors over large areas. *RSC Advances* 7 (63): 39420–39426.

147 Ho, D.H., Sun, Q., Kim, S.Y. et al. (2016). Stretchable and multimodal all graphene electronic skin. *Advanced Materials* 28 (13): 2601–2608.

148 Boutry, C.M., Kaizawa, Y., Schroeder, B.C. et al. (2018). A stretchable and biodegradable strain and pressure sensor for orthopaedic application. *Nature Electronics* 1 (5): 314–321.

149 Gerratt, A.P., Sommer, N., Lacour, S.P., and Billard, A. (2014). Stretchable capacitive tactile skin on humanoid robot fingers—First experiments and results. In *2014 IEEE-RAS International Conference on Humanoid Robots*. IEEE. pp. 238–245.

150 Zhang, Y., Liu, S., Miao, Y. et al. (2020). Highly stretchable and sensitive pressure sensor array based on icicle-shaped liquid metal film electrodes. *ACS Applied Materials & Interfaces* 12 (25): 27961–27970.

151 Ha, K.H., Zhang, W., Jang, H. et al. (2021). Highly sensitive capacitive pressure sensors over a wide pressure range enabled by the hybrid responses of a highly porous nanocomposite. *Advanced Materials* 33: 2103320.

152 Qu, X., Li, J., Han, Z. et al. (2023). Highly sensitive fiber pressure sensors over a wide pressure range enabled by resistive-capacitive hybrid response. *ACS Nano* 17 (15): 14904–14915.

153 Ha, K.-H., Li, Z., Kim, S. et al. (2021). Stretchable hybrid response pressure sensors. *Matter* 7 (5): 1895–1908. https://www.cell.com/matter/fulltext/S2590-2385(24)00164-4.

154 Lou, Z., Chen, S., Wang, L. et al. (2016). An ultra-sensitive and rapid response speed graphene pressure sensors for electronic skin and health monitoring. *Nano Energy* 23: 7–14.

155 Chen, W. and Yan, X. (2020). Progress in achieving high-performance piezoresistive and capacitive flexible pressure sensors: a review. *Journal of Materials Science and Technology* 43: 175–188.

156 Wang, H., Li, Z., Liu, Z. et al. (2022). Flexible capacitive pressure sensors for wearable electronics. *Journal of Materials Chemistry C* 10 (5): 1594–1605.

157 Huang, Y., Fan, X., Chen, S.C. et al. (2019). Emerging technologies of flexible pressure sensors: materials, modeling, devices, and manufacturing. *Advanced Functional Materials* 29 (12): 1808509.

158 Kumar, A. (2022). Recent progress in the fabrication and applications of flexible capacitive and resistive pressure sensors. *Sensors and Actuators A: Physical* 344: 113770.

159 Li, Z., Ha, K.H., Wang, Z. et al. (2022). Effects of AC frequency on the capacitance measurement of hybrid response pressure sensors. *Soft Matter* 18 (44): 8476–8485.

14

Active Mechanical Haptics Constructed with Curved Origami

Zhuang Zhang and Hanqing Jiang

Westlake University, School of Engineering, No. 600 Dunyu Road, Xihu District, Hangzhou, Zhejiang 310030, PR China

14.1 Introduction

The rapid emergence of the metaverse, fueled by the advancements in virtual reality (VR) and augmented reality (AR) technologies, is ushering in a transformative era. We are transitioning from a third-person perspective to an immersive, first-person experiential platform that vividly captures and replicates the features of the physical world [1–3]. This transformation is grounded in the creation of virtual environments enriched with authentic sensory perceptions, encompassing sight, hearing, and crucially, touch.

In this new era of media interaction, VR and AR technologies are poised to infiltrate a myriad of industry sectors, ranging from entertainment and communication to education, human–machine interaction, teleoperation, clinical therapy, and rehabilitation [4, 5]. This penetration is facilitated by advanced software, including communication technologies, various apps, and social networks, as well as hardware such as VR/AR and haptic devices, all of which serve to augment our interaction experiences [4, 5].

However, while state-of-the-art VR/AR devices have made strides in delivering stereoscopic visual and auditory sensory perceptions, they fall short in providing a tangible sensory dimension in the mechanical domain, particularly in terms of touch sensation [5–7]. There has been significant progress in introducing machine-triggered, passive haptic experiences to users. Yet, most of these experiences are limited to simple, hand-centered vibrations (vibro-haptics) [5–16] and cutaneous stimuli to the skin [17–19], or they involve the use of desktop surfaces [20–24] and multi-linkage mechanisms [25–31] to impose motion constraints. This stands in stark contrast to the physical world, where humans naturally perceive objects in terms of their hardness, softness, and even the breaking moments of fragile items through active touch via hands, feet, or other body parts [2].

Active mechanical haptics, encompassing both positive stiffness (perceiving hardness and softness) and negative stiffness (sensing breaking and falling

Mechanics of Flexible and Stretchable Electronics, First Edition. Edited by Yong Zhu and Nanshu Lu.

moments), are central to our first sensory perceptions in daily interactions with the physical world. They play a crucial role in exploring and processing information and in creating ontological metaphors [2, 32]. Therefore, in this chapter, we introduce mechanisms, materials, design principles, systems integration, and the corresponding sensory perceptions for human-initiated, body-centered, scalable, and wirelessly controlled haptic devices. These devices are uniquely enabled by curved origami modules, allowing users to actively generate and sense mechanical touching sensations with controllable stiffness, ranging from hard to soft and from positive to negative. Our aim is to produce human-triggered mechanical sensory perceptions, a crucial human sense for interacting with the physical universe, yet one that is rarely reproduced in the virtual world [2, 5]. These devices are poised to enhance VR/AR experiences beyond visual and auditory sensations, delivering a range of diverse and immersive experiences [33].

14.2 Stiffness Tuning via Curved Origami

Utilizing a circular arc-shaped cut line defined by an angle α, which is intricately linked to the circle's radius ρ through the relationship $\cos\alpha = \frac{b}{2\rho}$, where the rectangle's dimensions are given by $a \times b$, we successfully create a curved origami structure by folding the 2D plate along the curved crease (Figure 14.1a). The folding angle β at the crease is influenced by the material's plasticity during the folding process, resulting in smaller β values for more significant bending deformations of the curved panels and larger β values for lesser bending deformations.

To understand the mechanical behavior of the curved origami, including the curved crease, panels, and plasticity at the curved crease, we employ a variant of the nonstandard Miura origami (shown in Figure 14.1b). In this model, the angle γ is related to the plastic folding angle β, expressed as $\gamma(\beta)$. This relationship is crucial as it enables us to predict the stiffness of the curved origami and subsequently guide our design choices.

When a vertical load is applied to the origami, the curved origami exhibits varying force–displacement relationships. These variations are attributed to the interplay between bending deformation, contributing to positive stiffness, and folding deformation around the curved crease, resulting in negative stiffness Figure 14.1c presents the force–displacement relationships of a specific curved origami defined by $\alpha = 80°$ and $a/b = 1.25$, under different initial folding angles β, such as 60°, 90°, and 120°. Larger β values (e.g. 120°) result in a range that includes both positive and negative stiffness, while smaller β values (e.g. 60°) yield solely positive stiffness.

This implies that for a given curved origami (defined by angle α), one can adeptly modulate its stiffness by altering the folding angle β. This can be achieved through a cable-driven method, as depicted in Figure 14.1d, where an external force is actively applied atop the curved origami. Concurrently, a cable is integrated between the two panels to adjust angle β from a specific initial value (e.g. $\beta = 120°$ for a broader tuning range) by $\Delta\beta$ in real time.

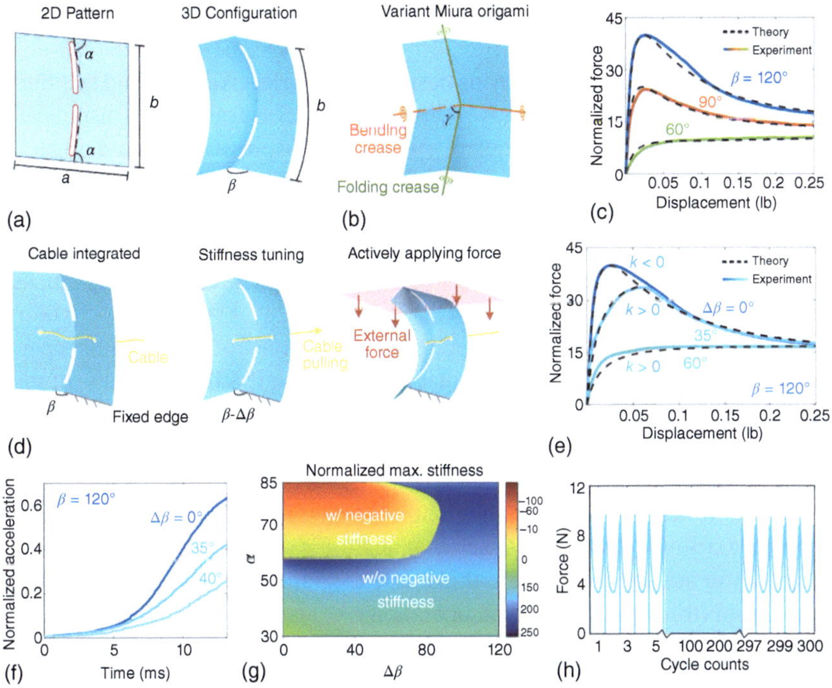

Figure 14.1 Working principle using curved origami to tune stiffness. (a) 2D pattern and 3D configuration of the curved origami module, where α, the angle between the horizontal edge and the tangent line at the end of the curved crease, denotes the normalized curvature of the crease; β, the angle between two bending panels, representing the plastic origami folding, (b) a variant Miura origami with straight creases and rigid panels for theoretical modeling of the curved origami. The solid lines represent mountain creases, and the dashed lines represent valley creases, (c) force–displacement relationship of curved origami modules with different folding angles β and a fixed curve $\alpha = 80°$, (d) working principle of the stiffness manipulation based on curved origami. Left: The initial state of the origami module with an integrated cable and a fixed edge; Middle: The cable is utilized to change the folding angle by $\Delta\beta$; Right: Upon actively pressing the curved origami vertically, the combined panel-bending and origami folding defines its stiffness, (e) force–displacement relationship of a specific module with $\alpha = 80°$ and $\beta = 120°$ and different angle $\Delta\beta$, (f) dropping acceleration of a dead load in negative stiffness domain for a specific module with $\alpha = 80°$ and $\beta = 120°$ and different angle $\Delta\beta$, (g) phase diagram of normalized secant stiffness for a curved origami module with a given folding angle $\beta = 120°$, varying crease curvatures $30° < \alpha < 85°$, and cable-driven folding angles $0° < \Delta\beta < 120°$. The red and blue zones represent parameters with negative and positive stiffness, respectively, (h) cyclic compression test of a curved origami module, suggesting good cyclability.

Contrary to the plasticity observed during the folding process that determines the folding angle β, the cable pulling results in an elastic folding process of the origami. This elasticity ensures that the origami instantly reverts to its original folding state (defined by angle β) once the cable is released. Figure 14.1e showcases the force–displacement relationships for a curved origami characterized by $\alpha = 80°$, $\beta = 120°$, and $a/b = 1.25$, under different cable pulling angles $\Delta\beta = 0°$, 35°, and 60°. This results in stiffness values spanning both positive and negative ranges, affirming the effectiveness of the cable-driven method in real-time stiffness modulation.

In terms of haptic perceptions, varying positive stiffness levels correspond to different degrees of hardness, while negative stiffness emulates sensations of crushing an object or experiencing a fall. Consequently, when actively pressing a curved origami exhibiting negative stiffness with a constant force, acceleration becomes a crucial metric for characterizing this unique feature. For the same curved origami depicted in Figure 14.1e, under $\beta = 120°$ and $\Delta\beta = 0°$, $35°$, and $40°$, the acceleration of a dead load (i.e. a constant force) can reach up to 0.6 G, with 1 G representing free fall, as illustrated in Figure 14.1f.

To guide the design of the curved origami for more pronounced mechanical perceptions, Figure 14.1g provides a phase diagram depicting the absolute value of the maximum stiffness as a function of the crease angle α and the tuning angle $\Delta\beta$, for a prescribed initial folding angle $\beta = 120°$. In scenarios where the origami exhibits both positive and negative stiffness ranges, only the maximum negative stiffness is considered. Conversely, when only the positive stiffness range is present, the maximum positive stiffness is selected. Given that angle α is a predetermined parameter defining the curved origami pattern, this phase diagram serves as a valuable tool, offering design (through angle α) and operational guidance (through tuning angle $\Delta\beta$) for utilizing curved origami in stiffness sensory perceptions.

Similar to the elastic deformation observed during cable pulling, the deformation experienced during cyclic pressing atop the curved origami also exhibits elasticity and demonstrates excellent cyclability, as shown in Figure 14.1h. Here, both negative and positive stiffness are distinctly observable in each cycle, further emphasizing the robustness and versatility of the curved origami structure in delivering diverse haptic experiences.

14.3 Theoretical Modeling and Analysis of Curved Origami

The curved origami belongs to the class of deformable origami, and thus, it has infinite degrees of freedom. Instead of using finite element methods to obtain the spatial distribution of stress and strain upon deformation, here we adopted an analytical analysis by bridging the curved origami (deformable origami) with a variant Miura origami (rigid origami), both in geometry and in mechanics, to obtain a simplified and more insightful solution.

14.3.1 Geometric Modeling

The folding of the curved origami involves plastic deformations at the creases; as a result, once the folding is finished, the curved origami maintains deformed states even without imposing external loadings. To approximate the curved origami by a variant Miura origami, we first need to ensure geometric similarity, i.e. in the loading-free condition, the variant Miura origami must present the equal apparent

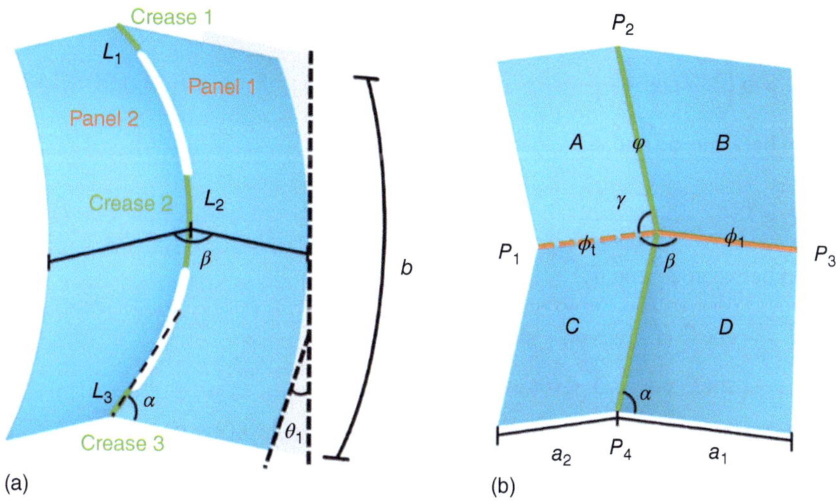

Figure 14.2 Illustration of curved origami and equivalent variant Miura pattern. (a) Illustration of curved origami, (b) illustration of the variant Miura pattern.

deformation, meaning that the initial deformation between two loaded edges and the angle between the panels for the curved origami and the variant Miura are the same.

In Figure 14.2b, ϕ_1, ϕ_2, φ are the acute angles (not necessarily the dihedral angles) between facets B and D, A and C, and A and B, respectively; γ, β, α are the angles between creases P_1 and P_2, P_3 and P_4, and creases P_4 and the lower loaded edge, respectively. It is worth emphasizing that ϕ_1, ϕ_2, φ these can be dihedral angles (when the dihedral angles are less than $\pi/2$) or their supplementary angles (when the dihedral angles are greater than $\pi/2$). For the traditional Miura origami, $\gamma = \alpha$, $\phi_1 = \phi_2$, and the four panels are coplanar as $\phi_1 = \phi_2 = 0$. However, for the variant Miura origami that is used to describe the behaviors of the present curved origami, $\gamma \neq \alpha$, $\phi_1 \neq \phi_2$, and four panels are not coplanar as $\phi_1 = 0$.

In the variant Miura origami, $\gamma > \alpha$. For the curved origami (shown in Figure 14.2a), panel 1 is almost flat when no external force applied to it, and the associated folding angle β is denoted as β_0. To satisfy the geometrical similarity, in the variant Miura origami, the angle β should also be β_0 as $\phi_1 = 0$. Through a simple geometrical analysis, this geometrical relationship is yielded, i.e. $\cos\gamma = -\cos\alpha\cos\beta_0$. When the curved origami and the variant Miura origami are subjected to the same displacement in the vertical direction, the angle φ in the variant Miura origami equals the rotation angles of creases 1 and 3 in the curved origami, and β in the variant Miura equals the rotation angle of the crease 2 in the curved origami.

Here, we endeavor to derive the relations between ϕ_1, ϕ_2, φ, β, and the vertical displacement u (the same one for both the curved origami and the variant Miura origami). The geometrical relations can be derived and listed:

Relation between ϕ_1 and u:

$$u = b\left(1 - \cos\frac{\phi_1}{2}\right) \quad (14.1)$$

Relation between ϕ_1 and ϕ_2:

$$\cos\frac{\phi_2}{2} = \frac{\cos\frac{\phi_1}{2}\sin\alpha}{\sin\gamma} \quad (14.2)$$

Relation between ϕ_1 and φ:

$$\begin{aligned}&\left(\frac{1}{\tan\alpha}\cos\frac{\phi_1}{2}(1-\cos\varphi) - \sin\frac{\phi_1}{2}\sin\varphi\right)\cos(\gamma-\alpha)\\ &-\left(\sin\frac{\phi_1}{2}\cos\frac{\phi_1}{2}(1-\cos\varphi) + \frac{1}{\tan\alpha}\sin\varphi\right)\\ &\times\sin(\gamma-\alpha)\sin\frac{\phi_1}{2} - \left(\cos^2\frac{\phi_1}{2}(1-\cos\varphi)+\cos\varphi\right)\sin(\gamma-\alpha)\cos\frac{\phi_1}{2} = 0\end{aligned} \quad (14.3)$$

Relation between β and φ:

$$\cos\beta = -\frac{\frac{\cos\gamma}{\cos\alpha} + \frac{\cos\alpha}{\cos\gamma} - \cos\alpha\cos\gamma(\tan^2\gamma + \tan^2\alpha - 2\tan\gamma\tan\alpha\cos\varphi)}{2} \quad (14.4)$$

14.3.2 Mechanics Modeling

The central task of mechanics modeling is to establish the mechanical similarity between the curved origami and the variant Miura origami. From the viewpoint of energy, the potential energies stored within these two configurations are the same for arbitrarily given apparent displacement. Here, we will first calculate the energy in the curved origami, i.e. the potential energy stored in the creases 1, 2, and 3, and the panels 1 and 2. In addition to the important geometrical parameters denoted in Figure 14.2, Young's modulus and the thickness of the panel are E and h, respectively.

Under vertical load on the top of the curved origami, the creases almost experience pure-bending deformations. The theory of one-dimensional deformable bodies can readily evaluate the equivalent stiffness of these creases. Given the width of the creases S and the bending stiffness of the panel EI, the equivalent bending stiffness of the crease can be defined as $k \triangleq \frac{EI}{S}$. The potential energy stored in the crease can be calculated by $\frac{1}{2}k\varphi^2$.

For the panels, we regard it as the pure bending deformation and calculate the equivalent stiffness. For a prescribed displacement along the vertical direction, we can readily calculate the relationship between the angle of pure bending (2θ) in the curved origami to its corresponding angle ϕ_1 and ϕ_2, i.e. $\frac{\sin\theta_i}{\theta_i} = \cos\frac{\phi_i}{2}$, with $i = 1$ or 2. Then, the potential energy stored in two panels can be easily calculated as $\frac{2EI}{b}\theta^2$.

In summary, for the curved origami, the potential energy stored in the creases and panels are expressed, respectively, as

$$\begin{cases} V_{\text{crease1}} = \dfrac{EL_1h^3}{24S}\varphi^2 \triangleq \dfrac{1}{2}k_1\varphi^2 \\ V_{\text{crease2}} = \frac{EL_2h^3}{24S}\beta^2 \triangleq \frac{1}{2}k_2\beta^2 \\ V_{\text{crease3}} = \frac{EL_3h^3}{24S}\varphi^2 \triangleq \frac{1}{2}k_3\varphi^2 \\ V_{\text{panel1}} = \frac{Ea_1h^3}{6b}\theta_1{}^2 \triangleq \frac{1}{2}K_1\theta_1{}^2 \\ V_{\text{panel2}} = \frac{Ea_2h^3}{6b}\theta_2{}^2 \triangleq \frac{1}{2}K_2\theta_2{}^2 \end{cases} \tag{14.5}$$

in which, k_i and K_i are the equivalent stiffness, L_i are the lengths of creases, h is the thickness of the panel, and a_i are the widths of two panels.

Here, we apply the principle of minimal potential energy to determine the force–displacement relations of the variant Miura origami, from which we can evaluate that of the curved origami. For a given static load F along the vertical direction with displacement u, the total potential energy, including the potential of the external force and the strain energy stored in torsional springs, is expressed as,

$$V = \frac{1}{2}k_1\Delta\varphi^2 + \frac{1}{2}k_3\Delta\varphi^2 + \frac{1}{2}k_2\Delta\beta^2 + \frac{1}{2}K_1\Delta\theta_1{}^2 + \frac{1}{2}K_2\Delta\theta_2{}^2 - Fu \tag{14.6}$$

in which all angles are functions of the vertical displacement u, $\Delta(\ldots)$ represents the change of each angle compared to their initial states. By setting $\frac{\partial V}{\partial u} = 0$, we obtain the following force–displacement relation:

$$F = \frac{(k_1 + k_3)\Delta\varphi\dfrac{\partial\varphi}{\partial\phi_1} + k_2\Delta\beta\dfrac{\partial\beta}{\partial\phi_1} + K_1\Delta\theta_1\dfrac{\partial\theta_1}{\partial\phi_1} + K_2\Delta\theta_2\dfrac{\partial\theta_2}{\partial\phi_1}}{\frac{1}{2}b\sin\dfrac{\phi_1}{2}} \tag{14.7}$$

By introducing the dimensionless parameters $\overline{k}_2 = \frac{k_2}{k_1}, \overline{k}_3 = \frac{k_3}{k_1}, \overline{K}_1 = \frac{K_1}{k_1}, \overline{K}_2 = \frac{K_2}{k_1}$, $\overline{F} = \frac{Fb}{k_1}, \overline{u} = \frac{u}{b}$, the dimensionless force displacement relation is expressed as,

$$\overline{F} = \frac{(1 + \overline{k}_3)\Delta\varphi\dfrac{\partial\varphi}{\partial\phi_1} + \overline{k}_2\Delta\beta\dfrac{\partial\beta}{\partial\phi_1} + \overline{K}_1\Delta\theta_1\dfrac{\partial\theta_1}{\partial\phi_1} + \overline{K}_2\Delta\theta_2\dfrac{\partial\theta_2}{\partial\phi_1}}{\frac{1}{2}\sin\dfrac{\phi_1}{2}} \tag{14.8}$$

in which the partial derivatives (i.e. $\frac{\partial\varphi}{\partial\phi_1}$) are calculated numerically from Eqs. (14.1) to (14.4) The first three terms in the numerator represent the contribution by creases and the last two terms are for the contribution by panels. As shown in Figure 14.2, the creases and the panels provide negative stiffness and positive stiffness, respectively. Obviously, we can design the mechanical responses of the curved origami on demand by appropriately allocating the creases and panels.

14.3.3 Active Tuning

By intentionally introducing a cable between two panels, the force–displacement relationship of the curved origami can be tuned. The aforementioned analysis can be readily adjusted to include this tuning factor. For one case with the tensional cable, the mechanical response can be divided into two successive stages: under a small load, the cable keeps in the tensional state (Stage I), while as the load approaches a certain critical value, the cable starts to be slack, and the mechanic's response comes into Stage II. In Stage 1, the dimensionless force–displacement relation is modified as,

$$\overline{F} = \frac{(1+\overline{k}_3)\Delta\overline{\varphi}\frac{\partial\varphi}{\partial\phi_1} + \overline{k}_2\Delta\overline{\beta}\frac{\partial\beta}{\partial\phi_1} + \overline{K}_1\Delta\theta_1\frac{\partial\theta_1}{\partial\phi_1} + \overline{K}_2\Delta\theta_2\frac{\partial\theta_2}{\partial\phi_1}}{\frac{1}{2}\sin\frac{\phi_1}{2}} \tag{14.9}$$

in which $\Delta\overline{\varphi}$, $\Delta\overline{\beta}$ are the angle changes caused by the vertical load, which are equal to the changes without cable pulling. φ, β take the values of the angles when the tensional cable is applied. In stage II, the force–displacement relation is just the one without cable pulling.

14.4 Closed-Loop Design and System Integration of Origami

Integrating the tactile feedback from the curved origami with synchronized visual cues from standard VR devices enables the creation of a fully immersive and interactive virtual environment, as depicted in Figure 14.3a. In this setup, users can tangibly interact with the virtual world, feeling the stiffness of objects they see in VR through user-initiated actions, utilizing haptic devices where the curved origami serves as a crucial component for adjusting stiffness. The virtual scenarios provide input to the haptic device, influencing the adjustment of the folding angle β by the integrated motor in $\Delta\beta$, to emulate the expected tactile response correlated with the VR visuals. Consequently, users can experience the mechanical stiffness of virtual objects in real time, whether through active hand-grasping or body-centric foot-stepping actions, which will be introduced later.

The user's interactions with the curved origami, such as pressing, result in changes in resistance. This change is captured using a layer of silver nanowires (AgNWs) deposited on the origami panel, serving as a sensing layer. These resistance changes are then transformed into voltage variations through an onboard microcontroller, as illustrated in Figure 14.3a,b. The voltage variations are wirelessly communicated to the virtual environment engine (e.g. Unity) on a PC and subsequently streamed to the VR device, enabling real-time adjustments in the virtual world, such as simulating the breaking of an ice surface when negative stiffness is triggered.

Figure 14.3c demonstrates the relative resistance variation ($\Delta R/R_0$) of the sensing layer during cyclic and active pressing on the curved origami, showcasing consistent repeatability and negligible influence from the folding angle $\Delta\beta$. This consistency is crucial for employing a universal algorithm (i.e. $\Delta\beta$ independent algorithm) for virtual environment modifications. Hence, active human interactions lead to changes in the virtual environment, correlating with the measured deformations through electrical resistance changes.

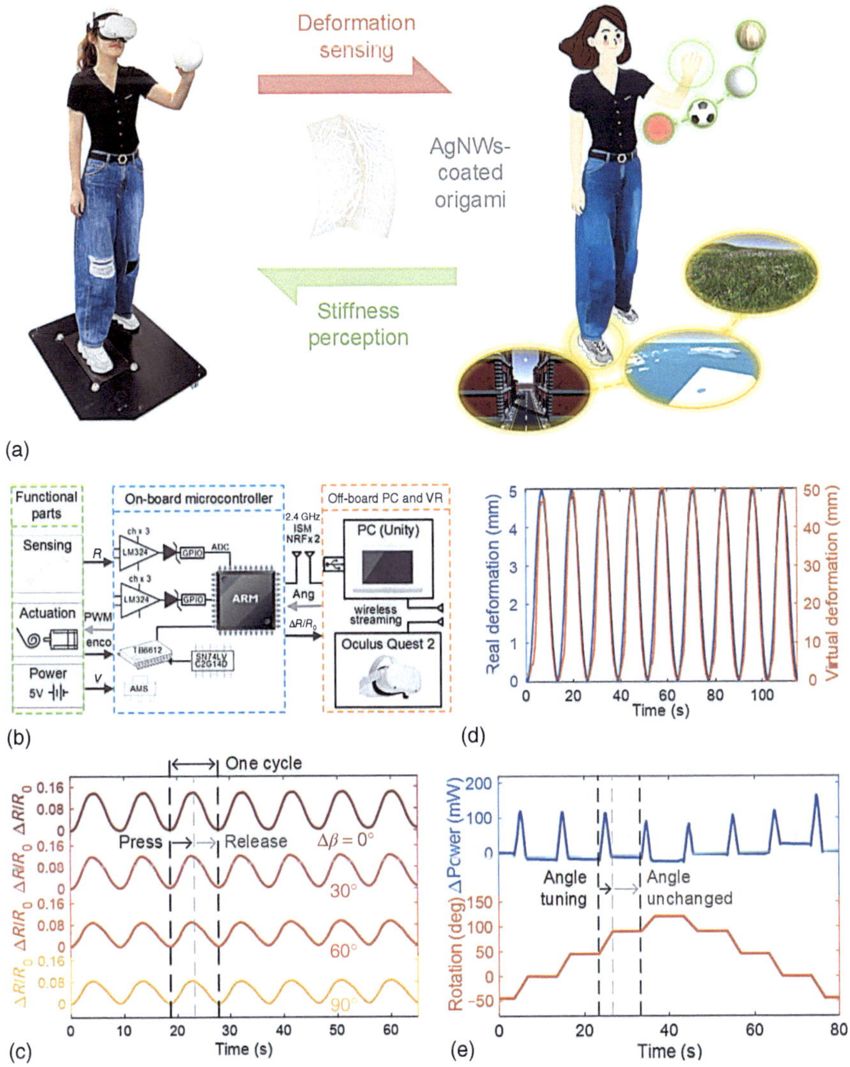

Figure 14.3 Closed-loop haptics based on curved origami. (a) Schematics of the AgNWs-coated curved origami as an interface between physical and virtual environments, (b) system integration and circuit diagram for the haptic devices constructed with curved origami, (c) resistance of the coated AgNWs sensing layers upon cyclic compressions at different folding angles of the curved origami with $\alpha = 80°$ and $\beta = 120°$, (d) deformation response of virtual objects triggered by actual deformation of the curved origami module upon active cyclic compression, demonstrating a highly synchronized deformation between the physical and virtual environments bridged by the curved origami, (e) power consumption of the actuation system during the haptic process.

Figure 14.3d confirms the device's stable synchronization between the virtual and real worlds through its wireless transmission system. It is important to note that the motor-driven cable integrated into the system is activated only when necessary, specifically when alterations in the curved origami's folding angle are required according to the virtual scenario. This design ensures that various stiffness perceptions are provided to the user without the need for complex control algorithms or additional actuation, resulting in an energy-efficient system for human-initiated active haptics, with a power dissipation of about 150 mW during stiffness tuning, as shown in Figure 14.3e.

14.5 In-hand Haptic Device

In the first demonstration, we introduce a haptic device designed to be held in hand, delivering active mechanical haptic feedback that spans a range of stiffness sensations, from positive to negative. The device is intricately designed, comprising four main subsystems (illustrated in Figures 14.4a and 14.5). The curved origami module serves as a bridge between the virtual environment and the user's tactile perceptions, ensuring a seamless integration of visual and tactile experiences (as depicted in Figure 14.4b). This integration allows for a variety of stiffness sensations to be provided in sync with the virtual objects viewed, with the user actively controlling the variations through button presses.

Four pivotal design principles guide the creation of the in-hand device: (i) User-initiated pressing should directly result in the curved origami's deformation; (ii) All five fingers must be able to perceive stiffness, mirroring everyday gripping actions; (iii) The device should be ergonomic, boasting a shape and weight that are easy to handle; and (iv) A wireless design is essential to ensure users are not hindered by cables. In light of these principles, we have crafted a spherical in-hand device featuring five origami buttons that are controlled in unison. This device encompasses four integral subsystems (Figures 14.4a and 14.5): (i) The five curved origami buttons, which facilitate the generation of feedback between the virtual and physical realms upon user interaction; (ii) A sophisticated actuation system capable of adjusting the curved origami's stiffness by converting motor rotations into cable movements; (iii) A suite of electronic control components responsible for managing the feedback loop, actuation, sensing, and ensuring wireless communication; and (iv) A supportive shell, which houses the buttons and is meticulously designed to align with finger positioning, electronic components, and ergonomic considerations for a natural gripping experience. The five origami buttons are strategically placed on the spherical shell, aligning with common hand gestures. The actuation system connects to their fixed bottom plates, while the moving top plates are guided by a slide, facilitating user interaction. Due to spatial constraints within the origami buttons, the cable is not inserted directly into the origami structures. Instead, we employ relative rotations between the origami buttons' fixed and moving plates to adjust folding angles and achieve stiffness tuning (Figure 14.6). This is made possible through a single input-multiple output (SI-MO) actuation strategy based on

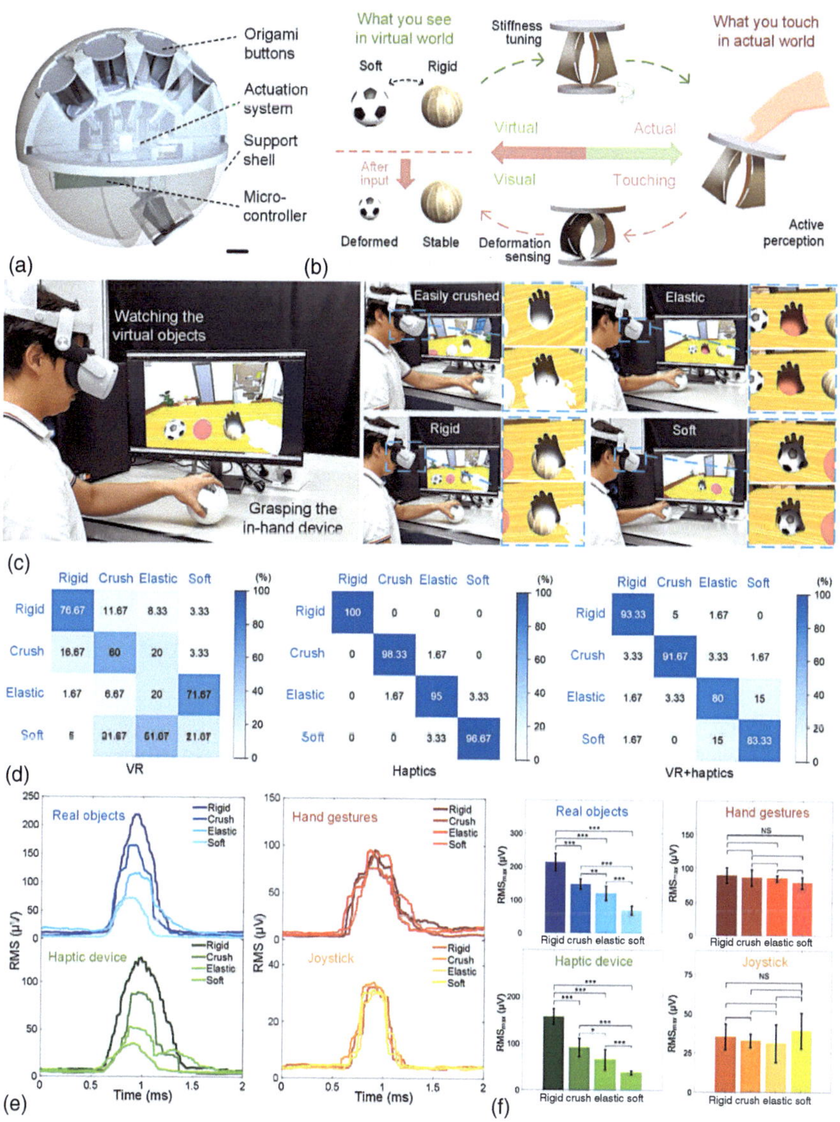

Figure 14.1 Haptic in-hand device and evaluation of its sensory perceptions. (a) Design of the in-hand haptic device with curved origami-based buttons. Scale bar, 10 mm, (b) schematics of the curved origami button as an interface to realize the closed-loop haptics, (c) a user perceives different stiffness of various objects in the virtual environment using the in-hand device and seeing their corresponding deformation, achieving *"what a user sees is what the user feels,"* (d) confusion matrices showing identification rates of 12 participants under three different interaction methods, (e) physiological signals (EMG) of the upper limb when the users grasp different real objects and virtual ones under three in-hand conditions, including the present haptic device, a joystick, and sole hand gestures, (f) evaluation of the EMG differences among the active grasps of different objects under four conditions. Asterisks denote significant effects at $^{*}P < 0.05$, $^{**}P < 0.01$, and $^{***}P < 0.001$.

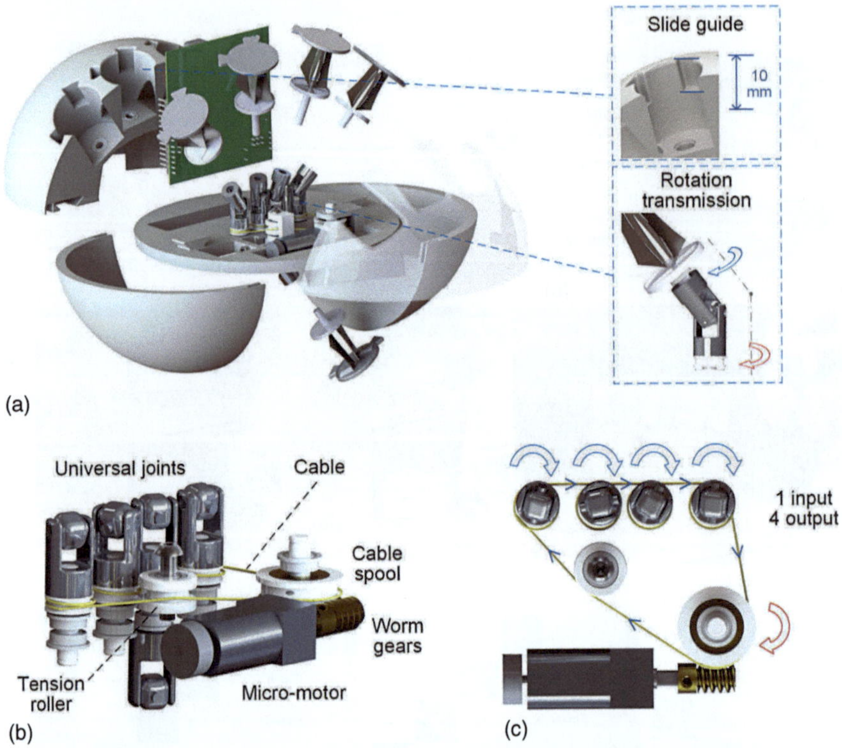

Figure 14.5 Mechanical structure of the in-hand haptic device. (a) Exploded-view schematic illustration of the spherical in-hand device with five synchronously controlled origami buttons. The slide guide is utilized to confine the compression range (10 mm) of the origami button. The universal joints are used for transmitting the rotations about the vertical axis between the top and bottom plates of curved origami modules, (b) schematic of the actuation system. An onboard micro-motor actuates the cable through worm gears. A tension roller is used for the pretension of the cable to avoid the slide between the cable and the rollers, (c) synchronous actuation of the origami buttons. An single input-multiple output (SI-MO) actuation strategy is constructed based on cable routing. The red arrow denotes the driving roller, and the blue ones denote follow-up rollers.

precise cable routing (Figure 14.5b,c). As the vertical orientations of the five buttons vary, we designed four rollers with axes parallel to the cable spool for planar cable routing, connecting them to the origami buttons via universal joints. To optimize space, the thumb and middle finger buttons share the same roller. This design ensures that a single micro-motor can simultaneously control the rotations of all five origami buttons' bottom plates, resulting in a device that is both compact and lightweight. Worm gears play a crucial role in this system, transmitting rotations from the micro-motor to the cable spool. Their self-locking nature ensures the device can withstand torque from the relative rotations between the origami's top and bottom plates, maintaining the desired stiffness levels and enhancing the system's energy efficiency. Beyond these mechanical elements, the device integrates electronic control components, including an onboard microcontroller (detailed in Figure 14.3b) and a compact 5 V battery, all housed within the supportive shell to achieve wireless operation.

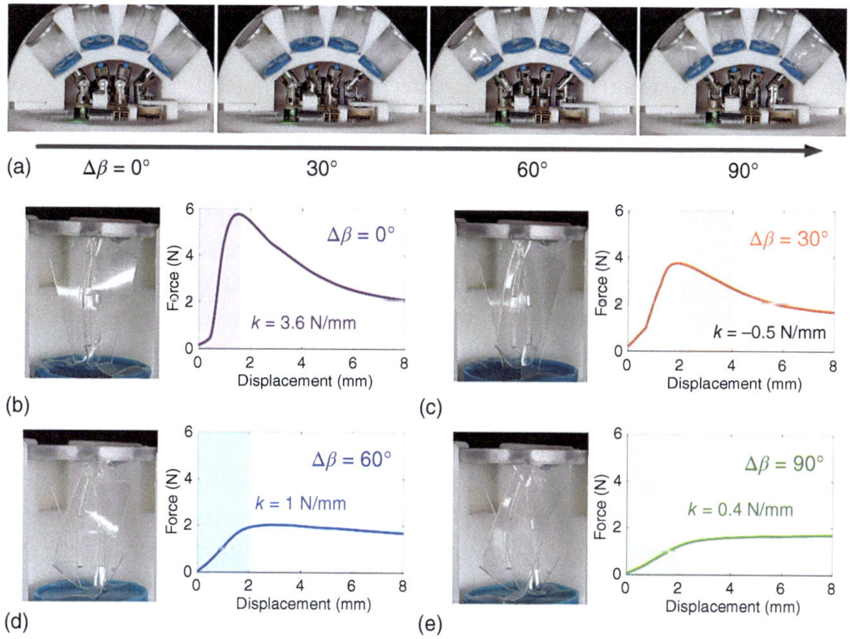

Figure 14.6 Stiffness manipulation of the in-hand device. (a) Synchronous control of the folding angles of the curved origami buttons, (b–e) optical images and the corresponding force–displacement relationships of a curved origami button with folding angle $\beta = 120°$ under different control angles $\Delta\beta = 0°, 30°, 60°$, and $90°$.

The device's supporting shells and cable rollers are crafted using an SLA 3D printer (Lite 800, Union Tech Inc.), employing photosensitive resin (Somos Imagine 8000 and Somos WaterClear Ultra 10122, DSM Desotech Inc.). The origami buttons' fixed and moving plates are 3D printed from polylactic acid (eSilk-PLA, eSun Inc.) using an FDM 3D printer (Pro3 plus, Raise3D Inc.). The device utilizes a Kevlar cable with a 0.8 mm diameter and incorporates standard metal parts for the universal joints, bearings, and worm gears. Currently, the in-hand device weighs approximately 600 g, with the 3D-printed support shell accounting for around 70% of this total.

14.6 Stiffness Perception via Active Pressing

In practical scenarios demonstrated in Figure 14.4c, users can grasp virtual objects of varying stiffness – from elastic to easily breakable, from soft to stiff – with the specific stiffness values of these objects detailed in Figure 14.6. The haptic device then provides the corresponding tactile feedback, allowing users to physically feel the stiffness of the objects they are interacting with in the virtual world. This bidirectional connection between the user and the virtual objects, facilitated by the curved origami module, not only makes the virtual objects tangible but also allows them to be manipulated in the virtual space through the user's active interactions, akin to handling real objects in the physical world. This results in a significantly enhanced immersive experience.

To quantitatively evaluate the haptic perceptions facilitated by our device, we conducted a study with 12 participants. They were tasked with identifying four spherical objects, each with distinct mechanical properties – labeled “rigid,” “elastic,” “soft,” and “crush” for varying stiffness levels – using three different methods: visual information from a VR headset, tactile feedback from our haptic device, and a combination of both. The results, detailed in the confusion matrices in Figure 14.4d, reveal a stark contrast in identification rates across the different methods. Sole reliance on visual information led to the lowest identification rates, especially for objects with similar stiffness levels. In contrast, the haptic device alone resulted in near-perfect identification rates, showcasing its precision in conveying stiffness information. Interestingly, when both visual and tactile feedback were combined, the identification rates slightly decreased, suggesting a potential dominance of tactile perception over visual cues in stiffness discernment.

To further objectify the evaluation of haptic perceptions, we recorded physiological signals electromyogram (EMG) from participants as they interacted with objects of varying stiffness using different methods, including our haptic device. This comprehensive evaluation, detailed in Figure 14.4e,f, underscores the effectiveness of our device in delivering distinguishable stiffness sensations, closely mirroring the responses elicited by real objects, and highlights the limitations of current interaction tools like joysticks and hand gestures in providing authentic haptic feedback.

14.7 Body-Centered Stepping Device

In the second demonstration, we introduce a stepping device that is centered around the lower limbs, incorporating larger-scale curved origami structures to facilitate whole-body movement and deliver corresponding stiffness sensations. This innovation takes the immersive, active mechanical haptic experience beyond the confines of hand-centric interactions. The device comprises four primary subsystems, as depicted in Figure 14.7a. The stepping device operates through a two-step transmission mechanism involving parallel worm gears and multi-knotted cables. This setup enables simultaneous adjustments to the stiffness of the curved origami tessellation, creating a variety of closed-loop haptic sensations. The user is thus immersed in an environment where the ground they step on in the virtual world is palpably represented by the stepping device (Figure 14.7b). Demonstrations of this immersive experience are provided in Figure 14.7c,d, where users navigate different virtual terrains (e.g. a solid avenue, a plush lawn, and a fragile ice surface with specific stiffness values) and receive tactile feedback corresponding to these environments through the stepping device.

For the stepping device, we have identified four crucial design considerations: (i) the device must facilitate direct deformation from the curved origami when users actively step on it; (ii) it should enable stiffness perception through body-level movements, akin to typical walking experiences; (iii) it needs to be robust enough to

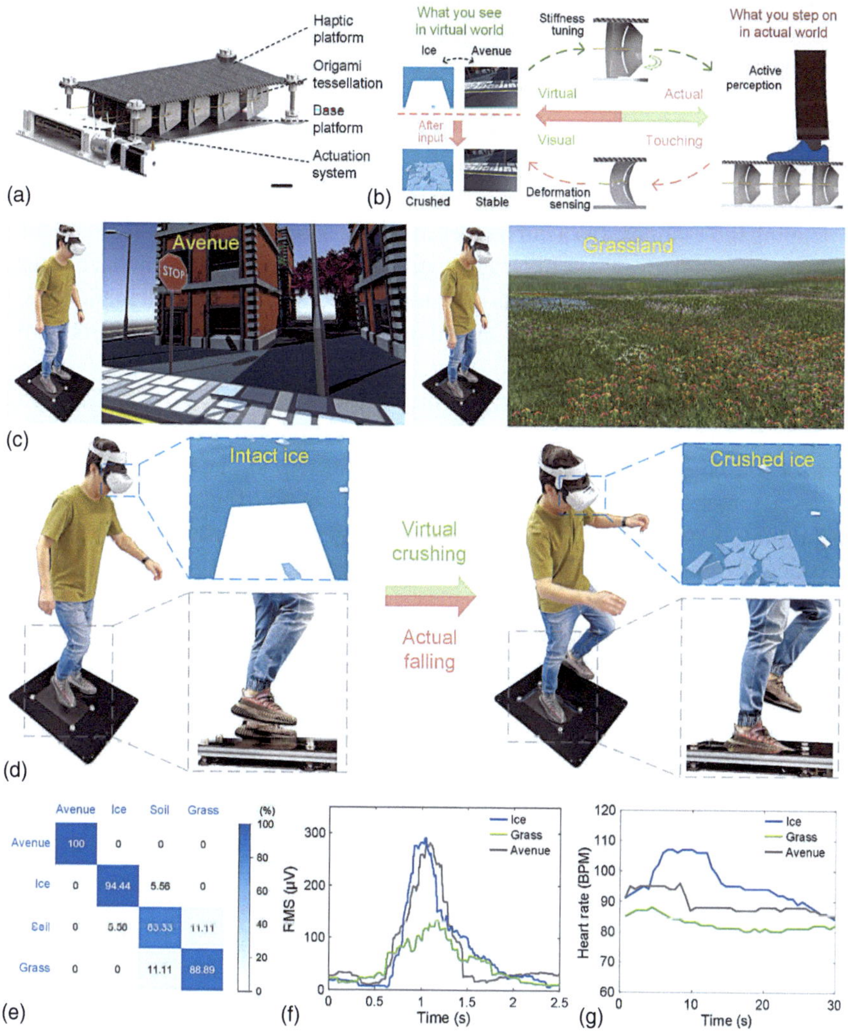

Figure 14.7 Haptic stepping device and evaluation of its sensory perceptions. (a) Design of the stepping device with curved origami tessellations. Scale bar, 50 mm, (b) schematics of the curved origami as an interface to realize the closed-loop haptics, (c) a user perceives different stiffness of various grounds in the virtual environment using the stepping device and seeing their corresponding deformation, achieving "*what a user is immersed in is what the user steps on,*" (d) A user capable of perceiving the sensation of falling using the present stepping device through hard stepping, synchronized by seeing a visual crushing of the ice surface in the virtual environment, (e) confusion matrix showing identification rates of 12 participants, (f) physiological signals (EMG) of the lower limb when the user steps on different virtual surfaces, (g) heart rates reflect psychological reactions when the user steps on different virtual surfaces.

support an adult's weight in a high stiffness state, yet flexible enough to deform easily in a low stiffness state; and (iv) the device's moving platform should be restricted to a single degree of freedom (vertical movement) to maintain haptic safety.

In adherence to these guidelines, we have conceptualized a stepping device featuring a flat, vertically deformable haptic platform, and a curved origami tessellation that is controlled synchronously (illustrated in Figure 14.7a and detailed in Figure 14.8a). The device encompasses four main subsystems: (i) a moving platform for stepping interactions; (ii) a curved origami tessellation that supports the user's weight and generates feedback upon active stepping; (iii) a base platform that houses the origami tessellation and supports the entire structure; and (iv) an actuation system that adjusts the stiffness of the curved origami by transmitting motor rotation to the origami folds via cables.

To guarantee that the haptic platform moves exclusively in a vertical direction during user engagement, we have affixed four cylindrical guides to the base and attached corresponding linear bearings to the haptic platform. The device features a dual-stage transmission system consisting of multi-head worm gears for transmitting rotation from the motor to four gears, which subsequently pull four independent cables. These cables drive the simultaneous folding of five curved origami modules through five knots located behind the moving panels (as illustrated in Figure 14.8b,c). This innovative design enables the synchronous manipulation of a 20-module origami tessellation arranged in a 4×5 matrix with just a single

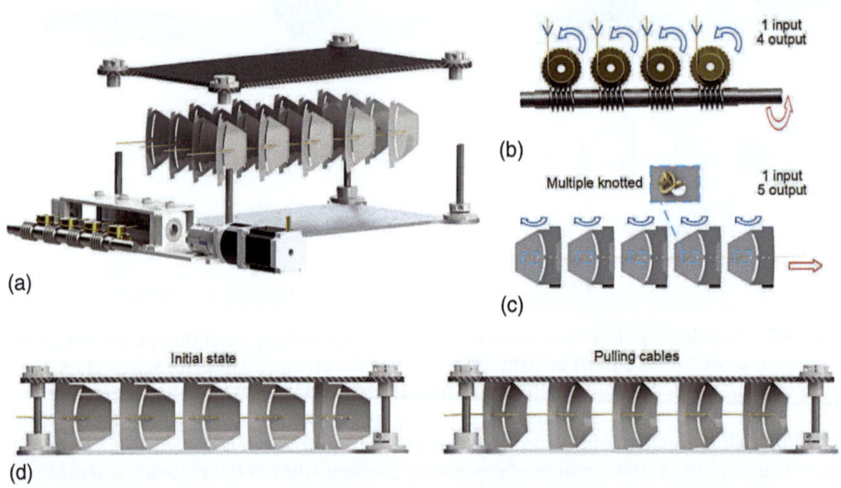

Figure 14.8 Mechanical structure of the stepping haptic device. (a) Exploded-view schematic illustration of the stepping device with synchronously controlled curved origami tessellation, (b) schematic of the multi-head worm gears transmission. The cable is actuated by a motor through worm gears. Four gears are synchronously controlled by one four-head worm, making four cables synchronously pulled/released by the motor, (c) schematic of the multi-knotted cable-driven transmission. Five knots are evenly made on each cable, rending the folding of five origami modules synchronously controlled by one cable. The red arrow denotes the input, and the blue ones denote the follow-ups, (d) synchronous actuation of the origami tessellation.

motor (Figure 14.8d), resulting in a streamlined, energy-efficient actuation system, reminiscent of the in-hand device.

The haptic platform is constructed from a 5-mm-thick carbon fiber plate, precisely shaped using an industrial engraving machine. The base platform, fabricated from aluminum alloy, is produced through computerized numerical control (CNC) machining. The Kevlar pulling cable, with a diameter of 0.8 mm, ensures a balance of strength and flexibility. Standard metal parts are utilized for the coupling, bearings, and cylindrical guides, while additional structural components are produced using polylactic acid (PLA+, eSun Inc.) with an FDM 3D printer (Pro3 plus, Raise3D Inc.).

14.8 Whole-Body Stiffness Perceptions

This bidirectional link between the user and the virtual environment ensures that the virtual world is not only perceptible to the user but also responsive to their interactions, such as active stepping, via a feedback mechanism. Among the various body-centric sensations facilitated by the device, experiences of falling or breaking objects (e.g. shattering an ice surface) are made possible through the implementation of negative stiffness. Such events are likely to leave a lasting impression on the user, as they often come unexpectedly, mirroring real-world experiences.

To assess the body-centric haptic perceptions provided by the device, a study was conducted with 12 participants, who were tasked with identifying four types of virtual grounds, each characterized by distinct mechanical properties. These included an "avenue" with high stiffness, "soil" with moderately low stiffness, "grass" with very low stiffness, and "ice" with a negative stiffness designed to simulate a crushing sensation (Figures 14.9 and 14.10). The results, presented in a confusion matrix (Figure 14.7e), reveal identification rates exceeding 83% for all scenarios, indicating the participant's ability to distinguish between the different haptic experiences offered by the body-centric stepping device.

To delve deeper into these sensations, we recorded EMG signals from the participants' legs as they navigated the virtual environment using a VR headset, all while stepping on the haptic device that offered synchronized stiffness feedback in line with the virtual terrain (depicted in Figure 14.7f). The EMG signals' RMS values revealed a distinct pattern across various scenarios: lower stiffness led to reduced EMG voltage, while crushing the ice surface resulted in a peak voltage comparable to stepping on a solid avenue, followed by a swift decline and a gradual decrease in voltage reduction. This intricate pattern reflects the muscles' dual response to both the exerted force and the traveled distance, showcasing the device's capability to faithfully recreate real-world sensations, especially the unique feeling of negative stiffness.

In addition to physiological responses, the stepping device also triggers psychological reactions, as illustrated in Figure 14.7g. Here, a user's heart rate is monitored during initial interactions with the device across three different stiffness scenarios.

Figure 14.9 Snapshots of the immersive whole-body haptic experiences on the stepping device with various virtual scenarios. (a) The user experienced low-value positive stiffness in the scenario of grassland, with his feet readily sinking into the virtual ground, (b) the user experienced negative stiffness in the scenario of an icy surface, with his whole body keeping still upon a light active stepping while experiencing a real falling upon a hard active stepping, (c) the user experienced high-value positive stiffness in the scenario of a rigid avenue, with his whole body keeping still upon active stepping.

While the heart rate remains stable when stepping on both rigid and soft surfaces, it spikes (~15 BPM) at the moment of crushing the ice surface, demonstrating the device's potential to elicit strong psychological reactions. This state of heightened alertness or fear takes a few seconds to dissipate, underscoring the device's efficacy in providing an immersive and emotionally resonant experience. In essence, this body-centric stepping haptic device not only enriches the immersive aspect of virtual environments but also allows users to experience a broad spectrum of sensory, physiological, and psychological responses commonly encountered in the real world but rarely replicated in current VR/AR and existing haptic devices.

14.9 Discussion

The innovative mechanical haptic technology introduced in this chapter leverages the curved origami structures, serving as a bridge between digital and tangible realms and bringing a user-centric, dynamic tactile interaction experience. This spectrum encompasses a wide array of stiffness sensations, extending from positive to negative. This paradigm shift from conventional machine-triggered, passive haptic systems to user-initiated interfaces offers a more authentic mimicry of real-world

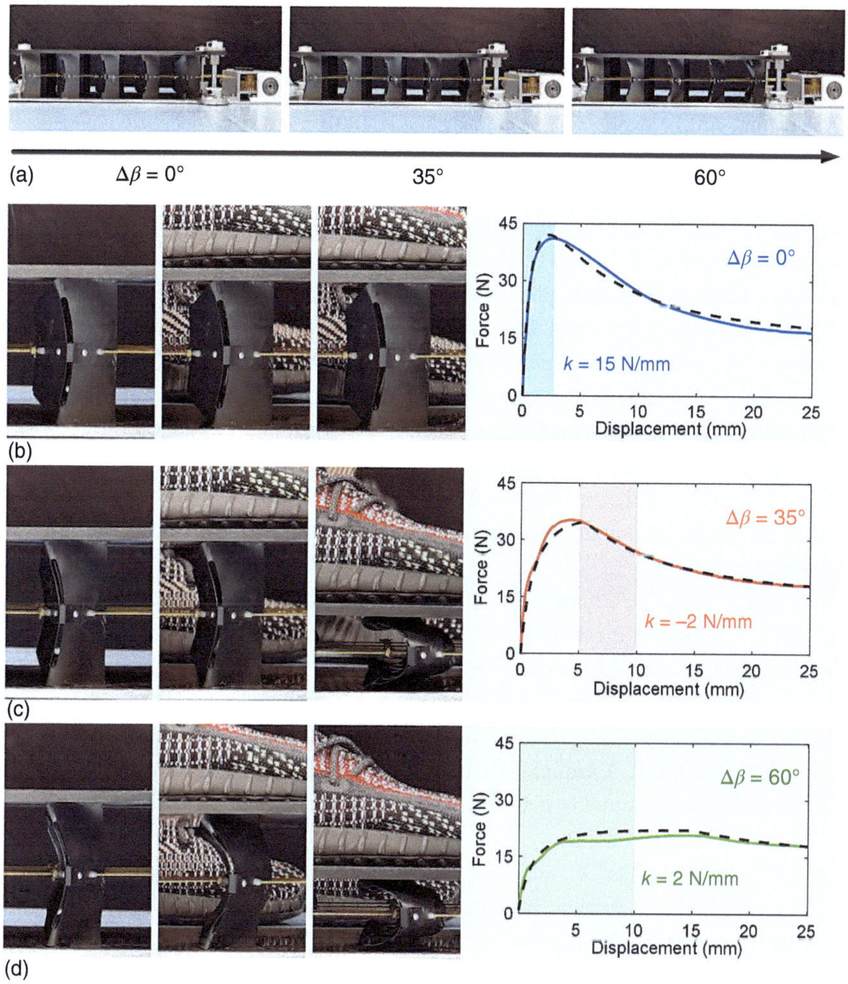

Figure 14.10 Stiffness manipulation of the stepping device. (a) Synchronous control of the folding angles of the curved origami tessellation, (b–d) Optical images of the stepping processes and the corresponding force–displacement relationships of a curved origami module inside the stepping device with folding angle $\beta = 120°$ under different control angles $\Delta\beta = 0°$, 35°, and 60°. Solid lines are measured results and dashed lines denote theoretical ones.

interactions. The inclusion of negative stiffness, along with various levels of positive stiffness, significantly amplifies the immersive quality of virtual environments. We provide an exhaustive exploration of the system's principles and characteristics, ensuring its replicability and seamless integration into various applications. Two prototype devices – a handheld device and a stepping device – have undergone rigorous testing, validating their ability to deliver a rich, first-person sensory experience within immersive virtual settings and highlighting the technology's potential for widespread adoption.

References

1 Sparkes, M. (2021). What is a metaverse. *New Scientist* 251 (3348): 18.

2 Ackerman, J.M., Nocera, C.C., and Bargh, J.A. (2010). Incidental haptic sensations influence social judgments and decisions. *Science* 328 (5986): 1712–1715.

3 Yu, X., Xie, Z., Yu, Y. et al. (2019). Skin-integrated wireless haptic interfaces for virtual and augmented reality. *Nature* 575 (7783): 473–479.

4 Biswas, S. and Visell, Y. (2021). Haptic perception, mechanics, and material technologies for virtual reality. *Advanced Functional Materials* 31 (39): 2008186.

5 Pyun, K.R., Rogers, J.A., and Ko, S.H. (2022). Materials and devices for immersive virtual reality. *Nature Reviews Materials* 7 (11): 841–843.

6 Li, S., Bai, H., Shepherd, R.F., and Zhao, H. (2019). Bio-inspired design and additive manufacturing of soft materials, machines, robots, and haptic interfaces. *Angewandte Chemie International Edition* 58 (33): 11182–11204.

7 Jung, Y.H., Kim, J.-H., and Rogers, J.A. (2021). Skin-integrated vibrohaptic interfaces for virtual and augmented reality. *Advanced Functional Materials* 31 (39): 2008805.

8 Zhu, M.L., Sun, Z.D., Zhang, Z.X. et al. (2020). Haptic-feedback smart glove as a creative human-machine interface (HMI) for virtual/augmented reality applications. *Science Advances* 6 (19): eaaz8693.

9 Liu, Y., Yiu, C., Song, Z. et al. (2022). Electronic skin as wireless human-machine interfaces for robotic VR. *Science Advances* 8 (2): eabl6700.

10 Jung, Y.H., Yoo, J.-Y., Vázquez-Guardado, A. et al. (2022). A wireless haptic interface for programmable patterns of touch across large areas of the skin. *Nature Electronics* 5: 1–12.

11 Sun, Z., Zhu, M., Shan, X., and Lee, C. (2022). Augmented tactile-perception and haptic-feedback rings as human-machine interfaces aiming for immersive interactions. *Nature Communications* 13 (1): 5224.

12 Kim, D., Kim, B., Shin, B. et al. (2022). Actuating compact wearable augmented reality devices by multifunctional artificial muscle. *Nature Communications* 13 (1): 4155.

13 Choi, I., Ofek, E., Benko, H. et al. (2018) CLAW: a multifunctional handheld haptic controller for grasping, touching, and triggering in virtual reality. *Proceedings of the 2018 CHI Conference on Human Factors in Computing Systems*, Montreal, QC, Canada. Montreal QC, Canada (21–26 April 2018): Association for Computing Machinery; Paper 654.

14 Scheggi, S., Meli, L., Pacchierotti, C., and Prattichizzo, D. (2015). Touch the virtual reality: using the leap motion controller for hand tracking and wearable tactile devices for immersive haptic rendering. *ACM SIGGRAPH 2015 Posters*, Los Angeles, California, Association for Computing Machinery: Los Angeles, California (09-13 August 2015); Article 31.

15 Giraud, F.H., Joshi, S., and Paik, J. (2022). Haptigami: a fingertip haptic interface with vibrotactile and 3-DoF cutaneous force feedback. *IEEE Transactions on Haptics* 15 (1): 131–141.

16 Pezent, E., Agarwal, P., Hartcher-O'Brien, J. et al. (2022). Design, control, and psychophysics of tasbi: a force-controlled multimodal haptic bracelet. *IEEE Transactions on Robotics* 38 (5): 2962–2978.

17 Yao, K.M., Zhou, J.K., Huang, Q.Y. et al. (2022). Encoding of tactile information in hand via skin-integrated wireless haptic interface. *Nature Machine Intelligence* 4 (10): 893–903.

18 Leroy, E., Hinchet, R., and Shea, H. (2020). Multimode hydraulically amplified electrostatic actuators for wearable haptics. *Advanced Materials* 32 (36): 2002564.

19 Chinello, F., Pacchierotti, C., Malvezzi, M., and Prattichizzo, D. (2018). A three revolute-revolute-spherical wearable fingertip cutaneous device for stiffness rendering. *IEEE Transactions on Haptics* 11 (1): 39–50.

20 Xu, H., Peshkin, M.A., and Colgate, J.E. (2019). UltraShiver: lateral force feedback on a bare fingertip via ultrasonic oscillation and electroadhesion. *IEEE Transactions on Haptics* 12 (4): 497–507.

21 Steed, A., Ofek, E., Sinclair, M., and Gonzalez-Franco, M. (2021). A mechatronic shape display based on auxetic materials. *Nature Communications* 12 (1): 4758.

22 Peng, Y., Serfass, C.M., Kawazoe, A. et al. (2021). Elastohydrodynamic friction of robotic and human fingers on soft micropatterned substrates. *Nature Materials* 20 (12): 1707–1711.

23 Choi, C., Ma, Y., Li, X. et al. (2022). Surface haptic rendering of virtual shapes through change in surface temperature. *Science Robotics* 7 (63): eabl4543.

24 Li, X., Ma, Y., Choi, C. et al. (2021). Nanotexture shape and surface energy impact on electroadhesive human–machine interface performance. *Advanced Materials* 33 (31): 2008337.

25 Mintchev, S., Salerno, M., Cherpillod, A. et al. (2019). A portable three-degrees-of-freedom force feedback origami robot for human–robot interactions. *Nature Machine Intelligence* 1 (12): 584–593.

26 Pacchierotti, C., Sinclair, S., Solazzi, M. et al. (2017). Wearable haptic systems for the fingertip and the hand: taxonomy, review, and perspectives. *IEEE Transactions on Haptics* 10 (4): 580–600.

27 Xiong, Q., Liang, X., Wei, D. et al. (2022). So-EAGlove: VR haptic glove rendering softness sensation with force-tunable electrostatic adhesive brakes. *IEEE Transactions on Robotics* 1–13.

28 Giri, G.S., Maddahi, Y., and Zareinia, K. (2021). An application-based review of haptics technology. *Robotics* 10 (1): 29.

29 Choi, I., Hawkes, E.W., Christensen, D.L. (2016). Wolverine: a wearable haptic interface for grasping in virtual reality. *2016 IEEE/RSJ International Conference on Intelligent Robots and Systems (IROS 2016)* (9–14 October 2016), 986–993.

30 Martinez, M. O., Campion, J., Gholami, T., Rittikaidachar, M. K., Barron, A. C., Okamura, A. M. 2017. Open source, modular, customizable, 3-D printed kinesthetic haptic devices. *2017 IEEE World Haptics Conference (WHC), 6–9 June 2017*; pp 142–147.

31 Jadhav, S., Majit, M.R.A., Shih, B. et al. (2022). Variable stiffness devices using fiber jamming for application in soft robotics and wearable haptics. *Soft Robotics* 9 (1): 173–186.

32 Ernst, M.O. and Banks, M.S. (2002). Humans integrate visual and haptic information in a statistically optimal fashion. *Nature* 415 (6870): 429–433.

33 Zhang, Z., Xu, Z., Emu, L. et al. (2023). Active mechanical haptics with high-fidelity perceptions for immersive virtual reality. *Nature Machine Intelligence* 5: 643–655.

15

Mechanics of Transient Electronics

Ankan Dutta[1,2] and Huanyu Cheng[1]

[1] *The Pennsylvania State University, Department of Engineering Science and Mechanics, Materials Research Institute, 212 Earth Engineering Sciences Building, University Park, PA 16802, USA*
[2] *The Pennsylvania State University, Center for Neural Engineering, University Park, PA 16802, USA*

15.1 Introduction

Although modern electronics are designed to be durable and function consistently over a long period of time, the introduction of transient electronics takes a different approach. These devices are meant to be destroyed intentionally in order to create unique opportunities. In addition, transient electronics dissolve at a predictable rate when exposed to water, releasing environmentally safe byproducts and making them suitable for various applications ranging from biodegradable electronics to therapeutic implants. For example, an electronic component can be temporarily implanted in a patient and then safely dissolve, eliminating the need for a second surgery. Ultimately, transient electronics can address the issue of safely disposing of electronics conveniently.

Transient electronics are characterized by their ability to dissolve into nontoxic substances when exposed to water. Consequently, dissolution is a significant area of study in this field, as evident from extensive research [1–11]. Early investigations focused on the partial dissolution of components by utilizing organic materials as substrates [12, 13]. For instance, organic thin-film transistors were created using paper made from cotton as a substrate [14]. Silk was also demonstrated to be a useful soluble substrate for implants within the body [15, 16]. However, the research was limited to the substrate, and the electronic devices remained insoluble.

Recently, research has resulted in the development of electronics that can completely dissolve. This is due to a significant discovery that semiconductor-grade monocrystalline silicon can dissolve in biofluids or water at physiological conditions with a predetermined lifespan relevant to biomedical applications [2]. As the rate of silicon hydrolysis to form silicic acid ($Si(OH)_4$) is extremely slow, thin forms of silicon devices were fabricated. A nanomembrane composed of silicon, with lateral dimensions similar to conventional circuits but with a thickness of only 70 nm, has been shown to dissolve in approximately 10 days [2]. Similarly, using similar chemistry, thin silicon dioxide (SiO_2) was chosen as a gate dielectric.

Mechanics of Flexible and Stretchable Electronics, First Edition. Edited by Yong Zhu and Nanshu Lu.

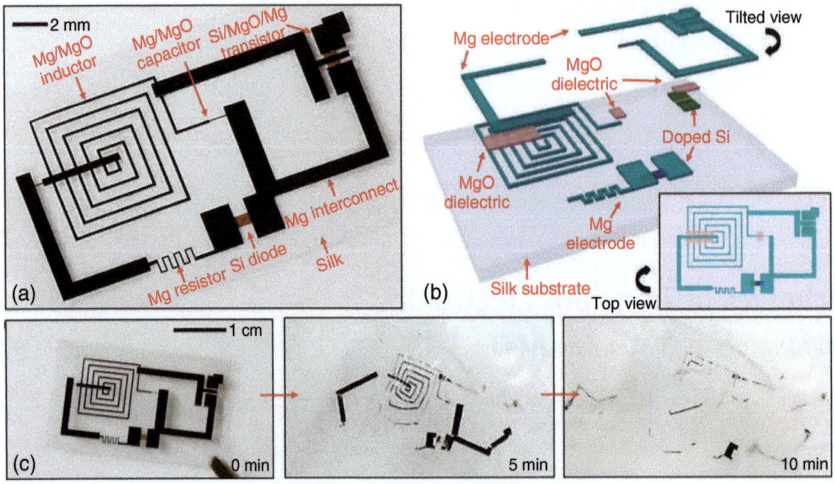

Figure 15.1 Proof-of-concept illustration for transient electronics, showcasing the primary materials and layout of the device structure. (a) The image depicting the device is on a thin silk substrate and includes multiple components such as transistors, diodes, inductors, capacitors, resistors, interconnects, and interlayer dielectrics. (b) The schematic illustration of the device is in an exploded view with a top view in the lower right inset. (c) Images demonstrate the time sequence at various dissolution stages in deionized water. Source: Ref. [2]. Reproduced with permission of American Association for the Advancement of Science - AAAS.

Combined with other dissolvable inorganic materials such as magnesium (Mg) and magnesium oxide (MgO) for conductors and the interlayer dielectric, silicon nanomembranes provide a fundamental way to construct transient electronic devices. Magnesium and magnesium oxide react spontaneously with water to produce biologically benign $Mg(OH)_2$. To demonstrate this concept, Figure 15.1a,b shows a schematic of a demonstration platform that uses silicon nanomembranes (Si NMs) for semiconductors, magnesium for conductors, magnesium oxide and silicon dioxide for dielectrics, and silk for substrate and packaging materials. These collectively configured devices dissolve and disintegrate when immersed in deionized (DI) water, as shown in Figure 15.1c.

The dissolution behavior of materials is dominated by surface reactions for reaction constants that are sufficiently large. However, the porosity of materials such as Mg, MgO, and SiO_2 has been found to be influential as it facilitates the diffusion of water through the material, which increases the dissolution rate by increasing the effective surface area. In the study of the dissolution of transient electronics, it is essential to consider the physical and chemical properties of materials and the ambient factors of an aqueous environment. Analytical models have been developed to better understand dissolution behaviors in transient electronics based on research into these factors and others. Such models can help design transient electronics. This review will discuss the hydrolysis of semiconducting materials, focusing on silicon nanomembranes, and then present the reactive diffusion model to account for the dissolution behavior of porous materials. The class of transient electronics, when combined with ideas from soft, tissue-like electronic devices, provides a promising means for minimally obtrusive health monitoring or care delivery.

15.2 Hydrolysis of Semiconducting Materials

Understanding the chemical kinetics of each functional material is crucial for establishing a practical set of materials, especially regarding the hydrolysis of semiconducting materials. The dissolution rates of semiconducting materials such as silicon, silicon-germanium, and germanium are considerably low at physiological pH levels and temperatures. The nanomembrane structure plays a vital role in minimizing the amount of semiconducting materials that need to be dissolved. For example, when monocrystalline silicon nanomembranes dissolve in phosphate-buffered saline (PBS) at biologically relevant temperatures, an intermediate oxidation product, SiO_2 or silicic acid $Si(OH)_4$, is formed through the equilibrium reaction: $Si + 4H_2O \rightleftharpoons Si(OH)_4 + 2H_2$. The dissolution rate depends on various factors such as the crystal structure, morphology, doping concentration of silicon, and the temperature and composition of solutions [1, 17–22].

The dissolution kinetics of silicon in various biofluids at different pH levels and temperatures were systematically characterized. To create patterned Si NMs of dimension $3\,\mu m \times 3\,\mu m \times 70\,nm$, a thermal oxide layer was first applied to a silicon wafer, followed by immersing it in 50 ml of aqueous buffer solutions. The thermal oxide dissolution rate was found to be negligible in comparison to that of silicon. The thicknesses of Si NMs were measured at specific time intervals, and then the sample was placed in a fresh buffer solution. The insignificant change in the dissolution rate for several days indicates that the dissolution rate was accurately measured. Similar setups were utilized to study SiGe and Ge [23, 24]. Atomic force microscope topographical images of a Si NM with an initial thickness of 70 nm collected at various stages of hydrolysis in biofluids (PBS with a pH of 7.4; 37 °C) are presented in Figure 15.2a, demonstrating its transient behavior.

The hydrolysis of monocrystalline silicon nanomembranes is constrained to the surface due to the compact arrangement of silicon atoms in crystal structures. The kinetics can be explained by a surface reaction with a constant dissolution rate, and this model implies that there is no diffusion of solution molecules into the silicon. Thus, a linearity between the measured thickness and time is observed, where the dissolution rates are represented by the slope [1, 2]. According to a previous empirical formula [19], an increase in the concentration of hydroxide ions $[OH^-]$ accelerates the dissolution rate in high pH solutions of KOH. Figure 15.2b shows that this formula can replicate the experimental trends for solutions at physiological pH levels if a different power law exponent is employed for $[OH^-]$ [1].

The dissolution rate of silicon nanomembranes is affected by the pH level and the type and concentration of ions present. Experiments have shown that the dissolution rate in bovine serum, which contains 0.12 M sodium chlorides and has a pH of 7.4 at 37 °C, is 20 times higher than that in PBS solution with a similar pH level and containing 0.14 M chlorides and 0.01 M phosphates. To investigate the

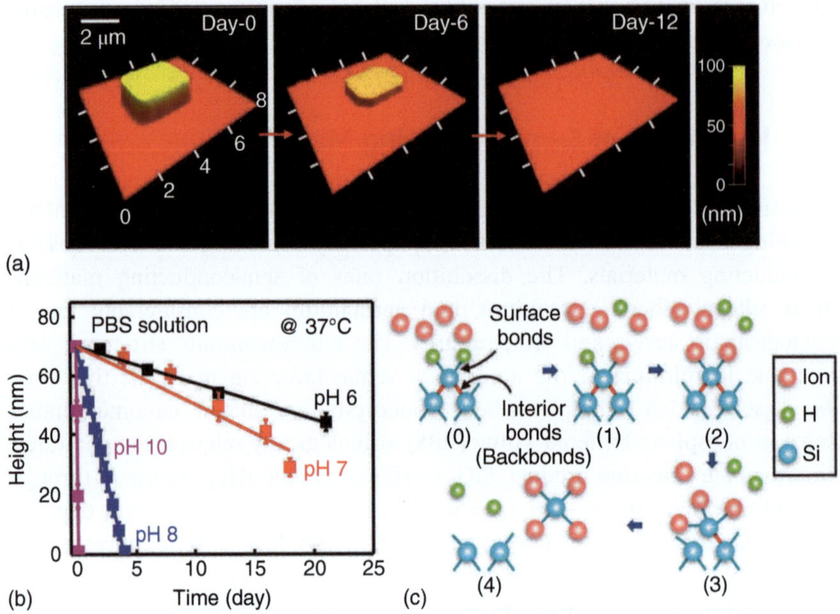

Figure 15.2 Experiments and theoretical and numerical analyses related to the dissolution of silicon. (a) Shows atomic force microscope topographical images of a silicon nanomembrane (Si NM) at various stages of hydrolysis in phosphate-buffered saline (PBS) at 37 °C. (b) Shows the theoretical and experimental dissolution of Si NMs at different pH levels (pH 6, black; pH 7, red; pH 8, blue; pH 10, purple) and physiological temperatures. The lines represent the theoretical predictions, and the symbols represent the experimental results. (c) Shows atomic configurations for each ion adsorption event in a density functional theory (DFT) simulation of the silicon dissolution process. These analyses provide insights into the dissolution behavior of silicon and can aid in developing analytical models for the design of transient electronics. Source: (a,b) Adapted from Ref. [24]. (c) Reproduced from Ref. [24]/with permission of John Wiley & Sons.

impact of anions commonly found in biological systems such as chlorides [Cl^-] and phosphates [HPO_4^{2-}], researchers studied the dependency of the dissolution rate on the ion concentration like phosphates and chlorides at 0.05–1 M. Experimental results indicated that the dissolution rates depend on a power law of [Cl^-] and [HPO_4^{2-}], similar to [OH^-] in high pH solutions of KOH. The process by which Si NMs dissolve layer by layer on the surface occurs because the OH^- nucleophilic attack on silicon surface bonds weakens the internal bonds of silicon. Other anions, such as Cl^- and HPO_4^{2-}, can also promote the dissolution of Si NMs by acting as nucleophiles. Atomic-level simulations have shown that the adsorption of OH^-, Cl^-, and HPO_4^{2-} on the silicon surface leads to the elongation of interior surface silicon bonds, suggesting a decrease in strength (Figure 15.2c). Anions that act as nucleophiles, such as Cl^- and HPO_4^{2-}, can engage with the silicon surface comparable to OH^-, helping dissolve Si NMs.

To create functional devices, it is necessary to understand the dissolution of doped silicon with different types and concentrations. Experimental results that measured the dissolution rates of Si NMs containing dopants, specifically phosphorus (P) or boron (B), indicate a substantial decline in dissolution rates beyond a certain concentration level, such as $1020/cm^3$. This observation aligns with previous research on silicon dissolution conducted at elevated pH levels and temperatures [18]. An empirical expression $R_{dissolution} = R_i/[1+(C/C_0)^4]$, describes this phenomenon where the dissolution rate remains constant (R_i) until a critical dopant concentration (C_0) is reached, after which the rate decreases sharply. The values of R_i and C_0 depend on the relationship between the dissolution rates of Si NMs and temperature, following the Arrhenius law, which exhibits varying activation energies. This empirical expression can also describe the dissolution behavior of doped silicon in physiologically relevant conditions [20].

15.3 Model of Reactive Diffusion for Transient Materials

Furthermore, for the chemical reactions happening on the surface, it is crucial to consider the diffusion that occurs within porous materials and the ensuing reaction between the diffused molecules and the surrounding porous material. Danckwerts [25] proposed a reactive diffusion model for studying the dissolution process of porous materials, which considers the diffusion of water and hydroxide ions into the porous materials effectively expanding the reactive surface area. The key parameters to consider in this model are water's diffusivity (D) within the porous material and the reaction constant (k) that regulates the interaction between water and the material. As the initial thickness h_0 is considerably smaller than both the width and length of the sample tested in the experiment, the dissolution behaviors can be effectively captured using a one-dimensional model (1D). Assuming that $y = 0$ is the bottom of the material layer (as shown in Figure 15.3a), the water concentration $w(y, t)$ at a specific location y and time t can be described by the reactive diffusion equation [25].

$$D\frac{\partial^2 w}{\partial y^2} - kw = \frac{\partial w}{\partial t}, 0 \leq y \leq h_0 \tag{15.1}$$

If the constant reaction k is negligible, Eq. (15.1) simplifies to the standard diffusion equation. The boundary conditions of Eq. (15.1) comprise a constant water concentration at the interface between the water and the porous material $w|_{y=h_0} = w_0$ and a zero water flux at the interface between the porous material and the substrate $\partial w/\partial y|_{y=0} = 0$. The initial condition is that the water concentration is zero $w|_{t=0} = 0$ $(0 \leq y < h_0)$. To transform the inhomogeneous boundary condition $w|_{y=h_0} = w_0$ into a homogeneous one, a new variable $\theta = w - w_0$ is introduced [6]. This results in an updated equation:

$$D\frac{\partial^2 \theta}{\partial y^2} - k\theta - \frac{\partial \theta}{\partial t} = kw_0, \quad 0 \leq y \leq h_0 \tag{15.2}$$

The problem at hand involves certain boundary and initial conditions, namely that $\theta|_{y=h_0} = 0$, $\partial\theta/\partial y|_{y=0} = 0$, and $\theta|_{t=0} = -w_0$ $(0 \leq y < h_0)$. Although Eq. (15.2) is not homogeneous, its general solution can be expressed as the sum of a particular solution θ_p and a homogeneous solution θ_h. The homogeneous solution θ_h satisfies the equation $D\,\partial^2\theta_h/\partial y^2 - k\theta_h - \partial\theta_h/\partial t = 0$ with homogeneous boundary conditions

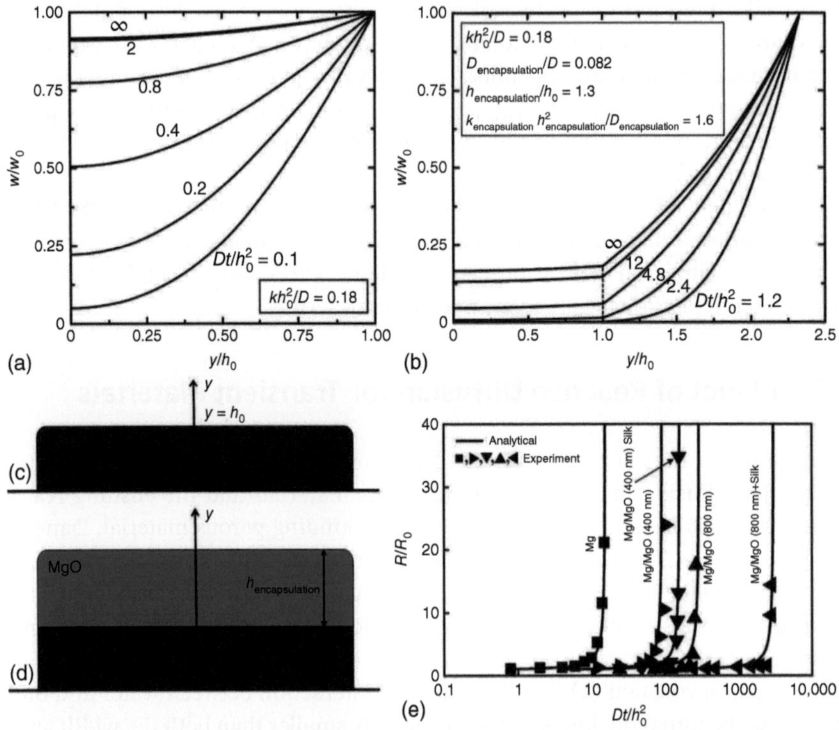

Figure 15.3 The expected distribution of water concentration and electrical resistance, as determined by models of reactive diffusion, is shown for a normalized time $Dt/h_0^2 = 0.1, 0.2, 0.4, 0.8, 2$, and ∞ in (a) an Mg layer ($kh_0^2/D = 0.18$) without an encapsulation layer, and (b) both an Mg conductor layer encapsulated by a MgO layer ($kh_0^2/D = 0.18$, $D_{\text{encapsulation}}/D = 0.082$, $h_{\text{encapsulation}}/h_0 = 1.3$, and $k_{\text{encapsulation}}h^2_{\text{encapsulation}}/D_{\text{encapsulation}} = 1.6$). (c) Single-layered structure and (d) bi-layered structure as a boundary condition to models of reactive diffusion. (e) Comparison of different encapsulation strategies between experimental and modeling results of the electric resistance of Mg and MgO. Source: Reproduced from Ref. [6]/with permission of John Wiley & Sons.

$\theta_h|_{y=h_0} = 0$ and $\partial\theta_h/\partial y|_{y=0} = 0$. By using the method of separation of variables, θ_h can be solved in the form of $\theta_h = Y(y)T(t)$. The homogeneous equation then leads to $DY''/Y - k = T'/T = -\lambda$, where λ is the eigenvalue that can be obtained from the boundary condition. The functions $T(t)$ and $Y(y)$ are expressed as $T = e^{-\lambda t}$ and $Y = A\sin\left[\sqrt{(\lambda-k)/D}y\right] + B\cos\left[\sqrt{(\lambda-k)/D}y\right]$, where A and B are constants to be determined from initial and boundary conditions. With $Y|_{y=h_0} = 0$ and $Y'|_{y=0} = 0$, which are obtained from the homogeneous boundary conditions, A is solved to be $A = 0$. The trigonometric equation $\cos\left[\sqrt{(\lambda-k)/D}h_0\right] = 0$ leads to the solution for eigenvalues $\lambda_n = [(2n-1)\pi/(2h_0)]^2 D + k$ $(n = 1, 2, 3, \cdots)$, which in turn gives the homogeneous solution.

$$\theta_h = \sum_{n=1}^{\infty}\left\{B_n e^{-\left[\left(\frac{2n-1}{2h_0}\pi\right)^2 D+k\right]t}\cos\left(\frac{2n-1}{2h_0}\pi y\right)\right\} \tag{15.3}$$

To find the constants B_n, the initial condition needs to be considered. The particular solution θ_p must satisfy the inhomogeneous equation and homogeneous boundary conditions. An appropriate choice for $\theta_p = w_0[\cos h(\sqrt{k/D}y)/\cos h(\sqrt{k/D}h_0) - 1]$. Substituting the general solution $\theta - \theta_h + \theta_p$ in the initial condition yields $B_n = 4(-1)^n (2n-1)\pi D w_0/\left[4kh_0^2 + (2n-1)^2\pi^2 D\right]$. Using this, the solution for Eq. (15.1) can be obtained.

$$w(y,t) = w_0 \left\{ \frac{\cos h\left(\sqrt{\frac{kh_0^2}{D}}\frac{y}{h_0}\right)}{\cos h\sqrt{\frac{kh_0^2}{D}}} + 2\sum_{n=1}^{\infty} \frac{(-1)^n\left(n-\frac{1}{2}\right)\pi}{\frac{kh_0^2}{D} + \left(n-\frac{1}{2}\right)^2\pi^2} e^{-\left[\frac{kh_0^2}{D}+\left(n-\frac{1}{2}\right)^2\pi^2\right]\frac{Dt}{h_0^2}} \times \cos\left[\left(n-\frac{1}{2}\right)\pi\frac{y}{h_0}\right] \right\} \tag{15.4}$$

The solution provided in the passage illustrates a scaling law $w/w_0 = w\left(y/h_0, Dt/h_0^2, kh_0^2/D\right)$, where the normalized water concentration w/w_0 depends on the normalized position y/h_0, normalized time Dt/h_0^2, and a non-dimensional parameter kh_0^2/D that scales with the ratio of reaction constant k to diffusivity D. Experimental measurements of Mg with an initial thickness of 300 nm suggest $D = 6.0\times10^{-12}$ cm^2/s and $k = 1.2\times10^{-3}$/s, which ranges within the reaction constants [26]. Figure 15.3b depicts the water concentration distribution for various normalized times, including $Dt/h_0^2 = 0.1$, 0.2, 0.4, 0.8, 2, and ∞. In this context, $Dt/h_0^2 = \infty$ corresponds to the steady-state limit of the water concentration in the Mg layer, given by $w(y, t \to \infty) = w_0 \cos h\left(\sqrt{kh_0^2/D}y/h_0\right)/\cos h\sqrt{kh_0^2/D}$.

Based on the definition of the reaction constant, the amount of water that reacted at a certain location per unit volume is kw. The quantity of dissolved porous material per unit volume at a specific location, where q water molecules react with one atom of the material, can be determined as $kwM/\left(qM_{H_2O}\right)$, where $M_{H_2O} = 18$ g/mol and M are the molar masses of water and porous material, respectively. By integrating $kwM/\left(qM_{H_2O}\right)$ over both time t and thickness y, the dissolved porous material amount can be obtained. Consequently, a scaling law for the remaining thickness (h) normalized by its initial thickness (h_0) before complete disappearance can be established, which is given by:

$$\frac{h}{h_0} = 1 - \frac{w_0 M}{q\rho M_{H_2O}}\frac{kh_0^2}{D}\left\{\frac{Dt}{h_0^2}\cdot\frac{\tan h\sqrt{\frac{kh_0^2}{D}}}{\sqrt{\frac{kh_0^2}{D}}} - 2\sum_{n=1}^{\infty}\frac{1-e^{-\left[\frac{kh_0^2}{D}+\left(n-\frac{1}{2}\right)^2\pi^2\right]\frac{Dt}{h_0^2}}}{\left[\frac{kh_0^2}{D}+\left(n-\frac{1}{2}\right)^2\pi^2\right]^2}\right\} \tag{15.5}$$

The mass density of the porous material is denoted by ρ. When k and D have values that are relevant to the use of transient electronics in biomedicine, the summation on the right side of Eq. (15.5) can be neglected in comparison to the other term, leading to an approximately linear expression for the thickness as $h/h_0 \approx 1 - t/t_c$. Here,

$t_c = \left(h_0/\sqrt{kD}\right)\left[q\rho M_{H_2O}/\left(w_0 M\right)\right]\left(\tan h\sqrt{kh_0^2/D}\right)^{-1}$ is the critical time taken by the transient material to completely physically disappear. The electrical dissolution rate for conductors can be determined from the linear approximation of the thickness as $v_{\text{dissolution}} = -dh/dt \approx h_0/t_c = \sqrt{kD}w_0 M/\left(q\rho M_{H_2O}\right)\tan h\sqrt{kh_0^2/D}$. The dissolution rate for 100, 300, and 500 nm-thick Mg layers is 0.044, 0.13, and 0.20 nm/s, respectively. These dissolution rates are of the same order of magnitude as those reported in previous experiments [27, 28].

In PBS, Mg reacts with two water molecules to form $Mg(OH)_2$ and H_2. Since water molecules are abundant in biofluids, their concentration is approximately 1 g/cm^3. The critical time it takes to dissolve Mg with a molar mass of 24 g/mol, a mass density of 1.738 g/cm^3, and an initial thickness of 300 nm is calculated to be 38 minutes, which agrees with the experimental measurement of 40 minutes. SiO_2 reacts with two water molecules to form $Si(OH)_4$, but the reaction is slower than Mg and takes days instead of minutes. The reaction constant between water and SiO_2 is much smaller than between water and Mg. The dissolution rates of SiO_2 range from 0.11 to 0.47 nm/h for an initial thickness between 35 and 100 nm and at temperatures between room and physiological temperatures. These rates are consistent with previous experiments on PECVD SiO_2 in water, which are higher than those for quartz.

Compared to measuring thickness intermittently, electrical properties can offer continuous measurements and enable evaluating a conductive material's dissolution behavior beneath a nonconductive layer. The width and length direction changes are relatively small compared to the thickness direction. Therefore, the electric resistance is inversely proportional to the remaining thickness, represented as $R = R_0 h_0/h \approx R_0/(1 - t/t_c)$, where R_0 is the initial resistance. Resistance changes reflect changes in thickness and influence related to nonuniform dissolution-induced porosity, pitting, and other nonuniformities. Figure 15.3c shows the normalized electric resistance R/R_0 of Mg as a function of normalized time t/t_c, with a chosen reaction constant k of 1.2×10^{-3}/s, diffusivity D of 6.0×10^{-12} cm^2/s, an initial thickness of 300 nm, and an initial resistance (per unit length) of 1.06 Ω/mm. It is important to note that a thin layer of MgO may exist on top of the Mg layer, which quickly reacts with water to form a crystalline hydroxide. Not as protective as noncrystalline films, but the single-layer dissolution model is still suitable for accounting for the hydrolysis of Mg.

The use of thin films made from other transient metals for transient electronics has also been investigated, with MOSFETs as an example. The analytical models discussed in the previous section are found to be valid for other bioresorbable metals such as molybdenum (Mo), zinc (Zn), Mg alloy, and tungsten (W). In addition, the model can accurately predict the observed dissolution behaviors in both deionized water and simulated body fluids, such as Hanks' solution, with a pH range of 5–8. Notably, the electrical dissolution rates in thin films can be vastly distinctive from the corrosion rates conventionally described for subsequent bulk materials. However, the model is not able to predict the dissolution behavior of iron (Fe) due to its nonuniform degradation and the low solubility of certain reaction products, such as Fe_2O_3 and Fe_3O_4.

Silicon oxides and silicon nitrides are essential materials in high-performance silicon-based electronics, serving as dielectrics and encapsulation layers. The dissolution rates of these materials depend on their physical and chemical properties, which are influenced by the deposition and growth methods and conditions. Density is a critical parameter that can help characterize these differences, with reduced density increasing porosity and reactive surface area and reducing the amount of material to be dissolved. The effective density of a porous material can be calculated based on the density of the fully dense material and the volumes of the porous material and air cavity. A modified reactive diffusion model has been developed to account for density variation [5].

Equation (15.1) is modified by using an effective diffusivity D_e instead of D. The relationship between D_e and the pore fraction in a porous medium is linear, where D_e for water is proportional to $V_{air}/(V_{air} + V_s) = (\rho_s - \rho_{eff})/\rho_s$. The effective diffusivity for each porous material can be obtained by directly measuring the densities of the porous materials under different deposition methods/conditions. At $t = 0$, water fills the air pores completely ($0 \leq y < h_0$) and is represented by $w|_{t=0} = w_0(\rho_s - \rho_{eff})/\rho_s$. The boundary conditions for Eq. (15.1) remain unchanged. The normalized thickness is obtained using the same method [5].

$$\frac{h}{h_0} = 1 - \frac{w_0 M}{q\rho_{eff} M_{H_2O}} \frac{kh_0^2}{D_e} \left\{ \frac{D_e t}{h_0^2} \frac{\tan h\sqrt{\frac{kh_0^2}{D_e}}}{\sqrt{\frac{kh_0^2}{D_e}}} - 2\sum_{n=1}^{\infty} C_n \frac{1 - e^{-\left[\frac{kh_0^2}{D_e} + \left(n - \frac{1}{2}\right)^2 \pi^2\right] \frac{D_e t}{h_0^2}}}{\left[\frac{kh_0^2}{D_e} + \left(n - \frac{1}{2}\right)^2 \pi^2\right]} \right\} \tag{15.6}$$

where C_n is $C_n = \left[kh_0^2/D_e + (n - 1/2)^2\pi^2\right]^{-1} + \left(\rho_{eff}/\rho_s - 1\right)/(n - 1/2)^2\pi^2$. The dissolution rate is then estimated as $-dh/dt = kh_0 w_0 M/(q\rho_{eff} M_{H_2O}) \tanh \sqrt{kh_0^2/D_e}/\sqrt{kh_0^2/D_e}$. By analyzing the densities and effective diffusivities, it was found that the reaction constants differ for silicon oxides or nitrides that are deposited under different conditions. This suggests that changes in diffusion and reactivity caused by density variations affect the dissolution rate. Additionally, hydrolysis can lead to the removal of nanoscale portions of the material as narrow regions of the porous matrix disappears. This discovery offers the possibility of developing customizable dissolution through geometric alterations. By combining this with patterned structures in the encapsulation layer or transient materials, the dissolution behavior of transient electronics can be actively controlled through changes in diffusivity and reactivity induced by stimuli such as swelling and thermal and mechanical loads.

15.4 Dissolution of the Device with Bi-layered Structures

Transient electronics used in biomedical applications must function reliably for a specific period before completely disappearing. However, all transient materials dissolve rapidly when exposed to biofluids. The lifetime of such devices is primarily determined by Mg interconnects due to their fast reaction with water. Although the device may function before completely breaking down, its performance is severely compromised. Therefore, it is crucial to investigate mechanisms that enable devices to operate within a predetermined lifetime.

Encapsulation layers or packaging materials can be applied to the device to increase its lifetime in a controlled manner. For example, MgO can act as an encapsulation layer for Mg. In this two-layered system (as shown in Figure 15.3d), the Mg and MgO layers have zero initial conditions at $t = 0$. The reactive diffusion Eq. (15.1) and zero water flux boundary condition at the bottom surface $y = 0$ remain valid for the Mg layer. Regarding the MgO encapsulation layer with an initial thickness of $h_{\text{encapsulation}}$, the reactive diffusion equation is modified [6].

$$D_{\text{encapsulation}}\frac{\partial^2 w}{\partial y^2} - k_{\text{encapsulation}}w = \frac{\partial w}{\partial t}\ \left(h_0 \le y \le h_0 + h_{\text{encapsulation}}\right) \tag{15.7}$$

where $D_{\text{encapsulation}}$ and $k_{\text{encapsulation}}$ are the diffusivities of water in MgO and the reaction constant between MgO and water, respectively. The MgO/water interface has a constant water concentration boundary condition represented by $w|_{y=h_0+h_{\text{encapsulation}}} = w_0$. Additionally, the MgO/Mg interface has continuity conditions for water concentration and flux, where $w|_{y=h_0-0} = w|_{y=h_0+0}$ and $D\partial w/\partial y|_{y=h_0-0} = D_{\text{encapsulation}}\partial w/\partial y\big|_{y=h_0+0}$. Similar to a single-layer system, the inhomogeneous boundary condition represents the water concentration as a sum of a homogeneous solution w_{h} and a particular solution w_{p}, i.e. $w = w_{\text{h}} + w_{\text{p}}$. The homogeneous solution w_{h} satisfies the homogeneous equation $D'\partial^2 w_{\text{h}}/\partial y^2 - k' w_{\text{h}} = \partial w_{\text{h}}/\partial t$, where $D' = D$ and $k' = k$ for $0 \le y \le h_0$ in the Mg layer, and $D' = D_{\text{encapsulation}}$ and $k' = k_{\text{encapsulation}}$ for $h_0 \le y \le h_0 + h_{\text{encapsulation}}$ in the MgO encapsulation. The boundary conditions also become homogeneous, i.e. $w_{\text{h}}|_{y=h_0+h_{\text{encapsulation}}} = 0$, $\partial w_{\text{h}}/\partial y|_{y=0} = 0$.

Using the separation of variables method, the homogeneous solution can be solved as $w_{\text{h}} = w_0\sum_{n=1}^{\infty} C_n e^{-\lambda_n t} f_n(y)$. The coefficients C_n are determined from the initial condition, and $\lambda_n (n = 1, 2, 3, \ldots)$ can be solved from the following eigen equation:

$$\tan\sqrt{\frac{\lambda - k}{D}h_0^2}\,\tan\sqrt{\frac{\lambda - k_{\text{encapsulation}}}{D_{\text{encapsulation}}}h_{\text{encapsulation}}^2} = \sqrt{\frac{\left(\lambda - k_{\text{encapsulation}}\right)D_{\text{encapsulation}}}{(\lambda - k)D}} \tag{15.8}$$

and $f_n(y)$ is written as

$$f_n(y) \equiv \begin{cases} \sin\left(\sqrt{\frac{\lambda_n - k_{\text{encapsulation}}}{D_{\text{encapsulation}}}}h_{\text{encapsulation}}\right)\cos\left(\sqrt{\frac{\lambda_n - k}{D}}y\right), 0 \le y \le h_0 \\ \cos\left(\sqrt{\frac{\lambda_n - k}{D}}h_0\right)\sin\left[\sqrt{\frac{\lambda_n - k_{\text{encapsulation}}}{D_{\text{encapsulation}}}}\left(h_0 + h_{\text{encapsulation}} - y\right)\right], \\ \quad h_0 \le y \le h_0 + h_{\text{encapsulation}} \end{cases} \tag{15.9}$$

By satisfying the reactive diffusion equation along with the inhomogeneous boundary condition, zero water flux at the bottom of the Mg layer, and continuity conditions, the particular solution w_p is solved as $w_p = w_0 g(y)$, where:

$$g(y) \equiv \begin{cases} G\cos h\left(\sqrt{\frac{k}{D}}y\right), 0 \le y \le h_0 \\ \cos h\left(\xi\frac{h_0+h_{\text{encapsulation}}-y}{h_{\text{encapsulation}}}\right) - H\sin h\left(\xi\frac{h_0+h_{\text{encapsulation}}-y}{h_{\text{encapsulation}}}\right), \\ h_0 \le y \le h_0 + h_{\text{encapsulation}} \end{cases} \tag{15.10}$$

$$G = \frac{1}{\sqrt{\frac{Dk}{D_{\text{encapsulation}}k_{\text{encapsulation}}}}\sin h\sqrt{\frac{kh_0^2}{D}}\sin h\xi + \cos h\sqrt{\frac{kh_0^2}{D}}\cos h\xi}$$

$$H = \frac{\sqrt{\frac{Dk}{D_{\text{encapsulation}}k_{\text{encapsulation}}}}\tan h\sqrt{\frac{kh_0^2}{D}} + \tan h\xi}{\sqrt{\frac{Dk}{D_{\text{encapsulation}}k_{\text{encapsulation}}}}\tan h\sqrt{\frac{kh_0^2}{D}}\tan h\xi + 1} \tag{15.11}$$

and $\xi = \sqrt{k_{\text{encapsulation}}h^2_{\text{encapsulation}}/D_{\text{encapsulation}}}$. With the expression of the water concentration $w = w_0\left[\sum_{n=1}^{\infty} C_n e^{-\lambda_n t} f_n(y) + g(y)\right]$, the zero initial condition leads to $\sum_{n=1}^{\infty} C_n f_n(y) + g(y) = 0$. Orthogonality of eigenfunctions requires $\int_0^{h_0+h_{\text{encapsulation}}} f_m(y) f_n(y)\,dy = 0$ (if $m \neq n$), which yields the coefficient C_n analytically as

$$C_n = -\frac{\int_0^{h_0+h_{\text{encapsulation}}} f_n(y) g(y)\,dy = 0}{\int_0^{h_0+h_{\text{encapsulation}}} f_n^2(y)\,dy = 0}$$

$$= \frac{-\frac{2}{\lambda_n}\sqrt{(\lambda_n - k_{\text{encapsulation}})D_{\text{encapsulation}}}\cos\left(\sqrt{\frac{\lambda_n-k}{D}}h_0\right)}{\left\{\begin{array}{l} h_0\sin^2\left(\sqrt{\frac{\lambda_n-k_{\text{encapsulation}}}{D_{\text{encapsulation}}}}h_{\text{encapsulation}}\right)\left[1+\frac{\sin\left(2\sqrt{\frac{\lambda_n-k}{D}}h_0\right)}{2\sqrt{\frac{\lambda_n-k}{D}}h_0}\right] \\ +h_{\text{encapsulation}}\cos^2\left(\sqrt{\frac{\lambda_n-k}{D}}h_0\right)\left[1-\frac{\sin\left(2\sqrt{\frac{\lambda_n-k_{\text{encapsulation}}}{D_{\text{encapsulation}}}}h_{\text{encapsulation}}\right)}{2\sqrt{\frac{\lambda_n-k_{\text{encapsulation}}}{D_{\text{encapsulation}}}}h_{\text{encapsulation}}}\right] \end{array}\right\}} \tag{15.12}$$

The solution to the reactive diffusion equation for bi-layered structures is completed using the expression in Eq. (15.12). In addition to depending on y/h_0, Dt/h_0^2, and kh_0^2/D, as in the single-layer solution, the normalized water concentration (w/w_0) also depends on the normalized reaction constant $k_{\text{encapsulation}}h^2_{\text{encapsulation}}/D_{\text{encapsulation}}$ of the encapsulation layer, the diffusivity ratio $D_{\text{encapsulation}}/D$, and the initial thickness ratio $h_{\text{encapsulation}}/h_0$. As the diffusivity ratio $D_{\text{encapsulation}}/D$ is much smaller than 1, the water concentration in Mg remains relatively uniform, while that in MgO does not. Compared to Figure 15.3a, the water concentration in Mg increases much more slowly, effectively extending the lifetime of Mg.

Similar to the approach described in the previous section, the normalized remaining thickness h/h_0 of the Mg layer can be obtained as:

$$\frac{h}{h_0} = 1 - \frac{w_0 M}{q\rho M_{H_2O}} k \left[Gt \cdot \frac{\sin h\sqrt{\frac{kh_0^2}{D}}}{\sqrt{\frac{kh_0^2}{D}}} + \sum_{n=1}^{\infty} \frac{C_n}{\lambda_n} \left(1 - e^{-\lambda_n t}\right) \times \frac{\sin\sqrt{\frac{\lambda_n - k}{D} h_0^2}}{\sqrt{\frac{\lambda_n - k}{D} h_0^2}} \sin\sqrt{\frac{\lambda_n - k_{\text{encapsulation}}}{D_{\text{encapsulation}}} h_{\text{encapsulation}}^2} \right] \tag{15.13}$$

where G is given in Eq. (15.11). It is difficult to measure the thickness change of Mg as it is located below the MgO encapsulation layer. However, since MgO is not conductive, the electric resistance of Mg (or the bi-layered structure) can be measured, from which the thickness can be calculated as $h = R_0 h_0 / R$. To study the thickness effect of MgO encapsulation, 400 nm-thick and 800 nm-thick MgO layers were studied on a 300 nm-thick Mg layer. As shown in Figure 15.3e, the experimental measurements match well with the theory predicting the relationship between normalized electrical resistance R/R_0 versus the time Dt/h_0^2. Comparing two different layouts of MgO + Mg structures shows that the dissolution time increases with the thickness of the encapsulation layer. The reason why the lifetime of Mg is extended and the dissolution time is controlled when a MgO encapsulation layer is used is that water diffuses much slower in MgO than in Mg.

For devices relevant to transient implants, the summation on the right-hand side of Eq. (15.13) can be neglected. In this case, the thickness of the Mg layer decreases linearly with time, and an approximate expression for the thickness is given as $h/h_0 \approx 1 - t/t'_c$, where t'_c is a characteristic dissolution time given by:

$$t'_c = \frac{t_c}{G\cos h\sqrt{\frac{kh_0^2}{D}}} = t_c \left(\sqrt{\frac{kD}{k_{\text{encapsulation}} D_{\text{encapsulation}}}} \tan h\sqrt{\frac{kh_0^2}{D}} \sin h\xi + \cos h\xi \right) \tag{15.14}$$

The critical time t'_c represents the time required for the Mg conductor layer to disappear completely physically and is also known as the critical time in the single-layer system. Based on the remaining thickness h, the dissolution rate can be determined as the negative derivative of h with respect to time, which is approximated as h_0 divided by the critical time t'_c. This dissolution rate is almost directly proportional to the initial thickness h_0, and for a sufficiently thin Mg layer, where kh_0^2/D is much smaller than 1, the expression for the dissolution rate can be further simplified as $v_{\text{dissolution}} = w_0 M / \left(q\rho M_{H_2O}\right) \left(kh_0 / \cos h\xi\right)$.

Encapsulation layers made of silicon oxides and nitrides can also be used in addition to their conventional use as gate and interlayer dielectrics, as they are known to be effective barrier materials for preventing the permeation of water [5]. However, defects such as pinholes are commonly found in these films, which can be a primary source of leakage for vapors or fluids. To address this, multilayer structures comprising different materials have been developed to eliminate defects cooperatively for

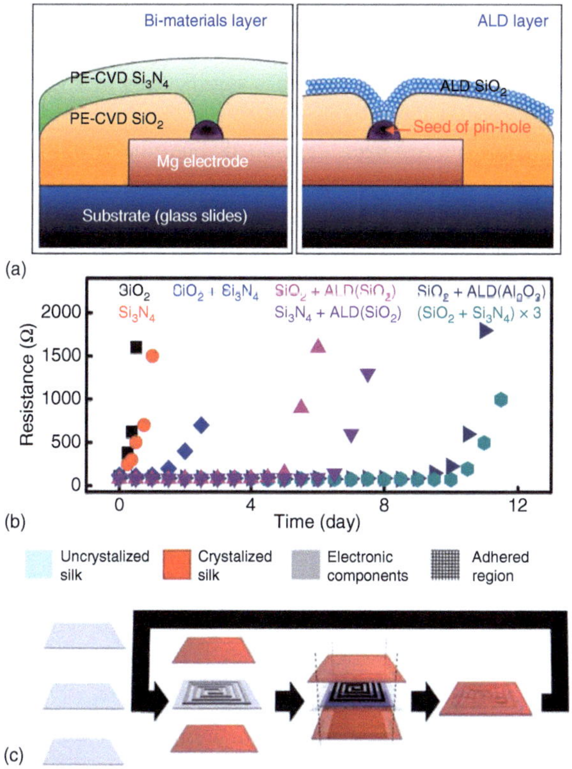

Figure 15.4 Multilayer structures leveraging encapsulation schemes. (a) The images show two different ways of encapsulating transient electronic devices. On the left, a bilayer of SiO_2/Si_3N_4 covers defects, such as pinholes. On the right, an ALD layer is used for the same purpose. (b) The experiment involved measuring the changes in resistance of 300 nm-thick Mg traces that were encapsulated using different methods while immersed in deionized water at room temperature. The encapsulation methods tested included a single layer of PECVD SiO_2 (1 μm, black), PECVD-LF Si_3N_4 (1 μm, red), and ALD SiO_2 (20 nm, orange). The experiment also included testing double-layer encapsulations, such as PECVD SiO_2/PECVD-LF Si_3N_4 (500/500 nm, blue), PECVD SiO_2/ALD SiO_2 (500/20 nm, magenta), and PECVD-LF Si_3N_4/ALD SiO_2 (500/20 nm, purple), as well as a triple layer encapsulation of PECVD SiO_2/PECVD-LF Si_3N_4 (200/200/200/200/100/100 nm, cyan). Source: (a,b) Reproduced from Ref. [5]/with permission of John Wiley & Sons. (c) The image illustrating the process for creating a multilayer silk pocket involves crystallizing the outer layers to make them water-insoluble while leaving the inner device substrate layer crystallized. Sealing the outer edges of the device with the crystallized layers creates a protective silk pocket around it to create the multilayer structure. Source: (c) Reproduced from Ref. [31]/with permission of American Chemical Society.

transient electronics [5, 29]. In addition to using a combination of multiple layers, atomic layer deposition (ALD) can also be employed to reduce defects and improve encapsulation performance, even with thin layers [30] (Figure 15.4a). Figure 15.4b shows the effectiveness of several encapsulation approaches, as demonstrated by the measured changes in resistance of a serpentine-shaped Mg trace with an initial thickness of 300 nm.

Silk overcoats have been used to further enhance the barrier properties of MgO and Mg layers in transient electronic devices, resulting in longer lifetimes [2]. Combining the MgO encapsulation with a silk overcoat has significantly increased device lifetime [6]. To take advantage of the multilayer structure, a new encapsulation method using multiple air pockets has been developed (Figure 15.4c) [31]. The method involves enclosing transient electronic devices on a silk substrate with silk films of varying crystalline and diffusion properties, which are then thermally sealed to create air pockets. The process can be repeated to provide additional silk pockets as needed. The silk protective layer collapses, and device degradation begins only when the air pocket collapses due to swelling in a wet environment [32].

These approaches to encapsulation result in a two-stage dissolution process in transient electronics. The first stage is defined by the encapsulation layers, which enable stable operation with minimal changes in electrical performance. The Mg

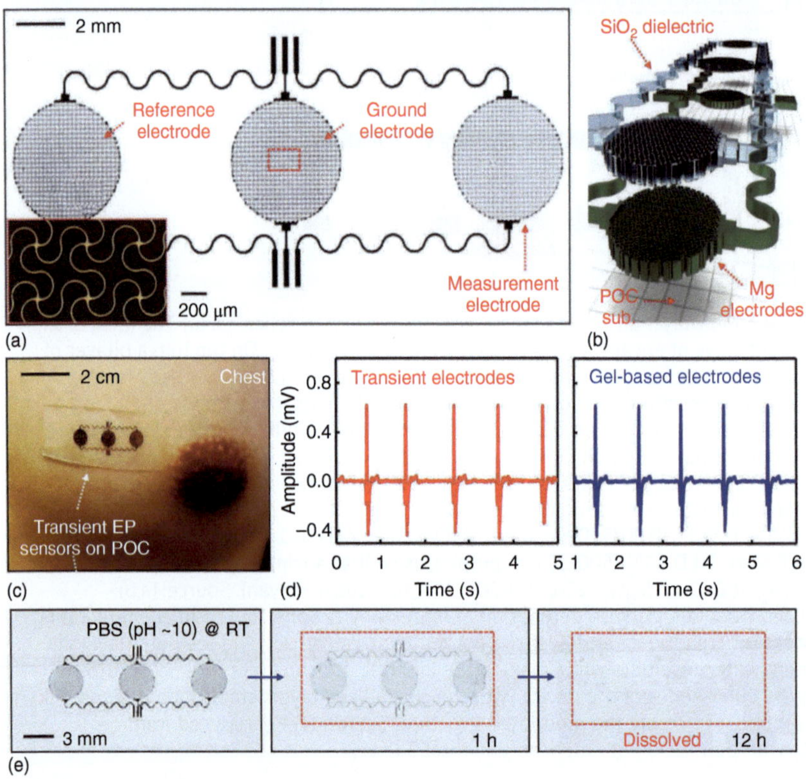

Figure 15.5 Transient electrophysiological sensors are designed for capacitive sensing in a stretchable prototype. (a) Illustration of a device and (inset) magnified view of serpentine electrode structures in the mesh layout and its (b) corresponding device. (c) Image showing the device mounted performing electrocardiogram (ECG) measurements on the chest. (d) Comparative ECG measurements using transient (red) and standard gel-based (blue) devices. (e) Images at different dissociation phases of a transient device in PBS (pH 10) at room temperature. Source: Reproduced from Ref. [3]/with permission of American Chemical Society.

layer defines the second stage, which causes the device to degrade rapidly. Effective encapsulation strategies can eliminate leakage pathways and significantly prolong stable operation time. However, to fully utilize the potential of transient electronics for implantable applications, it is necessary to achieve conformal contact with the body's organs. To achieve this, recent developments in transient medical devices have utilized the principles of stretchable electronics [33–40] through biodegradable elastomers [41–47].

As depicted in optical images and schematic illustrations (Figure 15.5a,b), a stretchable and transient electrophysiology sensor has been created. Thin layers of Mg (300 nm) and SiO_2 (100 nm), designed in the form of filamentary serpentine meshes [35, 48, 49], function as measurement, ground, and reference electrodes, as well as connecting leads. Using a biodegradable polymer [20, 50] between the Mg electrodes and the skin allows for capacitive sensing. The fidelity of the measurements is comparable to that of conventional gel electrodes, as demonstrated in the chest's high-quality electrocardiogram (ECG) measurements (Figure 15.5c,d). Figure 15.5e presents a set of images of a transient electrophysiology sensor at various dissolution stages in PBS (pH 10) at room temperature. The dissolution behavior of each component is consistent with separate studies of these materials discussed in the previous section, with Mg dissolving completely within hours and SiO_2 taking days or weeks to dissolve.

15.5 Conclusion

Transient electronics, a class of high-performance electronics that can disintegrate and dissolve in biofluids or water, have been extensively studied to control their dissolution behavior. Several discoveries have been made in this regard. First, researchers studied various materials, including semiconductors, and their application in designing transient electronics. Second, they developed a reactive diffusion model to predict how a component would dissolve in biofluids or water, considering factors such as the material's porosity. Additionally, the model was expanded to investigate reactive diffusion in a bi-layered structure, and analytical findings were acquired to link the crucial electrical property to models of reactive diffusion. These results can be used with established circuit simulators as a comprehensive design approach.

Since the dissolution of transient electronics is significantly influenced by the materials used, future material science research would be desirable. In addition to encapsulation, other strategies should be explored to control the dissolution behavior of transient electronics. Active control of the transience in devices is also an interesting direction for future development. Furthermore, combining ideas from soft, "tissue-like" devices with the possibilities offered by transient electronics opens up opportunities for biomedical applications. Overall, the research conducted thus far on the design of transient electronics has been extensive, and the potential use of this technology in the industry is apparent.

Acknowledgments

H.C. acknowledges the support provided by NIH (Award Nos. EB031629 and EB030140), NSF (Grant Nos. ECCS-1933072 and 2309323), and Penn State University. A.D. acknowledges the support provided by NIH Training Grant T32NS115667 and the Leighton Riess Graduate Fellowship at Penn State University.

References

1 Ko, G.-J., Kang, H., Han, W.B. et al. (2024). Materials and designs for extremely efficient encapsulation of soft, biodegradable electronics. *Advanced Functional Materials* 2403427. https://doi.org/10.1002/adfm.202403427.

2 Liu, J., Liu, N., Xu, Y. et al. (2024). Bioresorbable shape-adaptive structures for ultrasonic monitoring of deep-tissue homeostasis. *Science* 383: 1096–1103. https://doi.org/10.1126/science.adk9880.

3 Lee, J.Y., Shin, J., Kim, K. et al. (2023). Ultrathin crystalline silicon nano and micro membranes with high areal density for low-cost flexible electronics. *Small* 19: 2302597. https://doi.org/10.1002/smll.202302597.

4 Hwang, S.-W., Park, G., Edwards, C. et al. (2014). Dissolution chemistry and biocompatibility of single-crystalline silicon nanomembranes and associated materials for transient electronics. *ACS Nano* 8: 5843–5851.

5 Kang, S., Hwang, S.W., Cheng, H. et al. (2014). Dissolution behaviors and applications of silicon oxides and nitrides in transient electronics. *Advanced Functional Materials* 24: 4427–4434.

6 Li, R., Cheng, H., Su, Y. et al. (2013). An analytical model of reactive diffusion for transient electronics. *Advanced Functional Materials* 23: 3106–3114.

7 Han, W.B., Lee, J.H., Shin, J.W., and Hwang, S.W. (2020). Advanced materials and systems for biodegradable, transient electronics. *Advanced Materials* 32: e2002211.

8 Yang, Q., Liu, T.L., Xue, Y. et al. (2022). Ecoresorbable and bioresorbable microelectromechanical systems. *Nature Electronics* 5: 526–538.

9 Lee, G., Choi, Y.S., Yoon, H.J., and Rogers, J.A. (2020). Advances in physicochemically stimuli-responsive materials for on-demand transient electronic systems. *Matter* 3: 1031–1052.

10 Cheng, H. and Vepachedu, V. (2016). Recent development of transient electronics. *Theoretical and Applied Mechanics Letters* 6: 21–31.

11 Dutta, A. and Cheng, H. (2023). Pathway of transient electronics towards connected biomedical applications. *Nanoscale* 15: 4236–4249.

12 Bettinger, C.J. and Bao, Z. (2010). Organic thin-film transistors fabricated on resorbable biomaterial substrates. *Advanced Materials* 22: 651–655.

13 Irimia-Vladu, M., Troshin, P.A., Reisinger, M. et al. (2010). Biocompatible and biodegradable materials for organic field-effect transistors. *Advanced Functional Materials* 20: 4069–4076.

14 Eder, F., Klauk, H., Halik, M. et al. (2004). Organic electronics on paper. *Applied Physics Letters* 84: 2673–2675.

15 Kim, D.H., Kim, Y.S., Amsden, J. et al. (2009). Silicon electronics on silk as a path to bioresorbable, implantable devices. *Applied Physics Letters* 95: 133701.

16 Tao, H., Kim, Y.S., Amsden, J. et al. (2014). Silk-based resorbable electronic devices for remotely controlled therapy and in vivo infection abatement. *Proceedings of the National Academy of Sciences* 111: 17385–17389.

17 Rimstidt, J.D. and Barnes, H.L. (1980). The kinetics of silica-water reactions. *Geochimica et Cosmochimica Acta* 44: 1683–1699.

18 Seidel, H., Csepregi, L., Heuberger, A., and Baumgartel, H. (1990). Anisotropic etching of crystalline silicon in alkaline-solutions. 2. Influence of dopants. *Journal of the Electrochemical Society* 137: 3626–3632.

19 Seidel, H., Csepregi, L., Heuberger, A., and Baumgartel, H. (1990). Anisotropic etching of crystalline silicon in alkaline-solutions. 1. Orientation dependence and behavior of passivation layers. *Journal of the Electrochemical Society* 137: 3612–3626.

20 Hwang, S., Park, G., Cheng, H. et al. (2014). 25th Anniversary article: materials for high-performance biodegradable semiconductor devices. *Advanced Materials* 26: 1992–2000.

21 Worley, W.G. (1994). Dissolution kinetics and mechanisms in quartz-and grainite-water systems.

22 Wirth, G.S. and Gieskes, J.M. (1979). The initial kinetics of the dissolution of vitreous silica in aqueous media. *Journal of Colloid and Interface Science* 68: 492–500.

23 Kang, S.-K., Park, G., Kim, K. et al. (2015). Dissolution chemistry and biocompatibility of silicon-and germanium-based semiconductors for transient electronics. *ACS Applied Materials & Interfaces* 7: 9297–9305.

24 Yin, L., Farimani, A.B., Min, K. et al. (2015). Mechanisms for hydrolysis of silicon nanomembranes as used in bioresorbable electronics. *Advanced Materials* 27: 1857–1864.

25 Danckwerts, P.V. (1950). Absorption by simultaneous diffusion and chemical reaction. *Transactions of the Faraday Society* 46: 300.

26 Taub, I.A., Roberts, W., LaGambina, S., and Kustin, K. (2002). Mechanism of dihydrogen formation in the magnesium-water reaction⊥. *The Journal of Physical Chemistry. A* 106: 8070–8078.

27 Inoue, H., Sugahara, K., Yamamoto, A., and Tsubakino, H. (2002). Corrosion rate of magnesium and its alloys in buffered chloride solutions. *Corrosion Science* 44: 603–610.

28 Ng, W.F., Chiu, K.Y., and Cheng, F.T. (2010). Effect of pH on the *in vitro* corrosion rate of magnesium degradable implant material. *Materials Science and Engineering: C* 30: 898–903.

29 Rosink, J., Lifka, H., Rietjens, G.H., and Pierik, A. (2005). 34.1: Ultra-thin encapsulation for large-area OLED displays. *SID Symposium Digest of Technical Papers*, vol. 36, Wiley Online Library. 1272–1275. https://sid.onlinelibrary.wiley.com/doi/epdf/10.1889/1.2036236

30 Meyer, J., Görrn, P., Bertram, F. et al. (2009). Al_2O_3/ZrO_2 Nanolaminates as ultrahigh gas-diffusion barriers—a strategy for reliable encapsulation of organic electronics. *Advanced Materials* 21: 1845–1849.

31 Brenckle, M.A., Cheng, H., Hwang, S. et al. (2015). Modulated degradation of transient electronic devices through multilayer silk fibroin pockets. *ACS Applied Materials & Interfaces* 7: 19870–19875.

32 Lawrence, B.D., Wharram, S., Kluge, J.A. et al. (2010). Effect of hydration on silk film material properties. *Macromolecular Bioscience* 10: 393–403.

33 Cheng, H., Zhang, Y., Hwang, K.-C. et al. (2014). Buckling of a stiff thin film on a pre-strained bi-layer substrate. *International Journal of Solids and Structures* 51: 3113–3118.

34 Cheng, H., Wu, J., Li, M. et al. (2011). An analytical model of strain isolation for stretchable and flexible electronics. *Applied Physics Letters* 98: 61902.

35 Kim, R.H., Bae, M.H., Kim, D.G. et al. (2011). Stretchable, transparent graphene interconnects for arrays of microscale inorganic light emitting diodes on rubber substrates. *Nano Letters* 11: 3381–3886.

36 Rogers, J.A., Someya, T., and Huang, Y.G. (2010). Materials and mechanics for stretchable electronics. *Science* 327: 1603–1607.

37 Viventi, J., Kim, D.H., Moss, J.D. et al. (2010). A conformal, bio-interfaced class of silicon electronics for mapping cardiac electrophysiology. *Science Translational Medicine* 2: 24ra22.

38 Yang, L., Yan, J., Meng, C. et al. (2023). Vanadium oxide-doped laser-induced graphene multi-parameter sensor to decouple soil nitrogen loss and temperature. *Advanced Materials* 35: 2210322.

39 Niu, G., Wang, Z., Xue, Y. et al. (2022). Pencil-on-paper humidity sensor treated with NaCl solution for health monitoring and skin characterization. *Nano Letters* 22: 3.

40 Xue, Y., Wang, Z., Dutta, A. et al. (2023). Superhydrophobic, stretchable kirigami pencil-on-paper multifunctional device platform. *Chemical Engineering Journal* 465: 142774.

41 Hwang, S., Kang, S.K., Huang, X. et al. (2015). Materials for programmed, functional transformation in transient electronic systems. *Advanced Materials* 27: 47–52.

42 Chung, H.U., Rwei, A.Y., Hourlier-Fargette, A. et al. (2020). Skin-interfaced biosensors for advanced wireless physiological monitoring in neonatal and pediatric intensive-care units. *Nature Medicine* 26: 418–429.

43 Choi, Y.S., Jeong, H., Yin, R.T. et al. (2022). A transient, closed-loop network of wireless, body-integrated devices for autonomous electrotherapy. *Science* 376: 1006–1012.

44 Yang, Q. et al. (2021). Photocurable bioresorbable adhesives as functional interfaces between flexible bioelectronic devices and soft biological tissues. *Nature Materials* 20: 1559–1570.

45 Choi, Y.S., Wei, T., Yin, R.T. et al. (2021). Fully implantable and bioresorbable cardiac pacemakers without leads or batteries. *Nature Biotechnology* 39: 1228–1238.

46 Koo, J., MacEwan, M.R., Kang, S.K. et al. (2018). Wireless bioresorbable electronic system enables sustained nonpharmacological neuroregenerative therapy. *Nature Medicine* 24: 1830–1836.

47 Mickle, A.D., Won, S.M., Noh, K.N. et al. (2019). A wireless closed-loop system for optogenetic peripheral neuromodulation. *Nature* 565: 361–365.

48 Kim, D.H., Xiao, J.L., Song, J.Z. et al. (2010). Stretchable, curvilinear electronics based on inorganic materials. *Advanced Materials* 22: 2108–2124.

49 Kim, D.H., Ahn, J.H., Choi, W.M. et al. (2008). Stretchable and foldable silicon integrated circuits. *Science* 320: 507–511.

50 Yang, J., Webb, A.R., and Ameer, G.A. (2004). Novel citric acid-based biodegradable elastomers for tissue engineering. *Advanced Materials* 16: 511–516.

Index

y